Nonequilibrium Cooperative Phenomena in Physics and Related Fields

NATO ASI Series

Advanced Science Institutes Series

A series presenting the results of activities sponsored by the NATO Science Committee, which aims at the dissemination of advanced scientific and technological knowledge, with a view to strengthening links between scientific communities.

The series is published by an international board of publishers in conjunction with the NATO Scientific Affairs Division

A Life Sciences **B Physics**	Plenum Publishing Corporation New York and London
C Mathematical and Physical Sciences	D. Reidel Publishing Company Dordrecht, Boston, and Lancaster
D Behavioral and Social Sciences **E Engineering and Materials Sciences**	Martinus Nijhoff Publishers The Hague, Boston, and Lancaster
F Computer and Systems Sciences **G Ecological Sciences**	Springer-Verlag Berlin, Heidelberg, New York, and Tokyo

Recent Volumes in this Series

Volume 111—Monopole '83
edited by James L. Stone

Volume 112—Condensed Matter Research Using Neutrons: Today and Tomorrow
edited by Stephen W. Lovesey and Reinhard Scherm

Volume 113—The Electronic Structure of Complex Systems
edited by P. Phariseau and W. M. Temmerman

Volume 114—Energy Transfer Processes in Condensed Matter
edited by Baldassare Di Bartolo

Volume 115—Progress in Gauge Field Theory
edited by G. Lehman, G. t'Hooft, A. Jaffe, P. K. Mitter, I. Singer, and R. Stora

Volume 116—Nonequilibrium Cooperative Phenomena in Physics and Related Fields
edited by Manuel G. Velarde

Volume 117—Moment Formation in Solids
edited by W. J. L. Buyers

Series B: Physics

Nonequilibrium Cooperative Phenomena in Physics and Related Fields

Edited by
Manuel G. Velarde
Universidad Nacional de Educación a Distancia
Madrid, Spain

Plenum Press
New York and London
Published in cooperation with NATO Scientific Affairs Division

Proceedings of a NATO Advanced Study Institute on
Non-Equilibrium Cooperative Phenomena in Physics and Related Fields,
held August 1-11, 1983,
in El Escorial (Madrid), Spain

Library of Congress Cataloging in Publication Data

NATO Advanced Study Institute on Non-Equilibrium Cooperative Phenomena in Physics and Related Fields (1983: Escorial)
Nonequilibrium cooperative phenomena in physics and related fields.

(NATO ASI series. Series B, Physics; v. 116)
"Published in cooperation with NATO Scientific Affairs Division."
"Proceedings of a NATO Advanced Study Institute on Non-Equilibrium Cooperative Phenomena in Physics and Related Fields, held August 1-11, 1983, in El Escorial (Madrid), Spain"—Verso t.p.
Includes bibliographical references and indexes.
1. Phase transformations (Statistical physics)—Congresses. 2. Lasers—Congresses. 3. Quantum optics—Congresses.4. Nonequilibrium thermodynamics—Congresses. 5. Liquids—Congresses.6. Nonlinear theories—Congresses. I. Velarde, Manuel G. (Manuel García) II. North Atlantic Treaty Organization. Scientific Affairs Division. III. Title. IV. Series.
QC175.16.P5N38 1983 530.1′3 84-18143

ISBN 978-1-4684-8570-7 ISBN 978-1-4684-8568-4 (eBook)
DOI 10.1007/978-1-4684-8568-4

A Division of Plenum Publishing Corporation
233 Spring Street, New York, N.Y. 10013
Softcover reprint of the hardcover 1st edition 1984

In memory of Professor S. M. ULAM,
an enthusiastic supporter and scheduled lecturer for the ASI who at the last moment was unable to attend for pressing personal reasons.

PREFACE

This volume contains the lectures and invited seminars presented at the NATO Advanced Study Institute on NON-EQUILIBRIUM COOPERATIVE PHENOMENA IN PHYSICS AND RELATED FIELDS that was held at EL ESCORIAL (MADRID), SPAIN, on August 1-11, 1983.

Most nonlinear problems in dissipative systems, *i.e.*, most mathematical models in SYNERGETICS are highly transdisciplinary in practice and the list of lecturers and participants at the ASI reflects this diversity both in background and interest. The presentation of the material fell into two main categories: *tutorial lectures* on some basic ideas and methods, both experimental and theoretical, intended to lay a common base for all participants, and a series of more specific lectures and seminars, serving the purpose of exemplying selected but typical applications in their current state of development. Topics were chosen for their basic interest as well as for their potential for applications (laser, hydrodynamics, liquid crystals, EHD, combustion, thermoelasticity, etc.). We had more seminars and some of the oral presentations were supported or complemented with 16 mm films and on occasion with experimental demonstrations including a special seminar, a social one on broken symmetries in Art and Music. There is here no record of these non-standard activities. We had, indeed, quite a heavy load for which I was fully responsible. However, the reader and, above all, the participants at the ASI ought to be aware of the fact that in Spain, with over thirty five million people, in most universities and research institutes we do not possess adequate libraries, lab and workshop facilities nor do we have during the year the number of scientific seminars and visitors required for a healthy scientific life and the appropriate atmosphere for a competitive research. This contrasts with the standard of an average European or North-American institution.

I, like some other Spanish participants, wanted to act as a sponge with all pores open to the highest inflow of information, experimental data, ideas, exchange of views, etc. for there has been an alarming increase in the number of things one knows nothing about

and this approach of mine biased the ASI to certain extent. In this perspective it is my hope that, despite important limitations, a broad and reasonably representative coverage of the field has been achieved and that this volume may be a valuable aid to newcomers and practitioners in this fascinating realm of Science.

The speakers and thus the papers/chapters included in these *proceedings* correspond to fifteen invited lecturers and thirteen invited seminar speakers. Note that due to the diversity of the topics covered at the ASI, and in order to provide here a coherent and useful presentation some of the lecture notes have been either divided into separate contributions or else assembled into single comprehensive chapters. Moreover, the twenty nine chapters of these proceedings have been grouped into eight blocks. We start with two introductory chapters (part A) dealing with general aspects of synergetics and the theory of non-equilibrium phase transitions and cooperative phenomena that set the background and describe the rather unifying framework already available in the field. Then we have five chapters (part B) on quantum optics with emphasis on oscillatory instabilities and related problems in laser systems. Part C refers to the action of intense laser fields on nuclear processes. Part D contains five chapters on hydrodynamic instabilities where like in other parts of these proceedings there is a fair blending of phenomenology, heuristic arguments, experimental data, experimental techniques and theory. Part E deals in two chapters with experiments in systems where convection, diffusion and reaction operate simultaneously. Part F covers, in ther chapters experimental and theoretical aspects of combustion (premixed and diffusion flames). Non-Equilibrium phenomena in solids are discussed in the two chapters of part G. Recent advances in continuous and discrete nonlinear applied mathematics to physical and physicochemical problems are dealt with in the three chapters of part H. This is followed by part I, where in six chapters an account is presented of the most relevant results recently found in the stochastic analysis of nonlinear non-equilibrium phenomena (PDE, ODE, role of white and non-white noise, etc.)

Prof. F. Tito Arecchi and Prof. Marlan O. Scully, co-directors of the ASI, were kind enough to encourage and help me long before this School took place and throughout its development. Prof. Hermann Haken was a key and friendly supporter of our initiative. My gratitude goes to all three and also to the other members of the steering committee, to the lecturers and seminar speakers, and to my collaborators at UNED, who in one way or another provided invaluable help in my duties. I also wish to specially thank Mr. Luis de la Rasilla, from UNED, who through efficient action helped with the busing at arrival, departure and excursions offered to the ASI participants.

Finally, my appreciation also goes to the NATO Scientific Affairs Division for its economic support. Financial support also came from the Comisión Asesora de Investigación Científica y Técnica (Ministerio de Educación y Ciencia, Spain), IBM - Spain and the Universidad Nacional de Educación a Distancia (U.N.E.D., Madrid), that permitted covering travel and hostal expenses for a number of non NATO country participants.

Manuel G. Velarde

CONTENTS

PART A. INTRODUCTION: SYNERGETICS AND NON-EQUILIBRIUM PHASE TRANSITIONS

PART B. LASERS AND QUANTUM OPTICS

PART D. INSTABILITIES AND CONVECTION IN LIQUIDS AND LIQUID CRYSTALS

PART E. CONVECTION DIFFUSION AND REACTION

PART F. COMBUSTION

SYNERGETICS

H. Haken

Institut für theoretische Physik

Universität Stuttgart

1. INTRODUCTION

In my lectures I should like to give a brief outline of some basic ideas of synergetics. Then I shall present some of our recent results obtained by an application of our mathematical methods. The word SYNERGETICS is composed of two greek words and means COOPERATION. What we study in this field is the cooperation of individual parts of a system so that a self-organized formation of spatial, temporal, or functional structures on macroscopic scales becomes possible. In particular we shall ask whether there are general principles which govern self-organization irrespective of the nature of the individual subsystems which may be electrons, molecules, photons, biological cells, animals, or even humans. In these lectures I shall focus my attention on the application of the general ideas of synergetics to the natural sciences, and here especially physics. But it might be note worthy that such concepts have been applied to other branches of science as well, such as economy, sociology, and behavioral sciences. Since this kind of approach does not follow quite the traditional lines of science a few words of explanation might be in order.

Usually science is subdivided into its disciplines such as physics, chemistry, biology, up to sociology etc. These disciplines are then subdivided again, e.g.

physics into fluid dynamics, solid state physics, laser physics, plasma physics, high energy physics, etc. On the other hand some approaches are now known where the same phenomena, which may occur in different fields, are treated from a common point of view. An example, well known to physicists, is phase transitions which may occur in quite different fields. Another kind of study is concerned with solitons which can be observed in quite different fields, possibly even biology. Synergetics adopts a similar attitude by dealing in an interdisciplinary fashion with the above mentioned self-organized formation of macroscopic structures. Over the past one or two decades a great number of such phenomena could be found in quite different disciplines, and it has been possible to subsume many of the phenomena under the unifying concepts of synergetics. Furthermore it has been possible to predict entirely new phenomena due to the general principles which allow one to draw analogies between quite different substrates. In this way it has been possible to predict the existence of turbulent laser light which has now been found experimentally, or to establish analogies between the Soret instability and optical bistability.

Let me first list a few examples from physics where such phenomena occur. A rich field is provided by fluid dynamics. In the convection instability macroscopic rolls, or hexagons, or squares of velocity fields can be formed which at elevated Rayleigh numbers may show various kinds of oscillations at one or several frequencies, or chaotic motion. Laser light can show various transitions from incoherent to coherent waves, to ultrashort pulses, or to chaos. As in the case of fluid dynamics also certain kinds of period doublings have been observed. A variety of different kinds of oscillations can be produced by electronic devices. Electrons and holes in semiconductors may produce various kinds of oscillations or spatial patterns. Plasma physics contains a rich variety of phenomena of the sort considered here. In chemistry macroscopic temporal oscillations of concentrations as well as macroscopic waves of concentric or spiral type have been found. Flames exhibit a rich variety of spatial and temporal patterns. Because chemical processes lie at the basis of biological phenomena, models dealing with chemical pattern formation can be invoked to explain a number of morphogenetic phenomena. A unification of all these phenomena has become possible when we focus our attention on those situations which are the most interesting ones, namely when qualitative changes of structures occur, e.g. when a liquid starts forming a macroscopic pattern or a laser starts its coherent oscillations. Incidentally many of these transitions are strongly reminescent of phase transitions of systems in thermal equilibrium though the systems we are considering here are driven far away from thermal equilibrium. Because the field of synergetics is mushrooming it is impossible for me here to give a detailed account of its mathematical methods and experimental results. Therefore I should like to refer the reader to my books SYNERGETICS.AN INTRODUCTION [1] and to ADVANCED SYNERGETICS [2]. While most of the systems we shall be dealing within these notes belong to classical physics, adequate methods to cope with quantum systems have also been developed and a survey on a number of such methods is decribed in my book LASER THEORY [3].

2. OUTLINE OF THE GENERAL APPROACH

Let me take an example from physics. We may describe the behavior of a fluid at three different levels. At the microscopic level we deal with the motion of individual atoms or molecules. At the mesoscopic level we lump many molecules together into droplets so that we may speak of densities, temperature etc., but so that at this level no macroscopic structure is visible. At the macroscopic level we deal with the formation of structures e.g. rolls, hexagons etc. While e.g. in laser physics we directly proceed from the microscopic to the macroscopic level, in this lecture we shall adopt the following attitude. We assume that the transition from the microscopic to the mesoscopic level has been achieved by statistical mechanics or that adequate equations have been formulated at the mesoscopic level in a more or less phenomenological manner. An example is provided by the Navier Stokes equations, or by rate equations for chemical reactions. We then wish to study the evolution of patterns at the macroscopic level.

The state of the system is described by a set of variables $q_1 \dots q_n$ which we lump together into a state vector q. Because in general the processes depend on space and time, q is a function of x and t also. The following list gives a number of interpretations of the various components of q

numbers or densities of atoms or molecules	fluids, solidification chemical reactions
velocity fields	flames, lasers, plasmas
electromagnetic fields	electronic devices
electrons	solid state
firing rates of neurons	neural nets
numbers of specific cells	morphogenesis
monetary flows	economy
numbers of animals	ecology
number of people with specific opinion	formation of public opinion sociology

The processes may take place in various geometries e.g.in the plane, in threedimensional space, but also on a sphere. For instance pattern formation on spherical shells in biology have been studied by Velarde [4] or pattern formation in the atmosphere of planets by Busse and others. Also one may think of more complicated manifolds or even evolving manifolds. The concept of approach of synergetics rests on a number of paradigms, to use a word en vogue, namely

a) evolution equations
b) instability
c) slaving
d) order parameters
e) formation of structures
f) instability hierarchies

3. A BRIEF OUTLINE OF THE MATHEMATICAL APPROACH

a) Evolution equations

These equations deal with the temporal evolution of q, i.e. we have to study $\dot{q} = N(q)$. The r.h.s. is a nonlinear function of the components q_j, e.g. q_1^2, $q_1 q_2$ etc. The systems under consideration are dissipative i.e. they contain equations of the form

$$\dot{q}_1 = - \gamma q_1 + \dots \qquad (3.1)$$

They may contain transport terms describing

$$\begin{array}{ll} \text{convection:} & v\nabla v \\ \text{diffusion:} & \Delta \\ \text{waves:} & \Delta \end{array} \tag{3.2}$$

The systems are controlled from the outside, e.g. by changing the energy input. This control is described by control parameters, e.g. by α in the eq.

$$\dot{q} = (\alpha - \gamma)q + \dots \tag{3.3}$$

Finally, close to transition points of nonequilibrium phase transitions fluctuations play a decisive role. These fluctuations stem from fluctuating forces which represent the action of the microscopic "underworld" on the physical quantities q of the mesoscopic level. Lumping all the different terms together we are led to consider coupled nonlinear stochastic partial differential equations of the type

$$dq(x,t) = N(q,\nabla,x,\alpha,t)dt + dF \tag{3.4}$$

where we may use the Stratonovich calculus. Without fluctuations the equations reduce to

$$\dot{q} = N(q,\nabla,x,\alpha) \tag{3.5}$$

A special case treated in chemistry has the form

$$\dot{q} = R(q) + D\nabla^2 q \tag{3.6}$$

where the first term R describes the reactions whereas the second describes diffusion processes.
For sake of completeness we mention that as long as we deal with Markov processes we may also invoke other types of equations, e.g. the Chapman-Kolmogorov equation. Finally we mention that the methods we shall present below, including the slaving principle, possess a quantum mechanical analogue, where the evolution equations are replaced by Heisenberg's operator equations which contain damping terms and fluctuating operator forces.

b. Instability

We assume that we have found a solution of the nonlinear equations for given control parameters $\alpha=\alpha_0$. In practical cases such a solution may describe, for instance, a quiescent and homogeneous state, but our treatment may also include spatially inhomogeneous and oscillatory states. We denote the corresponding solution by q_0. When we change the control parameter that solution q_0 may loose its stability. To study the stability (or instability) we put

$$q(x,t,\alpha) = q_0(x,t,\alpha) + w(x,t,\alpha) \tag{3.7}$$

and insert it into (3.5). Assuming that w is a small quantity we may linearize (3.5) and study the resulting equations of the form

$$\dot{w} = L(q_0(x,t),\nabla,x,\alpha)w, \quad w = w(t) \tag{3.8}$$

If L is independent of t or depends on t periodically, or in a large class of systems in a quasiperiodic fashion, the solutions can be written in the form

$$w^{(j)}(t) = \exp(\lambda_j t)v^{(j)}(t) \tag{3.9}$$

where v(t) is bounded. Thus the global behavior of w is determined by the exponential function in (3.9). We call those solutions, whose real part of λ is positive, <u>unstable</u>, and those whose real part of λ is negative, <u>stable</u>. In order to solve the nonlinear equation (3.5) (or, more generally, its stochastic counterpart (3.4)) we make the hypothesis

$$q(x,t) = q_0(x,t,\Phi(t)) + \sum_j u_j(t)v^{(j)}(x,t,\Phi(t)) + \sum_k s_k(t)v^{(k)}(x,t,\Phi(t)) \tag{3.10}$$
(3.10)*

where Φ is a set of certain phase angles in case we deal with quasiperiodic motion. For details I refer the reader to my book ADVANCED SYNERGETICS. Here it may suffice to note that by inserting the hypothesis (3.10) into our original nonlinear equations (3.5) we find after some mathematical manipulations the following equations

---

* j and k run over the unstable and stable mode indices, respectively.

$$\dot{u}_j = \lambda_j u_j + N_j^{(u)}(u,\Phi,t,s), \quad (3.11)$$

$$\dot{s}_k = \lambda_k s_k + N_k^{(s)}(u,\Phi,t,s), \quad (3.12)$$

$$\dot{\Phi}_l = N_l^{(\Phi)}(u,\Phi,t,s). \quad (3.13)$$

Similarly, starting from (3.4) we obtain stochastic equations for u,s, Φ. Though in general one may not expect to simplify a problem by means of a transformation, the new equations (3.11)-(3.13) can be considerably simplified when a system is close to instability points, where the real parts of some λ's change their sign from negative to positive.

c) The slaving principle

For the situations just mentioned we have derived the slaving principle for stochastic differential equations and discrete noisy maps. The slaving principle states that we may express the amplitudes s of the damped modes by means of u and Φ at the same time, so that

$$s = f(u,\Phi,t) \quad (3.14)$$

We shall call u and Φ order parameters. We have studied numerous cases of dissipative systems and have found that in practically all of them there occur only few order parameters while there are still very many slaved modes. As a consequence we achieve an enormous reduction of the degrees of freedom because we may express all damped modes s by the order parameters. In this way we obtain a closed set of equations of the form

$$\dot{u} = N(u,\Phi,t), \quad (3.15)$$

$$\dot{\Phi} = N'(u,\Phi,t). \quad (3.16)$$

4. SOME SIMPLE EXAMPLES

So far our approach has been quite general and rather abstract. Close to instability points, the order parameter equations can be simplified considerably because in many cases u is a small quantity. This allows us to expand the r.h.s. of eq.(3.15) into a power series of u and to keep only the first few important terms. In many cases the dynamics is described by a single order parameter u. Since in a number of cases the coefficient of the quadratic term in u on the r.h.s. of (3.15) vanishes, for symmetry reasons the order parameter equation for u takes the simple form

$$\dot{u} = \lambda u - u^3 + F(t), \tag{4.1}$$

where we have included a fluctuating force F(t). For sake of simplicity we assume that F is independent of u. Since this equation is nonlinear and contains a stochastic force F(t) its solution is not easy. However, as I have shown nearly twenty years ago, its solution can easily be visualized by invoking an analogy with mechanics. To this end I interpret u as the coordinate of a particle moving in the u direction and I add an accelleration force mu. In this way I consider the equation

$$(m\ddot{u}) + \dot{u} = f(u) + F(t) \equiv - \frac{\partial V(u)}{\partial u} + F(t) \tag{4.2}$$

As indicated, the "force" $f(u) \equiv \lambda u - u^3$ can be derived from a potential V which is plotted in Fig.4.1. The motion of the particle in the potential V(u) described by F(t) and subject to random pushes can now be easily discussed and is left as an exercise to the reader. A number of important analogies with phase transitions of systems in thermal equilibrium can be established by use of the potential V. When the curve for $\lambda < 0$ is changed to $\lambda > 0$, the position u_0 becomes unstable and a new position must be taken by which the symmetry is broken. Thus we are dealing with a symmetry - breaking instability. Furthermore, when λ is increased starting from its negative region, the dashed curve becomes flatter and flatter, and the particle falls down the potential curve more and more slowly. This is the well known phenomenon of critical slowing down. Finally, when the curve becomes flatter, the fluctuating force F(t) becomes more efficient and we

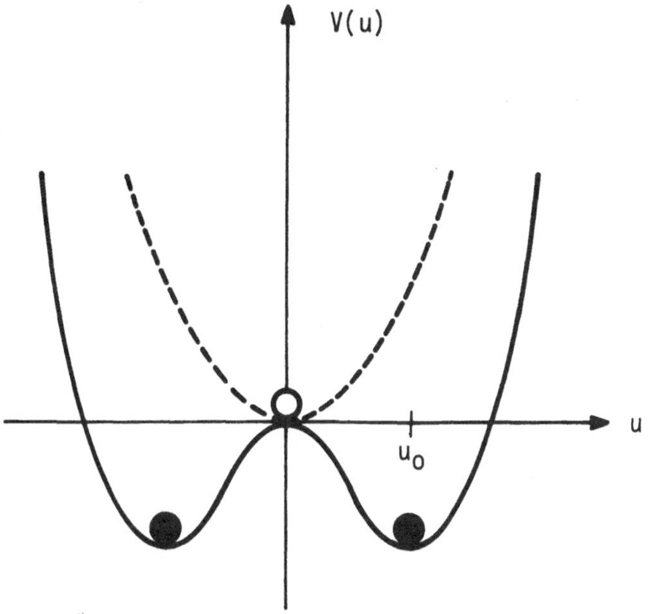

Fig.4.1 The potential V(u) as a function of u for $\lambda < 0$ (dashed line) and $\lambda > 0$ (solid line)

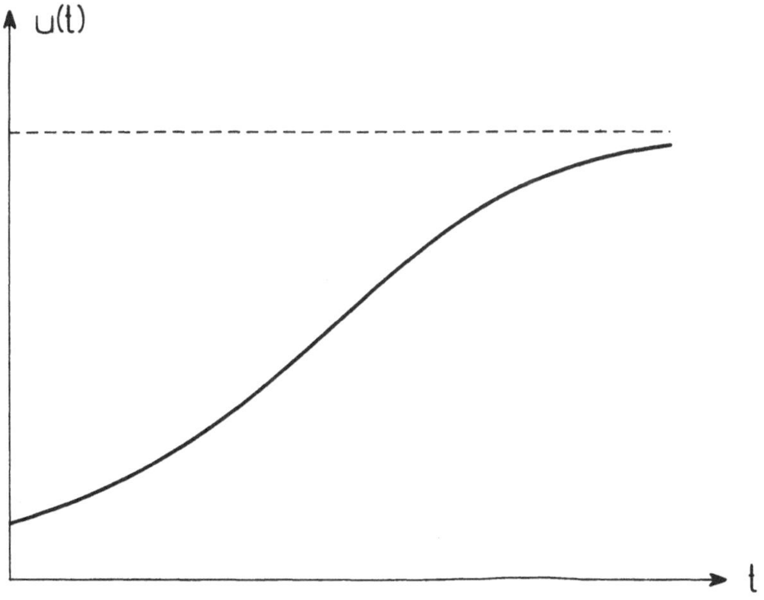

Fig.4.2

Increase of u(t) towards its stationary state

are dealing with critical fluctuations well known from phase transitions of systems in thermal equilibrium. As can be shown in a mathematically rigorous fashion from a formal point of view, V(u) plays the same role as the free energy of systems in thermal equilibrium. But we have to keep in mind that we are dealing here with systems driven far away from thermal equilibrium and that V(u) has quite a different physical meaning than a free energy and it would be a terrible confusion to attach that meaning to V. Without F(t) eq.(4.1) can be solved quite simply and leads to a behavior shown in Fig.4.2. In this way we can simply explain the

buildup of spatial patterns

To this end we assume that only one order parameter is present and that we deal with no phase angles Φ. In such a case the hypothesis (3.10) reduces to

$$q(x,t) = q_0 + u(t)v^{(1)}(x) + \sum_k s_k(t)v^{(k)}(x). \quad (4.3)$$

For a qualitative discussion we may neglect the last sum because it contains small quantities. We note that u(t) obeys eq.(4.1), whereas $v^{(1)}(x)$ describes a spatial pattern which results from a solution of the linearized equation (3.8). Fig.4.3 shows a plot of the first two terms of the solution q, which evolves in time. Thus our approach shows quite clearly how a spatial pattern may evolve in time. Note that in this way also more complicated cases in which several order parameters are present can be treated.

5. GENERALIZED GINZBURG-LANDAU EQUATIONS

When the dimensions of continuously extended systems are large compared to the fundamental length of developing patterns, the spectrum λ is practically continuous. In such a case particular mathematical difficulties arise because it is no more possible to distinguish clearly between undamped and damped modes. A way out of this difficulty can be found when we resort to the formation of wave packets. This in turn necessitates that the order parameters, which we shall call ξ, depend not only on time but now also on space (in a slowly varying fashion). Therefore our hypothesis reads

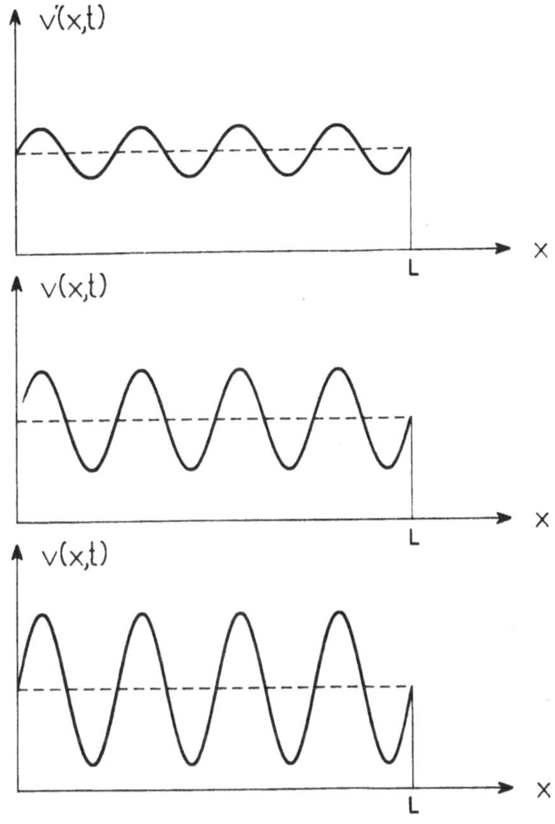

Fig.4.3 Build-up of spatial pattern

$$q(x,t) = q_0 + \sum_{k_c} \xi_{k_c}(x,t) v_{k_c}(x) + \sum \text{slaved modes} \quad (5.1)$$

where k_c runs over a discrete set of critical wave vectors at which the instabilities occur (compare fig. 5.1). For simplicity let us again consider a case in which no phase angles occur and let us furtheron be satisfied with an expansion of the nonlinear terms up to third order. The order parameter equations then acquire the form

$$\xi_{k_c}(x,t) = \lambda_{k_c}(\nabla)\xi_{k_c}(x,t) + \sum_{k_1,k_2} A\ldots\ \xi_{k_1}\xi_{k_2} \quad (5.2)$$

$$+ \sum_{k_1,k_2,k_3} B\ldots\ \xi_{k_1}\xi_{k_2}\xi_{k_3} + F_{k_c}.$$

I have called these equations, which I derived some time ago "Generalized Ginzburg-Landau-equations", because they are strongly reminescent of the famous Ginzburg-Landau-equations. But two important distinctions should be noted. While the original Ginzburg-Landau-equations refer to a system in thermal equilibrium my Generalized Ginzburg-Landau-equations refer to systems far from thermal equilibrium. Furthermore the original Ginzburg-Landau-equations were derived in a heuristic fashion, whereas here the Generalized Ginzburg-Landau-equations have been derived rigorously. Because of the double and triple sums these equations are quite clumsy. However, under well justified assumptions these equations can be simplified as I have shown recently. To this end I define a new function

$$\Psi(x,t) = \sum_{k_c} e^{ik_c x}\xi_{k_c}(x,t). \quad (5.3)$$

After a few elementary manipulations and under specific assumptions on λ, A and B eq.(5.2) can be cast into the form

$$\dot{\Psi}(x,t) = \left(a + b(k_0^2 - \nabla^2)^2\right)\Psi + a\Psi^2 + B\Psi^3 + F, \quad (5.4)$$

where I have chosen an explicit example for $\lambda(k)$ which refers to the eigenvalues of the convection instability.

We have solved this equation on a computer to study

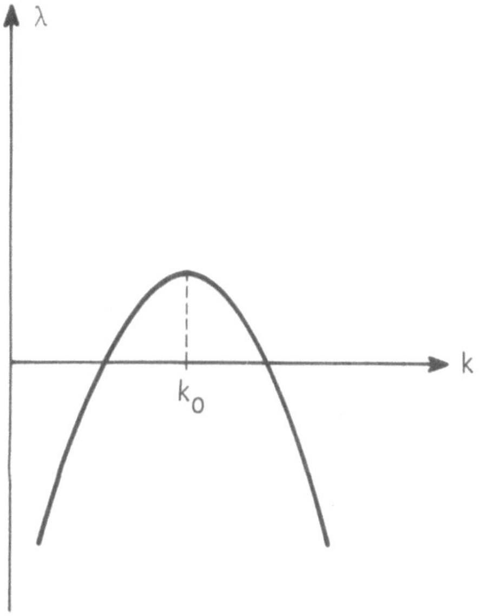

Fig. 5.1

Example for the dependence of λ on the wave number k. The instability occurs in the neighbourhood of k_0

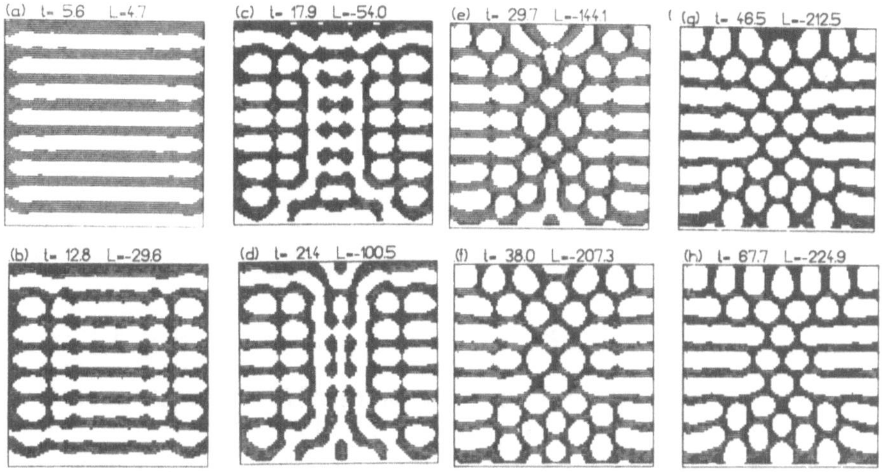

Fig. 5.2 In Fig. 5.2 a roll pattern is prescribed but the parameter values of the equation are chosen such that hexagons should be formed. The sequence a - h shows the formation of hexagons but the final state is reached only at infinitely large time (critical slowing dowm)

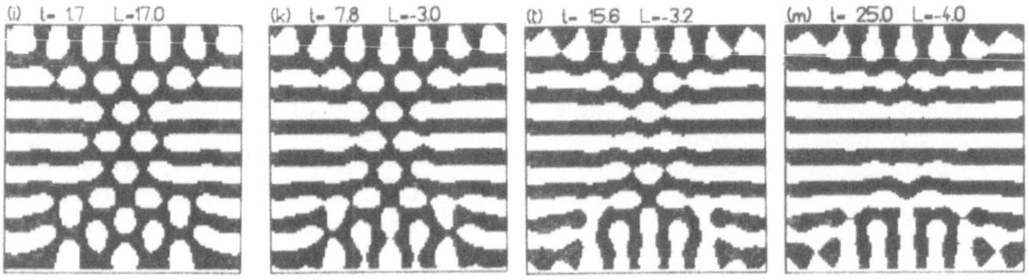

Fig. 5.3 shows the inverse process. Here the final pattern h) of Fig. 5.2 is taken as initial condition but the conditions are now set so that a roll pattern should evolve. The sequence i) to m) indeed shows this evolution

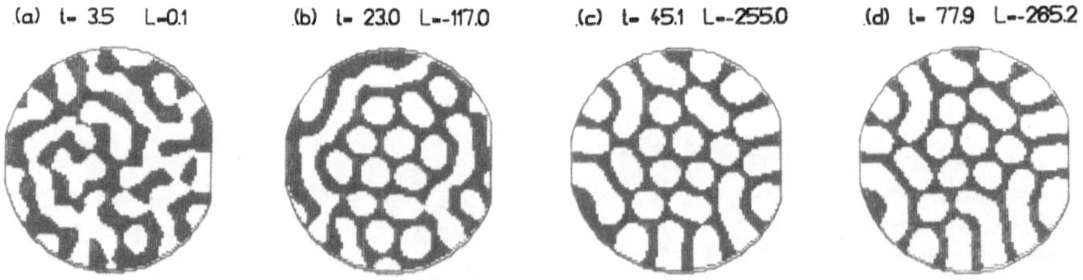

Fig. 5.4 shows the evolution of a pattern in circular geometry. A random dot pattern was prescribed at initial time Part a) shows initial pattern formation which then evolves through the states b) - c) to the final state d) exhibiting the coexistence of hexagonal cells with rolls

Fig. 6.1 Lines of constant vertical velocity in a plasma heated from below and subject to vertical constant magnetic field (the configuration is indicated in Fig. 6.1a)

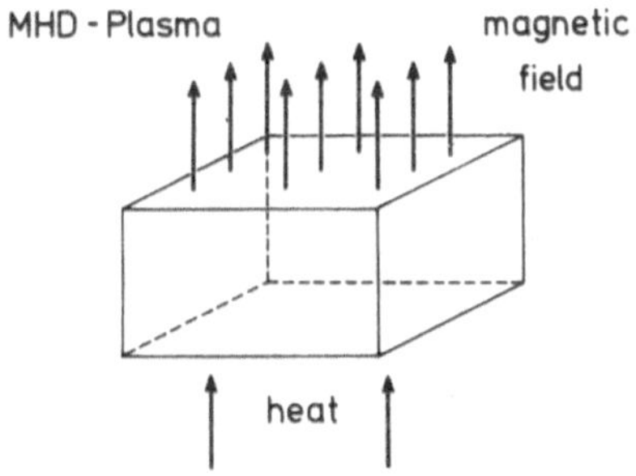

Fig. 6.1a Experimental set-up of MHD plasma

the temporal evolution of patterns. The results are shown in Fig. 5.2 for two kinds of geometries and various kinds of initial conditions.

6. SOME FURTHER APPLICATIONS

By means of the mathematical methods we have outlined above my coworkers and I have treated a number of explicit cases over the recent years. I present a few of them in order to demonstrate the applicability of the mathematical method I have briefly sketched in the beginning of my lectures.

a) Pattern formation of an MHD plasma which is heated from below and is subjected to a vertical constant magnetic field. The boundary conditions in the horizontal directions are chosen periodic. Several patterns could be found. In the single mode case rolls appear, well known from fluid dynamics. However, also two or three mode cases are possible. Typical velocity distributions are shown in Fig. 6.1.

b) Running waves in the positive column of a gas discharge in neon

By means of a nonlinear treatment it has been possible to derive the corresponding spatio-temporal pattern in good qualitative and semiqualitative agreement with experiments (cf.Fig. 6.2).

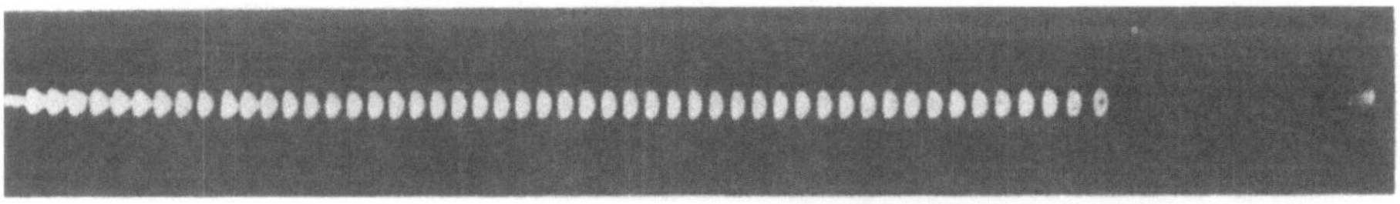

Fig. 6.2 Column in a gas discharge tube

c) Prepattern formation of the spiral wave pattern of a sunflower head

We have adopted reaction diffusion equations of the form (cf.3.6) and have used the Gierer-Meinhardt-model for the reaction terms. Since the specific form of the reaction terms is not so important we don´t discuss them here. However, it was assumed that the diffusion is space dependent. Results of the nonlinear analysis are shown in Fig.6.3. They clearly exhibit two counter rotating sets of spirals in good agreement with the observed fructification of the sunflower head.

d) A summary of spatial patterns:

By means of the methods outlined above we were able to derive a number of fundamental patterns which occur over and over again in the formation of microscopic spatial structures. Fig. 6.4 gives a survey on these.

7. HIERARCHIES OF TEMPORAL PATTERNS: FROM OSCILLATION TO CHAOS

The methods briefly described is section 3 allow us to study a series of transitions in which more and more complicated oscillations evolve. Let us start from a time independent state q . When we change the control parameter, this state may become unstable and an oscillation may start (Fig. 7.1). In mathematics this transition is called "Hopf-bifurcation". If one includes noise in the original equations, the evolving "limit cycle" is subject to phase diffusion and amplitude fluctuations, both of which can be treated in an elegant fashion by use of a Fokker-Planck equation. With further increase of the control parameter, the limit cycle can become unstable and motion at two basic frequencies can occur. If these two frequencies are irrational with respect to each other, the motion in q-space can be visualized as that on a torus (or, to use a popular wording, on a doughnut) (Fig.7.2). When the control parameter is changed further, this torus can become unstable and several things can happen. It has been suggested by Newhouse, Ruelle and Takens based on mathematical arguments that then the torus bifurcates into chaotic motion in the "generic" case. We have found, however, that this need not always be so but that the two-dimensional torus can bifurcate into a three-dimensional torus provided a KAM-condition and some

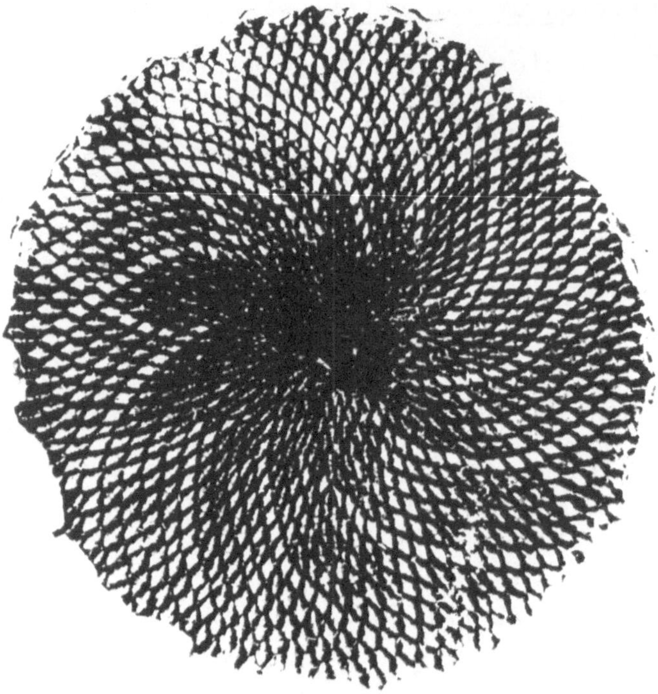

Fig.6.3a
Fructification of sunflower

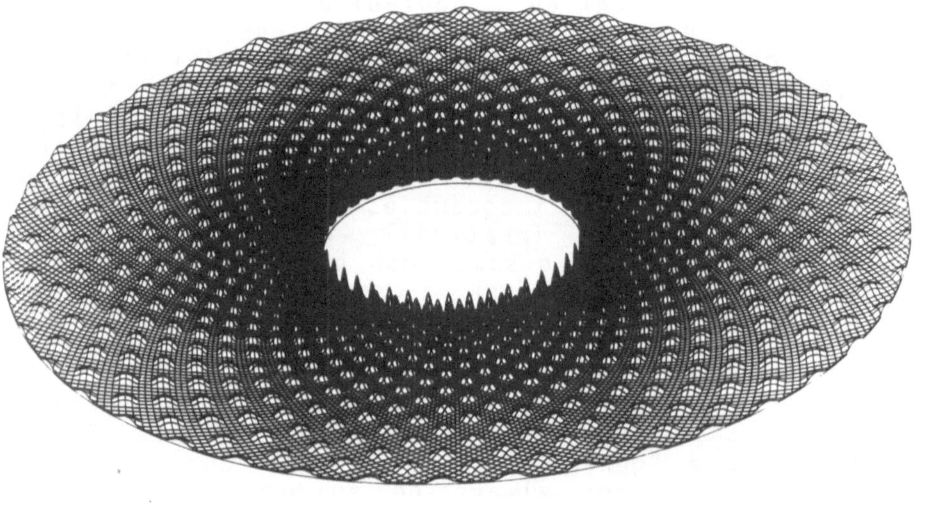

Fig. 6.3b Embossed map on model calculation on fructification of sunflower

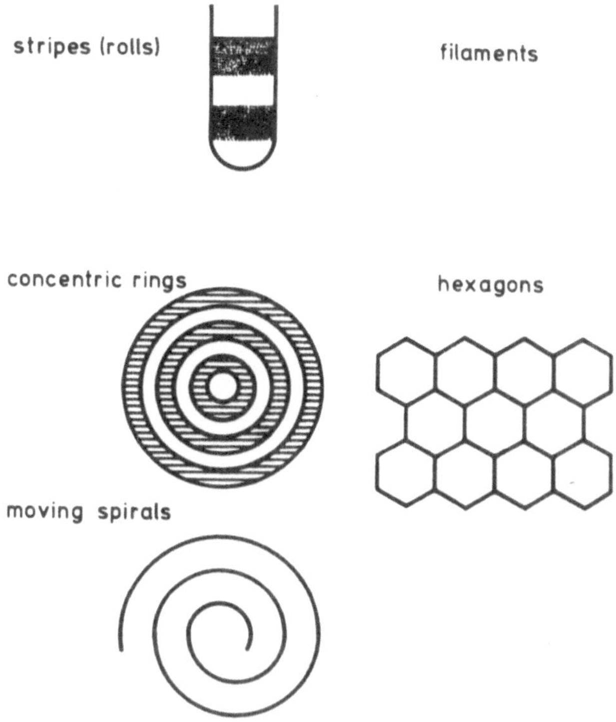

Fig. 6.4 Typical spatial patterns found in synergetic systems which can be calculated by the order parameter concept

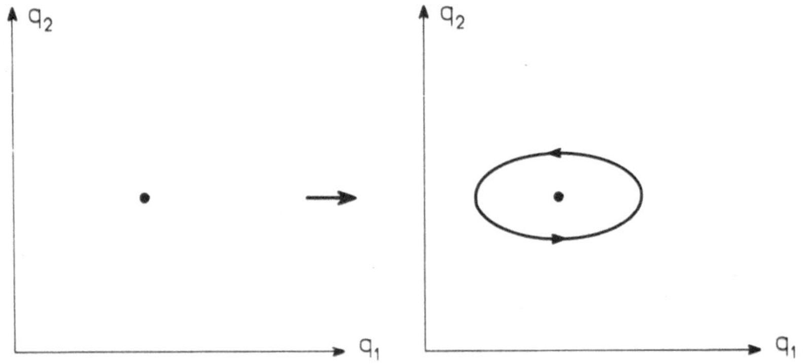

Fig. 7.1 Bifurcation of a stable fixed point in the q_1q_2 plane to a limit cycle

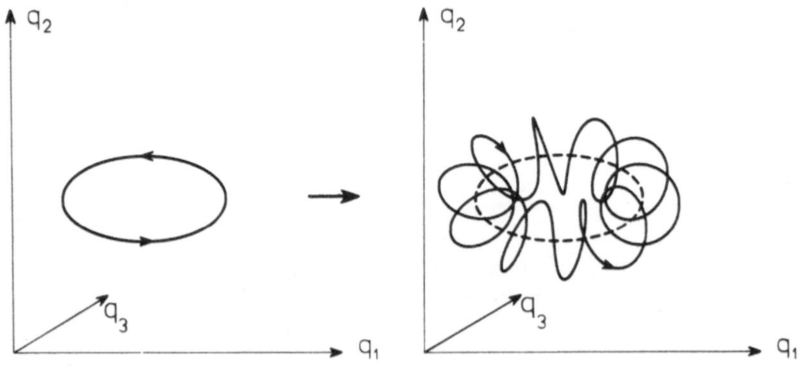

Fig. 7.2 Transition from a limit cycle in q-space into a limit cycle of a more complicated structure. If the corresponding two frequencies of the motions are irrational with respect to each other the solid curve fills a torus

technical conditions are fulfilled. It has turned out that the decision whether the one or the other case happens is a problem of measure theory and not a question of genericity. For sake of completeness we mention a third way leading to chaos namely via period doubling bifurcations (consult Fig.7.3).

In conclusion we give a classification of typical temporal patterns which occur again and again (compare Fig.7.4).

8. OUTLOOK

By means of the systematic approach of synergetics it has been possible to classify a number of spatial and temporal patterns which occur over and over again. In addition it has become possible to study the dynamics close to transition points in detail because the dynamics is governed by few order parameters only. A few words of future problems may be in order and I will list only three of them:

1) So far we have assumed that we start from a spatially homogeneous state. The whole approach works also if the original state is spatially inhomogeneous but time independent or time periodic. However, in order to solve the linearized equations and to derive the order parameter equations in most cases computer calculations may be needed.

2) When we go away from instability points, the patterns remain qualitatively the same as is known from numerous experiments. However, as it seems to me a rigorous theory far away from instability points is still lacking.

3) A rich field of further study is provided by chaos and it seems to me that we are just at the beginning of classifying and understanding chaotic motion.

In conclusion it might be worth pointing out that in the field of synergetics we need not only a further development of mathematical methods but also the corresponding experiments must be performed and a close interaction between experimentalists and theoreticians is needed. We believe that the approaches so far have not only given us fundamental insights into the way new patterns evolve at

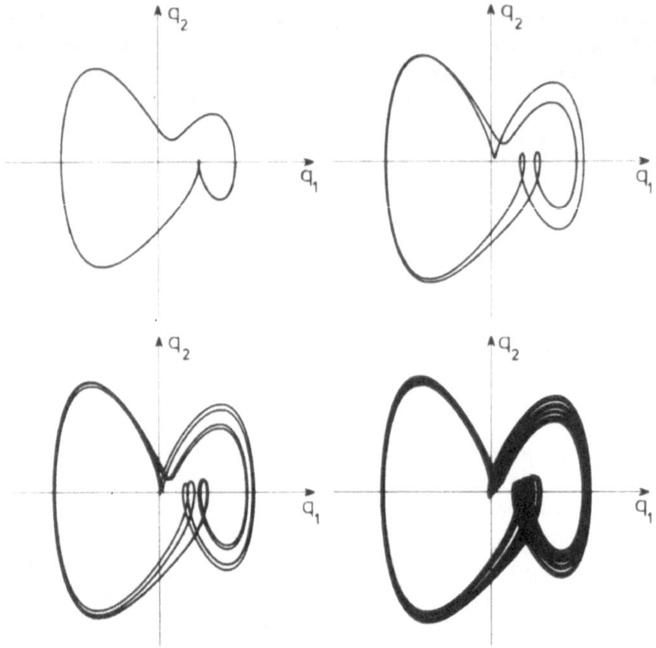

Fig. 7.3

A sequence of period doublings of the Duffing oscillator leading to chaos

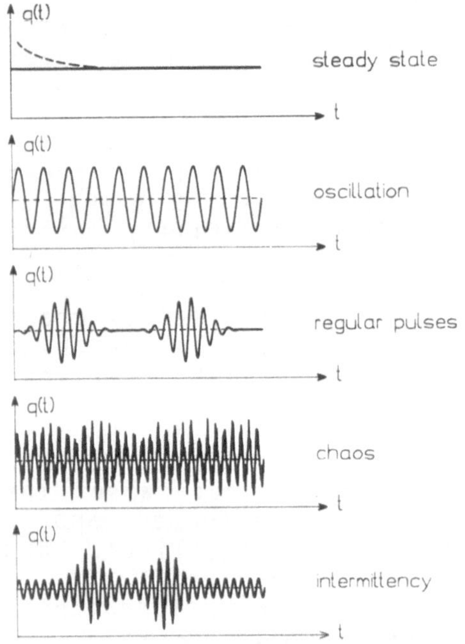

Fig. 7.4

Typical temporal patterns found in synergetic systems

instability points but have also led to a number of practical applications by exploiting analogies between different systems. These analogies become apparent through the order parameter equations. I am sure that this will lead to a development of new devices especially in solid state physics and quantum electronics.

References

1. H. Haken, "Synergetics. An Introduction. Nonequilibrium Phase Transitions and Self-Organization in Physics, Chemistry and Biology" Third Revised and Enlarged Edition Springer Verlag, Berlin, Heidelberg, New York, Tokyo 1983.
2. H. Haken, "Advanced Synergetics. Instability Hierarchies of Self-Organizing Systems and Devices" Springer Verlag Berlin, Heidelberg, New York, Tokyo 1983.
3. H. Haken, "Laser Theory" Encyclopedia of Physics Vol. XXV/2c, Springer Verlag Berlin, Heidelberg, New York 1970, reprinted 1983.
4. J.L. Ibanez, M.G. Velarde, J.Non-Equilib.Thermodyn. Vol.3, 63 (1978).

   Ch. Berding and H. Haken, J.Math.Biol. 14, 133 (1982).
5. An analogy between laser with saturable absorber and convection instability of fluids has been demonstrated by V. Degiorgio and L.A. Lugiato, Phys.Letters 77A, no 2.3 (1980)and
   M.G. Velarde in "Evolution of Order and Chaos in Physics, Chemistry, and Biology", ed. H. Haken Springer Verlag Berlin, Heidelberg, New York 1982 where further references are given.
6. Laser light chaos was predicted using the analogy with the Lorenz - model of fluid dynamics by H. Haken Phys.Letters 53A, no 1 (1975).

PHASE TRANSITION ANALOGIES: MAGNETS, LASERS AND FLUID FLOWS

J.Gea-Banacloche,[1] M.O. Scully[1,2] and M.G. Velarde[3]

(1) Dept. Physics & Astronomy, Univ. of New Mexico, U.S.A.
(2) Max Planck Inst. Modern Optics, Munchen, G.F.R.
(3) Fisica Fundamental, U.N.E.D., Madrid, Spain

1. INTRODUCTION

One of the early developments of the quantum theory of a laser was the uncovering of a striking similarity between the behavior of a laser around threshold and a second-order phase transition such as the order-disorder transitions of ferromagnetic and superfluid systems /1, 2/. The analogy was soon extended to a variety of phenomena in quantum optics, such as the symmetry broken laser (laser with injected signal /1/, lasers with saturable absorbers /3/, optical bistability /4/ and transient processes such as superfluorescence. Recently, the analogy has been extended in a bold though useful approach to hydrodynamic instabilities /5-8/. Such studies provided useful insights into the statistical mechanics of systems operating far from thermodynamic equilibrium and permitted the uncovering of fascinating phenomena in otherwise unrelated fields.

The physical reason for the similarity between an order-disorder phase transition and the passage of a laser oscillator above threshold can be understood in the context of a mean-field theory. In a ferromagnet, for instance, one may consider each individual magnetic dipole to interact with the macroscopic field created by all the other microscopic dipoles. In a laser each atom interacts with the macroscopic field radiated by all the other atoms in the cavity. When the temperature of the ferromagnet is lowered below a certain critical value (thus reducing the thermal agitation of the microscopic dipoles), a cooperative phenomenon takes place: dipoles far apart become "aware" of their respective orientations (via the common magnetic field they are immersed in) and align themselves parallel to each other, thereby creating a large, macroscopic, stable magnetization. In the same way in a laser, when the density of atoms in the excited state reaches a critical value (the laser threshold value), the phase and orientation of the radiation emitted by an individual atom is influenced by the "collective" field, and the atoms "cooperate" to radiate a macroscopic electromagnetic field.

It was this realization of large range correlation that led to the first suggestions of an analogy between lasers and other systems exhibiting second-order cooperative transitions, namely superfluids and superconductors. Martin /9/ drew up a table emphasizing the parallelism, in particular the existence of "coherent" states leading to interference phenomena in all these systems. Cumming and Johnston /10/ pointed out that a coherent state of the radiation field (which is a good approximation to the actual state of the field radiated by a laser) is formally identical to the BCS ground state of a superconductor: in fact, the radiation coherent state may be written as

$$|\alpha_{\vec{k}}\} > = C \prod_{\vec{k}} e^{\alpha_{\vec{k}} a^{+}_{\vec{k}}} | 0 >$$

where C is a normalization constant, the $\alpha_{\vec{k}}$ are (complex) mode amplitudes, and a^{+}_{k} is the creation operator of a photon in the k-th mode of the radiation field; while the BCS ground state may be written

$$|\Phi_0 > = C \prod_{\vec{k}} e^{g_{\vec{k}} b^{+}_{\vec{k}}} | 0 >$$

where C is a normalization constant,the $g_{\vec{k}} b^{+}_{\vec{k}}$ are complex numbers. and $b_{\vec{k}}$ is the creation operator of a Cooper pair of electrons having momenta $\vec{k}h$ and $-\vec{k}h$ (and opposite spins).

The laser threshold-phase transition analogy was independently proposed by DeGiorgio and Scully /1/ and Graham and Haken /2/; in the next sections we will review some of its main traits.

2. LASER THRESHOLD - SECOND-ORDER PHASE TRANSITION ANALOGY

The discussion in the preceding section suggests that the amplitude of the electric field E in a laser be chosen as the "order parameter", in the same way as the macroscopic magnetization in a ferromagnet. If one writes E as the real part of some complex amplitude α,

$$E = \frac{1}{2}(\alpha + \alpha^*)$$

it is possible /11/ to derive, from the general quantum-mechanical theory of a laser, and working in the coherent-state representation, the following Fokker-Planck equation

$$\frac{\partial P}{\partial t} = \frac{1}{2}\frac{\partial}{\partial \alpha}\{((A - C) - B|\alpha|^2)\alpha P\} + \text{c.c.} + A\frac{\partial^2 p}{\partial\alpha\partial\alpha^*} \tag{1}$$

where the function $P(\alpha, \alpha^*, t)$ is a "quasi-probability" distribution for the complex amplitues α, α^*. The coefficient A is the unsaturated gain in the active medium and is proportional to the population inversion σ ; B is the saturation parameter, also proportional to σ ; and C represents the cavity loss factor.

From (1) we derive an equation of motion for the expectation value of the field $< E >$:

$$< \dot{E} > = \frac{1}{2}(A - C) < E > - \frac{1}{2} B < E^3 > \tag{2}$$

If fluctuations in E are ignored(*), one may replace E by E , in which case equation (2) is identical to the equation derived in Lamb's semiclassical laser theory /8/. In this limit, then, the steady-state properties of the laser oscillator are given by the following "equation of state":

$$(A - C) < E > - B < E >^3 = 0 \tag{3}$$

Let now $A = a$, $B = b$, $C = a\sigma_t$, where σ_t is the treshold population inversion (the threshold condition to eq. (3), is $A = C$). Then the solutions of (3) are (see Fig.1)

(*) For non-equilibrium phase transitions and more specifically for hydrodynamic instabilities (Bénard problem, etc.) although the statistical mechanics is still in its infancy and a Landau (pseudo) free-energy (Lyapunov potential) is far from being justified there are, however, good reasons to validate this assumption: critical regions are generally so small and the influence of boundaries is so drastic that *non-classical* corrections do not appear as (experimentally) relevant /13/.

$$\langle E \rangle = 0 \qquad \text{if} \quad \sigma - \sigma_t < 0 \text{ (below threshold)}$$

$$\langle E \rangle = \left\{ \frac{a}{b} \left(\frac{\sigma - \sigma_t}{\sigma} \right) \right\}^{\frac{1}{2}} \qquad \text{if} \quad \sigma - \sigma_t > 0 \text{ (above threshold)} \tag{4}$$

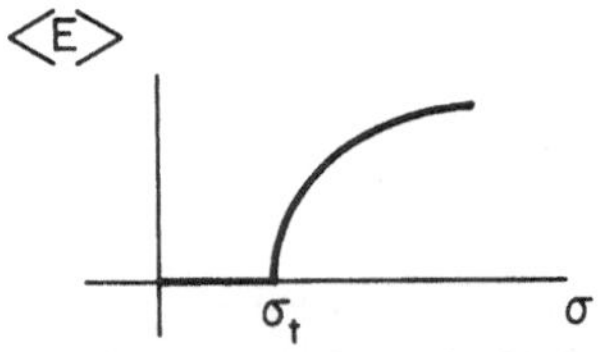

Fig. 1: The electric field amplitude versus the population inversion for the transition of a laser above threshold (second-order-like phase transition). The critical exponent is 1/2.

On the other hand, for a ferromagnet treated in the molecular-field theory, one obtains Weiss's equation of state

$$H = c(T - T_c) \langle M \rangle + dT \langle M \rangle^3 \tag{5}$$

where H is an external magnetic field, $\langle M \rangle$ is the average magnetization, T is the temperature, T_c is the critical temperature, and c and d are parameters. In the absence of an external field H, eq. (5) becomes formally identically to eq. (3), and has the solutions

$$\langle M \rangle = 0 \qquad \text{for } T \geqq T_c$$
$$\langle M \rangle = \left(\frac{c}{d}\right)^{\frac{1}{2}} \left\{\frac{T_c - T}{T}\right\}^{\frac{1}{2}} \qquad \text{for } T < T_c \tag{6}$$

The relationship between the order parameter η ($\langle E \rangle$ or $\langle M \rangle$) and the "reservoir variable" ε ($(\sigma - \sigma_t)/$ or $(T_c-T)/T$) near the critical temperature is in both cases of the form $\eta \sim \varepsilon^{\frac{1}{2}}$; the exponent $\frac{1}{2}$ is characteristic of mean-field theories.

At this point, one may wonder what happens if fluctuations in $\langle E \rangle$ are taken into account. The result turns out to be that E vanishes even above threshold. This is easily understood, since the quantum-mechanical average is an ensemble average, and the macroscopic phase of the laser oscillations above threshold, though well-defined, is arbitrary; hence on the average over a large number of lasers $\langle E \rangle$ would vanish, just as, in the absence of an external magnetic field, $\langle M \rangle$ would vanish on the average over many macrospoic ferromagnetic samples, since if H is zero the direction of M is arbitrary.

One can introduce in the laser problem a symmetry-breaking mechanism which is analogous to the external magnetic field in the case of a ferromagnet, and which leads to a non-vanishing ensemble average for E. This is simply an external injected signal (considered as externally controlled, and therefore described as a classical variable-not a quantum operator- in the theory).

It is worth pointing out that this example is not a merely academic one: the possibility of "locking" the phase of a laser to that of an injected signal coming from a master oscillator has very important and useful consequences. In particular, one could obtain in that way a whole array of laser oscillating in phase with each other, which would allow to concentrate very large amounts of energy in a common focus. This idea is discussed further in Ref./14/. If S is the amplitude of the injected signal, one obtains /1 / the following equation of state

$$(A - C) \langle E \rangle - B \langle E \rangle^3 + 2S = 0 \tag{7}$$

in place of (3); a comparison with (5) allows one to see that S plays indeed the same role as the external magnetic field H in a ferromagnet.

The quasi-probability distribution function $P(\alpha, \alpha^*, t)$ introduced in eq. (1) may be used to study fluctuations (and, in general, all the statistics of the laser light). For the general case of a laser with an injected signal, it is convenient to introduce the two real independent variables

$$x = \frac{1}{2} (\alpha + \alpha^*)$$

$$y = \frac{1}{2i} (\alpha + \alpha^*) \qquad (8)$$

Then the (stationary) quasi-probability distribution $P(x,y)$ may be written as

$$P (x , y) = N e^{-G(x,y)/K\sigma} \qquad (9)$$

where N is a normalization constant, $K = a/4$, and $G(x,y)$ may be written as

$$G (x,y) = - \frac{1}{4} a(\sigma - \sigma_x) (x^2 + y^2) + \frac{1}{8} b\sigma (x^2 + y^2) - Sx + G_o \qquad (10)$$

The form (9) shows explicitly that $P(x,y)$ is non-negative, and so it can be considered as a "true" probability distribution. The function $G(x,y)$ has the interesting property that, if P_x and P_y are the components of the macroscopic polarization that are respectively in phase and in quadrature with the injection signal S, one has

$$\dot{x} = P_x = - \frac{\partial G}{\partial x}$$

$$\dot{y} = P_y = - \frac{\partial G}{\partial y} \qquad (11)$$

On the other hand, according to the Landau theory of second-order phase transitions /15/, the probability density for the fluctuations in M may be written as

$$P(M) = N \ e^{-F(M)/kT} \tag{12}$$

where F(M) is the free-energy of the system, which according to Landau's hypothesis could be written in the form

$$F(M) = \frac{1}{2}c(T - T_c)M^2 + \frac{1}{4} dTM^4 - HM + F_0. \tag{13}$$

where F_0 is dependent only on T and H (but not M). The similarity between eq. (9)-(10) and (12)-(13) is apparent, and suggests the interpretation of G(x,y) as a type of thermodynamic energy function (which is also suggested by (11). The function G(x,y) is plotted in Fig.2 for y =0; in this form it is directly analogous to F(M) (eq. (13)). The effect of an external signal is shown in Fig. 2b.

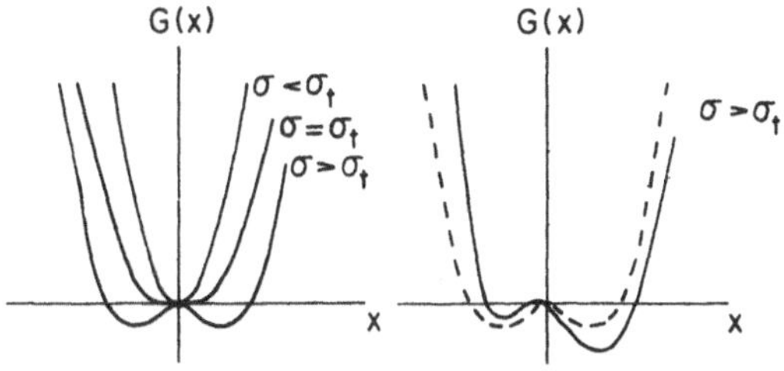

Fig. 2: The pseudo-free energy G(x,y) plotted as a function of x for y = 0. (a) without an injected signal, showing the transition that takes place as σ crosses the threshold value σ_t. The two minima for $\sigma > \sigma_t$ grow from the original minimum at x=0. (b) The effect of an injected signal (solid curve) compared with no injected signal (dashed curve); both curves above threshold.

The minima of F represent points of stable thermodynamic equilibrium; correspondingly and according to Eq. (10), the minima of G give the stable stationary solutions of (4)-(10). The phase transition is illustrated in Fig. 2a: as σ crosses its threshold value σ_t, the minimum at $x = 0$ becomes an maximum (meaning that the stationary solution $E = 0$ becomes unstable) and two minima are created, corresponding to the new stable solutions (which differ among them by a change in sign, i.e., a phase shift of Π radians; if the whole function $G(x,y)$ was plotted, one would obtain a surface of revolution with minima all around a circle in the $x - y$ plane, corresponding to all the possible phases of E). The symmetry is broken when an external signal is injected, as shown in Fig. 2a, leading to two minima of unequal depths, with the deepest one being the "most stable" (or "most probable", according to Eq. (9)).

3. LASER WITH SATURABLE ABSORBER- FIRST-ORDER PHASE TRANSITION

A laser with a saturable absorber inside the cavity may display a behavior resembling a first-order phase transition, just as the ordinary laser threshold resembles a second-order phase transition. The analogy was first pointed out by Scott et al./3/. The "pseudo-free energy" function G(E) for that system may be written as

$$G(E) = -\frac{1}{2}a(\sigma - \sigma_t)E + \frac{1}{2}a\sigma_s I_s \ln(1 + E^2/I_s) + \frac{1}{4}b\sigma E^4 \quad (14a)$$

$$\approx -\frac{1}{2}a(\sigma - \sigma_t - \sigma_s)E^2 + \frac{1}{4}(b\sigma - \frac{a\sigma_s}{I_s})E^4 + \ldots \quad (14b)$$

where the form (14b) is obtained by expanding the logarithm in (14a). Here σ_s and I_s are constants characteristic of the absorbing medium. Since there is no external signal, $G(x,y)$ is again rotationally symmetric in the $x - y$ plane; it can then be written as a function of $|E|$ alone, the magnitude of the dimensionless electric field. Fig. 3a shows its graph according to (14a). One can see that now secondary minima appear when the solution $E = 0$ is still stable (taht is, the system is still below threshold); it is natural to interpret these as *metastable* states. As σ is increased, these minima grow deeper and deeper until a threshold is reached at which all three states are equally stable; past that point, the solution $E = 0$ becomes metastable, and fluctuations may then cause the system to suddenly "switch on" to the new solution $E = E_0$. The behavior of the order parameter is then as plotted in Fig. 3b. The discontinuous jump in E is a consequence of the fact that in this case the new minina do not grow from the original minima at E = 0, as in the second-order phase transition, but instead appear at a finite value of E, $E = E_0$. This jump, as well as the hysteresis phenomena associated with the presence of metastable states, has been experimentally observed /16/.

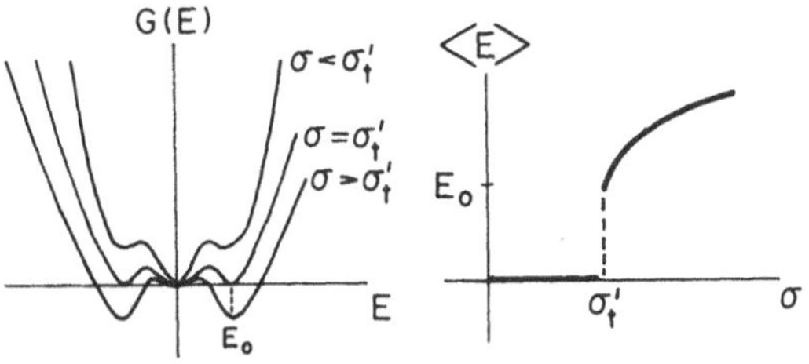

Fig.3: The first-order-like phase transition in a laser with saturable absorber. (a) shows the change in the pseudo-free energy G(E) as σ crosses the threshold value σ'_t (note that the new minima appear at a finite value of E). (b) shows the behavior of the electric field amplitude as σ is varied.

4. FROM LASERS TO FLUID FLOWS

Several authors /5-8, 17-22/ soon realized that the analogy discussed in the preceding sections could be extended to the field of convective instabilities, thus leading to an extraordinary cross-fertilization in the vast domain of nonlinear nonequilibrium cooperative phenomena. In fluid mechanics, however, steady or oscillatory states cannot, in general, be described in terms of a free energy or of a true potential (a Lyapunov functionsl). Nevertheless, one may describe convective flows by *ad hoc* polynomials whose form close enough to instability thresholds is very much like Eq. (13) or Eq. (14). Here the amplitude of the velocity field can be used as the order parameter and for certain cases a thermodynamic content has been given to the polynomial /17 - 22/.

From the purely heuristic viewpoint, Figure 4 depicts the kind of analogy that can be used when comparing a laser with saturable absorber and the two-component Bénard convection. The heuristic approach was supported by the discovery that under suitable, albeit not so restrictive approximations the time evolution of the respective order parameters in a laser (with or without absorber) and of convection (with or without competing agents) can be described with just the same mathematical equations /5,7,8/. Suppresion of the second component or of the absorber yields the kind of second-order, continuous transitions described in Section 2. With the absorber and the second-componnet or in a nematic liquid crystal, first-order, discontinuous transitions appear with spontaneous broken symmetries as well as second-order transitions for appropriate values of the parameters in each problem as the *ad hoc* polynomial is of order six. Multicritical points have also been predicted and some evidence of agreement between theory and experiment has been presented.

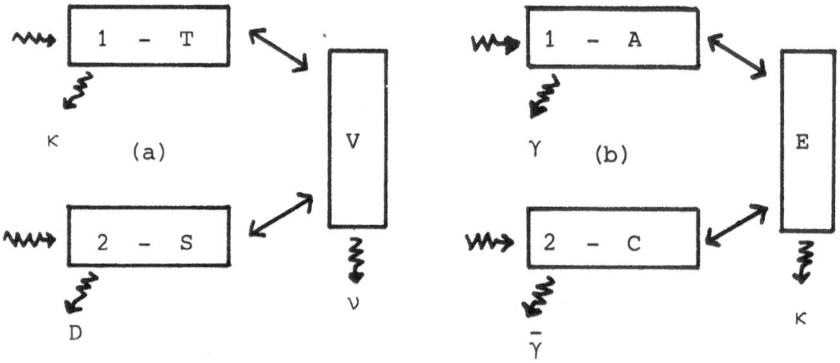

Fig.4. (a) Hydrodynamic instabilities (measured by velocity, V) and laser emission ($|E|$) when two agents operate: T, temperature and S, solute; A, active atoms and C, absorber. Thermal gradients and pumping rates are the control parameters. Straight arrows denote coupling between fields and wiggly arrows indicate characteristic decaying constants.

The mathematical equivalence between the laser and fluid flow models was also exploited by Haken /5,18/ and Velarde and Antoranz /8/ to describe the onset of chaos (aperiodic strange attractors) for various ranges of parameter values. For details about lasers with intracavity saturable absorbers see further on in this Book the chapters by Arimondo, and Velarde and Antoranz and for fluid flows with two competing agents see those by Zúñiga and Velarde, Riste, and Castellanos, Atten and Velarde.

REFERENCES

1. V. Degiorgio and M. O. Scully, Phys. Rev. *A2*, 1170 (1970).
2. R. Graham and H. Haken, Z. Physik *213* 31 (1970).
3. J.F. Scott, M. Sargent III and C.D. Cantrell, Opt. Commun. *15*, 13 (1975).
4. R. Bonifacio and L. A. Lugiato, Opt. Commun.*19*, 172 (1976).
5. H. Haken, Phys. Lett. *53A* , 77 (1975)
6. V. Degiorgio, Phys. Today, October issue, 42 (1976)
7. V. Degiorgio and L.A. Lugiato, Phys. Lett. *77A* , 167 (1980)
8. M.G. Velarde and J.C. Antoranz, Phys. Lett. *80A* , 220 (1980); Prog. Theor. Phys. *66* , 717 (1981). See also M.G. Velarde, in *Evolution of order and chaos*, edited by H. Haken (Springer-Verlag, Berlin, 1982) p. 132.
9. P. C. Martin, in *Low Temperature Physics*, edited by J. G. Daunt, D.O. Edwards, F.J. Milford, and M. Yaqub (Plenum, N. York, 1965) p. 9.
10. F. W. Cummins and J. R. Johnston, Phys. Rev. *151*, 105 (1966)
11. M. Lax and W. H. Louisell, IEEE J. Quantum Electron. *3* 67 (1967); R. Graham and H. Haken, Z. Physik *235*, 166 (1970).
12. W. E. Lamb, Jr., Phys. Rev. *134*, A1429 (1969).
13. C. Normand, Y. Pomeau and M.G. Velarde, Rev. Mod. Phys. *49*, 581 (1977)
14. W. W. Chow, M. O. Scully and E. W. Van Stryland, Opt. Commun. *15*, 6 (1975).
15. L. D. Landau and E. M. Lifshitz, *Statistical Physics* (Pergamon, London, 1958).
16. V.P. Chebotayev, I.M. Beterov and V. N. Lisitsyn, IEEE J. Quantum Electron. *QE-4* 788 (1968).
17. R. Graham, Phys. Rev. Lett *31* , 1479 (1973). See also in *Order and fluctuations in Equilibrium and Non-Equilibrium Statistical mechanics*, edited by G. Nicolis, G. Dewel and J. W. Turner (Wiley, N. York, 1981) pp. 235-73 and references therein.
18. H. Haken, *Synergetics* (3rd. edition, Springer-Verlag, Berlin, 1983). *Advanced Synergetics* (Springer-Verlag, 1983).
19. W.A, Smith, Phys. Rev. Lett. *32*, 1164 (1974).
20. J. Swift and P.C. Hohenberg, Phys. Rev. *A 15*, 319 (1977).
21. P. Glansdorff and I. Prigogine, *Thermodynamics of Structure, Stability and Fluctuations* (Wiley, N.York, 1971). See also G. Nicolis and I. Prigogine, *Self-organization in Nonequilibrium Systems* (Wiley, N. York, 1977).
22. J.E. Wesfreid, Y. Pomeau, M. Dubois, C. Normand and P. Bergé, J. Phys. (París) *39* , 725 (1978).

COLLECTIVE PHENOMENA IN QUANTUM OPTICS

F.T. Arecchi

Istituto Nazionale di Ottica and University of

Florence, Largo E. Fermi, 6- 50125 Firenze, Italy

INTRODUCTION

One of the recent chapters of physics has been the study of fluctuations and coherence in lasers: how and why 10^{20} atoms or molecules, rather than radiating e.m. field in a chaotic fashion, decide to "cooperate" to a single coherent field mode; then, for still higher excitation, how and why they organize in a complex pattern (many modes), each per-se highly coherent but with little correlations with one another. Such a "turbulent" stage has a "complexity" today represented by the word "chaos".

Quantum optics deals with lasers and laser-like phenomena. At an elementary level, they can be understood in terms of perturbation theory at lowest orders; that is in terms of competition between stimulated and spontaneous emission processes (Sec.1).

If one tries to build a nonperturbative picture, one is struck by the complexity of the problem. There are two conceptual escape ways: on one hand to increase the size of the system to infinity while keeping the density finite and the temperature uniform, and look for an asymptotic (thermodynamic) solution; on the other hand to drastically simplify the boundary conditions making it possible to excite only one or a few radiation modes. This was the original idea of Schawlow and Townes, when extending the Maser principle to optical frequencies by use of a Fabry-Perot cavity. This is also done in other classes of nonlinear field problems such as hydrodynamical instabilites, where one works with small "aspect ratios", that is, with cells with comparable sizes in all three dimensions, in order to excite few Fourier components of velocity and thus deal with a finite number of coupled equations. In quantum optics this procedure leads to a set of quantum equations still unsoluble. A further approximation is the so-called "semi-classical" one, leading to Maxwell-Bloch equations. These approximations are discussed in Sec. 2.

Sec. 3 shows the analogies between the study of bifurcations in nonlinear systems and the phase transitions in thermodynamic systems.

Sec. 4 introduces the experimental methods of photon statistics and show how the first fundamental instability of quantum optical devices, that is, the laser threshold can be studied in detail.

Finally Sec. 5 discusses a new features of quantum optical systems under current investigation. It is generally known that $n \geq 3$ degrees of freedom nonlinearly coupled may lead to multiperiodic or chaotic oscillatory behavior (turbulence). Since quantum optics, in the finite-boundary plus semiclassical approximations, is ruled by the 5 Maxwell-Bloch equations, one expects similar behavior in quantum optical devices. Often these instabilities are ruled out by time scale consideration. When the atomic variables have fast damping times, at any instant they are in quasi-equilibrium with the rather slow field amplitude, hence the evolution reduces to a one-equation dynamics (adiabatic elimination of atomic variables). That is why a gas laser, beyond threshold, assumes a smooth coherent behavior. But change the active medium, or add an external modulation as done for Q-switching or mode-locking: then one easily gets three-variables dynamics, sufficient to yield chaos, for particular values of the coupling constants. What was initially considered as a "bad" or "dirty" behavior (self-pulsing, irregular mode-locking) is nowadays studied as relevant phenomena.

REFERENCES

We give some general references to collective effects in quantum optics, considered as a chapter of non-equilibrium statistics mechanics:

1) Part I (Basic theory and laser physics) of "Laser Handbook", ed. by F.T. Arecchi and E.O. Schulz-Du bois, 1972, North Holland
2) H.Haken, "Synergetics -An Introduction", 2nd ed. 1978 Springer
3) "Order and Fluctuations on Equilibrium and Non-Equilibrium Statistical Mechanics", Proc. XVII Solvay Conf. on Physics, ed. by G. Nicolis, G. Dewel, and J.W. Turner, J.Wiley 1981.

These will be sufficient for Sec. 1 to 4. Sec. 5 will be provided with its own References.

1. PHYSICS OF STIMULATED EMISSION PROCESSES

In a rectangular e.m. cavity of sides X_1, X_2, X_3, volume $V = X_1X_2X_3$, the solution of the wave equation, with periodic boundary conditions, yields the plane wave expansion for the field

$$E(x,y,z,t) = \sum_k E(\vec{k},t)\, e^{i(k_1 x + k_2 y + k_3 z)} \tag{1.1}$$

where $k_i = n_i \cdot 2\pi / X_i$ $(i = 1,2,3 \,;\; n_i = 1,2, \ldots)$.
For each set of k_i we have a different field configuration,or mode.

The dispersion relation imposes a constraint between frequency ω_k and amplitude of the k vector

$$\omega_k = c\, k \tag{1.2}$$

In k space each mode occupies an elementary volume

$$\delta^3 k = (2\pi)^3 / V \tag{1.3}$$

In a spherical shell of radius k and thickness Δk there are

$$\Delta M = 2(4\pi k^2 \Delta k/\delta^3 k) = 8\pi(\nu^2 / c^3)\Delta\nu\, V \tag{1.4}$$

modes. The extra-factor 2 accounts for the two possible polarizations for each vector. Let the cavity contain radiators (atoms on the walls or inside).

The distinction between spontaneous and stimulated emission came in the 1917 Einstein's derivation of the black body formula. In the interaction between an atom and a field mode with n photons, the spontaneous rate A and the stimulated emission rate B are related by

$$A/B = \Delta M \tag{1.5}$$

This can be interpreted by representing the field modes as boxes and the excited atom as linked to all of them as in Fig.1. With the probabilities there indicated, the stimulated emission probability into the mode with n photons is larger than the total spontaneous emission over the empty ΔM modes (all those within the linewidth of the atomic emission when

$$n > \Delta M \tag{1.10}$$

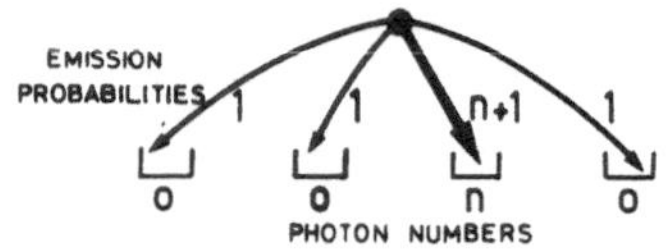

Fig.1. Decay channels of an excited atom into different field modes.

Let us call ΔN the atomic population difference between upper and lower states, P the rate of excitation (pump), n the photon number in the laser mode and

$$T_c = (L/c)\ (1/\theta) \tag{1.11}$$

the decay time of photons in the cavity made of two facing mirrors separated by a length L. T_c is equal to $1/\theta$ transit times, since the limited morror transmittivity $\theta = 1 - \rho < 1$ increases the number of transits. Condition (11) stems from considering photons as particles. It is a necessary, but not sufficient condition. Indeed, if we account for wave propagation and phase matching between forward and backward waves, the cavity is resonant for those frequencies corresponding to the standing wave condition

$$m\ \lambda/2 = L \qquad \text{(m integer)}$$

which amounts to a minimum frequency separation

$$\Delta\nu_{m,\ m+1} = c/2L \tag{1.12}$$

Only for these resonances the escape time is given by (11), otherwise it is much faster (just one transit time L/c).

The rate equations for photons and population inversion are then

$$dn/dt = B\,\Delta N\,n - n/T_c$$
$$d\Delta N/dt = P - B\,\Delta N\,n \qquad (1.13)$$

where we have neglected spontaneous processes. Solving them at equilibrium, the first gives

$$\Delta N = (B\,T_c)^{-1} \qquad (1.14)$$

and the second $P = B\,\Delta N\,n.$ (1.14)'

Combining the two with (10), the pump rate must be

$$P > \Delta M/T_c \qquad (1.15)$$

Let us introduce the concept of atomic cross section σ. The stimulated emission rate Bn can be written as

$$Bn = \sigma\,\phi \qquad (1.16)$$

where $\phi = cn/v$ is the photon flux and hence

$$\sigma = B\,V/c\ .$$

When the atomic line is broadened only by spontaneous emisssion process, put $\Delta\nu = A$ in Eq. (4) and have

$$\Delta M = (8\pi/\lambda^2)\ (V\,B/c)\ \Delta M$$

Hence

$$\sigma = \lambda^2/8\pi \qquad (1.17)$$

If thereis an extra broadening $\Delta\nu_a > A$ for collision process or other decays, σ reduces as

$$\sigma = (\lambda^2/8\pi)(A/\Delta\nu_a). \qquad (1.17)'$$

Cross section (17) holds for a bound electron, while for a free electron (fig.3) it is the square of the classical electron radius $r_o \approx 10^{-13}$ cm. On the other hand, writing the volume as V = S.L, condition (15) can be rewritten as

$$P > \theta \frac{S}{\sigma} \Delta\nu ,$$

which shows that the excitation rate is proportional to the ratio between the laser beam cross section S and the atomic cross section. Fig.2 shows why bound electrons are better than free electrons. However nowadays using high energy (~ 1 GeV) free electrons in a storage ring one can produce laser action down at wavelengths less than $\lambda = 1\ \mu$m.

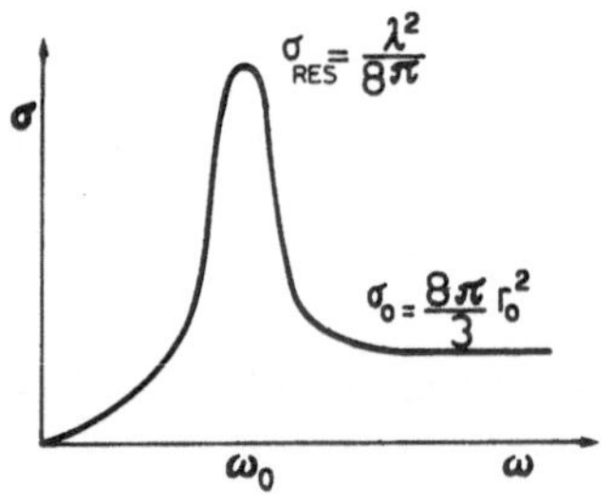

Fig. 2. Radiative electron cross-section versus frequency.

Fig.3 summarizes the different interactions and the spectral regions covered by lasers.

Once $n > \Delta M$ is fulfilled, that is, once the privileged mode has enough photons to neglect spontaneous decay channels, we must also take care for the cavity losses, and by (14) require that

$$B\,\Delta N \geq 1/T_c$$

This condition is represented in Fig.4 for two different ΔN. In the first case only one mode is above threshold, hence we have a single monochromatic frequency. In the second case we may have emission at three frequencies. Here we must introduce the fundamental difference between homogeneous and inhomogeneous linewidth. In the former case a monochromatic transition is broadened by cir-

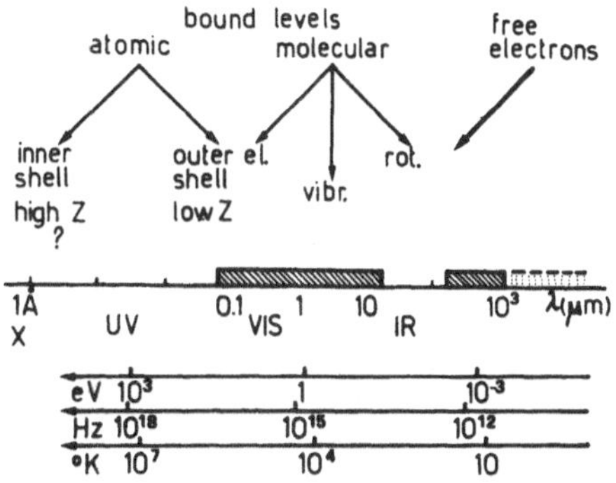

Fig . 3. Map of available lasers

cumstances which are equal for all atoms in the cavity (as spontaneous lifetime broadening in a gas,phonon interaction in a solid matrix). All atoms can contribute over the whole linewidth. Hence, once the mode nearest to the peak has been excited, as the associated field "sweeps" the cavity, it will "eat" all atomic contributions, forbidding the other modes form going above threshold.

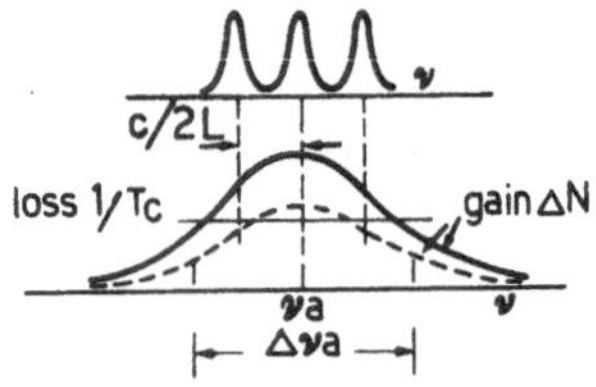

Fig. 4. One and three mode laser (dashed and solid gain respectively)

The inhomogeneous line broadening corresponds to different frequency locations of different atoms. This can be due, e.g., to Doppler shift in a gas where thermal agitation gives a distribution of velocities. Another inhomogeneity occurs in a crystal where active ions are exposed to a crystal field which changes from site to site. For an inhomogeneous line, different modes can go above threshold even without a standing wave pattern.

In general, if $c/2L$ is much smaller than the atomic linewidth there are many independent laser lines, without phase relations. If the different laser fields have fixed phase relations (mode locking), they act as the different Fourier components of a train of pulses, each lasting $1/\Delta\nu_a$ and separated by $2L/c$ (fig.5).

By using a Doppler broadened atomic line in a gas (like He-Ne, or in an A^+ laser) then

$$\Delta\nu_a \sim \nu/c \left(\frac{kT}{M}\right)^{\frac{1}{2}} \sim 10^9 \text{ Hz},$$

hence $t_{pulse} \simeq 1$ ns.

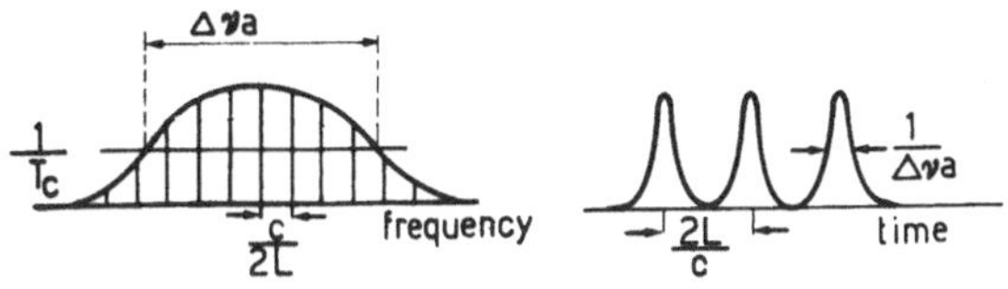

Fig.5. Mode locking operation

Using ions of a transition element embedded in a crystal or glass matrix, as Cr^{3+} in Al_2O_3 (ruby), or Nd^{3+} in glass, one may have large $\Delta\nu_a$. A large $\Delta\nu_a$ can also be achieved in the case of complex dye molecules in a liquid solution because of the overlapping among many vibrational and rotational levels. It is nowadays easy to achieve

$$\Delta\nu_a \sim 10^{13} \text{ Hz},$$

and hence

$$t_{pulse} \simeq 0.1 \text{ p sec}$$

Practical Lasers

Eq. (17)' gives the radiative cross section for a material with a spontaneous lifetime $\tau_{sp} = 1/A = 1/\Delta\nu_{sp}$ and a total linewidth $\Delta\nu_a$. The total gain per unit length $\alpha(cm^{-1})$ is given by

$$\alpha = \rho\,\sigma \tag{1.18}$$

where ρ is the density of inverted atoms. The minimal length ℓ of a lasing medium has to be of the order of $1/\alpha$. Once $\alpha\ell \sim 1$, laser action can be obtained even with a small feedback at the end of the rod.

In order to compare different media, we must briefly review the origins of the line broadening $\Delta\nu_a$. For sake of simplicity let us take a classical field, and evaluate the dipole source due to a radiating atom. It will be the expectation value of the dipole operator over the atomic state $|\psi\rangle$

$$\langle d \rangle = e \langle r \rangle = e \langle \psi | r | \psi \rangle \quad . \tag{1.19}$$

Now, let us consider an atomic transition from an s to a p state. It is easily seen from symmetry arguments that $\langle d \rangle = 0$ when $|\psi\rangle$ is the pure s or p state.

When interacting with a constant E field, an atom is driven back and forth in a reversible way as shown in fig. 6 at a rate

$$\Omega = \hbar / \mu E \tag{1.20}$$

where $$\mu = e \langle s | r | p \rangle \tag{1.21}$$

is the transition matrix element (taken for simplicity as a real number) and Ω is called the Rabi frequency. The instantaneous wavefunction $\psi(t)$ is given by

$$\psi(t) = a(t)| s\rangle + b(t)| p\rangle \tag{1.22}$$

where a(t) and b(t) have sinusoidal variations as shown in the figure.

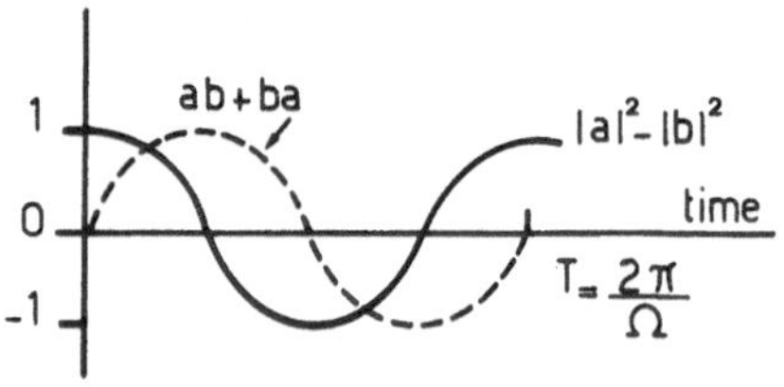

Fig. 6 . Population difference (solid line) and induced dipole (dashed line)

Correspondingly the induced dipole

$$\langle d \rangle = \mu (a^* b + ab^*) \qquad (1.23)$$

goes as the dashed line in fig.6. As shown in eq. (1.23) it depends on the phase of the wavefunction, whereas the population inversion $|b|^2 - |a|^2$ is phase independent. Both $|a|$ and $|b|$ return to equilibrium values $|a| = 1$, $|b| = 0$ by spontaneous emission processes. On a faster scale, phase destroying processes, as collisions, interrupt the coherent Rabi precession as shown in fig.7. The average interruption time is called T_2, and its reciprocal is the homogeneous linewidth $1/T_2 = \Delta \nu_a > \Delta \nu_{sp}$.

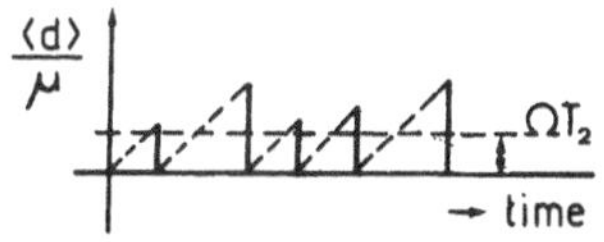

Fig. 7. Interruptions due to decay processes, giving a non zero polarization

Since $\langle d \rangle / \mu = \Omega T_2$ then the average dipole is

$$\langle d \rangle = \mu^2 E T_2 / \hbar = \mu^2 E / (\hbar \Delta \nu)$$

and the polarization (ρ being the atomic density)

$$P = \rho \langle d \rangle = \frac{\rho \mu^2}{\hbar \Delta \nu} \cdot E. \qquad (1.24)$$

We have thus shown a simple derivation of the susceptibility χ. If however there is an inhomogeneous broadening (as Doppler broadening in a gas) whose linewidth is bigger than $1/T_2$ (that is, $\Delta\nu_{inh} = 1/T_2^*$, $T_2^* < T_2$), then σ is evaluated with $\Delta\nu_{inh}$.

The Table summarizes the parameters of some lasers.

Table 1

	Ruby Nd	Dye	He-Ne	Ar^+	CO_2	Semi-conduct.
τ_{sp} (sec)	10^{-3}	10^{-8}	10^{-8}	10^{-8}	0.2	10^{-9}
T_2(sec)	10^{-12}	10^{-13}	10^{-8}	10^{-8}	$(4.4\times10^7/\text{torr})^{-1}$	10^{-14}
T_2*(sec)	10^{-10}	10^{-13}	10^{-10}	$\frac{1}{3}10^{-10}$	$(3.14\times10^8)^{-1}$	--
$\sigma(cm^2)$	3×10^{-20}	3×10^{-16}	10^{-12}	10^{-12}	10^{-15}	10^{-15}
$\rho(cm^{-3})$	10^{19}	10^{15}	10^9	10^{10}	10^{14}	10^{18}
$1/\alpha=\ell$(cm)	10	0.1	$10^{3(*)}$	$10^{2(*)}$	10	3×10^{-2}

(*) Good mirrors reduce the cavitylength to 10 - 100 cm

2. WHAT IS QUANTUM AND WHAT IS CLASSICAL IN LASER PHYSICS

The interaction model.

Let us consider an e.m. field and a set of atoms confined in a cavity. The interaction Hamiltonian is

$$H_{f-a} = - e/m \sum_i \vec{A}(x_i) \cdot \vec{p}_i$$

x_i being the position of atom i, $\vec{A}(x_i)$ the value of the vector potential at x_i, and p_i the momentum of the i-th electron.

Expand the field in plane waves

$$A(x;t) = \sum_k \left(\frac{\hbar}{2\varepsilon_o\omega_k V}\right)^{1/2} (a_k e^{-i(\omega_k t - kx)} + c.c.) \qquad (2.1)$$

(V = cavity volume; mks system used) and consider a_k and the conjugate a_k^+ as Bose operators:

$$\left\{ a_k , a_{k'}^+ \right\} = \delta_{kk'} \qquad (2.2)$$

For simplicity, we skip vector relations and give a scalar theory.

We consider the atoms as two-level atoms, so that the Hilbert space of a single atom is fully described by the identity operator I plus the three Pauli operators $\sigma^{\pm}$, σ_3,

$$\{ \sigma^+ , \sigma^- \} = \sigma_3 \quad , \{\sigma^{\pm}, \sigma_3 \} = \mp 2\sigma^{\pm} \qquad (2.3)$$

It is then a straightforward matter to obtain the following model Hamiltonian

$$H/h = \sum_k \omega_k a_k^+ a_k + \frac{\omega_o}{2} \sum_i \sigma_{3i} + \sum_{i,k} g_k (a_k \sigma_i^+ e^{ikx_i} + h.c.) \qquad (2.4)$$

where

$$g_k = (\omega_k / \hbar \varepsilon_o V)^{1/2} . \mu$$

$\mu = < 1|er|2 >$ being the dipole moment matrix element between the two atomic states coupled by the radiation at frequency ω_k.
We introduce the collective operators

$$J_k^{\pm} = \Sigma_i \sigma_i^{\pm} \exp(\pm i k x_i), \qquad J_2 = \Sigma_i \sigma_{3(i)} \qquad (2.5)$$

Only when the space dependence can be neglected, that is, for three cases

(a) point laser (cavity length $<< \lambda$)
(b) single mode case (one $k=k_o$ value),
(c) travelling plane wave field (again one k_o value),

one has closed commutation relation for $J^{\pm}$, J_z, and the Heisenberg equations of motion are obtained:

$$a = i \omega a - ig J^-, \quad \dot{J}^- = -i \omega_o J^- + iga J_z, \quad \dot{J}_z = ig(a^+ J^- - a J^+) \qquad (2.6)$$

They do not form a closed set, because one would need also an equation for the motion of bilinear operators such as Ja^+, etc

We study here the self-consistent field approximation (SCFA).

$$\text{SCFA:} \quad \langle a J^+ \rangle = \langle a \rangle \langle J^+ \rangle \tag{2.7}$$

The approximation is motivated by the following circumstances. First, if we shine a classical current on a quantum e.m. field initially in its vacuum state, then the field goes into a state with nonzero expectation value for the annihilation operator, $\langle a \rangle \neq 0$. Such a state (coherence state) which is minimum uncertainty, is very different from the energy eigen states used in the standard treatment of a harmonic oscillator, and obtained by the application of a finite number of creation operators to the vacuum state. Second, if we shine a classical field on a quantized atomic system, this latter one goes to a state which has similar coherent properties.

The two above approximations are semiclassical, insofar as only one of the coupled variables (field or atom respectively) are taken as quantum. In practice, once the quantum commutation relations have led us to Eqs. (2.6), a suitable approximation consists in considering them as C-number eqs. In such a case, the first is equivalent to Maxwell eq. for a single mode in Hamiltonian form, the other two are analogous to Bloch equations in Nuclear Magnetic resonance, and in vector form can be summarized as

$$\langle \dot{\vec{J}} \rangle = \vec{\Omega} \times \langle \vec{J} \rangle \tag{2.8}$$

where $\vec{\Omega} = (g \langle a \rangle, \ 0, \ \omega_0)$.

The semiclassical approach (only field or only atoms quantized) is summarized in the following scheme

$$\begin{array}{llll} \text{Classical current} & \xrightarrow{\text{Maxwell}} & \text{field state } \langle a \rangle \neq 0 & \searrow \\ & & & \text{coherence states} \\ & & & \nearrow \\ \text{Classical field} & \xrightarrow{\text{Bloch}} & \text{atomic state } \langle J^{\pm} \rangle \neq 0 & \end{array}$$

<u>The Maxwell-Bloch (M.B.) equations</u>

We approach the Bloch equations directly by the Schrödinger method rather than the Heisenberg method. In this way we attribute a physical meaning to the three components of $\vec{J}$.

The time-dependent wave function $\psi = a\,|s\rangle + b\,|p\rangle$ of a two-level atom, under a coherent field excitation, $E = \varepsilon \cos \omega t$ obeys the Schrödinger equation. We set up a linear combination

$$\vec{J} = \begin{pmatrix} a^* b + b^* a \\ i(b^* a - a^* b) \\ |a|^2 - |b|^2 \end{pmatrix} \tag{2.9}$$

of the two complex amplitudes a and b which weight the eigenstates $|s\rangle$ and $|p\rangle$ of the free-atom Hamiltonian. It obeys the vector equation (2.8). By evaluating $\langle\psi|\, er\, |\psi\rangle$, it is easy to see that the three components of J have the following meaning

J_1 is proportional to the in-phase polarization (dispersion),

J_2 is proportional to the out-of-phase polarization (absorption or gain),

J_3 is proportional to the population inversion.

We then write $\cos\omega t = \frac{1}{2}(e^{i\omega t} + e^{-i\omega t})$ and go to a frame rotating around z at angular velocity ω_0 (see Fig.2).

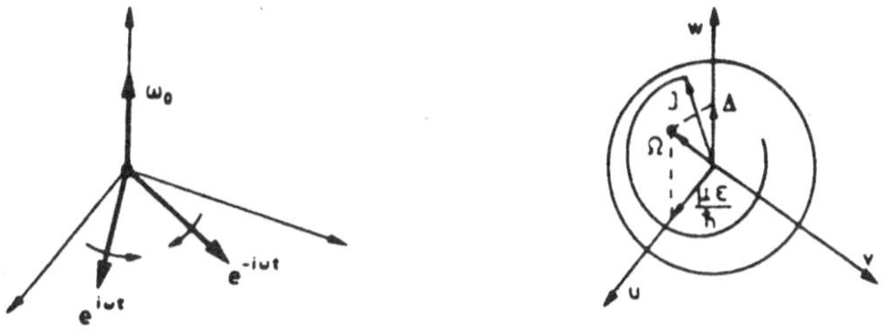

a b

Fig.8. a) splitting of $\cos\omega t$ into two counter-rotating components
b) precession of J around Ω. The end point of $\vec{J}$ moves on a sphere of radius N/2

$$\Omega \equiv \left(-\frac{\mu\varepsilon}{h},\ 0,\ \Delta = \omega - \omega_0 \right) \tag{2.10}$$

consider $\Delta = \omega - \omega_0 \ll \omega_0$ and hence neglect the effect of the fast component $e^{-i(\omega+\omega_0)t}$ (rotating wave approximation = RWA). The Bloch equations reduce to

$$\frac{d}{dt}\begin{pmatrix} u \\ v \\ w \end{pmatrix} = \begin{pmatrix} \Delta \cdot v & \\ -\Delta \cdot u & + \mu \varepsilon w \\ & - \mu \varepsilon v \end{pmatrix} \qquad (2.11)$$

This is the motion of a point on a spherical surface, as seen from the conservation relation

$$u^2 + v^2 + w^2 = \text{constant} = (N/2)^2 \qquad (2.12)$$

Indeed, at equilibrium, u=v=0. W= - N/2.
At resonance ($\Delta = 0$), we may take $u=0, \phi = 0$ always and reduce to

$$\dot{v} = \mu \varepsilon w$$

$$\dot{w} = - \mu \varepsilon v$$

with $v^2 + w^2 = 1$

The motion of $\vec{J}$ on the Bloch sphere is a motion on the (v,w) plane perpendicular to the ε direction. The Bloch eq. gives the motion of an atomic system shined by a field E (local effect). If we want to consider the reaction of the atoms back on a field propagating through the atomic system, we must couple Bloch with Maxwell equations. For a plane wave field

$$E = \varepsilon(z,t) \cos (\omega t - kz + \phi(zt) \qquad (2.14)$$

if ε and ϕ are slowly varying compared to optical period and wavelength, i.e., if

$$\frac{1}{\varepsilon}\frac{\partial \varepsilon}{\partial t} \ll \omega, \quad \frac{1}{\varepsilon}\frac{\partial \varepsilon}{\partial z} \ll k, \quad \frac{\partial \phi}{\partial t} \ll \omega, \quad \frac{\partial \phi}{\partial z} \ll k \qquad (2.15)$$

(slowly varying envelope approximation = SVEA), we can neglect the second derivates and have two coupled equations.

$$(\partial/\partial t + c\, \partial/\partial z)\varepsilon = (\omega N \mu / 2\varepsilon_o)\, v$$

$$(\partial/\partial t + c\, \partial/\partial z)\phi = (\omega N \mu / 2\varepsilon_o)\, u/\varepsilon \qquad (2.16)$$

Notice that v is the source term for amplitudes and u for phase.
Here N (cm^{-3}) is the density of atoms.

Let us now introduce phenomenological lifetimes T_2 for polarization components, T_1 for inversion ($T_2 \leq T_1$), and 1/K for the field, In case the material is confined in a region of size L without mirrors, $1/K$ is the transit time L/c.

Further, let us introduce, besides the atoms and field, a third partner, the 'pump', which in the absence of field yields a steady-state value of w

$$w = w_o$$

The Maxwell-Bloch equations become

$$\partial/\partial t \begin{pmatrix} u \\ v \\ w \end{pmatrix} = \begin{pmatrix} \Delta . v & & - u/T_2 \\ - \Delta . u & + \mu \varepsilon w & - v/T_2 \\ & - \mu \varepsilon v & -(w - w_o)/T_1 \end{pmatrix} \tag{2.17}$$

$$(\partial/\partial t + c \; \partial/\partial z)\varepsilon = (\omega N \mu/2 \varepsilon_o) v - k\varepsilon \tag{2.18}$$

In the absence of a pump, we must put

$$w_o = -1$$

because the atoms will spontaneously decay toward the South Pole of the Bloch sphere, with a lifetime T_1. A suitable way of re-writing MB eqs. is to normalize the field intensity in terms of photon number, and to write te Bloch vector for the totality N of atoms. For simplicity we refer to resonance ($\Delta = 0$), so that we have real field amplitude and one polarization component. Then MB eqs. become

$$\begin{aligned} (\partial/\partial t + c \; \partial/\partial z) a &= g S - k a \\ \partial/\partial t \; S &= 2g a \Delta N - \gamma_\perp S \\ \partial/\partial t \; \Delta N &= -2g a S - \gamma_\parallel (\Delta N - N/2) \end{aligned} \tag{2.19}$$

where

$$g = (\omega\mu^2 / 2\hbar \varepsilon_o V)^{1/2} \tag{2.20}$$

is the coupling constant and

$$k = 1/T_c \;, \quad \gamma_\perp = 1/T_2 \;, \quad \gamma_\parallel = 1/T_1$$

are field and atom loss rates. The normalization is such that $a^2=n$ is the photon number, and ΔN the number of the inverted atoms in volume V. Also, N is a source term; for an absorbing medium in the ground state (no pumping) it must be changed into $-N_{tot}$. We neglect here the space derivative $\partial/\partial z$. Care can be taken by transforming to a suitable moving frame in the cases of an amplifying and of an unexcited medium.

The order of magnitude, for dilute gas of atoms with allowed transitions in the visible, and for $V = 1\ cm^3$, is

$$g \sim 10^4\ s^{-1}, \qquad \gamma_\perp \cong \gamma_\parallel = \gamma \sim 10^8\ s^{-1},$$

and

$$k \sim 10^7\ s^{-1}, \quad \text{or} \quad 10^{10}\ s^{-1}$$

depending on whether the gas is in a laser cavity or is distributed over a length of some centimeters, without mirrors at the end.

The numbers given above suggest two distinct time scales, when $k \ll \gamma$. The fast atomic variables S, ΔN relax toward their equilibrium values, while the slow variable a changes little. Hence a is the order parameter in terms of which the coupled system can be described.

Single Mode Laser

When $\gamma \gg k$, eqs. for S and ΔN can be solved at steady state in terms of field amplitude, giving

$$S = (gN/\gamma)\,(a/1+x^2) \cong g\frac{N}{\gamma}\,a\,(1-x^2) \tag{2.21}$$

where we have introduced the ratio of Rabi frequency to atomic loss rate

$$x = 2\,g\,a/\gamma$$

Since $g/\gamma \sim 10^{-4}$, it takes $a^2 \sim 10^8$ photons to have $x = 1$. Hence for $n < 10^8$ photons the saturation term in the polarization can be approximated as a cubic correction. Replacing S in the first of eqs. (19), the field equation becomes

$$\dot{x} = k(C\frac{x}{1+x^2} - x) \cong k((C-1)x - C\,x^3) \tag{2.22}$$

where C is a cooperation parameter given by

$$C = g^2N/\gamma\,K \tag{2.23}$$

The steady, stable solution x^2 goes smoothly from zero to a nonzero value as C goes from below to above 1 (fig.9).

Optical Bistability

Take the same configuration (atoms within a cavity), leaving the atoms in the ground state (this changes the sign of S) and injecting an external field $\underline{a}$, which amounts to changing the first of eqs. (2.19) into

$$\dot{a} = gS - k(a - \alpha) \tag{2.24}$$

Introducing $y = 2g\alpha/\gamma$ gives for the steady solution

$$y = x + C\,x/1+x^2 \tag{2.25}$$

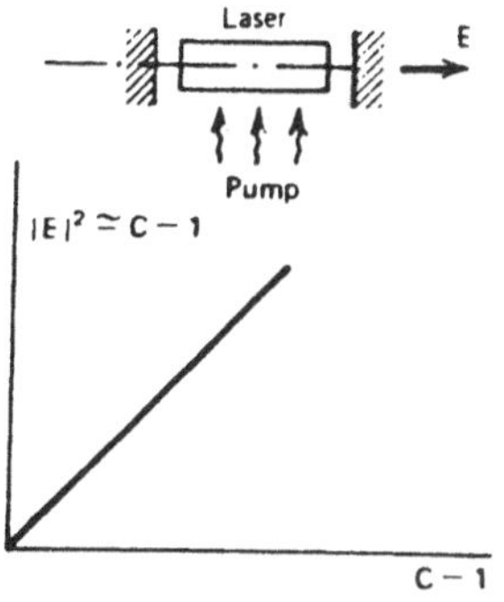

Fig.9 . Plot of the square of the order parameter(intensity of the laser field) versus the cooperation number((c-1)is proportional to the difference between gains and losses.

Plots of transmitted versus injected field are given in fig. 10 for increasing atomic densities, that is, for increasing C values. For $C = 0$ the cavity is tuned so that transmitted and incident fields are equal. For $C < 1$ the system is in the linear absorption regime. For $C \gg 1$ the system jumps into the saturated regime, where atoms become transparent, and hence there is a return to full transmission.

By decreasing the impinging field, one gets a hysteresis cycle peculiar of a first-order phase transition.

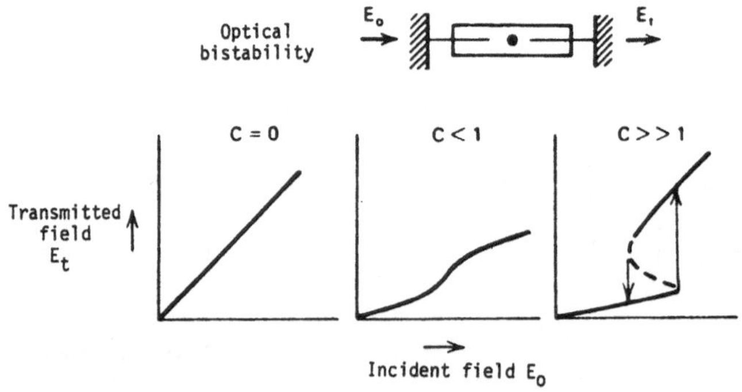

Fig. 10. Plot of transmitted field versus impinging field in an optical bistable device, for three different values of cooperation number C

3. BIFURCATIONS IN NON-EQUILIBRIUM SYSTEMS

Analogies with phase transitions

In this Section we show the analogies between phase transitions in a thermodynamic system (with boundary conditions at infinity) and bifurcations in the stationary solution of a pumped system, that is, a nonlinear finite system with excitations at the boundary . The analogies are displayed in the Table.

Table 2

		Phase transitions in Quantum Optics	
		Thermodynamic	Non equilibrium (pumped)
steady	2nd order	- order - disorder - ferro/paramagnetic H = 0 - liquid gas (at critical point)	- laser threshold
steady	1st order	- ferro/paramagnetic H = 0 - liquid gas (away from a critical point	- laser + saturable absorber - optical bistability
TRANSIENT		(spinodal decomposition)	- laser transient - superfluorescence

The three main features of the critical point of a phase transition are listed in fig. 11 for the case of a para-to-ferromagnetic transition. $\langle M \rangle$ is the average magnetization, or order parameter, which is zero below and increases above the critical point ; $\langle \Delta M^2 \rangle$ is the variance in the fluctuations of M, which diverges at the critical point; and τ is the lifetime of the correlations in the fluctuations $\langle \Delta M(o) \; \Delta M(t) \rangle$ of the order parameter, which also diverges, giving a zero spectral width.

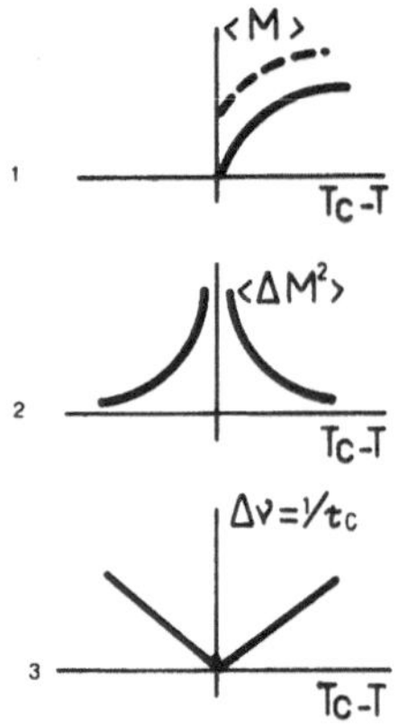

Fig.11. Solid lines:main features of 2nd order phase transitions.(Dashed line:1st order phase transition in the presence of a field)

1. long range order
2. critical fluctuations
3. critical slowing down

The features of Fig.11 are qualitatively explained by the Landau model. Let us forget space variations and consider a single-domain (zero-dimensional system) characterized by a dynamical variable q. The free energy can be written as the following expansion

$$F = F(0,T) + \alpha/2 \; q^2 + \beta/4 \; q^4 \qquad (3.1)$$

F(0,T) is slowly varying function of T, β is taken as a positive constant, and α changes with T as

$$\alpha = a(T - T_c) \qquad (a > 0) \qquad (3.2)$$

The thermodynamic probability P(q) is given by the relation

$$P(q) = N e^{-F/k_B T} \qquad (k_B = \text{Boltzmann constant}) \qquad (3.3)$$

and is maximum where F is minimum.

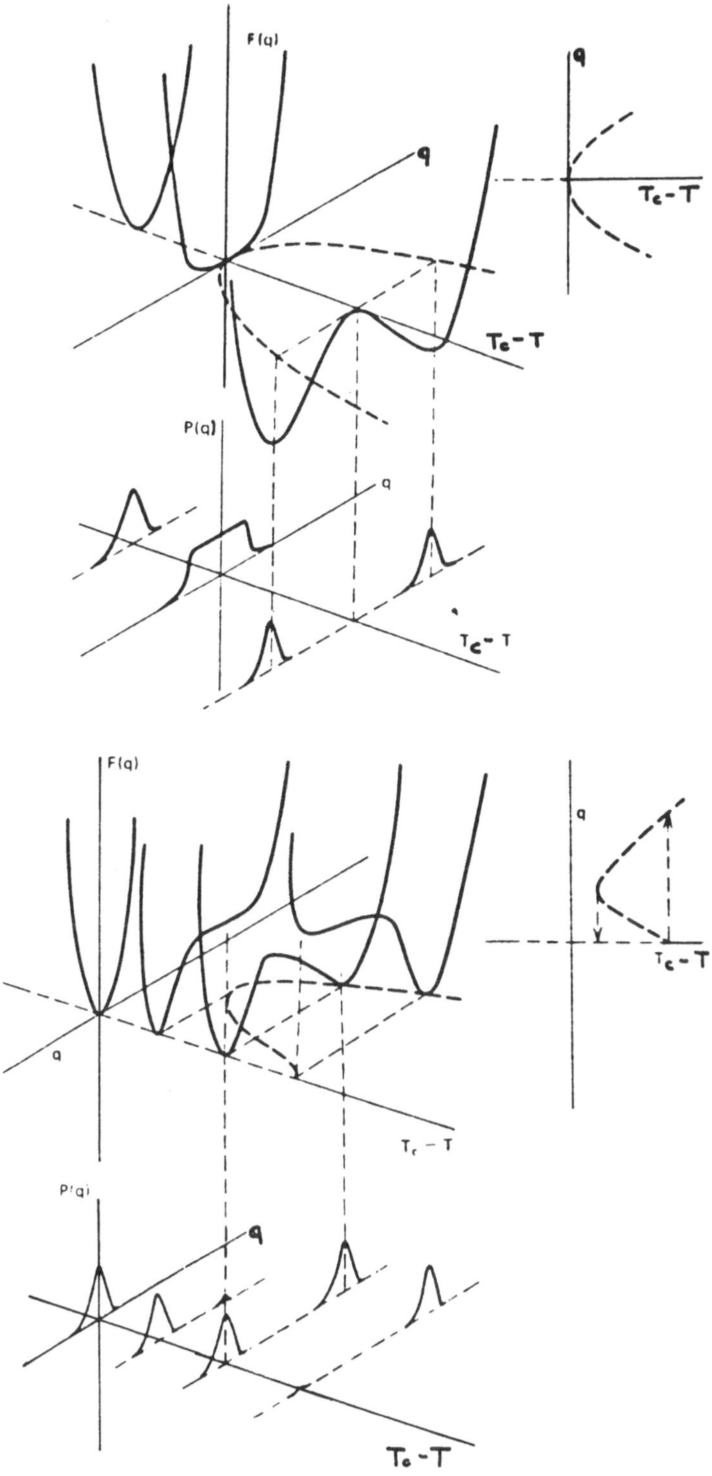

Fig.12.
Free energy and probability for 2nd and 1st order phase transition.

Fig. 11 shows how F and P change versus q and T. If we further introduce and odd term as

$$F - F_o = \alpha/2 \; q^2 + \gamma/3 \; q^3 + \beta/4 \; q^4 \qquad (3.4)$$

we have 1st order phase transitions as shown in fig. 12.

Let us now consider the nonlinear dynamical equation of a Q.O. device as the laser or the bistable system. Including a noise to account for interaction with the Universe (beside the main atoms-field interaction studied by the M.B. eqs.) we have a nonlinear Langevin eq.

$$\dot{q} = K(q) + \xi(t) \quad ; \quad \langle \xi(t)\xi(t')\rangle = D\,\delta(t-t') \qquad (3.5)$$

and consequently a nonlinear F.P. eq. for the probability $P(q,t)$

$$\partial P/\partial t + \partial/\partial q \{ K(q) P - D/2 \; \partial P/\partial q \} = 0 \qquad (3.6)$$

The steady solution is very simply

$$P(q) = N\, e^{-V(q)/D/2} \qquad (3.7)$$

where

$$V(q) = -\int K(q) dq.$$

It is easily seen that (3.7) is similar to (3.3) with $k_B T$ replaced by $D/2$ and the free energy replaced by the potential $V(q)$ of the nonlinear force $K(q)$. Hence fig. 12 may represent respectively the laser and the O.B.

4. PHOTON STATISTICS AND THE LASER THRESHOLD

The Photon Statistics (PS)

Consider a photodetector illuminated by a light beam. By means of an electronic gate lasting for a time T, the number n of photons annihilated at the photo-surface in T is counted. The random variable n has a statistical distribution p(n) that can be determined by iterating the procedure for a large number of samples.

In fig. 13 we show an experimental plot of the statistical distribution of photocounts p(n) versus the number of counts

$$n = \eta \langle I \rangle \cdot T, \qquad (4.1)$$

where $< I >$ is the average intensity, T the gating time and η a constant accounting for the quantum efficiency of the detector plus other instrumental factors. The curves refer to three physical cases that are indistinguishable from the point of view of classical optics. Indeed, in the three cases we have about the same average photon number $<n>$, the same diffraction-limited plane wave, and the same line width $\Delta\omega$, filtered out in such a way that

$$\tau = 1/\Delta\omega \gg T$$

From the PS point of view the three radiation fields are dramatically different .

The three fields L,G,S correspond to the following cases. Field L comes from a stabilized single mode laser well above threshold. Moment analysis shows that it is well approximated by a Poisson distribution, with a variance

$$< \Delta n^2 > = < n > \tag{4.2}$$

Field G (Gaussian) is obtained by scattering L over a collection of microparticles in Brownian motion, being sure that the correlation time is longer than T, and then putting up a diffraction-limited plane wave at a given angle. Moment analysis shows that it is a Bose-Einstein distribution with a variance

$$< \Delta n^2 > = < n > + < n >^2 . \tag{4.3}$$

Field S is the superposition of L (with $< n_1 >$ average photons) and G (with $< n_2 >$ average photons) over the same spatial mode. The associated variance is

$$< \Delta n^2 > = < n_1 > + < n_2 > + < n_2 >^2 + 2 < n_1 > < n_2 > . \tag{4.4}$$

A heuristic view of the photodetection process explains the results given above, without resorting to the theory. If the field is uniform, as we expect for a stabilized laser, the photons, being particles with zero mass, cannot be localized; hence there is no a priori correlation between two annihilation events at two different points either in space or in time. The photocounts from a single detector whose average number is proportional to the square field and the measuring time T

$$< n > = |E|^2 = T \eta$$

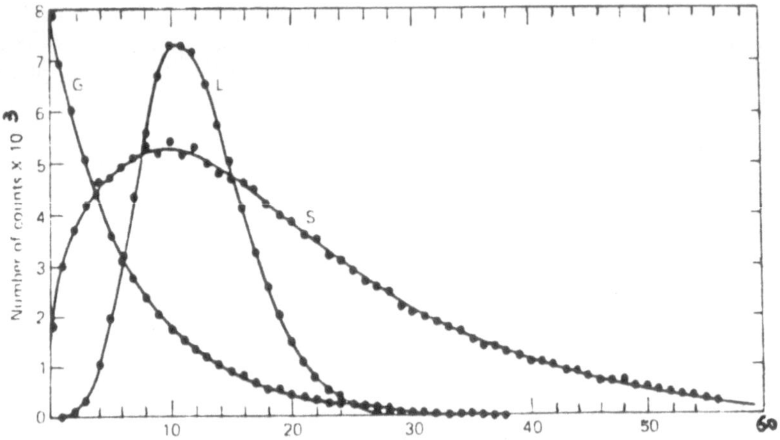

Fig. 13. Photocount distributions of three radiation fields. L=Laser field; G=Gaussian field; S=linear superposition with L and G onto the same space mode.

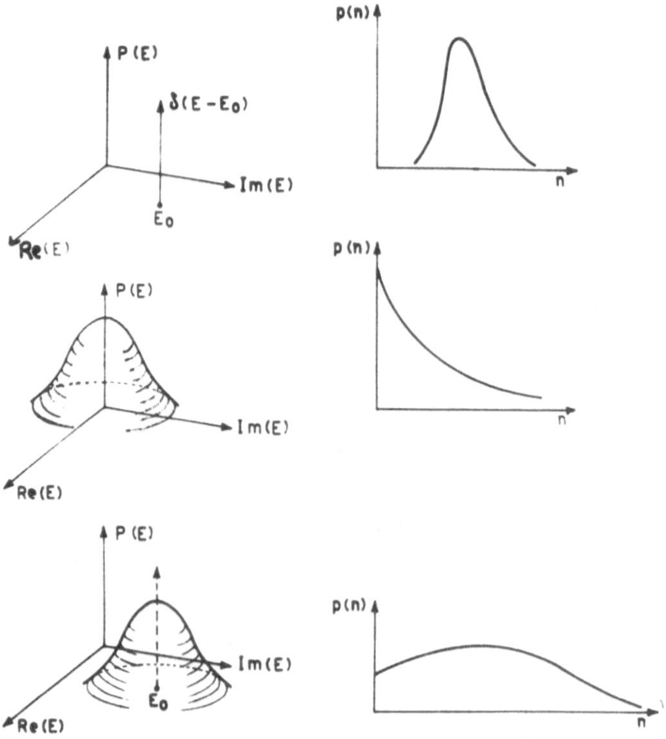

Fig. 14. Field and photon statistical distributions for an ideal coherence field (no fluctuation), for a thermal equilibrium field (Gaussian with zero average), and for the superposition of the two shifted Gaussian. They model the physical cases L, G, and S of fig.13

(η is the quantum efficiency of the detector) must then be distributed as a Poissonian, that is,

$$P_0(n|E) = \langle n \rangle^n \; e^{-\langle n \rangle}/n! \qquad (4.5)$$

This is shown in Fig. 14.

If now the complex field is randomly distributed with a statistics $P_1(E,t)$, and each measurement lasts for a time T much smaller than the coherence time τ (in order to have a constant field within each sample), we must average the detector statistics over the field statistics

$$p(n,T,t) = P_0(n|E)\, P_1(E,t) d^2 E \qquad (4.6)$$

In fig. 14 the results are shown pictorially for the three cases of fig.13.

By the central limit theorem a collection of random scatterers gives a Gaussian field statistics. By eq. (4.6) a Gaussian $P_1(E,t)$ will give rise to a Bose-Einstein distribution. Similarly, the superposition of a Gaussian and a zero-fluctuation distribution gives a convolution statistics for the field, which is a shifted Gaussian, whose associated photon statistics has the variance (4.3).

The Laser Threshold

From the description of Sec. 3 we expect at threshold the appearance od a non-zero-order parameter, a large increase in fluctuations, and a critical slowing down. To correct that picture we must add the following:

1. The laser equation is nonlienar; hence there is neither divergence in the fluctuations nor zero line width. The infinities are smoothed by the nonliearity.
2. The laser field is a complex parameter that should be described in modulus and phase

$$E = /E/\, e^{i\phi}$$

The photon statistics destroys phase information because $n \propto |E|^2$. Phase information is recovered by performing an interference experiment with two independent lasers, so that the output intensity has phase information. Indeed

$$n = |E_1 + E_2|^2 = I_1 + I_2 + 2 \,\mathrm{Re}\; E_1^* E_2 \,.$$

For comparing experiments and theory we use the second reduced factorial moment of the photocount distribution

$$H_2 = \langle n(n-1) \rangle / \langle n \rangle^2 - 1 \tag{4.7}$$

which goes from one (Gaussian field distribution, well below threshold) to zero for an amplitude-stabilized field (well above threshold), and the third reduced factorial moment

$$H_3 = \langle n(n-1)(n-2) \rangle / \langle n \rangle^3 - 1 \tag{4.8}$$

which goes from five (Gaussian distribution) to zero (amplitude-stabilized field) (fig.15).

From the stationary solution of the dynamic laser equations, the distribution of photocounts and the associated factorial moments can be derived. It is evident from the figures that the agreement between experiments and theory is very good.

Measurements of the power spectrum $S(\omega)$ of the intensity fluctuations for a He-Ne laser, both below and above threshold, are reported in fig. 16, showing the critical slowing down.

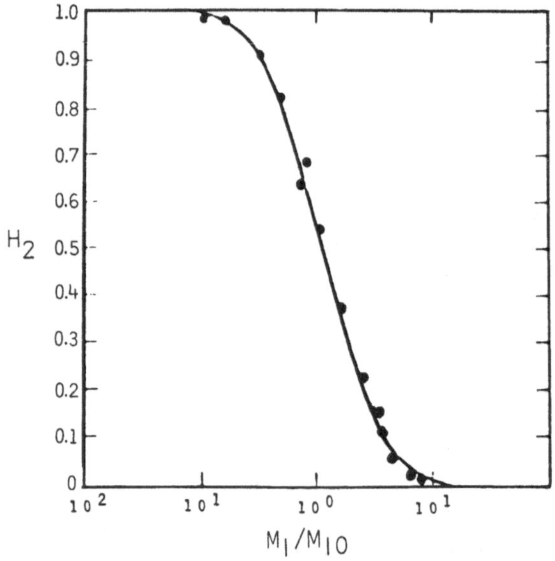

Fig.15.a. Measured and theoretical values of the reduced 2nd-order factorial moment of the photon distribution vs. the intensity M_1 normalized to threshold value M

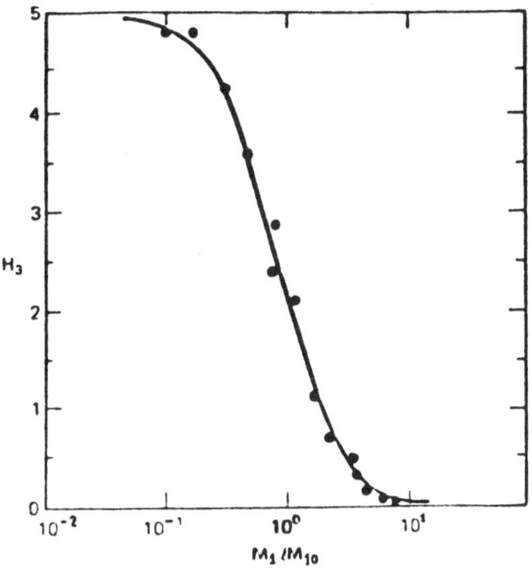

Fig.15.b Measured and theoretical values of the reduced 3rd-order factorial moment of the photon distribution.

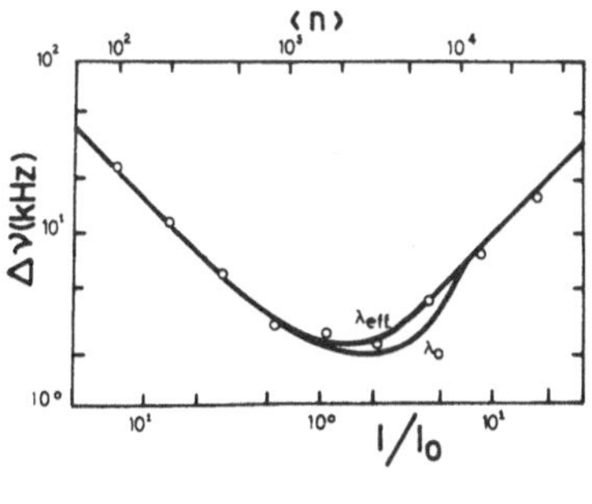

Fig. 16. "effective" line width λ_{eff} of the laser intensity fluctuations versus the intensity I, normalized to the threshold value I_o. The horizontal axis is also calibrated in values of the average photon number $\langle n \rangle$ inside the cavity

5. TURBULENCE IN QUANTUM OPTICS

Nonlinear Dynamics and Turbulence

Let us consider a dynamic system, ruled by the equation

$$\dot{x} = F(x \; ; \; \mu) \tag{5.1}$$

where x is an n-dimensional vector, F a nonlinear function and μ an m-dimensional control parameter.

We study the equilibrium solutions

$$F\,(x \; ; \; \mu) = 0 \tag{5.2}$$

for different μ . For some critical μ , the stationary solution may switch from stable to unstable (bifurcations). An example is the laser threshold. For $\mu \equiv C = 1$ the branch $x_1 = 0$ becomes unstable, and it appears a new stable branch $x_2 = C - 1$.

We are interested in how bifurcations lead some systems to turbulence, or chaos. We call turbulence the appearance of a continuous power spectrum, which corresponds to unpredictable features.

By Wiener-Khintchine theorem the power spectrum $S(\omega)$ of a dynamical variable $x(t)$ is the Fourier transform of its first order correlation function, i.e.,

$$S(\omega) = \mathcal{F}\{\langle x(o)\, x(t)\rangle\}$$

Notice that the above definiton of turbulence does not imply knowledge of higher order correlation functions, hence its statistical relevance is rather limited.

In 1963 Lorenz (MIT) [(1)] showed that three coupled nonlinear eqs. are enough to reach chaos. Now, three eqs. can mimic field eqs. with drastic cut-offs due to the boundary conditions, as said in the Introduction. It can be shown (Chandrasekhar)[(2)] that in case of a fluid heated from below the interplay between the velocity field and the temperature field can be reduced to three coupled equations between one velocity mode and two temperature modes.

If in eq. (1) we take $X = (x,y,z)$ we have a three-dimensional phase space. After a transient, the trajectories become closed loops (periodic or stable attractors)[(3)] corresponding to power spectra made of discrete lines. But for crucial values of the control parameters μ , one has a strange attractor, that is, an intricate loop which never appears to repeat itself, and correspondingly a continuous spectrum.

Eqs. (5.3) are the Lorenz eqs. with parameter values suitable to give a strange attractor.

$$\begin{aligned} \dot{x} &= -10x + 10y \\ \dot{y} &= -xz + 28x - y \\ \dot{z} &= xy - 8/3\, z \end{aligned} \tag{5.3}$$

Notice that Lorenz chaos is a deterministic chaos, because we do not have noise sources .

Eqs. (5.3) are like Maxwell-Bloch. If, instead of adiabatic elimination, we take field (x) polarization (y) and inversion (z) with almost equal damping times, we should obtain turbulence in a single mode laser [4].

Such a 3-mode chaos has not yet been verified in quantum optics . Equivalent to three coupled eqs. is a ystem of two 1st order eqs. (or one 2nd order eq.) plus an external modulation . An example is the driven Duffing oscillator

$$\ddot{x} + \gamma \dot{x} + \omega_o^2 x - \beta x^3 = A \cos \omega t \tag{5.4}$$

which can be experimentally realized with the electronic circuit of fig. 17 [5]

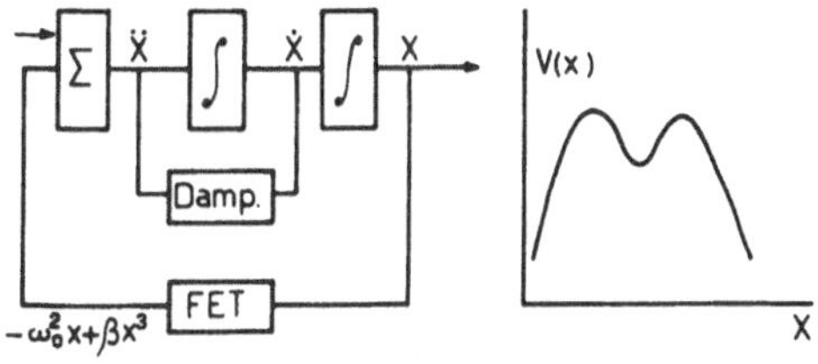

Fig.17. Electronic circuit obeying the Duffing equation and associated one-minimum potential.

Eq. (5.4) is equivalent to 3 coupled eqs.

$$\dot{x} = y$$
$$\dot{y} = -\gamma y - \omega_o^2 x + \beta x^3 + A\cos z$$
$$\dot{z} = \omega \qquad (5.4)'$$

The potential corresponds to a single minimum. For different control parameters μ (either modulation amplitude A or frequency ω) it may give a sequence of subharmonic bifurcation leading eventually to chaos as shown qualitatively

The sequence of subharmonic bifurcations corresponds to the successive appearance of periods T = 2 , 2 T, 4 T ,... $2^n T$ in the output. If we call μ_n the value of the parameter at which the period $2^n T$ appears then the following relation is verified

$$\delta = (\mu_{n+1} - \mu_n)/(\mu_{n+2} - \mu_{n+1}) \quad n \gg 1 \quad 4.669... \qquad (5.5)$$

This number has been shown by Feigenbaum[6] to be universal.

Generalized Multistability

So far we have sketched how, by varying the control parameter a system undergoes a sequence of bifurcations eventually leading to chaos. Noise is not essential (deterministic chaos), but if we add it, the number of subharmonic bifurcations before chaos become smaller and smaller. This can be put in terms of a scaling law where the variance of the external noise appears somewhat as a modification of the control parameter. Let us now explore what I call "generalized multistability".

Depending on the initial conditions, we have two or more independent attractors . Increase μ until they both get strange. Now, addition of a random noise may trigger jumps from one to the other. These jumps may be considered as a superchaos insofar as they couple two strange attractors, otherwise independent. The jump over different attractors occurs in a time which is "secular" (very long) compared to the inner correlation time within each attractor correspondingly we have a low frequency enhancement, which can be approximated by a power law $f^{-\alpha}$ ($\alpha = 0.6 \div 2$) over a few decades in frequency. After the first evidence of the jumping phenomenon, a similar effect was observed in a Q-modulated CO_2 laser [7] . It corresponds to a set of 2 coupled rate eqs., with time dependent cavity losses k(t), that is, calling Δ the population inversion and n the photon number, to

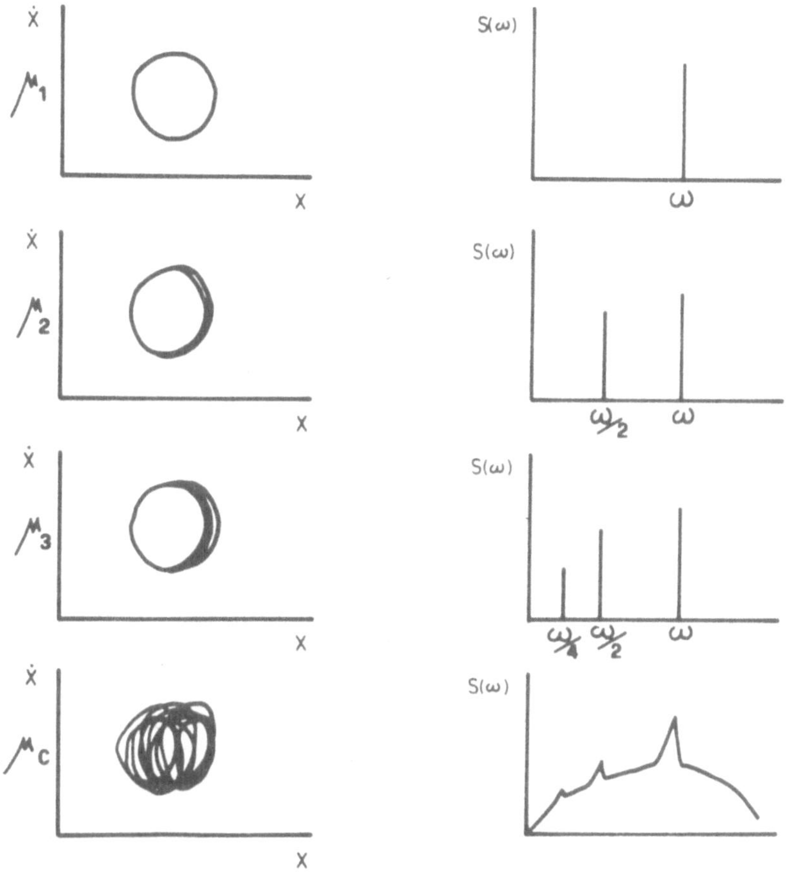

Fig. 18. Phase space plots $(\dot{x}, x)$ and power spectra $S(\omega)$ for different control parameters

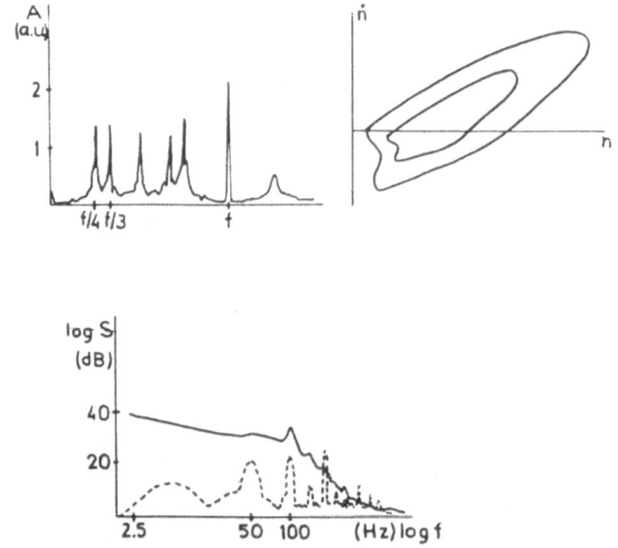

Fig.19. Bistability in a CO_2 laser with loss modulation a–b) coexistence of two attractors (period 3 and 4 respectively)
c) comparison between the low frequency cut-off when the two attractors are stable and the low frequency divergence when the two attractors are strange

$$\dot{\Delta} = R - 2Gn\Delta - \gamma_{\parallel}\Delta$$

$$\dot{n} = Gn\Delta - k(t)\,n \tag{5.6}$$

where $k(t) = k_o\,(1 + m\cos\omega t)$.

Fig.19 shows bistability, that is, the simultaneous coexistence of two attractors corresponding respectively to f/4 and f/3 subharmonic. Increasing the modulation depth m, the attractors become strange and one has the spectral divergence as in fig.19c so that the extension of turbulence considerations to more than one attractor (multistable situations) seems a successful conjecture (Fig.19).

REFERENCES

1. E. Lorenz, Jour. Atmos. Sc., 20 :130-141 (1963)
2. S. Chandrasekhar, Hydrodynamic Stability, Oxford U.P., London (1961)
3. D. Ruelle and F. Takens, Comm. Math. Phys., 20 : 163 (1971)
4. H. Haken, Phys. Lett., 53A : 77 (1975)
5. F.T. Arecchi and F. Lisi, Phys. Rev. Lett., 49:94 (1982)
6. M.J. Feigenbaum, J. Stat. Phys., 19 :25 (1978 ; 21:669 (1979)
7. F.T. Arecchi, R. Meucci, G. Puccioni, J. Tredicce, Phys. Rev. Lett., 49 : 1217 (1982)

OPTICAL BISTABILITY AND RELATED TOPICS

L.A. Lugiato
Dipartamento di Física, University di Milano
Milano 20133, Italy

L.M. Narducci
Physics Department, Drexel University
Philadelphia, PA 19104 U.S.A.

1. INTRODUCTION

The last decade has witnessed a rapidly growing interest for the general subject of cooperative phenomena in nonequilibrium systems [1,2]. The discovery that such dynamical fireworks as period doubling bifurcations and chaos give evidence of universal features [3,4] has stimulated extensive studies and produced encouraging advances. It is generally known that Quantum Optics has played a major role in the development of Synergetics [1]. Optics Bistability, in particular, has been a center of attraction because of its rich phenomenology and potential usefulness in technological applications. Actually, the field of quantum optics has uncovered more than one system capable of bistable behavior, such as, for example, the laser with a saturable absorber[5,8]. However, the term optical bistability is now commonly used to identify a very specific system; as such, this will be the focus of our attention.

To set the stage for our subsequent discussion, consider the typical layout of an optically bistable device. A coherent, monochromatic field E_I generated by a laser operating in a stationary regime is injected in an optical cavity, such as the Fabry-Perot shown in Fig. 1a. The cavity is adjusted to be resonant or nearly-

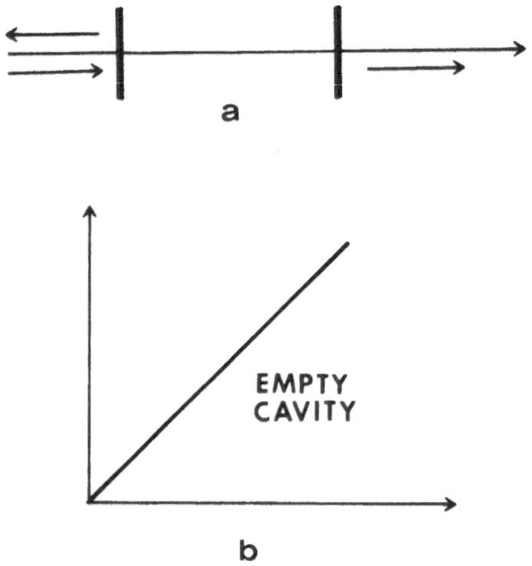

Fig. 1 (a) Empty Fabry-Perot cavity. I_T *is the incident intensity;* I_T *and* I_R *are the transmitted and reflected intensities, respectively. (b) Transmitted intensity versus incident intensity for an empty cavity.*

resonant with the incident field. Obviously, when the cavity is empty, the intensity $I_T = |E_T|^2$ of the transmitted field is proportional to the incident intensity $I_I = E_I^2$ (Fig. 1b), and the constant of proportionality depends on the mirror's transmission coefficient, on the degree of resonance between the incident radiation and one of the empty cavity modes, as well as, in practice, on the quality of the optics that makes up the interferometer.

The physical effects of interest for us arise when the cavity is filled with an absorbing medium, which can also be resonant or nearly-resonant with the incident light (Fig. 2a). In this case, the steady state behavior of the system is governed by the parameter

$$C = \frac{\alpha L}{2T} \tag{1}$$

where α is the amplitude linear absorption coefficient per unit length, L is the longitudinal dimension of the atomic sample, and

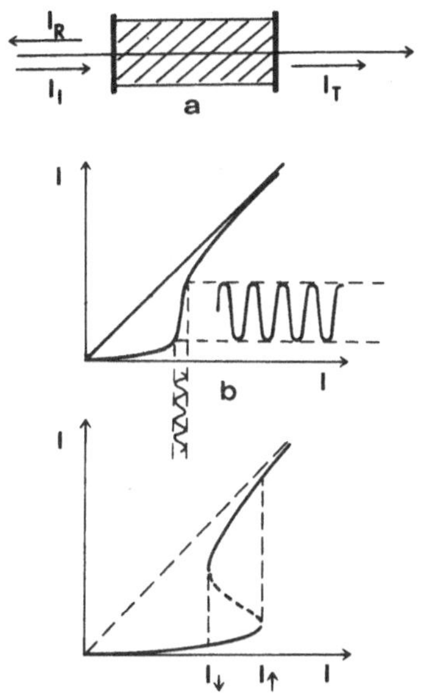

Fig. 2 (a) Fabry-Perot cavity filled by a nonlinear medium; (b) Optical transistor operation: a small modulation of the input intensity is amplified in the range where the differential gain is larger than unity; (c) Bistable operation: the dashed segment with negative slope is unstable; the range $I_\downarrow < I < I_\uparrow$ is bistable.

T is the intensity transmission coefficient of the mirror. For small values of C, the steady state transmission curve that relates the output with the input intensity is single-valued, and generally such that its slope dI_T/dI_I is smaller than unity. On increasing the value of C, a region of the transmission curve develops "differential gain" in the sense that dI_T/dI_I becomes greater than one (Fig. 2b). Under these conditions, the system works as an optical transistor because a slow intensity modulation imposed on the incident field gets amplified at the output. If C is increased even further, the steady state transmission curve becomes S-shaped (Fig. 2c). The segment with negative slope is unstable, while, typically (but not always) the upper and lower branches are stable. One can then find a range of values of the incident intensity for which the system displays a bistable character. With this setting, if we slowly increase the incident intensity from zero to a value beyond the bistable region, and then return to the starting point, the system traces a hysteresis cycle with low and high transmission branches. The two levels of transmission correspond to the logical states "0" and "1", thus suggesting the possibility that bistable devices may eventually form the heart of optical memories.

The threshold value of the parameter C for which bistable action develops depends on several physical parameters such as the degree of resonance between the incident field, the nonlinear medium and the cavity, the atomic linewidth, the type of resonator and so on. As we shall see, the emergence of bistable behavior is essentially tied to two main ingredients: one is the nonlinear nature of the atom-field interaction, and the other is the feedback action of the mirrors.

The existence of optical bistability was predicted in 1969 by Szöke and collaborators[9], but active investigations on this subject did not really begin until after the first experimental observation by Gibbs, McCall and Venkatesan in a sodium filled Fabry-Perot interferometer[10]. Shortly thereafter, Bonifacio and Lugiato formulated a first-principle, analytical description of optical bistability[11] which prompted very active research in two distinct directions, the first mainly concerned with the technological goal of producing practical, miniaturized, room temperature devices, and the second directed to the theoretical understanding of fundamental issues. In fact, the steady state hysteresis cycle of optical bistability exhibits striking analogies to first-order phase transitions in equilibrium systems. Furthermore, by controlling the parameters of the system, one can induce the emergence of spontaneous pulsations in the transmitted intensity. Depending on the choice of the parameters, the output oscillations can be highly periodic (regular self-pulsing behavior) or completely irregular (chaotic behavior, or optical turbulence).

For reasons of space, we have had to omit many interesting details: the reader can retrieve much additional information in the review papers by Gibbs et al.[12], by Abraham and Smith[13], by Arecchi and Salieri[14], by Lugiato[15,16] and in the papers on optical bistability contained in Ref. (17).

2. *A PRACTICAL OPTICAL BISTABLE DEVICE: RECENT ADVANCES*

Optical bistability in all-optical systems has been observed under widely varying conditions and experimental settings, from microwave to visible frequencies, from cavities of macroscopic size (several tens of centimeters) to miniaturized, micron-sized wafers. From a practical point of view, the most interesting situation concerns the smallest possible cavities. In fact, the ideal bistable device to be used as the basic element in a logical system ought to display the following attributes:

i. *Miniaturization.* Both the diameter and the length of the cavity should be in the micron range. In this case, high packing density can be achieved while keeping the cross-talk between neighboring units down to a negligible level, and at the same time, the cavity round-trip time is of the order of 10^{-2} picoseconds.

ii. *Fast response.* Both the switch-up and the switch-down time (i.e., the time required for the system to jump from the lower to the upper branch and viceversa) should be of the order of one picosecond.

iii. *Low energy requirement.* In order to minimize the energy consumption and the cooling requirements, the holding intensity should be less than one mW/μm^2.

iv. *The device should operate at room temperature.*

While these stringent conditions have not been entirely satisfied by the existing prototypes, great progress has been made in recent times using semiconductor materials such as GaAs[18] and InSb[19]. The situation is likely to improve rapidly even as these notes are being prepared. One of the devices constructed by Gibbs, McCall and collaborators at the Bell Telephone Laboratories (Murray Hill, NJ) consisted of a sample of GaAs, with a thickness of 4.1 μm, sandwiched between two AlGaAs layers (Fig. 3a) and reflecting coatings with a transmittivity of 10%. This system was supported by a 150 μm thick substrate of GaAs with a hole of 1-2 mm etched on it, to let the incident radiation through (Fig. 3b). Optical bistability, primarily due to the dispersive component of the medium's index of refraction was observed over a temperature range from 5 to 120°K for an incident wavelength of about 0.8 μm. The holding power of this device was about 1mW/μm^2 and the switching times were smaller than 40 nsec. More recently, bistable operation at room temperature has been reported with a GaAs-AlGaAs multiple quantum well structure[20].

The nonlinear Fabry-Perot constructed by Smith, Miller and collaborators[13] using InSb consisted of an uncoated plane-parallel crystal of dimensions 5x5 mm^2 by 560 μm. Bistability was observed at 5°K using as the input source a CO laser with an output wavelength of about 5 μm (Fig. 4). The holding intensity for this device was only 15 $\mu W/\mu m^2$. For InSb also, room temperature bistability has been reported recently[21].

3. *THEORY OF OPTICAL BISTABILITY IN A RING CAVITY*

A convenient setting for the analysis of a bistable optical system is a ring cavity because the incident light can be forced to propagate in only one direction. As shown in Fig. 5, a sample of length L and volume V containing N>>1 two-level atoms is placed in a ring cavity of total length $\mathcal{L}$. The incident field amplitude

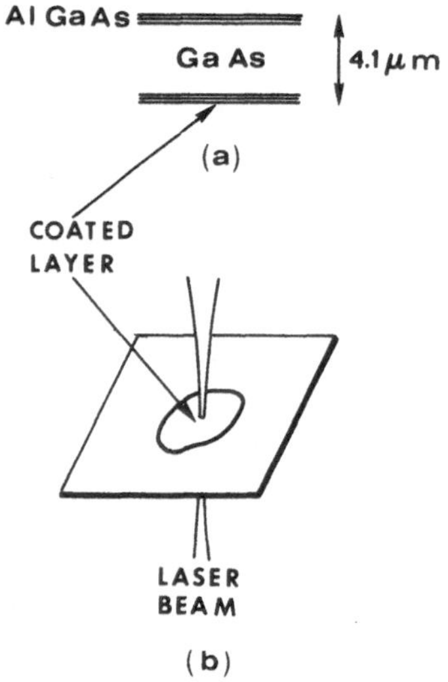

Fig. 3 (a) Miniaturized all-optical bistable system; a 4μm wafer of GaAs is sandwiched between two layers of AlGaAs. (b) GaAs substrate with a hole etched in the center to let the laser beam through.

is labelled $\mathcal{E}_I$, while $\mathcal{E}_T$ and $\mathcal{E}_R$ denote the transmitted and reflected amplitudes respectively. The upper mirrors have a reflectivity coefficient R = 1-T while the lower mirrors are assumed to have 100% reflectivity. We denote by $\mathcal{E}(z,t)$ and $\mathcal{P}(z,t)$ the electric field amplitude and the macroscopic atomic polarization inside the resonator. The field obeys the boundary conditions

$$\mathcal{E}(0,t) = \sqrt{T}\,\mathcal{E}_I + R\,\mathcal{E}(L,t-\Delta t) \tag{2}$$

where T is the transmittivity coefficient of mirrors 1 and 2 and Δt is the light transit time from mirror 2 to mirror 1

$$\Delta t = \frac{\mathcal{L}-L}{C} \tag{3}$$

The second term in Eq. (2) is responsible for the feedback mechanism which, as we mentioned, is essential for the emergence of bistability. We now let ω_0 and $K_0 = \omega_0/C$ denote the radian frequency and the wave

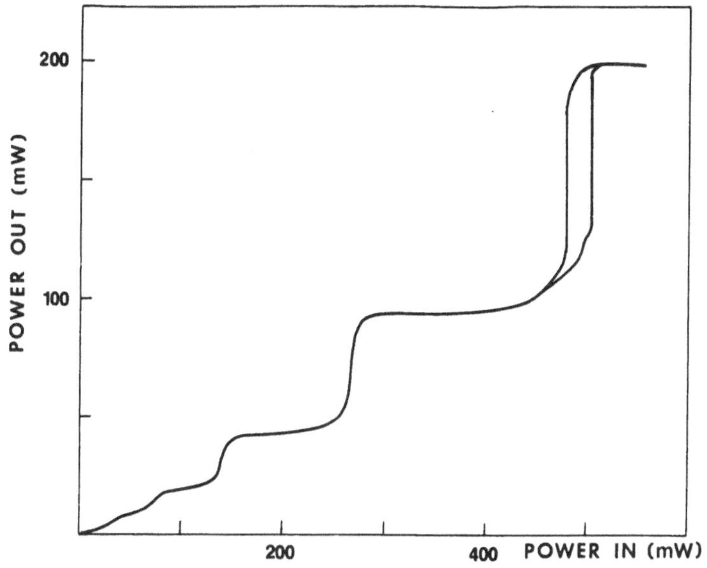

Fig. 4 Bistable action in InSb.

vector of the light field and set

$$\mathcal{E}(z,t) = E(z,t) \exp[-i(\omega_0 t - \kappa_0 z)] + c.c. \tag{4a}$$

$$\begin{Bmatrix} \mathcal{E}_I \\ \mathcal{E}_T \end{Bmatrix} (t) = \begin{Bmatrix} E_I \\ E_T \end{Bmatrix} (t) \exp(-i\,\omega_0 t) + c.c. \tag{4b}$$

$$\mathcal{P}(z,t) = P(z,t) \exp[-i(\omega_0 t - \kappa_0 z)] + c.c. \tag{4c}$$

where E(z,t) and P(z,t) are the slowly varying field and polarization envelopes, respectively. After substituting Eq. (4a) into Eq. (2), we obtain

$$E(0,t) = \sqrt{T}\, E_I + R\, e^{-i\delta_0} E(L,t-\Delta t) \tag{2'}$$

where δ_0 is the cavity detuning parameter

$$\delta_0 = \frac{\omega_c - \omega_0}{c/\mathcal{L}} \tag{5}$$

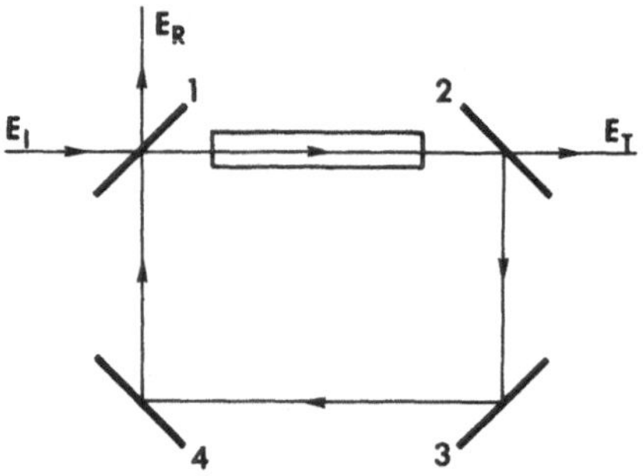

Fig. 5 Schematic layout of a ring cavity. $\mathcal{E}_I$, $\mathcal{E}_T$ and $\mathcal{E}_R$ are the incident, transmitted and reflected field amplitudes, respectively.

and ω_c is the empty cavity resonance which lies nearest to ω_0 (we recall that in a ring cavity the eigenfrequencies are given by $2\pi cm/\mathcal{L}$, m=1,2,...). In addition to Eq. (2'), we have the obvious relation

$$E_T(t) = \sqrt{T}\; E(L,t) \tag{6}$$

As the incident field propagates through the sample, the medium reacts back on it. The coupled atom-field dynamics is described by the so-called Maxwell-Bloch equations[22]. In the case of homogeneously broadened, two-level atomic systems, these equations take the form

$$\frac{\partial E}{\partial t} + c\,\frac{\partial E}{\partial z} = -\,g\,P \tag{7a}$$

$$\frac{\partial P}{\partial t} = \frac{\mu}{\hbar}\,E\,D - \gamma_\perp\,(1+i\Delta)\,P \tag{7b}$$

$$\frac{\partial D}{\partial t} = \frac{-\mu}{2\hbar}\,(E^*P+EP^*) - \gamma_\parallel\,(D-N/2) \tag{7c}$$

where D is half the population difference of the lower and upper level, μ is the modulus of the atomic dipole moment of the atoms, g is the coupling constant

$$g = \frac{4\pi\omega_0}{V}\mu \tag{8}$$

$\gamma_{\parallel}$ and $\gamma_{\perp}$ are the longitudinal and transverse atomic relaxation rates, and in particular $\gamma_{\perp}$ coincides with the atomic linewidth. The parameter Δ measures the atomic detuning

$$\Delta = \frac{\omega_a - \omega_0}{\gamma_{\perp}} \tag{9}$$

and ω_a labels the atomic transition frequency. Δ controls the dispersive effects: in particular, for $\Delta=0$ no dispersion in present and the bistability effect is called purely absorptive. In the presence of dispersion, it is common practice to refer to the effect as dispersive bistability, even though the absorptive part of the atomic polarization may be entirely nonnegligible.

3.1 STEADY-STATE BEHAVIOR IN ABSORPTIVE OPTICAL BISTABILITY WITH ZERO CAVITY DETUNING[23]

This is the simplest situation that one can encounter from a mathematical point of view. When $\Delta=0$, Eqs. (7b,c) in steady state ($\partial/\partial t = 0$) yield

$$P(z) = \frac{N}{2}\sqrt{\frac{\gamma_{\parallel}}{\gamma_{\perp}}}\ \frac{F(z)}{1+F^2(z)}$$
$$D(z) = \frac{N}{2}\ \frac{1}{1+F^2(z)} \tag{10}$$

where F(z) is the scaled field amplitude

$$F(z) = \frac{\mu E(z)}{\hbar\sqrt{\gamma_{\perp}\gamma_{\parallel}}} \tag{11}$$

If we now substitute Eq. (10) into Eq. (7a), the stationary field equation takes the form

$$\frac{dF}{dz} = -\alpha\frac{F}{1+F^2} \tag{12}$$

where α is the linear absorption coefficient per unit length (at the atomic line center)

$$\alpha = \frac{\mu g N}{2\hbar c\gamma_{\perp}} \tag{13}$$

In terms of the normalized incident and transmitted field y and x

$$
\begin{aligned}
y &= \mu E_I/\hbar\sqrt{\gamma_\perp \gamma_\parallel T} \\
x &= F(L) = \mu E_T/\hbar\sqrt{\gamma_\perp \gamma_\parallel T}
\end{aligned}
\tag{14}
$$

the boundary conditions (2') with $\delta_o = 0$, can be rewritten in the form

$$F(0) = Ty + Rx \tag{15}$$

and Eq. (12) can be integrated at once with the result

$$\ell n(F(0)/x) + \frac{1}{2}[F^2(0) - x^2] = \alpha L \tag{16}$$

Finally, on combining Eqs. (15) and (16) we obtain the exact relation between the transmitted field x and the incident field y

$$\ell n\left[1 + T\left(\frac{y}{x} - 1\right)\right] - \frac{x^2}{2}\left\{\left[1 + T\left(\frac{y}{x} - 1\right)\right]^2 - 1\right\} = \alpha L \tag{17}$$

Figure 6 shows the transmission curve $x = x(y)$ for $C \equiv \alpha L/2T = 10$ and several values of αL and T. An important feature is that, if T becomes too large, the bistable behavior disappears. In particular this is always so for T=1, regardless of the values of the other system parameters. This result shows that in order to achieve bistability, one needs not only a nonlinear medium, but also feedback action (i.e., good mirrors).

3.2 THE EFFECT OF DISPERSION - KERR MEDIUM

Consider first the case of an empty cavity in steady state. Because in this case we have $E(0) = E(L) = E_T/\sqrt{T}$, Eq. (2') yields the well known Airy transmission function

$$\mathcal{T} \equiv \frac{I_T}{I_I} = \left[1 + \frac{4R}{T^2}\sin^2\frac{\delta_o}{2}\right]^{-1} \tag{18}$$

The dependence of $\mathcal{T}$ on $\delta_o\, c/\mathcal{L}$ displays the usual resonances shown in Figure 7. The peaks of the resonances coincide with the empty cavity eigenfrequencies. Each resonance has a width

$$\kappa = cT/\mathcal{L} \tag{19}$$

which is commonly denoted as cavity linewidth. When the cavity is filled with a nonlinear medium, it is useful to represent the internal field in terms of its modulus and phase:

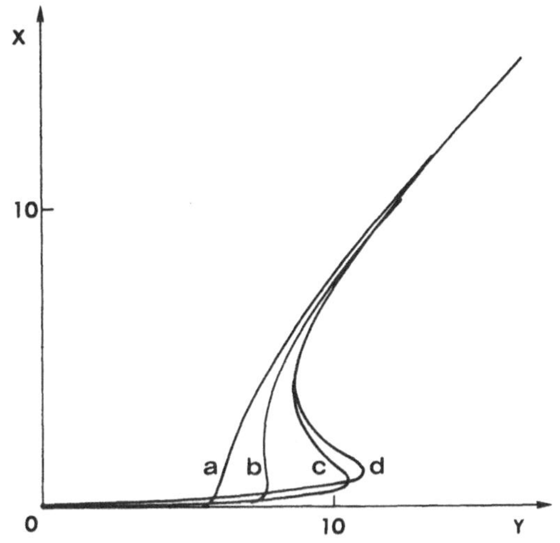

Fig. 6 Steady state transmitted field x as a function of the input field y for $\Delta=\delta_0=0$. All the segments with negative slope are unstable. The parameter $C = \alpha L/2T$ is held fixed and equal to 10. (a) $\alpha L=20$, $T=1$, (b) $\alpha L=10$, $T=0.5$, (c) $\alpha L=2$, $T=0.1$, (d) mean field limit $\alpha L \to 0$, $T \to 0$ with $C=10$.

$$E(z,t) = \rho(z,t)\,\exp(i\phi(z,t)) \tag{20}$$

In steady state, the space dependence of ρ and ϕ can be easily derived from Eqs. (7)

$$\frac{d\rho}{dz} = -\,\alpha\,\rho\chi_a(\rho^2) \tag{21a}$$

$$\frac{d\phi}{dz} = -\,\alpha\,\chi_d(\rho^2) \tag{21b}$$

where χ_a and χ_d represent the absorptive and dispersive parts of the nonlinear dielectric susceptibility of the medium, i.e.,

$$\chi_a = \frac{1}{1+\Delta^2+\rho^2}\,,\ \chi_d = \Delta\chi_a \tag{22}$$

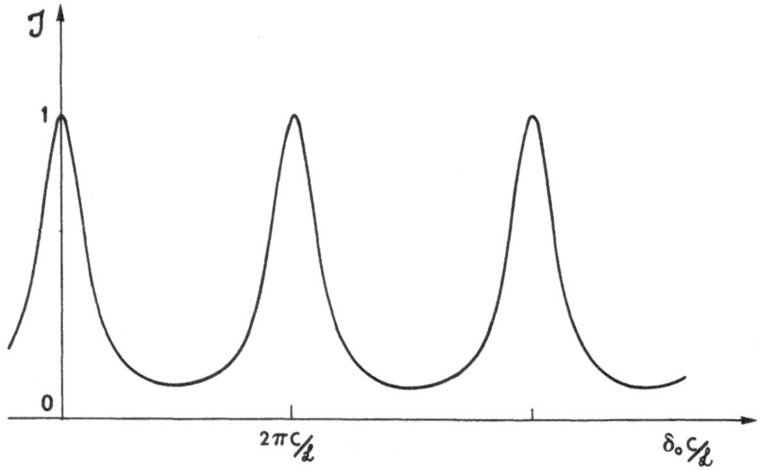

Fig. 7 The transmission function of the empty cavity is plotted as a function of the cavity detuning parameter $\delta_0 c/\mathcal{L} = \omega_c - \omega_0$.

Equation (21a) leads to a well known physical consequence in the limit when the field amplitude is sufficiently small ($\rho^2 \ll 1+\Delta^2$). After approximating χ_a with $(1+\Delta^2)^{-1}$, Eq. (21a) can be integrated at once to yield

$$\rho(z) = \rho(0)\exp(-\bar{\alpha} z), \quad \bar{\alpha} \equiv \frac{\alpha}{1+\Delta^2} \tag{23}$$

where $\bar{\alpha}$ is the out of resonance linear absorption coefficient. This result is known in optics at the Beer absorption law.

More generally, the boundary conditions (2') and Eq. (6) lead to the following general expression for the transmission of a ring cavity in steady state

$$\mathcal{T} \equiv \frac{I_T}{I_T} = \frac{T^2}{(\eta-R)^2 + 4R\eta \sin^2[\frac{1}{2}(\delta_0-\Delta\phi)]} \tag{24}$$

where

$$\eta \equiv \frac{\rho(0)}{\rho(L)} \geqslant 1$$

$$\Delta\phi \equiv \phi(L) - \phi(0) \qquad (25)$$

In the limit of an empty cavity, i.e., when $\alpha=0$, Eqs. (21) lead to $\eta=1$ and $\Delta\phi=0$, so that the transmission function (24) reduces to the form given by Eq. (18), as expected. Note also that, in the small field limit, η is given by $\exp(\bar{\alpha} L)$.

It is important to observe at this point that Eqs. (21) are much more general than one might surmise from the present discussion. If the nonlinear medium is not made up of homogeneously broadened two-level atoms, the nonlinear dielectric susceptibility functions are no longer given by Eq. (22) and depend, of course, on the specific medium. The crucial point, however, is that, in all instances, η and $\Delta\phi$ *depend* on the transmitted intensity, i.e., $\eta = \eta(I_T)$ and $\Delta\phi = \Delta\phi\,(I_T)$. In our case, these functions can be calculated explicitly by solving the differential equations (21) while taking Eq. (6) into account. Thus, the interferometer transmission function (24) provides an explicit expression for the incident intensity I_I as a function of I_T

$$I_I = \mathcal{f}(I_T)$$

$$\mathcal{f} = \frac{I_T}{T^2}\left\{[\eta(I_T)-R]^2 + 4R\eta(I_T)\ \ \sin^2\left[\frac{1}{2}\ (\delta_0-\Delta\phi(I_T))\right]\right\} \qquad (26)$$

The function $\mathcal{f}$ is single-valued by definition; however, if we plot I_I as a function of I_T, and exchange the horizontal with the vertical axis, we see that I_T can be a multivalued function of I_I (Fig.8). In this case, the s-shaped form of the transmitted intensity curve gives the clue for the existence of bistability.

When the atomic detuning is large ($\Delta^2 >> 1+\rho^2$), the absorptive component of the susceptibility becomes negligible with respect to the dispersive part, and bistability becomes, properly, of the dispersive type. As a concrete example of the analysis developed in this subsection, we focus on the case of dispersive bistability in a Kerr medium in which

$$\chi_a=0,\ \chi_d = c_1+c_2\ \rho^2 \qquad (27)$$

From eqs. (25), (21) and (6) we easily obtain

$$\eta=1,\ \Delta\phi= a_1-a_2\ I_T \qquad (28)$$

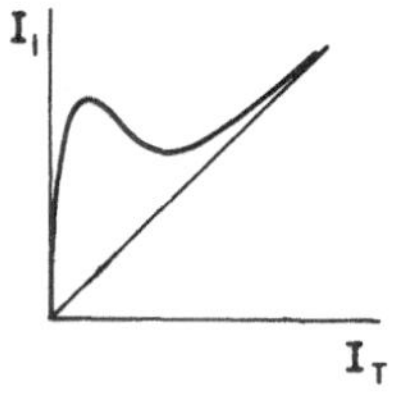

Fig. 8 Incident intensity plotted as a function of the transmitted intensity.

where $a_1 = -\alpha\Delta L c_1$, $a_2 = \alpha L\Delta c_2/T$ (*Note:* the appearance of the absorption coefficient should not be interpreted as a contradiction of the statement $\chi_a=0$; the Kerr limit can be obtained by detuning the incident field from the atomic resonance to the point that the absorption at the selected frequency is negligible, while, of course, the line center absorption coefficient is different from zero). In this case, Eq. (24) leads to the following transmission function

$$\mathcal{J} \equiv \frac{I_T}{I_I} = \frac{1}{1+\frac{4R}{T^2}\sin^2\left\{\frac{1}{2}\left[(\delta_0-a_1)+a_2\, I_T\right]\right\}} \tag{29}$$

If we plot $\mathcal{J}$ as a function of $I' \equiv a_2\, I_T$, we obtain Figure 9, a curve that coincides with the empty cavity curve ($\mathcal{J}$ vs. δ_0, Fig.7) apart from a translation of the horizontal axis by an amount $-a_1$. The stationary solutions of the system can be found with the help of a simple graphical procedure devised by Felber and Marburger[24] which consists of looking for the intersects of Eq. (29) with the

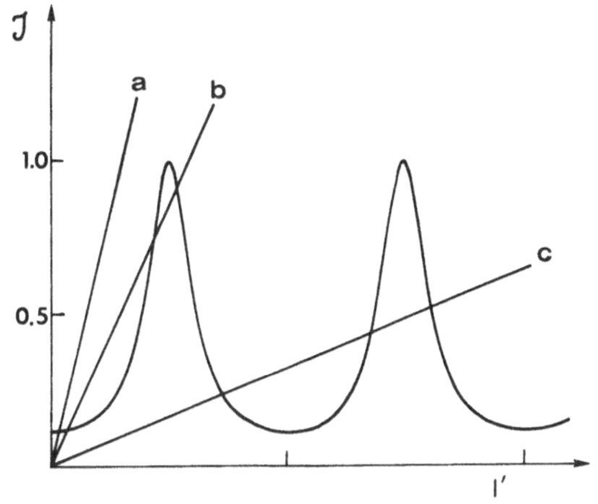

Fig. 9 Graphical search of the stationary solutions for a Kerr medium. I' is defined as $a_2 I_T$ (see text).

straight lines $\mathcal{J} = I_T/I_I = I'/a_2 I_T$ whose slope is inversely proportional to the incident intensity. For small values of I_I, one has only one intersect (line a). On increasing the strength of the incident intensity, three intersections develop (line 6) which correspond to a bistable situation because the intermediate solution is unstable. Hence, if one plots I_T as a function of I_I, one finds a hysteresis cycle. For larger values of I_I, one obtains multiple solutions (line c) which lead to multi-stability and multiple limit cycles. When T approaches unity, the curve (29) flattens out and bistability disappears.

4. *THE MEAN FIELD MODEL OF OPTICAL BISTABILITY*

The description of our system takes a particularly simple form in the limit of small absorption, small transmittivity coefficient and small cavity detuning

$$\alpha L \ll 1, \quad T \ll 1, \quad \delta_0 \ll 1 \tag{30a}$$

as long as the following ratios remain finite, but otherwise arbitrary

$$C \equiv \frac{\alpha L}{2T} \qquad \text{constant and arbitrary} \tag{30b}$$

$$\theta \equiv \frac{\delta_o}{T}$$

This situation is called "mean field limit" because when L approaches zero, the internal field becomes spacially uniform, so that its value at the output mirror coincides with its mean value in space. Note that here the words "mean field" have a different meaning from that which is commonly understood in the context of equilibrium phase transitions (e.g., in the Landau theory).

In the limit (30), the time evolution is governed by the so-called "mean field model", first proposed by Bonifacio and Lugiato[11].

$$\kappa^{-1} \dot{x} = -i\theta x - (x-y) - 2Cp \tag{31a}$$

$$\gamma_\perp^{-1} \dot{p} = xd - (1+i\Delta)p \tag{31b}$$

$$\gamma_\parallel^{-1} \dot{d} = -\frac{1}{2}(xp^* + x^*p) - d+1 \tag{31c}$$

The dot denotes derivative with respect to time and

$$p = \left(\sqrt{\frac{\gamma_\parallel}{\gamma_\perp}}\,\frac{N}{2}\right)^{-1} P, \quad d = \left(\frac{N}{2}\right)^{-1} D$$

The steady state solution of Eqs. (31) is easily calculated and it leads to the state equation

$$y^2 = |x|^2 \left\{(1+2C\chi_a(|x|^2))^2 + (\theta - 2C\chi_d(|x|^2))^2\right\} \tag{32}$$

where χ_a and χ_d are given by Eq. (22). If, in particular, one sets $\theta = \Delta = 0$, it is easy to prove that the state equation reduces to

$$y = x\left(1+\frac{2C}{1+x^2}\right) \tag{33}$$

which coincides with the limit of Eq. (17) when $\alpha L \to 0$, $T \to 0$ and $C = \alpha L/2T$ = constant. This result is displayed graphically in Figure 6.

The limit $\delta_0 << 1$ implies that the operation of the system is restricted to the resonant cavity mode at frequency ω_c. Hence, the mean field model is a *single-mode* model. It has played an important role in quite a number of investigations on the transient properties of Optical Bistability.

5. *SELF-PULSING AND CHAOTIC BEHAVIOR*

In this section we discuss the matter of *instabilities* in Optical Bistability. As usual, a stationary state (or a stationary regime) is said to be unstable when, as a result of a perturbation, the system departs exponentially from its initial configuration. Of course, an unstable state is never realized, in practice, because even the slightest perturbation removes the system from it.

The existence of optical bistability is itself a manifestation of the existence of an instability. In fact, on approaching the upper bistability threshold $I_{\uparrow}$ from below (see Fig. 2c) the low transmission state becomes unstable and precisely for this reason the system jumps to the high transmission branch. The same process in reverse is responsible for the downward transition at the lower bistability threshold $I_{\downarrow}$.

It is well known, on the other hand, that nonlinear systems usually display sequences of successive instabilities where a steady state becomes unstable, thus driving the system to another stable steady state, which in turn becomes unstable, and so on. This sequence can be identified by varying the control parameters of the system (for example, in optical bistability, the intensity of the incident field or the length of the cavity). Typically, after a few instabilities the system runs out of stationary states and develops an oscillatory behavior with the emergence of spontaneous pulsations. The oscillations can be regular, i.e., perfectly periodic in time, as well as completely irregular and aperiodic. In the latter case, one speaks of chaotic or turbulent behavior.

Two points are of basic importance:

(a) *The oscillations are not induced by external manipulations. They arise even when all the external parameters are constant in time. It is not unexpected, for example, that pulsed behavior should be produced in a bistable system by modulating the cavity length; this is a rather trivial fact. It is a different story, instead, when time-independent external parameters produce spontaneous pulsations because this is indicative of the existence of self-organization from within the system itself (for this reason, a behavior of this type is called self-pulsing).*

(b) *In the chaotic case, the time-dependent output intensity is often reminiscent of the type of behavior that may be expected of a system which has been contaminated by a noise source. In the situations of interest to us, chaotic behavior persists even without any external noise perturbation; because its origin resides in the differential equations themselves it is often called* *deterministic chaos*.

Chaotic self-pulsing and the various routes by which this effect is produced have been the object of very active research in recent years, although our understanding of these phenomena is still far from being satisfactory. The best understood and most frequently occurring route to chaos in quantum optical systems is the so-called "period-doubling route"[3,4]. This begins with the system in a stable oscillatory state characterized by a fundamental frequency $\bar{\omega}$ (in point of fact, $\bar{\omega}$ usually varies slowly as a function of the external parameters, but this is not important for this discussion). As one varies a selected control parameter, at some point the oscillatory state becomes unstable, and the system approaches a new stable state of oscillation whose period is twice as long as the previous one. In this case, the power spectrum of the observable variables develops a component at the subharmonic frequency $\bar{\omega}/2$. By further changing the control parameter, a subsequent instability develops which then leads to a new periodic oscillation whose period is four times as large as the original one. This process goes on indefinitely (again, we must stress that because of the slow dependence of $\bar{\omega}$ on the control parameters, it is not exactly true that the period of, say, the nth bifurcation is 2^n times longer than the original period. What is rigorously true, however, is that the power spectrum of the nth bifurcation displays a string of subharmonic frequencies components $\bar{\omega}_n/2^n$ where $\bar{\omega}_n$ is the fundamental oscillation frequency of the new stable solution). The values of the control parameter at which period doubling is observed get closer and closer according to a geometrical law, such that the ratio between the intervals in which period 2^n and period 2^{n+1} are observed approaches the value 4.6692..., as n approaches infinity. This value is universal in the sense that it characterizes all period doubling sequences. Hence, very rapidly, the sequence reaches an accumulation point beyond which the system exhibits chaotic behavior. In the chaotic regime, the power spectrum is no longer a simple line spectrum, but it displays a continuous background.

As shown in this section, optical bistability displays both periodic and chaotic self-pulsing. We shall distinguish two cases according to the mechanism which is responsible for the emergence of an instability in the stationary state. The optical cavity is characterized by an infinite number of possible excitation frequencies (modes); accordingly the behavior of the system can be described in terms of *mode amplitudes*. In the two cases to be discussed, the amplitude of the resonant mode becomes unstable (case 1); in the second case, the instability arises from the off-resonance modes.

6. INSTABILITY MODES

6.1 RESONANT-MODE INSTABILITY

This type of instability can arise only in the dispersive case[25,26]. The stability analysis of the mean field model (31) shows[26] that under appropriate conditions, a sizeable segment of the high transmission branch becomes unstable (see Fig. 10). Roughly speaking, the mechanism of the instability can be described as follows. In the dispersive situation, the atomic and the incident field frequency are mismatched. When the mismatch is sufficiently large, the system is no longer able to adapt itself to the driving field; the stationary state becomes unstable and the output begins to develop oscillations in the form of an undamped sequence of pulses. From a practical point of view, this behavior is very

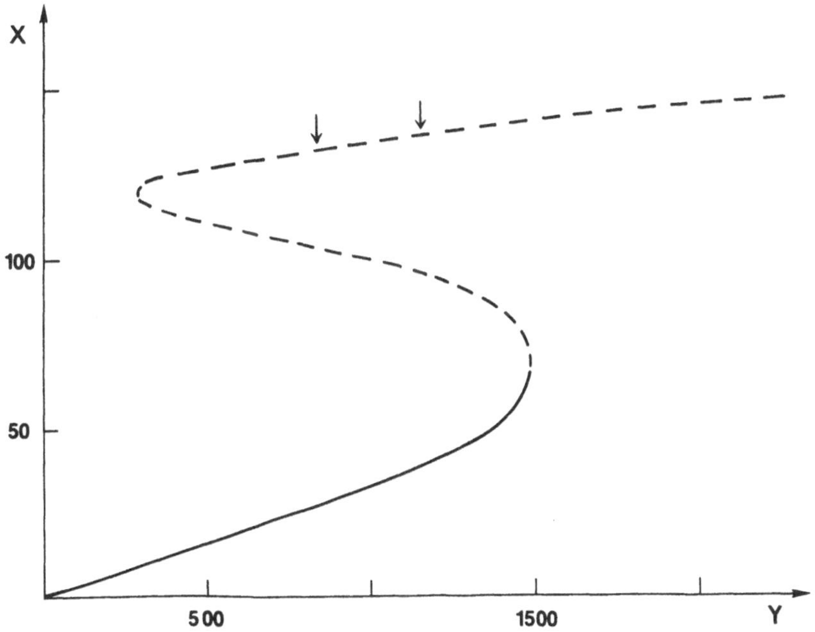

Fig. 10 Steady state transmission curve for C=70,000, Δ=374 and θ=340. The dashed segment is unstable. The arrows mark the region of the upper branch where chaotic behavior is found.

interesting, as it suggests the possibility that an optical device may convert a coherent constant intensity beam into coherent pulsed light.

Consider now what happens when the system jumps from the lower to the higher transmission branch. Since, in the case illustrated in Figure 10, the higher branch is unstable at the switching point, one expects immediately the appearence of a regular sequence of oscillations (Fig. 11a). Another way of looking at the same behavior is to consider the plane of the real and imaginary parts of the normalized transmitted field. In this case, the trajectory in this plane is a simple limit cycle (Fig. 11b). If one now decreases the incident field strength (the control parameter), the system undergoes a period doubling bifurcation (Fig. 11c) which corresponds to the trajectory shown in Fig. 11d. This trajectory arises from a kind of fission process of the limit cycle into two distinct parts. A further decrease of y leads to the appearence of period four (Figs. 11e,f) and so on. Finally, the system enters the chaotic domain (Figs. 10,11g,h) with the time trace displaying no remnant of periodicity.

6.2 OFF-RESONANCE MODE INSTABILITY

This phenomenon arises when two cavity modes, other than the resonant one, fall below the power broadened absorption line of the atomic medium. This condition can be fulfilled, for example, by increasing the length of the cavity, thus decreasing the intermode spacing. In this situation, under suitable conditions, the off-resonant mode amplitudes become unstable, while the resonant mode remains stable (in the sense of linear stability) analysis.

Consider first the simplest absorptive case $\Delta = \theta = 0$. The external parameters of practical interest are the incident field y and the total length $\mathcal{L}$ of the ring cavity. Instead of y and $\mathcal{L}$, it is more convenient to use x, which is linked to y by Eq. (33), and $\tilde{\alpha}/\gamma_{\perp} = 2\pi c/\mathcal{L}\gamma_{\perp}$ which is the spacing of two adjacent cavity modes in units of atomic linewidth $\gamma_{\perp}$. In Figure 12, we display the plane of the control parameters $\tilde{\alpha}/\gamma_{\perp}$ and x. The stationary state of the high transmission branch becomes unstable when the operating point lies in the hatched part of the plane, which will be called instability region.

The mechanism that produces the instability in this case is different from the one discussed in the previous subsection. It is, instead, similar to what one finds in the so-called saturation spectroscopy[27] where an atomic sample is illuminated by a strong coherent stationary field which saturates the medium, and sampled by a weak probe beam (see Fig. 13). In this case, one finds that for suitable ranges of values of the frequency detuning between

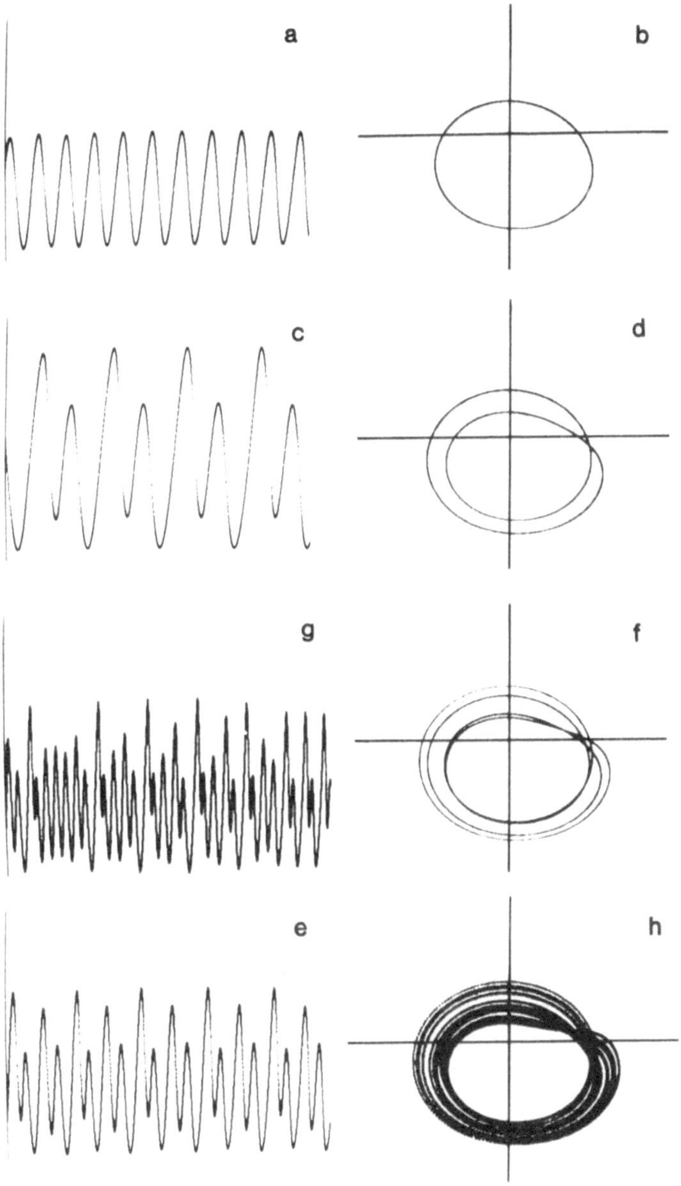

Fig. 11 Self-pulsing oscillations and corresponding phase-space trajectories in the plane of the real and imaginary parts of the normalized electric field for solutions of the type: period 1 (a,b), period 2 (c,d), period 4 (e,f), and chaotic (g,h), respectively. The values of the external field are y=2000, 1350, 1225 and 950 for the four sets of solutions.

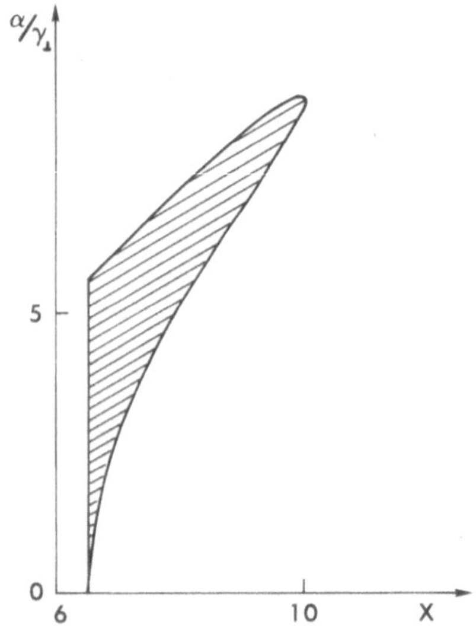

Fig. 12 Instability region in the plane of the control parameters $\tilde{\alpha}/\gamma_{\perp}$ *and x.*

saturating and probe beam, the probe experiences gain instead of loss (as one might expect, because the sample under study does not have a population inversion). The same happens in our system: the resonant cavity mode saturates the medium, while the neighboring modes play the role of the probe fields. Under appropriate conditions, some of the sidebands may experience gain, and when the gain becomes larger than the loss the sidebands become unstable. In some sense, the absorbing medium behaves as a laser with respect to the sidebands even without a built-in population inversion.

When the steady state in the higher branch is unstable, the system can exhibit two different kinds of evolutions[27,28]. In the first case, the system approaches a self-pulsing behavior; as we see from Fig. 14a, if the system is initially slightly displaced from steady state, it will begin to develop an oscillatory behavior. The amplitude of the oscillations increases with time, until a stationary regime is reaches, as evidenced by the flat shape of the self-pulsing envelope. The self-pulsing frequency in this case is equal to $\tilde{\alpha}$ which corresponds to a period equal to the transit time,

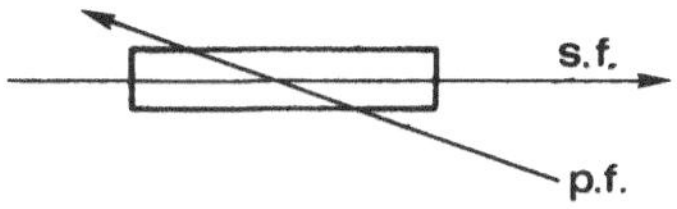

Fig. 13 Typical set-up of a saturation spectroscopy experiment. sf = saturating field; pf = probe field.

$\mathcal{L}/c$, of the photons in the cavity. The second possibility is that the system, after a transient self-pulsing action, simply precipitates to the low transmission state corresponding to the same value of the incident field. This possibility is illustrated in Fig. 14b from which we see that the oscillations are amplified, at first, but that eventually they die off as the system precipitates to the low transmission state.

The analytical treatment of this self-pulsing behavior, which is a multi-mode phenomenon, as it involves not only the resonant but also some of the sidebands, has been carried out in terms of a procedure called the "dressed mode theory of Optical Bistability"[29,30]. In the case of only two unstable sidebands (i.e., only three resonator modes altogether) this treatment is capable of handling the behavior of the system in terms of a simple two-dimensional description. The only two relevant variables are *(i) the half-amplitude of the oscillations, ρ, and (ii) the displacement, σ, of the mean value of the oscillations from the unstable steady state value x_{st}.* The upper and lower envelopes of the oscillations $x_{upper}(t)$ and $x_{lower}(t)$ are obtained as follows:

$$x_{\substack{upper\\lower}}(t) = x_{st} + \sigma(t) \pm \rho(t) \tag{34}$$

Hence, the time evolution can be described in equivalent ways by the envelope of the oscillations or by a trajectory in the phase-space (ρ,σ).

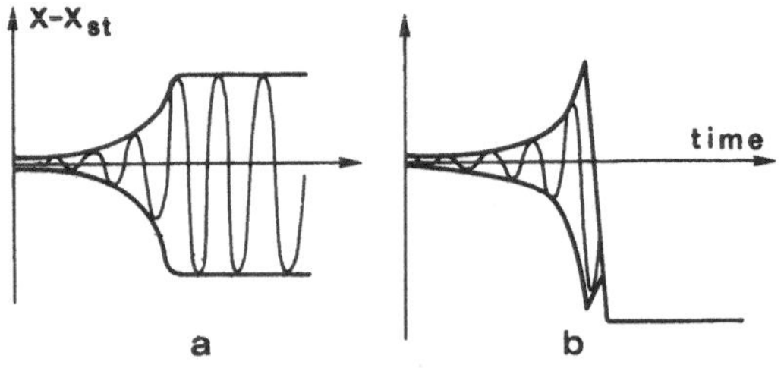

Fig. 14 Self-pulsing (a) and precipitation (b). x_{st} *is the value of the normalized transmitted field at steady state in correspondence to the high transmission state.*

Using the dressed mode theory, we can follow the behavior of the self-pulsing state over its entire domain of existence. This domain is shown in Figure 15 and is quite a bit wider than the instability region ABE. It can be subdivided into a "soft excitation" domain ABD, and a "hard excitation" domain BCC'D. The region ADE, instead, is the precipitation domain. The region bounded by the lines ABD corresponds to values of the control parameters $(x,\tilde{\alpha}/\gamma_{\perp})$ for which the self-pulsing state is stable, while the stationary state is unstable. Hence, a small initial deviation from steady state (soft excitation) is enough to push the system into the self-pulsing state. In the domain BCC'D, one encounters coexisting stationary and self-pulsing states, both being stable. Exactly which of the two stable states will be occupied by the system depends on the initial conditions; thus, if the initial fluctuation away from steady state is small, the system simply returns to the stationary state. The self-pulsing state is reached, instead, after a sufficiently large initial fluctuation (hard excitation).

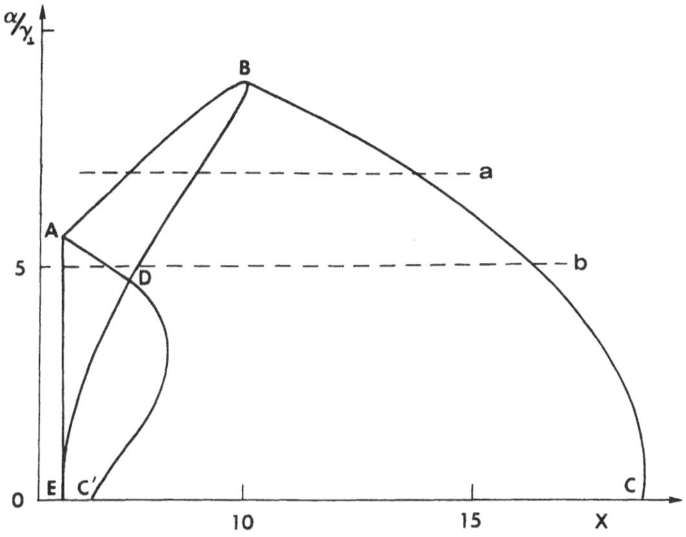

Fig. 15 Instability region, precipitation region and domain of existence of the stable self-pulsing solution for C=20, $\Delta=\theta=0$ in the mean field limit.

This situation leads to the appearence of a hysteresis cycle of new type. Figure 16 shows the half amplitude of the oscillations for long times, $\rho(t{=}\infty)$, upon varying the incident field along the horizontal line a of Fig. 15. On entering the instability region from the left, a stable self-pulsing state develops. As the incident field is increased, the amplitude of oscillation also increases, and, in fact, it continues to do so even outside the instability region, until the system returns discontinuously to the stationary state in the high transmission branch. If one now decreases the amplitude of the incident field, the system continues to operate in the stationary high transmission branch until we reach the right boundary of the instability region, where it jumps discontinuously to the self-pulsing regime with a finite amplitude.

The diagram of Figure 16 can be viewed as representing a second and first order phase transition simultaneously. The second order transition occurs on the left boundary of the instability domain. The first order transition is tied to the hysteresis cycle that

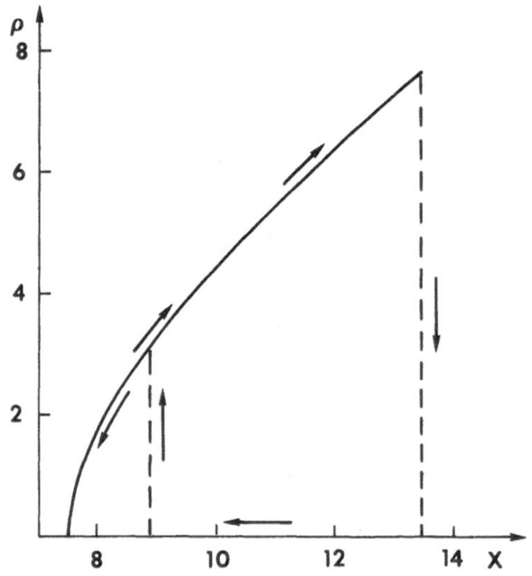

Fig. 16 The half-amplitude of the oscillations in the self-pulsing state is plotted as a function of the stationary value x of the transmitted field in the high transmission branch for C=20, $\tilde{\alpha}/\gamma_{\perp}=7$ *(line a of Fig. 15). The arrows indicate the behavior of the system when the incident field is decreased or increased.*

begins at the right boundary domain. It is worth stressing that in this case the bistability involves stationary and self-pulsing states, and *not just* stationary states as in the operation of usual bistable systems. This hysteresis cycle enriches the phenomenology of optical bistability, especially because the new self-pulsing branch is accessible from the usual steady states by suitably varying the external parameters.

An interesting situation occurs in the neighborhood of the line AD (Fig. 15) that separates the self-pulsing from the precipitation domains. In order to describe this behavior, consider a continuous variation of the cavity length along the line b of Fig. 15. On approaching the line AD from the right, the time-dependent envelope begins to develop considerable oscillations (Fig. 17a). This behavior is usually called "breathing". The breathing pattern observed in our case, however, lasts only a finite amount of time

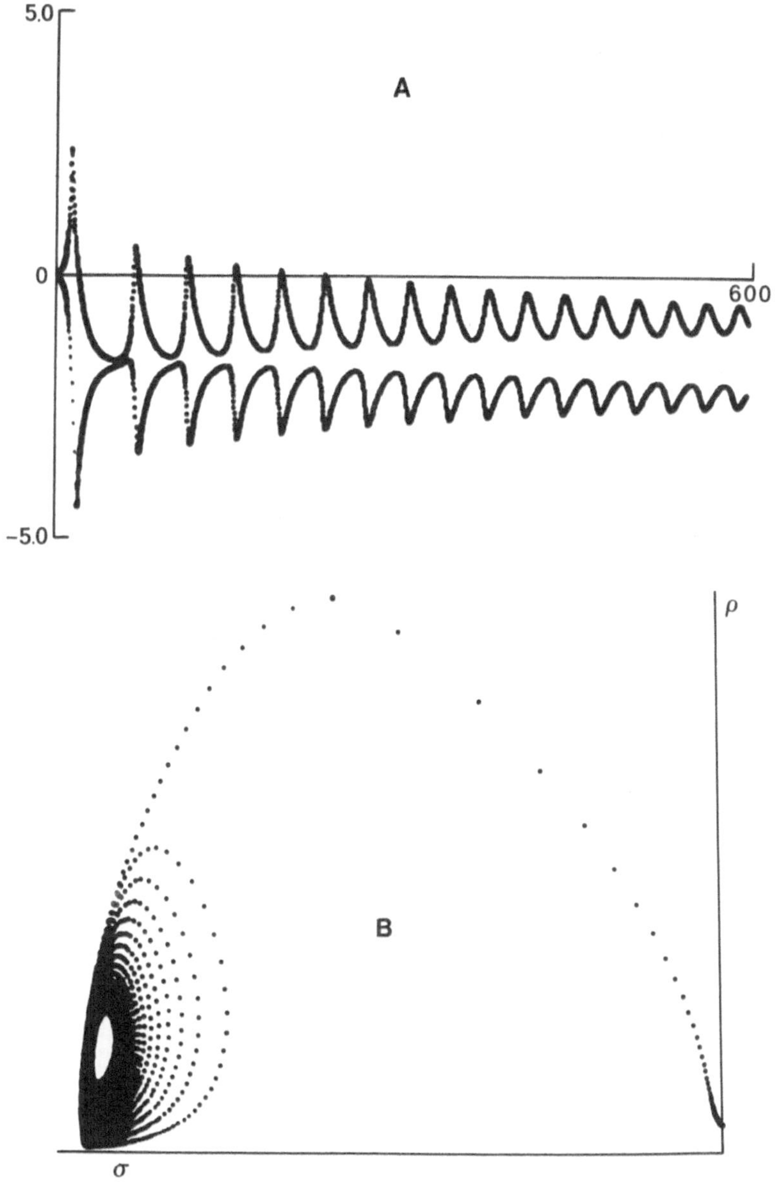

Fig. 17(a,b) Envelope breathing and phase trajectory in the (ρ,σ) plane for C=20, $\tilde{\alpha}/\gamma_{\perp}$=5, and x = 6.867. (continued)

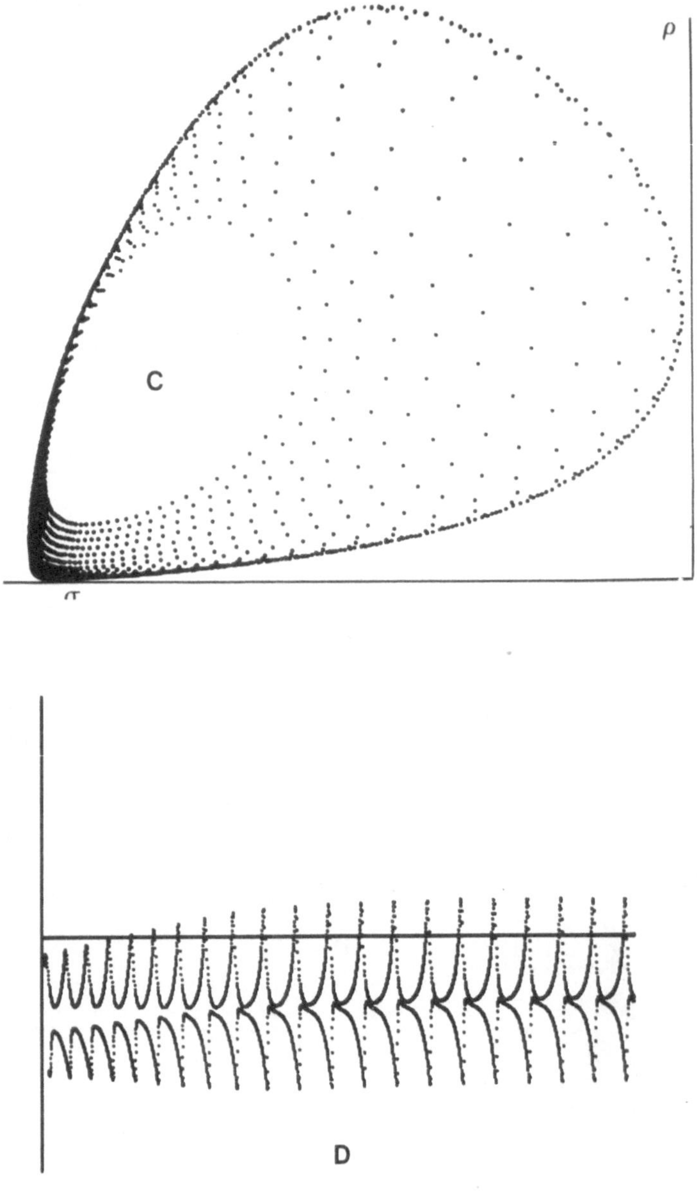

Fig. 17(c,d) Envelope breathing and limit cycle in the (ρ,σ) plane for C=20, $\tilde{\alpha}/\gamma_{\perp}=5$, and $x = 6.8669$. The limit cycle and the corresponding time dependent envelope have been obtained by integrating the equations of motion backward in time.

before eventually the envelope settles down to a steady asymptotic value. Figure 17b shows the same phenomenon, but from the point of view of the phase-space variables (ρ,σ). The trajectory in phase space spirals towards the point corresponding to the self-pulsing state, which behaves as a stable focus. After crossing the line AD, the focus becomes unstable through a Hopf bifurcation. In this case, one expects the appearance of a limit cycle in the phase space, which can be stable or unstable, depending on the nature of the Hopf bifurcation. The limit cycle that accompanies the crossing of the boundary line AD is unstable, i.e., it repels nearby trajectories. In order to display its presence, we have used a trick; by integrating the time evolution backward in time, the repeller becomes an attractor. Figure 17c shows the backward approach to the limit cycle in the (ρ,σ) phase-plane. Similarly, Figure 17d shows the backward time evolution of the envelope of oscillations of the transmitted field. Now for long times we have a perfectly periodic breathing regime in which the envelope marks time as a "regular" clock (but backwards in time). Of course, unstable limit cycles, just as unstable steady states, cannot be observed directly.

ACKNOWLEDGEMENTS

The research described in this review was partially supported by the Italian National Research Council (CNR), by a contract with the U.S. Army Research Office and by a grant from the Martin-Marietta Research Laboratories. We are grateful to Mrs. Debbie DeLise-Hughes for her skillful handling of this manuscript.

REFERENCES

1. H. Haken, *Synergetics - An Introduction*, Springer-Verlag, Berlin, 1977.
2. G. Nicolis and I. Prigogine, Self-Organization in Non-Equilibrium Systems: From Dissipative Structures to Order Through Fluctuations", Wiley and Sons, NY 1977.
3. S. Grossman and J. Thomae, Z. Naturforsch. 32a, 1353 (1977).
4. M.J. Feigenbaum, J. Stat. Phys. 19, 25 (1978) and 21, 669 (1979).
5. L.A. Lugiato, P. Mandel, S.T. Dembinski, and A. Kossakowski, Phys. Rev. A18, 238 (1978).
6. J.C. Antoranz, J. Gea, M. Velarde, Phys. Rev. Lett. 47, 1895 (1981); J.C. Antoranz, L.L. Bonilla, J. Gea, M. Velarde, Phys. Rev. Lett. 49, 35 (1982).
7. T. Erneux and P. Mandel, Z. Physik 44B, 353 and 365 (1982).
8. E. Arimondo, F. Casagrande, L.A. Lugiato and P. Glorieux, Appl. Phys. 30B, 57 (1983).

9. A. Szöke, V. Daneu, J. Goldhar and N.A. Kurnit, Appl. Phys. Lett. 15, 376 (1969).
10. H.M. Gibbs, S.L. McCall and T.N.C. Venkatesan, Phys. Rev. Lett. 36, 113 (1976).
11. R. Bonifacio and L.A. Lugiato, Opt. Comm. 19, 172 (1976).
12. H.M. Gibbs, S.L. McCall and T.N.C. Venkatesan, Opt. Eng. 19, 463 (1980).
13. E. Abraham, S.D. Smith, Rept. Progr. Phys. 45, 815 (1982).
14. F.T. Arecchi and P. Salieri, Physics Bull. 33, 20 (1982).
15. L.A. Lugiato, in *Progress in Optics*, Vol. XXI, edited by E. Wolf, North-Holland (in press).
16. L.A. Lugiato, Contemporary Physics (in press).
17. R. Bonifacio, Editor, *Dissipative Systems in Quantum Optics: Resonance Fluorescence, Optical Bistability, Superfluorescence*, Springer-Verlag, Berlin (1982).
18. H.M. Gibbs, S.L. McCall, T.N.C. Venkatesan, A.C. Gossard, A. Passner, and W. Wiegmann, Appl. Phys. Lett. 35, 451 (1979); S.S. Tarng, K. Tai, J.L. Jewell, H.M. Gibbs, A.C. Gossard, S.L. McCall, A. Passner, T.N.C. Venkatesan, and W. Weigmann, Appl. Phys. Lett. 40, 205 (1982).
19. D.A.B. Miller, S.D. Smith, C.T. Seaton, IEEE J. Quant. Elect. QE17, 312 (1981).
20. H.M. Gibbs, S.S. Tarng, J.L. Jewell, D.A. Weinberger, K.C. Tai, A.C. Gossard, S.L. McCall, A. Passner and W. Wiegmann, Appl. Phys. Lett. 41, 221 (1982).
21. S.D. Smith, Topical Meeting on Optical Bistability, Rochester (1981).
22. L. Allen and J.H. Eberly, *Optical Resonance and Two-Level Atoms*, Wiley and Sons, NY 1975.
23. R. Bonifacio and L.A. Lugiato, Lett. al. Nuovo Cimento, 21, 505 (1978).
24. F.S. Felber and J.H. Marburger, Appl. Phys. Lett. 28, 731 (1976).
25. K. Ikeda and O. Akimoto, Phys. Rev. Lett. 45, 709 (1980).
26. L.A. Lugiato, L.M. Narducci, D.K. Bandy and C.A. Pennise, Opt. Comm. 43, 281 (1982).
27. M. Gronchi, V. Benza, L.A. Lugiato, P. Meystre and M. Sargent III, Phys. Rev. A24, 1419 (1981).
28. R. Bonifacio and L.A. Lugiato, Lett. al Nuovo Cimento 21, 510 (1978).
29. L.A. Lugiato, V. Benza, L.M. Narducci and J.D. Farina, Zeit. Phys. B49, 351 (1983).
30. H. Haken, Zeit. Phys. B21, 105 (1975); ibid. B22, 69 (1975).
31. J.E. Marsden and M. McCracken, *The Hopf Bifurcation and its Applications*, Springer-Verlag, Berlin (1976).

LASER WITH INTRACAVITY ABSORBER: Q-SWITCHING, MULTISTABLE NONLINEAR OSCILLATIONS AND CHAOS

M.G. Velarde and J.C. Antoranz

Departamento de Física Fundamental, U.N.E.D.

Apartado 50.487, Madrid, Spain

1. INTRODUCTION

The laser problem refers to a highly simplified mathematical model obtained from a more refined description given by LUGIATO *et al.* |1,2|. We have

$$dE/dt = -\kappa E + NV + \bar{N}\,\bar{V} \tag{1}$$

$$dV/dt = -\bar{\gamma}_{\perp} V + |g|^2\, D\, E \tag{2}$$

$$d\bar{V}/dt = -\bar{\gamma}_{\perp} \bar{V} + |g|^2\, \bar{D}\, E \tag{3}$$

$$dD/dt = -\gamma_{\parallel} D - 4\, V\, E + \gamma_{\parallel}\, \sigma \tag{4}$$

$$d\bar{D}/dt = -\bar{\gamma}_{\parallel} \bar{D} - 4\, \bar{V}\, E + \ddot{\gamma}_{\parallel}\, \bar{\sigma} \tag{5}$$

where N is the number of two-level excitable (active)atoms. E is the electric field amplitude. V is the polarization of the atoms. D is the perturbed atomic inversion (population inversion). g is the field-matter (active or passive) coupling constant. κ is the damping constant of the field in the cavity . σ is the unsaturated inversion. $\gamma_{\parallel}$ and $\gamma_{\perp}$ are the longitudinal and transverse relaxation constants, respectively (their inverse are a measure of the population and dipole decay times, respectively). the bar over a quantity indicates the corresponding variable for the *passive* atoms (*absorbing medium*).

2. PHASE DIAGRAM

For universality of the description we shall make use of dimensionless quantities. We pose

$E = -(\gamma_{\|}\gamma_{\perp})^{1/2} a/|g|2$, $V = -\sigma|g|(\gamma_{\|}/4\gamma_{\perp})^{1/2} p$, $\bar{V} = -\bar{\sigma}|g|(\gamma_{\|}/4\gamma_{\perp})^{1/2}\bar{p}$,

$D = \sigma(1-d)$, $\bar{D} = \bar{\sigma}(1-\bar{d})$, $t' = \gamma_{\perp}t$, $\omega = \gamma_{\|}/\gamma_{\perp}$,

$r_1 = \bar{\gamma}_{\perp}/\gamma_{\perp}$, $r_2 = \bar{\gamma}_{\|}/\gamma_{\|}$, $\rho = \kappa/\gamma_{\perp}$, $A = N|g|^2\sigma/\kappa\gamma_{\perp}$ and

$C = 1 - \bar{N}|g|^2\bar{\sigma}/\kappa\bar{\gamma}_{\perp}$. Note that $\kappa = c(1-R)/L$, where R is the reflectivity of the mirrors and L is the length of the cavity. c is the velocity of light. A and C (or more precisely $C-1$) are the "control" parameters which hide the pumping rates of *active* and *passive* atoms, respectively. It is clear that the problem has too many parameters that can be considered free to vary. In dimensionless form we have

$$da/dt = \rho\{-a + Ap + r_1(1-C)\bar{p}\} \tag{6}$$

$$dp/dt = a(1-d) - p \tag{7}$$

$$d\bar{p}/dt = a(1-\bar{d}) - \bar{p}r_1 \tag{8}$$

$$dd/dt = \omega(-d + ap) \tag{9}$$

$$d\bar{d}/dt = \omega(-r_2\bar{d} + a\bar{p}) \quad , \tag{10}$$

where $\{a, p, \bar{p}, d, \bar{d}\}$ are the new "field amplitudes" of the model-problem and the remaining are parameters . As no confusion is expected we have replaced t' by t.

To illustrate some of the results found we now set some of the parameters to given values. We set $r_2 = 1$, *i.e.*, we take the population decay time the same for both active and passive atoms which is consistent with the assumption of resonance between the emitting and absorbing transitions. We fix $\omega = 0.01$, *i.e.*, we take the dipole decay time to be two orders of magnitude smaller than the other time constant ($T_1 = 1/\gamma_{\|}$, $T_2 = 1/\gamma_{\perp}$, $T_2 \ll T_1$). We also take $\rho = 0.1$ and $\omega < \rho$. Other parameter ranges have been studied in the literature |3, 4, 5| . r_1 can be let free to vary but for purposes of illustration we shall also fix to a value , say $r_1 = 0.4$.

Fig. 1 illustrates the results of the linear stability analysis of the trivial motionless state or the *no lasing* state. This figure also contains the locus of *finite amplitude* instability (*meta-stability)*. Note that the stability diagram is restricted to the range of parameter values where oscillatory modes are expected which corresponds to the case where the absorber plays its *role of absorber*. Later on we shall discuss the case where the absorber becomes active and cooperates in the lasing process.

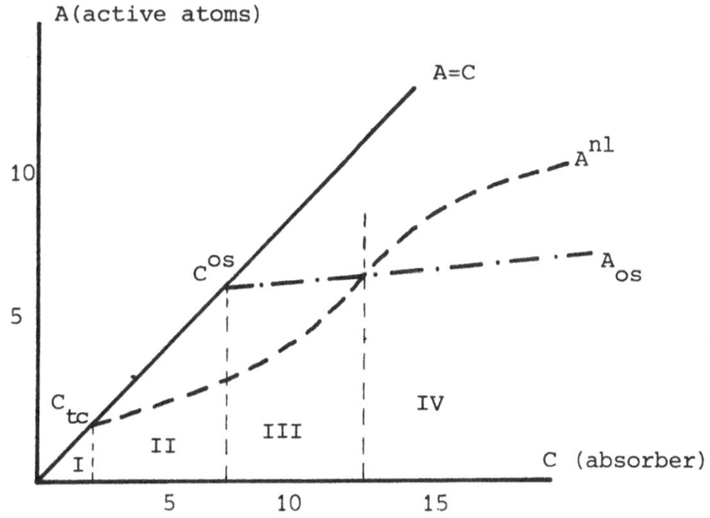

Fig.1. Phase diagram with primary bifurcations of a system with two competing agents (laser with absorber). I: soft steady state bifurcates; II: hard steady bifurcates; III: generally hard steady state bifurcation precedes soft oscillation and IV: generally, soft oscillation precedes hard steady state.

3. SOFT OSCILLATORY LASING

In region $C > C^{os}$ (Fig.1, region IV) the system is expected to pass directly from no lasing to oscillatory lasing |6|. In mathematical terms the system possesses a Hopf bifurcation to the correct side, *i.e.*, a supercritical branch of oscillations bifurcates softly with amplitude and period related to the distance to the bifurcation (instability point). Figure 2 illustrates the kind of oscillation obtained.

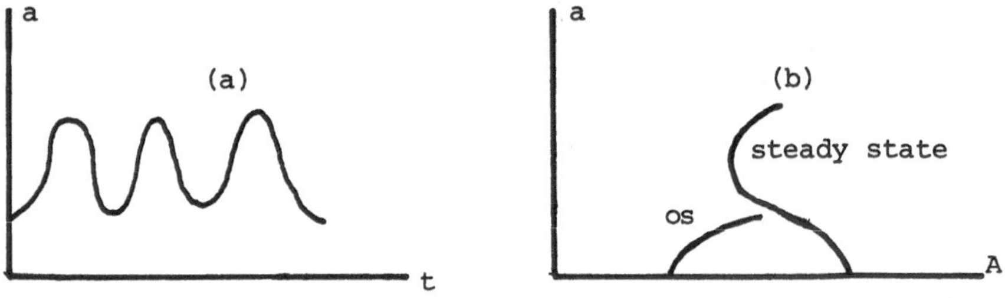

Fig.2.(a) Nonlinear soft-excited oscillation that bifurcates from $a = 0$ prior to another nonlinear steady state (b) that also bifurcates from $a = 0$ (Fig.1, region IV)

4. Q - SWITCHING AND COEXISTENCE BETWEEN OSCILLATIONS AND CW LASING

Without absorber the threshold for laser behavior is at $A = 1$ and the steady state laser intensity grows with $(A-1)$. The absorbing medium is pumped as an emitter if C is smaller than unity, with threshold at $C = 1 - 1/r_1$. Thus we now restrict consideration to the case C larger than unity. Then, for

$$(1 - r_1)^{-1} \equiv C_{tc} \leqq C \leqq (\rho + r_1)/\rho(1 - r_1) \equiv C_{osc}$$

the zero-field solution is stable up to $A = C$ where a branch of steady solutions bifurcates subcritically like in a *hard - mode (discontinuous, first- order)* transition. This brings a region of A-values where *two steady states are available to the system,* since the middle branch (Fig. 2b) is always unstable. One usually expects that past A^{nl} the nonlinear steady branch would be stable, at least for A not too different from A^{nl}. This nonlinear steady branch can be obtained analytically and a linear stability analysis of such branch can be carried out. It appears that the nonlinear steady branch is unstable for the range of parameters indicated in Fig. 3. Then at a value A_u there is a right-hand bifurcation from this nonlinear branch. Using Floquet theory |7| and the two-time scale method |4, 8| it appears that the bifurcation is to the *wrong side, i.e., it is subcritical* thus bringing bifurcation of an *unstable limit cycle,* whereas the nonlinear steady state is stable for values of A larger than A_u. For illustrative purposes, note that this prediction is valid in the neighborhood of $\rho = 0.1$ and r_1 smaller than unity. Such value of ρ corresponds to a one meter cavity in the laser with some 3% losses if we take $\gamma \sim 10^{-8}$ s^{-1}. The condition r_1 smaller than unity and T_2 larger than T_2, is consistent with having an active medium with higher gas pressure than that in the absorber.

The divergence of the system of differential equations is always negative and there must be an attractor in the solutions. As neither of the two steady states is stable below A_u and no softly excited limit cycle is expected we wonder about the evolution of the above mentioned limit cycle. We have explored the region $C < A < A_u$ by means of the Poincaré map. Then a relaxation oscillation type of limit cycle was found (Fig.4). Its width is of the order of the photon lifetime in the cavity, *i.e.*, $1/\kappa(1/\rho$ in dimensionless units) and its period is of the order of the decay time of the excited state $T_1(1/\omega$ in dimensionless units). These predictions agree with experimental results obtained by Arimondo (see his lecture Notes) and by Arecchi. In the latter case the experiment does not refer to a case with intracavity absorber but the system corresponds rather well to a case of two competing processes in the cavity, very much like the gain-absorber competition.

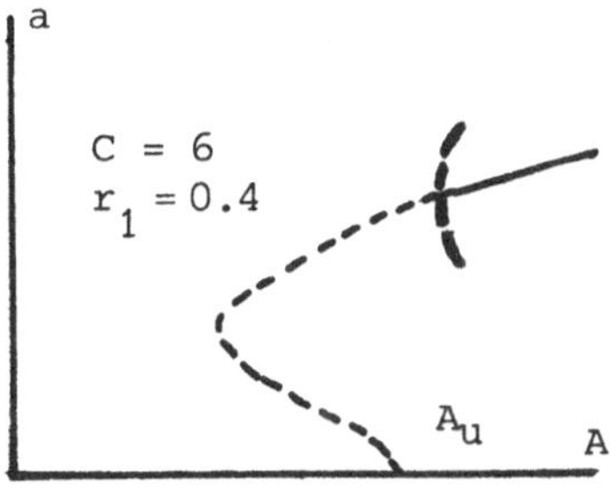

Fig.3 Bifurcation of hard-excited oscillations (pulses) from the steady state bifurcated from $a = 0$ in Fig.1, region II.

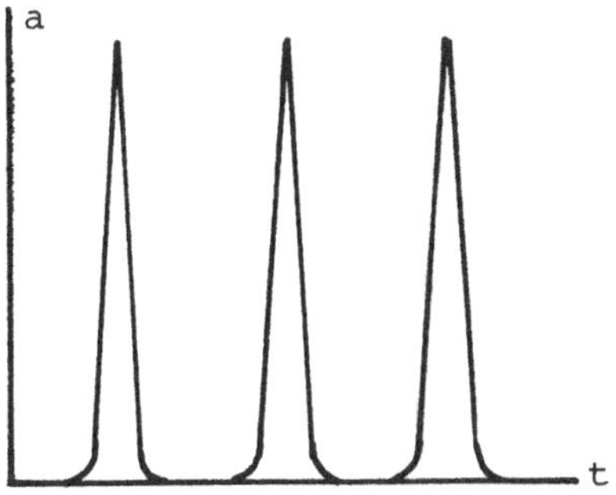

Fig.4 A sketch of the pulses bifurcated at $A = A_u$ when for $A < A_u$ they become stable. Note the difference between these oscillations (secondary bifurcation) and the oscillations in Fig.2 (primary bifurcation).

Note that at $\bar{d} = 1$ the absorber becomes transparent with equal number of atoms in the excited and ground states. It actually becomes *active* for a short interval of time, and only there *cooperates* with the gain cell. The pulse peak intensity decreases with increasing value of the constraint, A. Its value is an order of magnitude higher than the corresponding value at the unstable steady state. The period decays with increasing values of A. On the other hand when A tends to A_u from below the pulses broaden and tend more to a smooth oscillation keeping, however, a nonvanishing amplitude and finite period at A_u. When A approaches C from above, the limit cycle period rises drastically, the minimum intensity tends to zero, and the peak intensity remains essentially constant. All these features are characteristic of a *saddle-loop* at A = C. At values of A around C the oscillation tends to a single pulse with infinite rising and decaying times but, however, finite width and height.

5. COEXISTENCE OF VARIOUS OSCILLATIONS INCLUDING Q-SWITCHING

A more complete mathematical description of the problem discussed in the preceding sections is obtained when we consider the field amplitudes together with their phases. Thus we jump from five to eight differential equations. We have

$$da_r/dt = \rho\{-a_r + Ap_r + r_1(1-C)\bar{p}_r\} \tag{11.a}$$

$$da_i/dt = \rho\{-a_i + Ap_i + r_1(1-c)\bar{p}_i\} \tag{11.b}$$

$$dp_r/dt = a_r(1-d) - p_r \tag{12.a}$$

$$dp_i/dt = a_i(1-d) - p_i \tag{12.b}$$

$$d\bar{p}_r/dt = a_r(1-\bar{d}) - r_1\bar{p}_r \tag{13.a}$$

$$d\bar{p}_i/dt = a_i(1-\bar{d}) - r_1\bar{p}_i \tag{13.b}$$

$$dd/dt = \omega(-d+a_r p_r + a_i p_i) \tag{14}$$

$$d\bar{d}/dt = \omega(-r_2\bar{d}+ a_r\bar{p}_r + a_i\bar{p}_i)\ , \tag{15}$$

where the subscript "r" and "i" have the obvious meaning of real and imaginary parts of the variable defined in Section 2. Note that Eqs.(11)-(15) provide non-trivial phase effect, *i.e.*, non-vanishing values of a_i and p_i, only when the atomic system is prepared in a coherent superposition state at the initial time. A more realistic case would be to consider the action of an external field or a detuning between the atoms and the field.

As in the preceding Section we set $r_2 = 1$, $\omega = 0.01$, and $\rho = 0.1$. Thus we again have $\omega < \rho$. Then the steady solutions of (11)-(15) are either the *emissionless (no lasing)* state ($a_r = a_i = 0$) or a nonlinear emission with non-vanishing $a, a_r^2 + a_i^2 = X^2 \neq 0$, where X is any of the positive roots of $X^4 + X^2(1-A+r_1C) + r_1(C-A) = 0$. We have non-vanishing roots only for C larger than C_{tc} and below A_{nl} the only solution available is $X=0$. Thus A^{nl} corresponds to the appearance of four solutions in the algebraic equation. We also have $d = X^2/(1+X^2)$, $\bar{d} = X^2/(r_1 + X^2)$, $p_r = a_r/(1+X^2)$, $p_i = a_i/(1+X^2)$, $\bar{p}_r = a_r/(r_1+X^2)$, and $\bar{p}_i = a_i/(r_1 + X^2)$.

The emissionless state is linearly unstable when A is larger or equal than C or

$$A \geq (\rho + r_1)\}1 + r_1 + \rho(1 + r_1 C) / \rho(1+\rho) \equiv A^{os},$$

Along $A = C$ there is *exchange of stabilities* (transitions between steady states in the jargon used above). The transition here is soft for C smaller than C_{tc} and hard at $A = A^{nl}$ for C larger than C_{tc}. At $C = C_{os}$ (defined in Section 3) and all along $A = A_{os}$ there is *overstability*. Actually this overstability corresponds to a soft limit cycle bifurcation with a pair of complex semisimple eigenvalues both of multiplicity two. Their imaginary part is μ_o with $\mu_o^2 = r_1\{\rho C(1-r_1) - (\rho + r_1)/(1+\rho)$. Thus it happens that the linear stability diagram of the problem with complex amplitudes (11)-(15) is exactly the same as the diagram depicted in Fig. 1 for (1)-(5). The discussion that follows refers to $A^{os} < A^{nl}$.

At values of A slightly above A_{os} two stable limit cycles bifurcate from the emissionless state. They have been constructed using a method due to kielhöfer |9,10| (for a similar calculation using a different method see |11|). One of the limit cycles (LC1 in Fig.5) has constant phase and corresponds to the oscillatory solution described in Section 3 |6|. Thus LC1 is the same whether or not we consider the phases in (1)-(5) or (11)-(15) in this Section. The other limit cycle (LC2 in Fig. 5) has linearly growing phase and *does not appear* in the model (1)-(5). That LC1 and LC2 both bifurcate stable has been verified analytically by means of Floquet theory and numerically by using the Poincaré map. Also with the Poincaré map we have been able to locate the end-points where these limit cycles become unstable. Both yield to tori which bifurcate to the wrong side. For LC2 this has been verified by means of the two-time scale method . The coordinates of its corresponding fixed point in the Poincaré map are known |11| . Then we have constructed the bifurcated limit cycle in such plane and we have seen that it bifurcates to the wrong side, *i.e.*, the limit cycle bifurcating in the Poincaré map shows up *unstable* which indicates the appearance of an unstable torus in the eighth-dimensional space. The analytical result has been checked by means of a direct integration of (11)-(15). We have observed that there is a passage from the fixed point to an outwardly spiraling orbit in the Poincaré map. A similar behavior appears in the Poincaré map of LC1 although for this oscillation we have not been able to establish this property analytically.

As in Section 4 we have also studied the stability of the nonlinear steady lasing state (A^{nl} in Fig. 1). We have again found that this state is unstable for values of A smaller or equal to A_u and stable otherwise. The actual value of A_u depends on the values given to the remaining parameters of the problem (11)-(15). However, this value A_u is *exactly* the same as the value found with the system without the phases (1)-(5). Thus the phases in (11)-(15) play no relevant role in the stability of the nonlinear steady lasing state. They play indeed an interesting role in the evolution of the emissionless state as they permit the appearance of LC2. This fact was already discovered by different authors |11, 12|. In Ref. 12 there is also a phase diagram like that shown in our Fig. 1.

With respect to the hard-excited limit cycle found it is very much like the passive *Q- switching* described in Section 4 although they differ in some non-trivial properties . The pulses here found for (11)-(15) have peak intensity that *increases with increasing pumping rate (hidden in A), which is the opposite case to the Q-switching described in Section 4 where C was smaller than* C_{os} . Note that in Section 4 the Q-switching was the only available oscillatory solution due to the chosen range of parameter values . In both cases the maximum of the pulses is attained before the minimum in the emitter's curve is reached (Refs. 8 and 9). For the case of the present Section we have coexistence of limit cycles not only between soflty excited oscillations but also between a solf- and a hard-excited cycle.

At $A \leqq A_1$ the oscillation LCQ disappears whereas LC2 disappears at $A \geqq A_1$. The latter cycle yields to an unstable torus that we *conjecture* dies at LCQ. Hopefully the major qualitative findings with (11)-(15) remain valid for the model with the addition of an external field or a detuning between the field and the atoms.

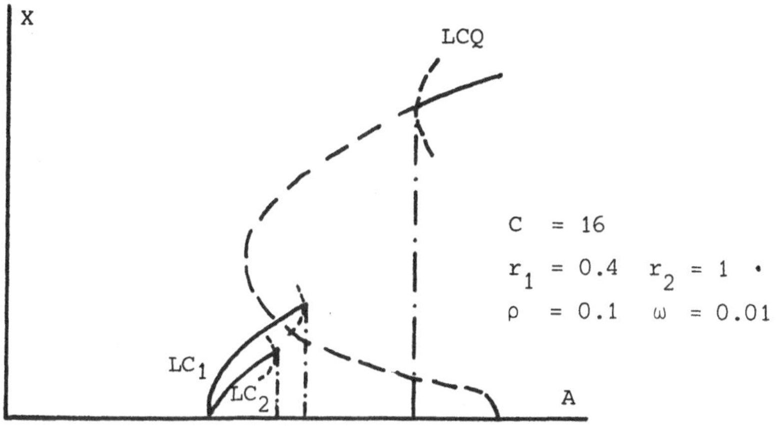

Fig.5. Oscillations that bifurcate from either X = 0 or from the nonlinear steady state bifurcated from X = 0 in Fig. 1, region IV. Bistable oscillatory regimes are described in the main text.

6. A LORENZ-LIKE STRANGE (APERIODIC) SOLUTION AND ITS FRACTAL DIMENSION

In region $C < 0$ of Figure 1 strange attractors appear like in the original Lorenz model. Strange attractors are erratic (chaotic, turbulent) solutions which share with turbulent flows in hydrodynamics the basic properties of power spectra with broadband noise, decaying correlation functions and sensitivity to changes in initial conditions (instability in phase space) thus leading to lack of predictability in the deterministic time evolution of the system. One possible way of locating such erratic solutions is to follow the evolution of the Lyapunov exponents of the system. Recipes exist in the literature. We have used a modified version of that given by Shimada and Nagashima |3| together with a fourth-order Runge-Kutta integration of the differential equations (1)-(5). The relevant quantity is just the first- Lyapunov exponent. The second one vanishes which corresponds to the facts that the differential system is autonomous and the basin of attraction is a finite domain.

Fig 6 depicts the evolution of the Lyapunov exponent. The dramatic change from negative to positive values corresponds to the transition from a steady state to the *strange attractor*. Listed in Table 1 are typical values of the Lyapunov exponents for two different values of the relevant bifurcation parameter and the following values:

$$\chi = 148.04, \quad N = \bar{N} = 10^4, \quad \sigma = 2.221,$$
$$\bar{\sigma} = 0.00107, \quad \gamma_\perp = 14.8, \quad \bar{\gamma}_\perp = 0.148,$$
$$\gamma_\parallel = 39.47, \quad \bar{\gamma}_\parallel = 0.39.$$

The two columns correspond to $|g|^2 = 2.35$ and $|g|^2 = 3.10$, respectively. $\varepsilon = A - A_c$, where $A = N|g|^2\sigma/\chi\gamma_\perp$ and A_c is the critical value which corresponds to the onset of steady transition. $A_c = C$, where $C = 1 - \bar{N}|g|^2\bar{\sigma}/\chi\bar{\gamma}_\perp$. Listed in Table 1 are the fractal dimensions of the two strange attractors computed using a formula proposed by Yorke and collaborators |14|. Note that the strange attractor in region $C < 0$ (not shown in Fig 1) appears in a region where the intracavity absorber plays an *active* role, *i.e.*, it behaves like a second gain cell thus sustaining and amplifying the lasing signal emitted by the first set of active atoms The fractal dimension, D, has been obtained using a formula attributed to Yorke,

$$D = j + \sum_{i=1}^{j} \lambda_i / |\lambda_{j+1}|$$

where j is the last Lyapunov exponent that permits

$$\sum_{i=1}^{j} \lambda_i \geq 0.$$

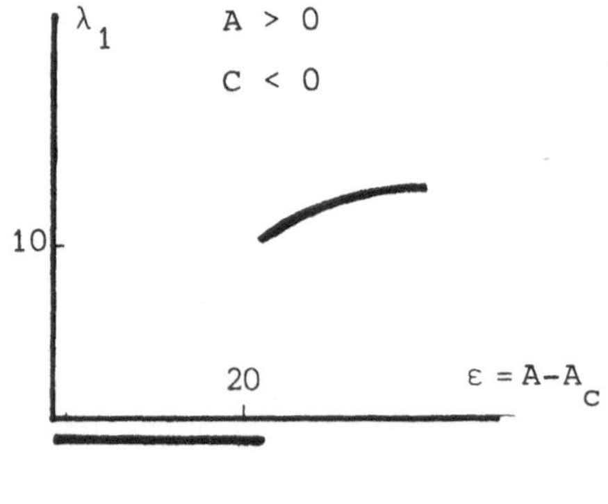

Fig.6. Evolution of the first and relevant Lyapunov exponent as a function of the pumping rate ($\varepsilon \sim 22$) past the first bifurcation. We have bifurcation to a strange attractor from a nonlinear CW steady state.

Table 1. Two examples of the strange attractor determined by the corresponding five lyapunov exponents, λ_i.D is the fractal dimension [14].

ε	24	32.0
λ_1	11	14.5
λ_2	0	0
λ_3	-0.3	-0.3
λ_4	-0.3	-0.3
λ_5	-213.3	-216.8
D	4.04	4.06

7. A FEIGENBAUM PERIOD-DOUBLING CASCADE, THE SUBSEQUENT STRANGE (APERIODIC) ATTRACTOR AND ITS FRACTAL DIMENSION

We have also found that the soft-excited limit cycle described in Section 3 does indeed become unstable to time-periodic disturbances in a Feigenbaum cascade. For illustration we here choose the parameters as follows: $\omega = 8/3$, $\rho = 10$, $r_1 = r_2 = 0.4$ and $C = 16$. Then the soft Hoph bifurcation to limit cycle occurs at $A = 7.785$. A subsequent bifurcation to a cycle with period twice the former is at $A = 7.793$. Further period-doublings appear at $A = 7.7953$, 7.79546, 7.79546, 7.79550, and so on. Past $A = 7.7955$ there is erratic behavior. At $A = 7.796$ there is a strange attractor. Such type of attractor exists for a finite range of values of A until the value 7.8 is attained. Then in accordance with our computer calculations the system escapes the erratic behavior and goes to a fixed point. We have not yet studied neither in detail nor accurately the region of chaos. However, for the elements that we have obtained of Feigenbaum's cascade we are confident in the results. We have used a Poincaré map ($\bar{p} = 0$ in the laser variables). Tables 2 and 3 illustrate the results found. It should be noted that Yorke's fractal dimension is a little over two in the neighborhood of accumulation point. We wonder if upon increasing the distance from this point an increase up to a value higher than four is obtained. We are working on this problem at the moment as well as on the stability of the hard-excited (Q-switched) limit cycles that might lead also to chaos through intermittency.

Table II. Lyapunov exponents for a strange attractor of (1)-(5) and the fractal dimension according to formulas provided by Yorke |14| and Mori |15| in the case of a transition from a (soft-excited) limit cycle to chaos via a sequence of period-doubling instabilities (A = 7.798)

λ_1	0.116	
λ_2	0	
λ_3	-0.798	
λ_4	-1.696	
λ_5	-12.75	
D_M	2.02	(two plus the fractal)
D_Y	2.14	(two plus the fractal)

Table III. Direct (a) and inverse (b, in the chaotic region) cascades of period-doubling (n indicates the n-th doubling) for $\rho = 1$, $\omega = 8/3$, $r_1 = r_2 = 0.4$ and $c = 16$

(a)		(b)	
n	A	n	A
1	7.833	5	7.84804
2	7.8453	4	7.84812
3	7.84725	3	7.84842
4	7.84784	2	7.84965
5	7.84797		

ACKNOWLEDGMENTS

This research has been sponsored by the Stiftung Volkswagenwerk.

REFERENCES

1. L.A. Lugiato, P. Mandel, S.T. Dembinski and A. Kossakowski , Phys. Rev. A *18* (1978), 238.
2. V. Degiorgio and L.A. Lugiato, Phys. Lett. A *77*(1980), 167.
3. J.C. Antoranz and M.G. Velarde, Optics Commun. *38* (1981), 61.
4. M.G. Velarde, in Nonlinear Phenomena at Phase Transitions and Instabilities (T. Riste, editor), Plenum Press, New York, 1981.
5. H. Knapp, H. Risken and H.D. Wollmer, Appl. Phys. *15*(1978),265.
6. M.G. Velarde and J.C. Antoranz, Phys. Lett. A *80*(1980), 220.
7. G. Ioss and D.D. Joseph, Elementary Stability and Bifurcation Theory, Springer-Verlag, New York, 1980 (Chapters V and VII).
8. L.L. Bonilla and M.G. Velarde, J. Math. Phys. *20* (1979),2692.
9. J.C. Antoranz, L.L. Bonilla, J. Gea and M.G. Velarde, Phys. Rev. Lett.*49* (1982), 35.
10. H. Kielhöfer, Arch. Rat. Mech. Anal. *69* (1979),53
11. T. Erneux and P. Mandel, Z. Phys. B *44*(1981), 353.
12. S.T. Dembinski, A Kossakowski, P. Peplφwski, L.A. Lugiato and P. Mandel, Phys. Lett. A *68*(1978), 20.
13. I. Shimada and T. Nagashima, Prog. Theor. Phys. *61* (1979),1605.
14. P. Frederickson, J.L. Kaplan and J.A. Yorke, preprint .
15. H. Mori, Prog. Theor. Phys. *63* (1980), 1044.

NON-LINEAR OPERATION OF CO_2 LASERS WITH INTRACAVITY SATURABLE ABSORBERS

Ennio Arimondo

Istituto di Fisica Sperimentale
Universitá di Napoli and Gruppo Nazionale
Struttura della Materia, Pisa, Italy

INTRODUCTION

The laser operation is an important case of a non-linear system and a large effort of the quantum optics has been devoted to the interpretation of the non-linear features, also from the point of view of specific applications. In a simple system the non-linearities are in the response of the amplifying medium. In a laser with an intracavity saturable absorber, where the non-linear response of the absorber is present, new phenomena occur. The behaviour of a laser containing an intracavity saturable absorber was investigated initially a long time ago (1). However recently a large interest in the phenomena arose owing to the observation of optical bistability and owing to the development of precise theoretical models. In the experiments involving the operation of an infrared CO_2 laser containing a saturable molecular absorber, it is quite easy to realize a single frequency operations. Thus the analysis presented here will be restricted to this case,and the laser instabilities produced by mode or frequency competition will be not considered. The experimental apparatus, the experimental results and a theoretical analysis will be presented.

EXPERIMENTAL APPARATUS

The typical intracavity apparatus was composed by a CO_2 cell, with a discharge length ℓ = 1.6 m contained in a infrared cavity, L = 3.9 m. A grating at one end of the cavity forced the laser operation on a single line, while a 95% reflectivity mirror at the other end formed the optical cavity.

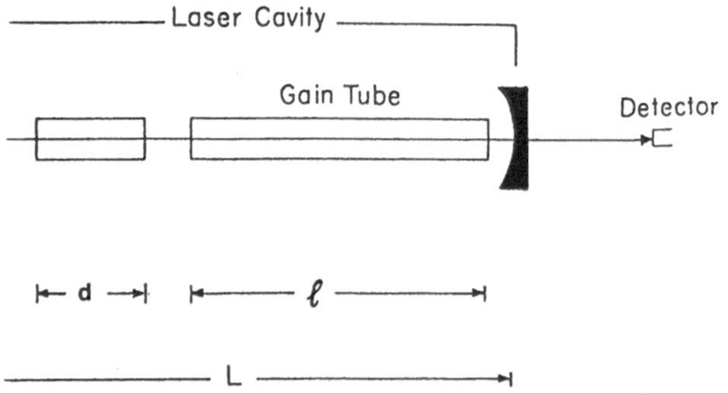

Fig. 1. Schematic diagram of the experimental set-up.

Three irises, at the end and the center of the cavity, restricted the operation to a single transverse mode. The absorbing medium was contained in a cell, length d = 0.8 m, located near the grating end of the cavity (Fig.1). The laser output power was monitored by a PbSnTe ir detector coupled to the input of a boxcar integrator through a wide-band amplifier, the total rise time of the detection system being approximatively 0.5 μs. Particular care was taken to have very stable operation on a long time scale, because thermal drifts and hysteresis may affect the observations.

The laser operation may occur either in a continuous wave (c.w) steady state regime or with undamped spiking oscillations, a self--pulsing behaviour known as passive Q-switching regime. In the passive Q-switching regime the time averaged output power and the period of oscillations were measured versus the parameters defining the laser operation, i.e. the amplifier current or pressure, the absorber pressure and the laser cavity detuning. In the c.w regime the laser output power and the laser frequency were monitored. Furthermore in order to complete the analyst in this regime a small perturbation was applied to the laser operation and the recovery towards the new steady through damped oscillations was monitored. Two methods have been used to produce a perturbation of the cavity absorption. In the Stark splitting technique a dc electric field was applied to the absorber to produce a splitting of the absorber transitions, and a change in the absorption coefficient. In the radio-frequency double-resonance method a radio-frequency field acts on the lower or upper level of the absorber infrared transitions and modifies the number of molecules interacting with the infrared field. Identical transient responses of the laser were observed with the two modulation techniques.

EXPERIMENTAL RESULTS

If a saturable absorber is introduced into the cavity of a CO_2 laser, either the laser output power is drastically reduced, or the self-pulsing Q-switching behaviour occurs. The Q-switching appears when the laser operates at a low pumping parameter near the threshold value, i.e. at low current or pressure in the amplifier cell. At larger values of the pumping parameter, as reached increasing the amplifier current or pressure, the Q-switching regime terminates and the c.w operation is recovered.

Most CO_2 intracavity observations have been performed on SF_6 gas which presents a large absorption coefficient for several 10 μm region laser lines, but also the methyl halides compounds produce large non-linear behaviour[2]. Figure 2 reports a typical pulse observed on the CO_2 laser output for operation on the 10P(20) line at 20 Torr pressure (partial pressures CO_2 15%, N_2 10%, He 75%) and 5 mA current, when 40 mTorr SF_6 gas was introduced in the absorber cell. Similar pulse shapes were observed by previous observations of passive Q-switching[3]. The width of the first pulse was in 0.1 - 2 μs range (deformed by the amplifier time in the record of Figure 2). The pulse repetition was in the 10 - 100 μs range. The pulses reported in the Figure presents a tail, whose length depends on the frequency tuning of the laser. The source of such a tail has not been identified, however the experimental observations support the role played by the detuning effects of the laser frequency relative to the absorber frequency or the simultaneous absorption of the SF_6 gas on a second transition starting from the upper level of that producing the passive Q-switching.

Because in recent theoretical analyses[4] the Q-switching pulse have been specifically investigated, width, intensity and period of the pulses have been measured for the 10P(22)CO_2 laser line in presence of SF_6 intracavity absorption with pressure in the 20 - -150 m Torr range.

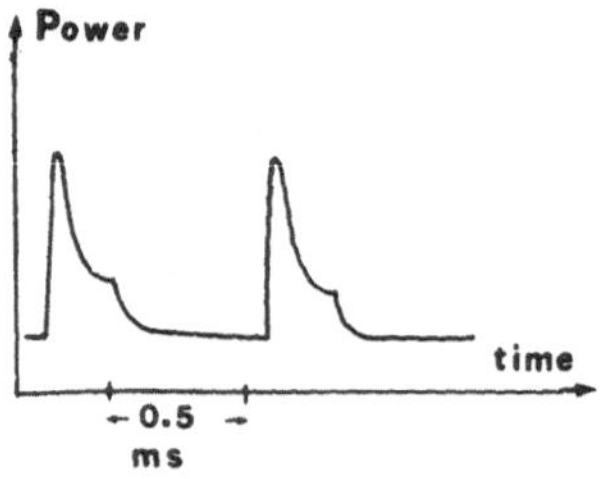

Figure 2 - Record of the Q-switch pulses on the output power of a 10P(20) CO_2 laser containing 40 mTorr SF_6 absorber

The pulse width did not change with the pumping inversion in the amplifier and with the laser frequency in agreement with the prediction that the pulse width is equal to the inverse of the photon decay rate in the cavity, denoted by K. For the pulse height it resulted that laser frequency and pumping inversion produced changes smaller than 10% and 5% respectively, while the iris diameter affected the pulse heigth up to 25%. For the pulse period it resulted that a dramatic rise occurred whenever a transition from the passive Q-switching to laser of operation is approached, i.e. a typical slowing down phenomenon at a bifurcation with an unstable solution. Such a behaviour is illustrated in Figures 3 and 4 for the dependence of the period pulse on the amplifier current, i.e the pumping inversion, and the laser frequency detuning respectively. In figure 3 the critical slowing down at low currents is observed for different SF_6 pressures, while the extension of the Q-switching region changes with the pressure. In figure 4 an increase in the pulse period is observed or both sides of the frequency tuning, with a longer period for laser tuning approaching the center of the SF_6 absorption. The pulse period is of the order of the decay time of the amplifier excited state[4], but figure 3 shows that when the decay time decreases increasing the amplifier pressure

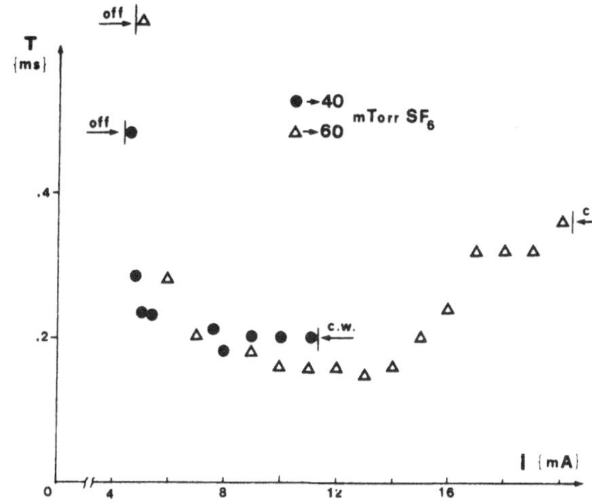

Figure 3. Dependence of the Q-switch period on the amplifier current for the 10P(22) CO_2 operation at several SF_6 absorber pressures

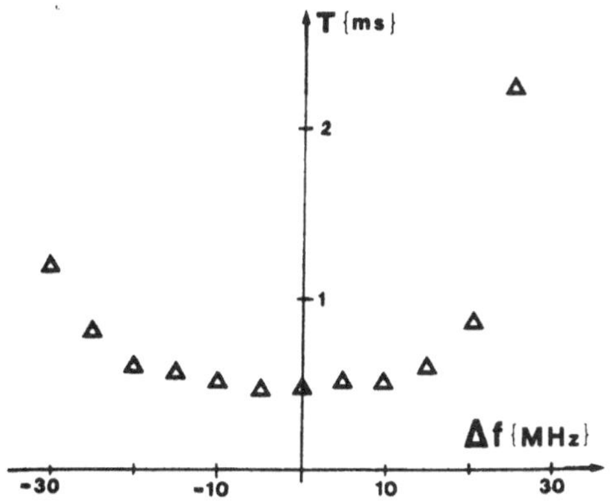

Figure 4. Dependence of the Q-switch period on the laser detuning, for the 10P(22)CO_2 operation with 70 mTorr SF_6 pressure. Laser frequency was measured by beating with a waveguide CO_2 laser. The center of the SF_6 absorption is at Δf = 50 MHz.

the pulse period does not decrease because the amplifier inversion is simultaneously increased.

As a final point in understanding the behaviour of a Q-switched laser it should be noticed that the averaged laser output power increases with the increase of the amplifier pumping parameter, because the pulse width remains constant but the pulse period increases. Furthermore at the transition from the Q-switching regime to the c.w. operation the laser output power changes in a continuous way (see Figure 5) proving that all the stored energy is released either in theQ-switching pulses or in the c.w. operation.

Additional information on the c.w. regime was obtained monitoring the transient recovery to the steady state following a perturbation of the intracavity absorption, as described in the previous paragraph. This transient behaviour presented damped oscillations, whose period and damping time were measured, as reported in Figure 6, for the case of 25 mTorr CH_3I absorption on the 10P(32) laser line in a region of parameters near the Q-switching operation. It may be noticed that the pulse repetition period and the oscillation period have a continuity at the transition point proving that in both cases the energy transfer between amplifier medium and absorber is involved. The damping time of the oscillation period diverges as the Q-switching region is approached, thereby showing the critical slowing down behaviour that is typical of the approach to an instability. This instability is the very cause of the Q-switching regime, where the oscillations are no longer damped out.

Finally the optical bistability phenomenon produced by an intracavity saturable absorber on the laser output power has been monitored[5]. At first the output power of the laser in the absence of the intracavity absorber was measured a function of the discharge current keeping the pressure constant in the discharge tube in the ratio CO_2 : N_2 : He, 1.8 : 2.4 : 7.8 Torr. The dashed line of Fig. 7 reports the measured power (in mW) versus the current (in mA). The relaxation rates of the CO_2 amplifier, depending mainly on the amplifier pressure, are not influenced by the discharge current. The amplifier population inversion increase with the discharge current in the range of the measurements; a linear dependence law of the inversion on the current is satisfied.

Finally the absorber cell was filled with 25 mm Torr pressure SF_6 gas that has several absorption lines coincident with the CO_2P(16) laser line,and the laser power was measured versus the current. The observed behavior is represented by the continuous line of Figure 7 with a bistable behavior extending over a large range of the monitored region.

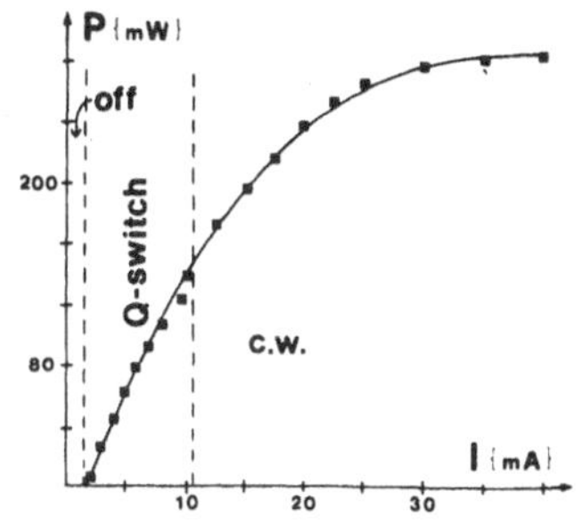

Figure 5. Plot of the averaged laser power versus the current amplifier for the 10P(32) CO_2 laser operation containing 60 mTorr CH_3I absorber plus 160 mTorr He buffer gas. The continuous behaviour of the power in passing from the Q-switch region to the c.w regime should be noticed. The line through the experimental points is not a least square fit.

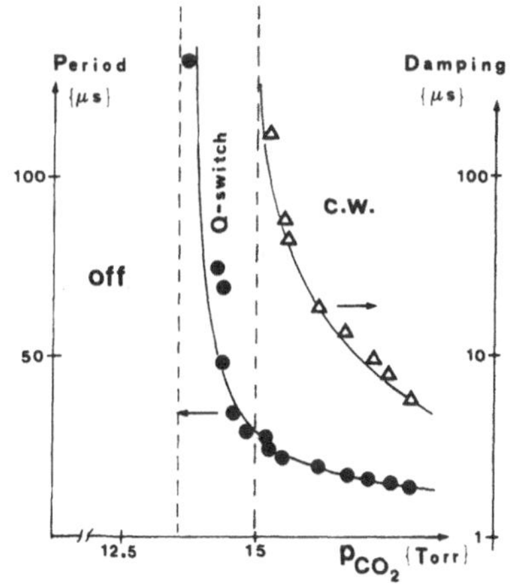

Figure 6. Plot of the oscillator period (● and upper left scale) and the damping time (Δ and upper right scale) versus the amplifier perssure for a 10P(32) CO_2. The lines through the experimental points are not a least square fit and have been drawn for convenience of reading.

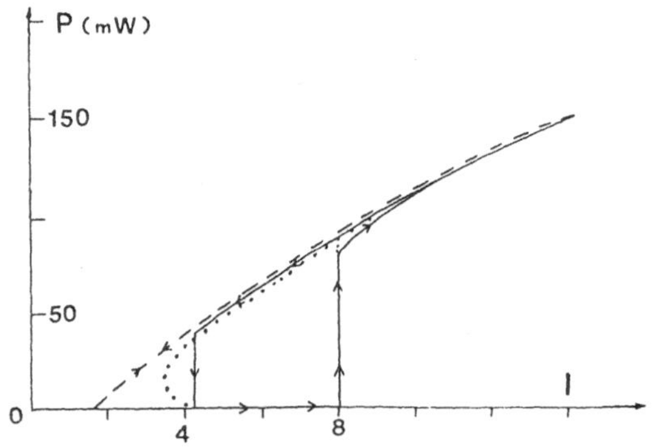

Figure 7. Experimental and theoretical curves for the optical bistability in a 25 mm Torr SF_6 sample. The dashed line represents the observed behavior for the laser without absorber and the continuous lines reports the behavior recorded with intracavity absorber. Dotted line is a theoretical fit.

THEORETICAL MODEL

The physical situation concerns a laser containing two cells inside its resonant cavity, the first cell with the amplifying atoms pumped to a positive population inversion, the second one containing the absorbing species. Two-level or four-level models, homogeneous or inhomogeneous broadening may be introduced into the description.

The Maxwell-Bloch equations for the slowly varying envelope E of the cavity electric field interacting with homogeneously broadened ensembles of atoms in the amplifier and in the absorber read in the c.g.s. unit system[6]:

$$\frac{dE}{dt} + K\,E = -\,4\pi\omega_o\,(\mu\,\zeta\,U + \bar{\mu}\,\bar{\zeta}\,\bar{U})$$

$$\left(\frac{d}{dt} + \gamma_\perp\right) U = (\omega - \omega_o)\, V$$

$$\left(\frac{d}{dt} + \gamma_\perp\right) V = (\omega_o - \omega)\, U + \frac{\mu}{2\hbar} E D$$

$$\left(\frac{d}{dt} + \bar{\gamma}_\perp\right) \bar{U} = (\bar{\omega} - \omega_o)\, \bar{V}$$

$$\left(\frac{d}{dt} + \bar{\gamma}_\perp\right) \bar{V} = (\omega_o - \bar{\omega})\, \bar{U} + \frac{\bar{\mu}}{2\hbar} E \bar{D} \tag{1}$$

$$\left(\frac{d}{dt} + \gamma_\parallel\right) D = -2\frac{\mu}{\hbar} E V - \gamma_\parallel N$$

$$\left(\frac{d}{dt} + \bar{\gamma}_\parallel\right) \bar{D} = -2\frac{\bar{\mu}}{\hbar} E \bar{V} + \bar{\gamma}_\parallel \bar{N}$$

where all barred quantities refer to the absorber and have the same meaning as the corresponding parameters of the lasing medium. $U + iV$ is the macroscopic atomic polorization per unit volume, D is the difference between the population of the lower and upper levels, N the population inversion created in the amplifier μ is the modulus of the dipole moment of the atoms, ω_o the frequency of the incident field and ω the transition frequency of the amplifier atoms. $\gamma_\parallel$ and $\gamma_\perp$ are the inverse of the longitudinal and transversal relaxation times of the amplifier atoms and $K = cT/2L$ the decay rate of the empty cavity electric field as function of the mirror transmittivity T and the cavity total length L. Finally $\zeta = \ell/L$ is the ratio of the length ℓ of the amplifier cell to the cavity length and $\bar{\zeta} = d/L$ the corresponding ratio for the absorber cell.

The stationary state defined by the condition that all time derivative vanish in Eqs.(1), leads to the following equation for the stationary intensity of the field:

$$2KI_{st}^{\frac{1}{2}}\left(1 - \frac{A}{1 + I_{st}} + \frac{\bar{A}}{1 + aI_{st}}\right) = 0 \tag{2}$$

where the normalized intensity I is given by

$$I = \frac{\mu^2}{\hbar^2} \quad \frac{E^2}{\gamma_\perp \; \gamma_\parallel \; (1 + \Delta^2)}$$

with $\Delta = (\omega_\circ - \omega) / \gamma_\perp$ (3)

The pumping parameters of the amplifying cell results

$$A = \frac{\zeta \, c \, \alpha}{K(1+\Delta^2)} = \frac{2\pi\omega_\circ\mu^2 \, N \, \zeta}{\hbar\gamma_\perp K(1+\Delta^2)} \tag{4}$$

where α is the unsaturated absorption coefficient on resonance, and $A = 1$ is the threshold condition for laser oscillation with no absorber.

The pumping parameter $\bar{A}$ of the absorber, defined by a similar equation envolving barred quantities, assume also positive values, according to the definitions of Eqs. (1). Finally the relative saturability a is defined by

$$a = \frac{\mu^2}{\gamma_\perp \; \gamma_\parallel(1+\Delta^2)} \times \frac{\bar{\gamma}_\perp \; \bar{\gamma}_\parallel(1+\bar{\Delta}^2)}{\bar{\mu}^2} \tag{5}$$

Equation (2) has been analyzed in detail in[6] It admit the following roots

$$I_{st} = 0$$

$$I_{st} = I_\pm = (1/2a) \left\{ a(A-1) - (\bar{A}+1) \right. \tag{6}$$
$$\left. \pm \left(\left(a(A-1) - (\bar{A}+1) \right)^2 - 4a(\bar{A}+1-A) \right)^{\frac{1}{2}} \right\}$$

The reality of I_+ depends on the interplay among the relevant paramenters A, $\bar{A}$, a. In the case $0 < a < 1$, the admissible roots of Eq. (6) are $I = 0$ if $A < \bar{A} + 1$ and $I = 0$, $I = I_+$ if $A > \bar{A} + 1$. For $a > 1$ and $\bar{A} < \bar{A}_{tr} = 1/(a-1)$ a similar behaviour occurs. In the case of $a > 1$ and $\bar{A} > \bar{A}_{tr}$ the admissible roots of Eq.(2) are

$$I = 0 \quad \text{if} \quad aA < X_+$$

$$I = 0\,,\; I = I_-,\; I = I_+ \quad \text{if} \quad X_+ < aA < a\,(\bar{A} + 1) \qquad (7)$$

$$I = 0\,,\; I = I_+ \quad \text{if} \quad A > \bar{A} + 1$$

where $X_+ = a + \bar{A} - 1 + 2\sqrt{\bar{A}(a-1)}$. Hence there is a region of A values in the (A,I) plane where the three roots[6] of Eq. (2) coesist, as shown in Fig.8 . The linear stability analysis reveals the root $I = I_-$ is always unstable while the root I = 0 is unstable for $A > \bar{A} + 1$. Hence when the stationary solution I_+ is stable the sýstem undergoes a hysteresis cycle and bistability is observed in the laser output power versus the pumping paramenter A, as reported in the experimental results of the previous paragraph (see Figure 7).

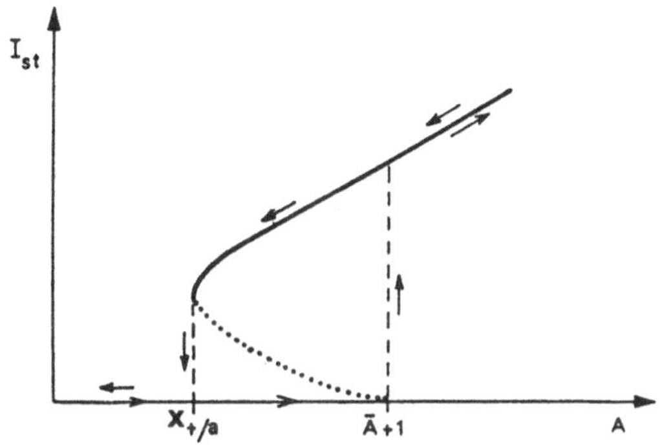

Figure 8. Plot of the emitted light I vs the pump parameter A for $a > 1$ and $\bar{A} > \bar{A}_{tr}$

In a plot in the $(A, \bar{A})$ plane the $A = \bar{A} + 1$ and $a\,A = X_+$ lines define regions of emissionles or bistability states, more specifically for $a\,A < X_+$ an emissionless state occurs, for $\bar{A} > \bar{A}_{tr}$ and $X_+ < aA < a(\bar{A} + 1)$ a bistable regime may occur.

Additional solutions have been found and analyzed in a series of paper [9,10] but they will be discussed in the framework of their stability.

If an inhomogeneous broadening with Doppler width u (and $\bar{u}$) is introduced in the description of the amplifier and absorber media, an equation similar to Eq. (2) is obtained [7,8]. In effect the absorption coefficient α_{in} for the inhomogeneous broadening in the Doppler limit and for detuning larger than the power broadened homogeneous linewidth may be written:

$$\alpha_{in} = \sqrt{\pi}\ \frac{2\pi\omega_o\mu^2 N}{\hbar uc}\ \exp\left(-(\omega_o - \omega)^2/u^2\right) \qquad (8)$$

The stationary state is then determined by

$$2K\ I^{\frac{1}{2}}\left(1 - \frac{A}{(1 + \frac{1}{2} I)^{\frac{1}{2}}} + \frac{\bar{A}}{(1 + \frac{1}{2} a\ I)^{\frac{1}{2}}}\right) = 0 \qquad (9)$$

where $A = \zeta c\alpha_{in} / K$ and $\bar{A} = \bar{\zeta} c\bar{\alpha}_{in} / K$ (10)

Unfortunately the analysis of Eq. (9) may be carried out only numerically, so it was not used in detailed comparison with the experiment.

In order to complete the analysis of a laser system with an intracavity saturable absorber it is required to check the stability of the solution, that means control for each solution of the system of Eq. (1) that introducing a small perturbation such perturbation is damped out and the solution values are recovered. It is important to notice that a similar procedure was introduced in the course of the experimental tests to derive the damping of the laser system (see Figure 6). In particular it is important to control for the existence of hard-mode excitation, like a first--order transition, where several steady states are available to the system, or control for Hopf bifurcations, where oscillatory states are allowed.

A stability analysis of the system of Eqs.(1) was performed in [4] and the oscillatory states of passive Q-switching regime have been predicted, and the pulse shapes have been constructed. However that analysis has been restricted to a small range of parameters owing to the amount of numerical work required to analyse the solution.

The stability analysis of refs [4,9,10] has shown that a new class of oscillatory solutions, whose existence depends on dispersion relations of the laser with saturable absorber, occurs when $\gamma_\perp > \bar{\gamma}_\perp$ and

$$\bar{A} > \gamma_\perp (K + \bar{\gamma}_\perp) \;/\; K(\gamma_\perp - \bar{\gamma}_\perp) - 1 \qquad (11)$$

As it will appear in the following when numerical parameters are introduced for the case of CO_2 lasers containing saturable absorber, such a range of values for $\bar{A}$ cannot be reached so that this new class of oscillatory solutions is not observable in the experiments considered here.

Furthermore it results that in the CO_2 laser medium as well in the molecular absorbers the transverse relaxation rate $\gamma_\perp$ is much larger than the longitudinal relaxation rate $\gamma_\parallel$. Thus the adiabatic elimination of the polarization components in Eqs.(1) may be applied for both the amplifier and absorber media and the rate equation approximation, where the population differences D and $\bar{D}$ only are dealt with, is obtained. Within this approximation the following set of equations is obtained:

$$\left(\frac{d}{dt} + 2K\right) I = 2I(\zeta G D + \overline{\zeta G D})$$

$$\left(\frac{d}{dt} + \gamma_\parallel\right) D = -\gamma_\parallel (N + D\, I) \qquad (12)$$

$$\left(\frac{d}{dt} + \bar{\gamma}_\parallel\right) \bar{D} = \bar{\gamma}_\parallel (\bar{N} - \bar{D}\, I)$$

where the gain parameter G (or $\bar{G}$) is given by

$$G = \frac{2\pi\omega_o \frac{\mu^2}{\hbar} \gamma_\perp}{\gamma_\perp^2 + (\omega - \omega_o)^2} \qquad (13)$$

For what concerns the steady state the above equations lead to the stationary equation (2) for the normalized field intensity I_{st}, while for the oscillatory.Solutions the class of solutions discussed above and based on the frequency detuning effects cannot be reproduced by the rate equation approximation.

From the linear stability analysis the following equation applies[2]:

$$\frac{2KI}{(1+I)\ (1+aI)} \left(\frac{\gamma_{\parallel}}{\bar{\gamma}_{\parallel}} A - \frac{\bar{\gamma}_{\parallel}}{\gamma_{\parallel}} a \bar{A} \right) + \gamma_{\parallel}(1+I) + \bar{\gamma}_{\parallel}(1+I) < 0 \qquad (14)$$

I being the nontrivial stationary solution of the system of Eqs.(12) When $K >> \gamma_{\parallel}, \bar{\gamma}_{\parallel}$ and I is not too small the last two terms in (14) can be neglected and one obtains the approximate instability condition

$$A < (\bar{\gamma}_{\parallel} / \gamma_{\parallel})^2 a \bar{A} \qquad (15)$$

Let us now consider the situation $a < 1$, which the system is monostable and the solution I_+ exist for $A > \bar{A} + 1$. It is clear from (15) that I_+ can become ustable only provided that $\bar{\gamma}_{\parallel} > \gamma_{\parallel}$. This prediction definitely contrasts with experimental findings in weak absorbers, that show PQS for $a < 1$ and $\bar{\gamma}_{\parallel} < \gamma_{\parallel}$.

This fact led several authors to conclude that the two-level model is inadeguate to describe the passive Q-switching, and to introduce a four-level model for both the amplifier and absorber media[2].

In this model which two energy levels with populations M_{j1}, M_{j2} ($\bar{M}_{j1}$, $\bar{M}_{j2}$) of the active (passive) medium resonant with the field mode belong to two vibrational bands. They are coupled also to all other rotational levels present in the same vibrational bands and surrounding the resonant levels (Fig.9). The coupling is via rotational relaxation processes, with rates γ_R and γ_R'. These further levels are all treated on the same footing, so that one has essentially a four-level model of the LSA. In the rate-equation regime the dynamics is described by a system of nine coupled equations for the photon number density ϕ, the population densities of the resonant levels M_{j1}, M_{j2}, $\bar{M}_{j1}$, $\bar{M}_{j2}$ and those of the rotational sets M_1, M_2, $\bar{M}_1$, $\bar{M}_2$. It results that if we introduce the following definitions

$$S = \delta \frac{2G}{\gamma_{\parallel}} \frac{\gamma_R'}{\gamma_R}; \qquad a = \frac{\bar{S}}{S}; \quad I = S\phi \qquad (16)$$

$$A = \frac{\zeta GN}{2K\gamma_{\parallel}} \frac{\gamma'_R}{\gamma_R} \qquad \delta = 1 + \frac{\gamma_{\parallel}}{\gamma'_R} \tag{16}$$

the stationary solution for the normalized field intensity I coincides with the Eq. (2) obtained within the two-level model.

For the linear stability analysis in ref. 2 an adiabatic elimination of the rotational variables $M_{j2} - M_{j1}$ and $\bar{M}_{j2} - \bar{M}_{j1}$ was introduced, because the rotational relaxation rates γ_R are typically 10^7 sec^{-1}, much larger than the vibrational relaxation rates $\gamma_{\parallel}$ and $\bar{\gamma}_{\parallel}$ typically in the 10^4 sec^{-1} range. As a consequence

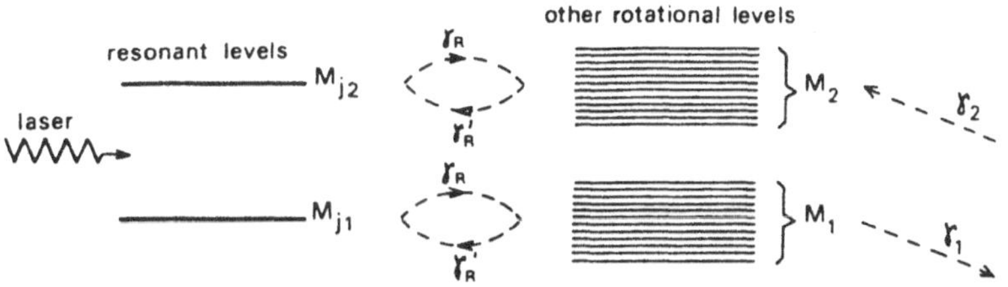

Figure 9. Scheme of the four-level model. The resonant level J_1, J_2 and the other rotational levels 1,2 form the vibrational states involved in the interaction of the lasing medium with the radiation mode. The corresponding populations M_{j1} and $M_{1,2}$ change also due to the incoherent processes ruled by the transition rates γ_R, γ'_R and the pump (decay) rate $\gamma_{\parallel}$. A similar picture applies to the absorber.

a set of three equations is derived, in general more complicated than that obtained in the two-level model, but where the linear stability analysis can be completed.

In this framework the condition for the occurrence of the passive Q-switching results:

$$\frac{\gamma_{\parallel}}{\gamma'_R} \frac{A}{\delta + \frac{\gamma_{\parallel}}{\gamma'_R} I} \frac{I}{1 + I} < \frac{\bar{\gamma}_{\parallel} a}{\bar{\gamma}'_R} \frac{A}{\bar{\delta} + \frac{\bar{\gamma}_{\parallel}}{\bar{\gamma}_R} a I} \frac{1}{1 + aI} \tag{17}$$

This relation simplifies in two limiting cases; at low field intensity I << 1 it becomes

$$A < \frac{\gamma_R}{\bar{\gamma}_R} \frac{\bar{G}}{G} \bar{A} = \zeta \bar{A} \tag{18}$$

at large saturation I >> 1 it leads to

$$A < \frac{\bar{A}}{a} \tag{19}$$

Both these relations are straight lines in the A, $\bar{A}$ phase plane, allowing a simple analysis of the bistability and passive Q-switching conditions for a laser with intracavity saturable absorber. Figure 10 reports such an analysis supposing that I << 1 and Eq. (18) represents the passive Q-switching condition.

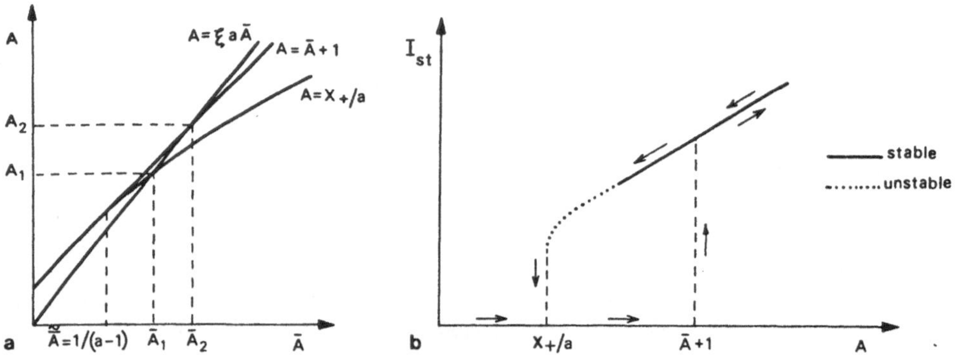

Figure 10. In a) phase plane (A,$\bar{A}$) of the system; in b) plot of the emitted light I versus pump parameter A for $\bar{A}_1 < \bar{A} < \bar{A}_2$

Fig. 10 b reports a typical plot of I versus A for a fixed $\bar{A}$ value included between $\bar{A}_1$ and $\bar{A}_2$. Such a plot represents the simultaneous occurrence of a bistability diagram with passive Q-switching, as observed in some experimental records with strong molecular absorbers[2,11]. Improvement to the solutions of the four-level model of the laser with saturable absorber have been presented in[12].

In ref.[8] where a rate equation approach to the inhomogeneous broadening was introduced, the mechanism of the saturation of a velocity set and transfer to other molecules was an additional non--linear mechanism leading to oscillatory instabilities in the operation of a laser with saturable absorber. This model is similar to the four-level model discussed above but a detailed comparison of the results obtained in the two approach has never been completed.

When a quantitative analysis of the experiments is started several drawbacks appear; the rate equasion approach cannot reproduce the Q-switching pulse shape, and the results derived from the numerical solution of Eqs.(1) are scarse. Thus all the experimental results on Q-switch pulse duration , height and period cannot be adeguately fitted to a theoretical model this the Q-switching threshold and the output laser power were described by the rate equation model, and fits of the experimental results have been presented in[2].

As an example of the parameters involved in the numerical analysis we report the values appropriate to the fit of the experimental results on the CH_3I absorption of the CO_2 laser line shown in Fig. 5 :

$$A = 0.745\ p \qquad \bar{A} \doteq 0.13 \qquad a = 5\ p$$

$$\gamma_R = 8 \cdot 10^6\ p\ sec^{-1} \qquad \gamma'_R = 3 \cdot 10^5\ p\ sec^{-1}$$

$$\gamma = 5 \cdot 10^3\ sec^{-1} \qquad \bar{\gamma}_R = 2.5 \cdot 10^5\ sec^{-1}$$

$$\bar{\gamma} = 7 \cdot 10^4\ sec^{-1} \qquad \bar{\gamma}'_R = 5 \cdot 10^2\ sec^{-1}$$

$$2K = 1 \quad 10^7 sec^{-1}$$

with p the pressure in Torr of the amplifier. On the basis of these values the Q-switching threshold and the output laser power may be reproduce satisfactorily within the rate-equation approach of the four-level model. However such an agreement for experimental results as well for those of refs[7,12] cannot be perfect owing to the hypothesis of homogeneous broadening for both the amplifier and absorber media. This hypotesis is valid for the CO_2 amplifier where the pressure broadening is larger than the Dóppler broadening, but is not valid fro the absorber where the Doppler broadening dominates.

Another difficulty encountered in the numerical analysis of the passive Q-switching data is thatvarying the pressure p of the amplifier, the pumping parameters A and the saturation parameter a are simultaneously varied. Thus in the bistability measurements of[5] a different approach was used and the current in the amplifier was modified (see Fig. 7). The saturation parameter of the amplifier depends on the relaxation rates only and is constant as function of the discharge current, while the pumping parameter A increases linearly with the discharge current.

However the quantitative analysis of the bistability results of Fig. 7 has required the modification of the theoretical model to include different experimental features. First of all the experiments deal with Faber- Perot cavities, while the theory is developed for a ring cavity. The presence of a standing wave laser field with nodes and antinodes instead of an uniform ranning wave modifies the range of parameters where optical bistability occurs. In ref [5] the standing wave effects have been approximatively included through a spatial average over the wavelength--dimension variations. Furthermore in the theoretical models of optical bistability the hypothesis of small gain in the amplifying medium and small losses in the laser cavity is usually introduced. In practice in CO_2 lasers with molecular absorbers the hypothesis of small gain and losser cannot be applied and the exponential increase or decrease of the laser intensity over the amplifier or absorber lengths should be included in the theory. As a consequence the state equation (2) for the laser with saturable absorber has been substituted by an equation that has been solved through a graphical analysis[5]. For the experimental results of Fig.(7) the dotted line reported in the figure represents the results obtained from the state equation when the pumping parmeter A was determined from the output power measured without absorber, the saturability parameter a was fixed to 60 and the A parameter was to the A + 1 value of the upper transition pôint. These values of the parameters for the SF_6 absorption are in good agreement with the values derived in independent measurements.

CONCLUSION

In conclusion if most features appearing in the lasers with saturable absorbers are well interpreted theoretically any further improvement and mainly an accurate fit of the experimental observations will require a large amount of numerical work. The instabilities of the nonlasing state, produced by dispersive phenomena, have been predicted theoretically, but have not been observed experimentally. From the experimental point of view the role of the frequency detuning of the absorber, amplifier and cavity has been never investigated in detail, so that further experimental work to investigate it should be preformed.

The material presented in this work is based on published and unpublished work in collaboration with F. Casagrande, B.M. Dinelli, P. Glorieux, L. Lugiato and E. Menchi to whom I am greatley indebted.

REFERENCES

1. O.R. Wood and S.E. Schwarz, Appl. Phys. Lett. 11 88-89 (1967)

2. E. Arimondo, F. Casagrande, L. Lugiato and P. Glorieux, Appl. Phys. B30 57 - 77 (1983)

3. J.Dupré, F. Meyer, C. Meyer, Rev. Phys. Appl. (Paris) 10 285 - - 293 (1975)

4. J.C. Antoranz, J. Gea and M.G. Velarde, Phys. Rev. Lett. 47 1895 - 8 (1981); J.C. Antoranz, L.L.Bonilla, J. Gea and M.G. Velarde, Phys. Rev. Lett. 49 35 - 8 (1982)

5. E. Arimondo and B.M. Dinelli, Optics Comm. 44 277 - 282 (1983)

6. L.A. Lugiato, P. Mandel, S.T. Dembinski and A. Kossakowski, Phys. Rev. A18 238 - 254 (1978)

7. R. Salomaa and S. Stenholm, Phys. Rev. A8 2695 - 2711 (1973)

8. Yu.V. Brzhazovškii, L.S. Vasilenko, S.G. Rautian, G.S. Popova and V.P. Chebotayev, Soviet Physics JETP 34 265 - 270 (1972)

9. S.T. Dembinski, A. Kossakowski, L.A. Lugiato, P.Mandel, P. Peplowski, Phys. Lett. 68A 20 - 2 (1978)
F. Mrugala and P. Peplowski, Z. Phys. B 38 359 - 364 (1980)

10. T. Erneux and P. Mandel, Z. Phys. B 44 353 - 63, 365 - 374 (1981)

11. A. Jacques and P. Glorieux, Opt. Commun. 40 455 - 460 (1982)

12. M.L. Asquini and F.Casagrande, Nuovo Cimento D, in press

13. H.J. Carmichael, Optica Acta 27 147 (1980)
J.A. Hermann, Optica Acta 27 159-70 (1980)

STRUCTURALLY STABLE BIFURCATIONS IN OPTICAL BISTABILITY

D. Armbruster and G. Dangelmayr

Institute for Information Sciences
University of Tübingen, Köstlinstr. 6
D - 7400 Tübingen, F.R.G.

1. INTRODUCTION

The equations encountered in optical bistability are solvable analytically only for limiting cases (mean field limit, pure absorptive case etc.). Here the question arises: are these solutions and their types of bifurcations persistent when moving away from the limits? An answer to this question of structural stability is given by imperfect bifurcation theory (developed by Golubitsky & Schaeffer)[1]. The aim of imperfect bifurcation theory is the classification of degenerate bifurcation problems. Bifurcation means a qualitative change of the behaviour of the solutions of nonlinear equations (p.d.e., o.d.e., algebraic) when a distinguished physical parameter is varied. This corresponds to the fact that experimental graphs are usually plotted against a single parameter (in our case this is the incident field E_I). Classification is done by means of an equivalence relation, which puts all bifurcation problems that behave qualitatively similar (i.e., exhibit the same stability changes) into the same class described by a polynomial normal form. Two classes differ by the number and type of perturbations which change the bifurcation qualitatively. These perturbations are called unfoldings and their number is the codimension of the problem. If partial information on the bifurcation of a system in special cases is available, then the information on the whole system can be completed by determining a singularity (i.e., a degenerate bifurcation point) with finite codimension, which acts as an organizing center for the bifurcation processes. The idea of an organizing center is based on the fact that the apparently global structure of a system is just a blown up version of the local structure. If one can find an organizing center, then a full unfolding of it yields all structurally stable bifurcation diagrams which govern the process, thus uncovering new phenomena in addition to those already known. Using these techniques we [2] have performed a classification of bifurcation problems with a reflection symmetry in 2 dimensions: The bifurcation equations with that symmetry have the form

$$G(x,\lambda,y) \begin{pmatrix} a(x,\lambda,y^2) \\ yb(x,\lambda,y^2) \end{pmatrix} = 0 \qquad (1)$$

where y is the amplitude of the solution which is odd with respect to the Z(2)-symmetry, x is the coordinate along the symmetry axis and λ is the bifurcation parameter. G is the result of a Lyapunov-Schmidt reduction or may describe the stationary solutions for the o.d.e. resulting from Haken's slaving or a center manifold reduction. A major application so far has been the interaction of Hopf and steady-state bifurcations [3], where y describes the amplitude of the periodic solution, and x the amplitude of the steady-state solution. Applying this to optical bistability leads to an organizing center for optical bistability and self-pulsing[4], reviewed in Section 2. Another application can be found in the so-called polarization switching[5], reviewed in Section 3.

2. SELF-PULSING AND OPTICAL BISTABILITY

Here, analytical results are most advanced in the mean field limit of pure absorption where we have the diagram of Fig. 1. It shows for a fixed cavity length, which determines the frequency difference α_1 of the first cavity sidemode and the resonance frequency, those parts of the high transmission branch, denoted by X, which are stable or unstable respectively. For details see Lugiato's lecture in this volume. In order to facilitate comparison with the bifurcation diagrams below we show in Figs. 2(a) and (b) the scans a and b as plots of the transmitted intensity X as a function of the incident intensity Y. The dashed curve represents the average intensity of the self-pulsing solution. Note that there are first and second order transitions at the high and low end of the self-pulsing branch respectively. The scan d exhibits singular behavior: For scans with α_1/γ at higher values than scan a, the points B,C,D approach each other until, for scan d at $\alpha_1=\alpha_{min}=\alpha_{max}$, the self-pulsing solution reduces to a point A on the high transmission branch. This point A is a degenerate Hopf-bifurcation point and describes the simultaneous occurance of a first and a second order transition to self-pulsing. As can be inferred from the classification of degenerate Hopf-bifurcations by Golubitsky & Langford [6], this point must have two unfolding parameters in order to produce all structurally stable diagrams around it. It is called an H(7)-degenerate Hopf-bifurcation. All structurally stable unfolded diagrams for this degeneracy are shown in Fig. 4, diagrams 1-3, 6-7. We recognize the two diagrams for pure absorptive behavior, viz. diagrams 4.1 and 4.3. The other three diagrams are new and represent two first order or two second order transitions to self-pulsing, respectively, and an isola solution.

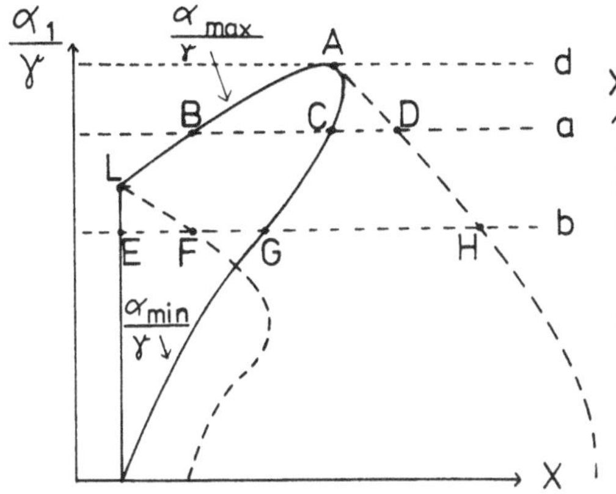

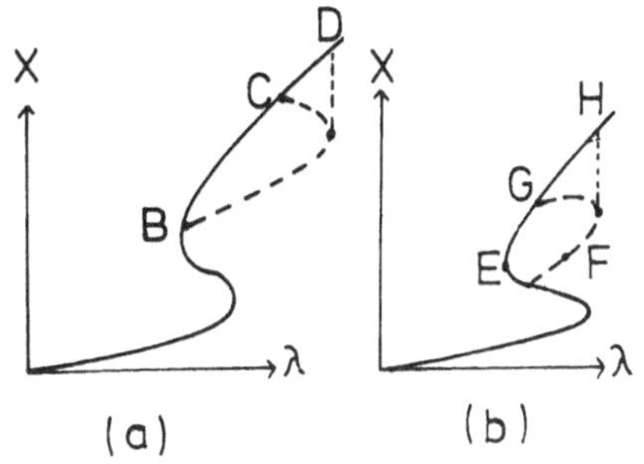

Fig. 1: Instability regions for self-pulsing in the pure absorptive limit. The dashed lines A-H and L-F represent first order transitions to self-pulsing and bifurcations to tori, respectively.

Fig. 2: Bifurcation diagrams corresponding to the scans a,b in Fig. 1. The dashed lines describe the mean value of the self-pulsing intensity while the full line represents the cw-transmission.

Note that for the self-pulsing isola the cw-transmission branch is not unstable. We can regard α_1 or the cavity length $\mathcal{L}$ as one of the unfolding parameters. Since there are no more free parameters in the pure absorptive limit we expect that cavity detuning Θ and/or atomic mistuning Δ will play the role of the other unfolding parameter. In addition to the singularities of the self-pulsing solutions, the cw-transmission curve itself can also possess a singularity, namely a hysteresis point where the curve has a vertical tangent. It has been shown by Agrawal & Carmichael[7] that this point exists for a wide range of C, Δ and Θ. Further partial information about the whole system is the existence of self-pulsing solutions without bistability in the dispersive case (Lugiato[8]). So, although for pure absorption the H(7) point always lies at $C > C_{Hysterese} = 4$, we suggest that the most general interaction of self-pulsing and bistability is governed by an unfolding of an organizing center which incorporates the existence of an H(7) singularity into a hysteresis point. The corresponding normal form has codimension 4 and its unfolding is

$$G = \begin{pmatrix} -(x^3 + y^2 - \lambda + \beta x + \alpha) \\ -y(x^2 + \lambda^2 + \gamma x + \delta) \end{pmatrix} \qquad (2)$$

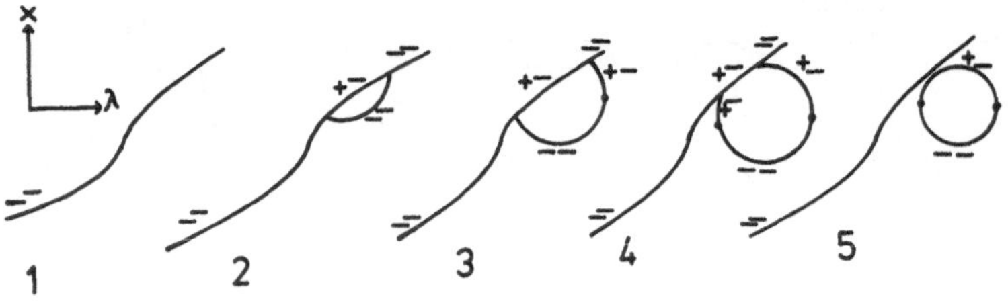

Fig. 3: Bifurcation diagrams without bistability. Here, (±,±) denote the signs of the real parts of the eigenvalues of the Jacobean of G.

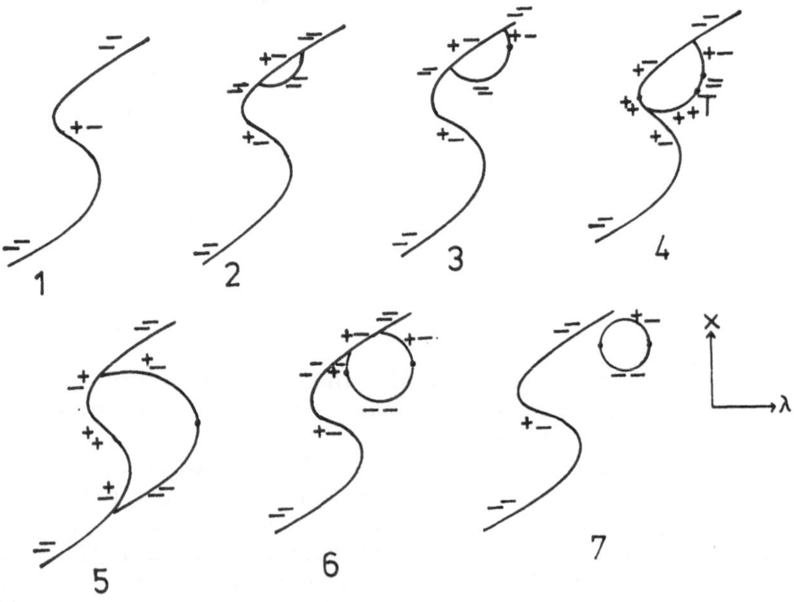

Fig. 4: Bifurcation diagrams with bistability.

Figs. 3 and 4 show the most interesting diagrams that are structurally stable and result from G=0. Since stability changes are initiated in the (x,λ)-plane, we only plot the projection of the periodic solution. Stability changes are at limit points of the self-pulsing solution with respect to λ and occur not necessarily at the maximum pulse amplitude, described by y_{max}. Since all the so far known

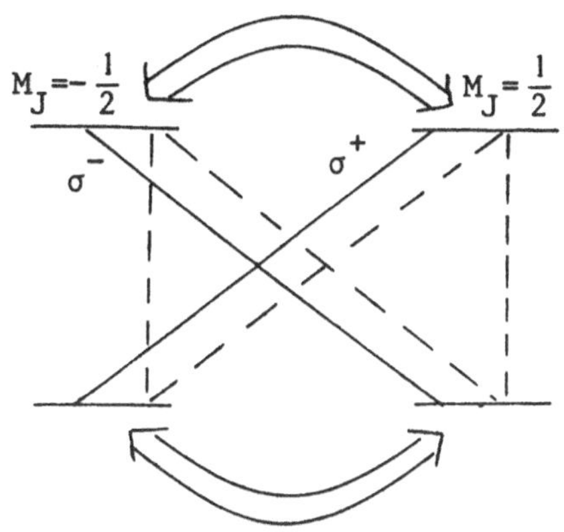

Fig. 5: Spontaneous decay (--) and collisional transfer (⇐⇒) for the model atom, pumped by σ^+ and σ^-.

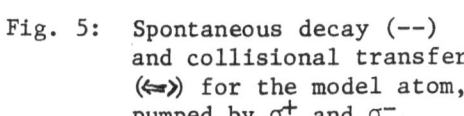

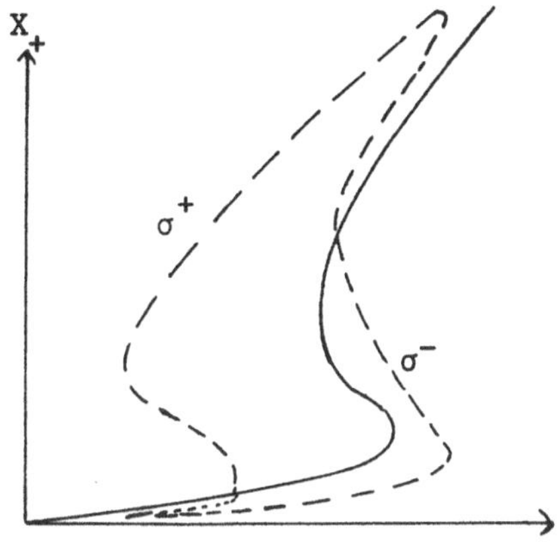

Fig. 6: Projection of typical solutions of (3) onto (X_+,Y)-plane.

bifurcations to self-pulsing in optical bistability can be found in these diagrams, one expects that these pictures fill the gaps between the known diagrams for pure absorption and for the dispersion dominated case, giving now an over-all description of the interaction between self-pulsing and optical bistability.

3. POLARIZATION SWITCHING

The physical setting is the following: We inject a coherent laser beam into a ring cavity. In contrast to the usual optical bistability we consider not only two-level atoms but a four state $J=1/2$ to $J=1/2$ transition model atomic system coupled by collisions and radiative interactions (Fig. 5). Hamilton et al.[9] have shown that for exact resonance between the laser frequency, the atomic transition and a cavity resonance (i.e., for the pure absorptive limit) we find for a linearly polarized input a switching to a circular polarized output σ^+ or σ^-. Of crucial importance are the collisional parameter β and the cooperativity parameter C, the former depending on the values of the spontaneous emission rates and the collisional transfer rates for the upper and lower states while the latter depends essentially on the length of the cavity L and the transmittivity T of the mirrors. In the mean field limit Hamilton et al. showed that we have the following relation between the linear input field Y and the two circular polarized outputs X_+ and X_-.

$$Y = X_\pm[1 + 2C\{1+8X_\pm^2 + \frac{(2\beta-4)(X_\mp^2-X_\pm^2)}{1+4\beta X_\pm^2}\}^{-1}] \tag{3}$$

A discussion of these equations leads to the following results: There is always a linearly polarized solution (full line) which exhibits bistability for $C>4$ and monotonic behavior for $C<4$. In addition, depending on β, there may be two stable branches connecting the low intensity branch of the linearly polarized solution to the high intensity branch by means of an almost circular polarized solution (dashed, cf. Fig. 6). These two branches are symmetric with respect to the diagonal $X_+=X_-$. Note the first order transitions to the circular polarized state - hence the term polarisation switching. Clearly the system has a reflection symmetry with respect to σ_+ and σ_- solutions, and with $x=X_++X_-$, $y=X_+-X_-$, $\lambda=Y$, (3) takes the form of Equ. (1). With the classification of Z(2)-symmetric bifurcation problems we also found conditions for an arbitrary bifurcation problem H to be equivalent to a certain normal form. These conditions involve Taylor coefficients of H at a degenerate point. Here the most degenerate point is the interaction of a hystereses point and the coalescing secondary bifurcation points. It turns out that for $C=4$ and $\beta=2$ the equations (3) fulfill all degeneracy conditions to be contact equivalent to

$$\begin{pmatrix} -x^3+\lambda-9xy^2 \\ -y(x^2+y^2) \end{pmatrix} = 0 \tag{4}$$

which has codimension 4. A detailed discussion (see [5]) of the bifurcation diagrams associated with a full unfolding of (4) yields 31 different diagrams. Some of them are shown in Fig. 7. The essential new features in these diagrams are:

1. We can have one-sided polarization switching showing a switching behaviour for low (high) intensity and a continuous transition to linearly polarized light for high (low) intensities (diagram 1-3).
2. We can have a transition to circular polarized light (dashed in Fig. 7) among any part of the bistability curve for linearly polarized light (full line) (diagrams 2,4-9).
3. We can have self-pulsing: Starting with diagram 10 and changing parameters to get via diagram 11 to diagram 12 shows that the branch running from A to B should be stable. On the other hand at B the bifurcation is from a completely unstable branch (++), so the branch AB should also be completely unstable (++). The question is resolved by a tertiary bifurcation point T which is a Hopf bifurcation point converting the (++) signature into a (--) signature. This Hopf bifurcation indicates the beginning of a self-pulsing solution.

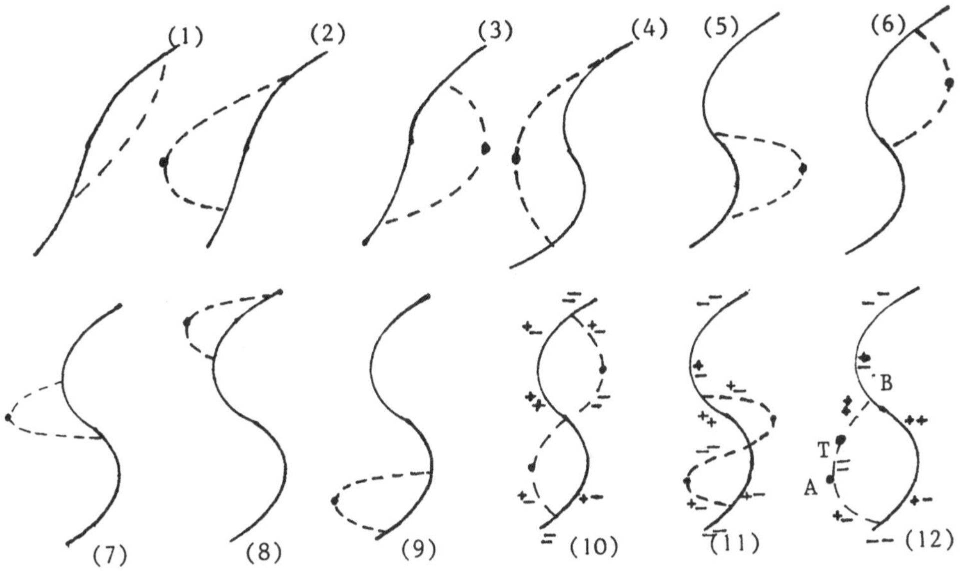

Fig. 7: Projections of bifurcation diagrams, associated with an unfolding of (4), onto (x,λ)-plane.

REFERENCES:

1. M. Golubitsky and D. Schaeffer, Comm. Pure & Appl. Math. 32:21 (1979).
2. G. Dangelmayr and D. Armbruster, Proc. London Math. Soc. (3), 46:517 (1983).
3. D. Armbruster, G. Dangelmayr and W. Güttinger, Imperfection sensitivity of interacting Hopf and steady-state bifurcations, preprint Tübingen (1982), submitted for publication.
4. D. Armbruster, An organizing centre for optical bistability and self-pulsing, submitted for publication.
5. D. Armbruster and G. Dangelmayr, An organizing centre for polarization switching, to be published.
6. M. Golubitsky and W.F. Langford, J. Diff. Equs. 41:375 (1981).
7. G.P. Agrawal and H.J. Carmichael, Phys.Rev.A. 19:2074 (1979).
8. L.A. Lugiato, Optics Comm. 33:108 (1980).
9. M.W. Hamilton, R.J. Ballagh and W.J. Sandle, Z. Phys. B 49:263 (1982).

ALLOWED NUCLEAR BETA DECAY IN AN INTENSE LASER FIELD Δ

R. R. Schlicher*, W. Becker†, and M. O. Scully *†

†Institute for Modern Optics
University of New Mexico
Albuquerque, New Mexico 87131

*Max-Planck Institut für Quantenoptik
D-8046 Garching bei München
West Germany

INTRODUCTION

It is the purpose of this lecture to study the influence of a strong external electromagnetic field on the spectrum of the emitted particles in a nuclear beta decay and in particular on the lifetime of the radioactive nucleus. The possibility of manipulating nuclear lifetimes [1] in a laboratory is very exciting and could bring important applications. The first experimental proof of this effect is more than 30 years old [2,3]. Experimentally accessible are those nuclear decay modes which involve an interaction between the nucleus and the atomic electrons such as internal conversion [3] and orbital electron capture [4] which is closely related to beta decay. By changing the chemical environment of the atom, by applying high pressure technology, by optical excitation with strong fields, by ionization and implantation, etc. the electronic structure of the atom can be modified, which results in changes of the nuclear lifetime up to a few parts in 10^2 [4].

However, it always seemed hopeless to influence the majority of the nuclear decay modes which take place without any interaction between the decaying nucleus and its environment. All imaginable fields in a laboratory are weak compared to the strong interaction or the Coulomb field at the surface of a nucleus.

ΔSupported in part by the Air Force Office of Scientific Research

Hence, changes of these nuclear decay rates have only been investigated under conditions which are of interest in astrophysics. In particular, nuclear beta decay was studied (i) under the condition of a statistical equilibrium in stellar interiors where excited nuclear states are thermally populated [5], (ii) for single photon absorption by the emitted electron from a Planck spectrum at stellar temperatures of the order of 10^8 °K [6], and (iii) for a strong uniform constant magnetic field [7] as it exists on the surface of a pulsar.

Due to the ongoing recent progress in the development of high power lasers it is nowadays also possible to produce extraordinarily strong fields in the laboratory. For example, in the beam of a Nd-glass laser, which produces TW pulses, the intensity can be (after focusing the beam down to, say, ten wavelengths) of the order of 10^{18} W/cm^2, corresponding to a field strength of about 10^{10} V/cm. These experimental facilities suggest the theoretical treatment of nuclear decays in the presence of strong electromagnetic plane wave fields.

The effects of intense plane electromagnetic waves on different quantum processes were already investigated two decades ago [8]. The interest soon focused on the decays of elementary particle like muons and pions [9] or neutrons [10,11] under the influence of a monochromatic external field. A common feature of Refs. 7-9 and 11 is the result that the influence of the external field on the <u>total</u> decay rate depends on the ratio between the field strength and the so called critical field strength $E_c = m^2c^3/e\hbar$. This is the limit for the applicability of classical electrodynamics, beyond which quantum effects are dominant [12]. This result seems to indicate that the influence of an external field of optical frequency on the lifetime of an elementary particle is a pure quantum effect which is very small as long as E is small compared to E_c. For electrons the critical field strength is about $1.3\text{x}10^{16}$ V/cm and therefore still far out of the range of present laser systems.

On the other hand, Ref. 10 predicts a measurable change of the lifetime of free neutrons in presently feasible fields. Furthermore, during the last few years nuclear decays in the presence of intense plane wave fields have also been investigated. These calculations predicted appreciable enhancements of the total decay rates in the field of available lasers, both for nuclear gamma decay [13] and nuclear beta decay [14], and also recently for forbidden nuclear beta decay [15].

In these lecture notes we will follow Ref. 14 in describing the decay process. This implies the use of a modified version of the Keldysh approximation [16] which was introduced for the theoretical description of laser-ionization in the so called electric field gauge. This approximation includes two steps: (i) the interaction between the bound system and the field is neglected, i.e. we consider the nucleus to be unaffected by the external field; (ii) in the final state the interaction between the emitted particle and the residual bound system is neglected, i.e. we neglect the Coulomb interaction between the emitted electron and the residual nucleus. The quality of the first approximation will be demonstrated in the next Section. With the second approximation we neglect a usually small effect [17]. In the framework of this approximation only the electron emitted in the nuclear beta decay couples to the external field. This is why we choose nuclear beta decay as the most promising decay mode: the smaller the mass of the emitted charged particle, the stronger its coupling to the external field and hence the stronger the laser impact on the decay process.

There are two general reasons why we might expect a change in the nuclear lifetime in the framework of this model: (i) it is well known for ordinary beta decay that taking into account the Coulomb corrections between electron and nucleus changes not only the spectrum of the electrons, but also the nuclear lifetime [17]. Hence, including the interaction of the electron with the laser field could result in the same effect; (ii) as we will show below the most likely energy transfer from, say, a Nd-laser with an intensity of $I \sim 10^{18}$ W/cm^2 to an electron emitted in a typical nuclear beta decay is of the order of MeV! Such a large energy increase leads to a much larger phase space of the emitted electrons which should result in a considerably faster decay of the nucleus.

In this lecture we shall restrict the discussion to (i) the nonrelativistic theory and (ii) to electromagnetic fields of circular polarization. This will simplify the calculations so much that they can easily be followed in detail. Strictly speaking, the nonrelativistic limit holds only if the mechanical momentum $\vec{p}-e\vec{A}$ of the electron in the field is small compared to mc. This implies a limiting condition for the kinetical electron energy E outside the field, $E = p^2/2m \ll mc^2$ and for the strength of the external field $e|A|/mc = \nu \ll 1$. This approximation seems to be unrealistic for many nuclei whose energy release in the decay is larger than the electron rest mass as well as for the fields which we intend to consider. Hence, in principle, the Dirac theory is required for the description of the electrons. Actually, relativistic effects show up in both the electron spectrum and the total decay rate. However, all the essential physical features which are introduced by the interaction with the laser field are included in the nonrelativistic theory in a very instructive way. Furthermore, as we shall demonstrate, the change of the lifetime in the presence of an external field is fairly well described by the nonrelativistic theory for all real nuclei and arbitrary field strengths, not only in the nonrelativistic limit $|\vec{p}-e\vec{A}| \ll mc$. Hence, we will present here only the nonrelativistic theory. However, all the numerical results shown are gained from the relativistic beta decay theory for allowed transitions with Dirac wave functions and V-A interaction. This theory will be published elsewhere.

We will first derive the wave functions of the different particles involved in the nuclear beta decay. This is based on Ref. 18. Whhat follows will be completely independent of Ref. 18. Next the partial transition rates will be calculated in the electric field gauge and in the radiation gauge. We will then evaluate the electron distributions and finally we will discuss the nuclear lifetime.

WAVEFUNCTIONS

For the calculation of the wavefunctions of the different particles involved in a nuclear beta decay we have to recall some points discussed in Ref. 18. We can obtain the Hamiltonian of a particle with charge e interacting with an external electromagnetic field from the Hamiltonian $H = p^2/2m + V(\vec{r})$ of a particle in a potential V with the help of the substitution

$$\vec{p} \rightarrow \vec{p} - e\,\vec{A}^g(\vec{r},t)\ , \qquad H \rightarrow H^g - e\,U^g(\vec{r},t).$$

This procedure yields the Hamiltonian

$$H^g = \frac{1}{2m}(\vec{p} - e\vec{A}^g(\vec{r},t))^2 + e\,U^g(\vec{r},t) + V(\vec{r})\ . \tag{2.1}$$

The vector potential $\vec{A}^g(\vec{r},t)$ and the scalar potential $U^g(\vec{r},t)$ are related to the electric field $\vec{E}(\vec{r},t)$ and the magnetic field $\vec{B}(\vec{r},t)$ by

$$\vec{E}(\vec{r},t) = -\vec{\nabla}U^g(\vec{r},t) - \frac{\partial}{\partial t}\vec{A}^g(\vec{r},t)\ ,$$

$$\vec{B}(\vec{r},t) = \vec{\nabla}\times\vec{A}^g(\vec{r},t). \tag{2.2}$$

The index g denotes the gauge freedom in the potentials. $\vec{E}$ and $\vec{B}$ remain unchanged if we transform from a gauge $(\vec{A}^g, \vec{U}^g)$ to a gauge $(\vec{A}^{g'}, \vec{U}^{g'})$ according to

$$\vec{A}^g(\vec{r},t) \rightarrow \vec{A}^{g'}(\vec{r},t) = \vec{A}^g(\vec{r},t) + \vec{\nabla}\chi(\vec{r},t)$$

$$U^g(\vec{r},t) \rightarrow U^{g'}(\vec{r},t) = U^g(\vec{r},t) - \frac{\partial}{\partial t}\chi(\vec{r},t). \tag{2.3}$$

where χ is an arbitrary function of $\vec{r}$ and t .

Also, the state vector $|\Psi^g\rangle$, which is a solution of the Schrödinger equation

$$i\hbar \frac{\partial}{\partial t} | \Psi^g \rangle = H^g | \Psi^g \rangle , \qquad (2.4)$$

depends on the gauge. The wave function in a gauge g is transformed to another gauge g' by the unitary transformation

$$\Psi^g(\vec{r},t) \rightarrow \Psi^{g'}(\vec{r},t) = e^{ie\,\chi(\vec{r},t)/\hbar}\, \Psi^g(\vec{r},t) \qquad (2.5)$$

The Hamiltonian has to be distinguished from the energy operator

$$\varepsilon^g = \frac{1}{2m} (\vec{p} - e \vec{A}(\vec{r},t)^2) . \qquad (2.6)$$

This (unperturbed) energy operator is a physical quantity in the sense of Ref. 19, since it transforms under a gauge transformation like

$$\varepsilon^{g'} = e^{ie\,\chi/\hbar}\, \varepsilon^g\, e^{-ie\,\chi/\hbar} .$$

In contrast, the Hamiltonian is an unphysical quantity since it transforms like

$$H^{g'} = e^{ie\,\chi/\hbar}\, H^g\, e^{-ie\,\chi/\hbar} - e \frac{\partial}{\partial t} \chi .$$

This distinction is important for the definition of an unperturbed state as discussed in Ref. 18.

Throughout these lecture notes we will describe the laser field by a uniform electric field $\vec{E}(t)$, neglecting the magnetic field. This long wavelength or dipole approximation is well justified for nuclear beta decay since the wavelength of visible light is about 5×10^{-5} cm, whereas the radius of a nucleus is of the order of 5×10^{-13} cm. Hence, when calculating matrix elements of the nucleus-laser interaction the external field can be considered as constant over the integration area. Within this long wavelength approximation two gauges are most frequently used: the electric field gauge (E-gauge) with

$$\vec{A}^E(t) = 0, \quad U^E(\vec{r},t) = -e\,\vec{r}\,\vec{E}(t) \qquad (2.7)$$

and the radiation gauge (R-gauge) with

$$\vec{A}^R(t) = -\int_{t_0}^{t} d\tau\, \vec{E}(\tau) \;, \qquad U^R(\vec{r},t) = 0 \;. \tag{2.8}$$

Both gauges are related by the gauge transformation

$$\chi R \to E(\vec{r},t) = -\vec{A}^R(\vec{r},t)\vec{r} = \vec{r}\int_{t_0}^{t} d\tau\, \vec{E}(\tau) \;. \tag{2.9}$$

For the nuclear wave functions to be used for the calculation of the beta-decay we shall adopt a simple one-particle shell model description: the nucleus is divided into a "valence" nucleon and an inert core which is affected neither by the field nor by the decay but generates the potential $V(\vec{r})$ for the valence nucleon which decays. The valence nucleon has an effective charge e_N and mass M. Both do not have to be specified; we have just to recall that because of charge conservation the effective charges of the initial and final valence nucleon will satisfy the relation $e_{N,i} - e_{N,f} = \pm e$ for beta$^{\mp}$-decay ($e = -|e|$ denotes the electron charge).

It is convenient to use the E-gauge. The nuclear wave-function $\Psi_N(r,t)$ is then a solution of the Schrödinger equation

$$i\hbar \frac{\partial}{\partial t} \Psi_N(\vec{r},t) = \left(\frac{\vec{p}^2}{2m} + V(\vec{r}) - e_N\vec{E}(t)\vec{r}\right)\Psi_N(\vec{r},t) \;. \tag{2.10}$$

The interaction term $-e_N\vec{E}\vec{r}$ is at the nuclear surface of the order of $10^{-2} - 10^{-1}$eV if we apply a field of 10^{10} V/cm. This is completely negligible compared to the nuclear binding potential or to the Coulomb potential at the nuclear surface, which are of the order of MeV. A noticeable modification of a bound nuclear state would generally require field strengths close to the critical one. We shall therefore assume that the initial and final nuclear states do not interact with the external field. This is the first part of the Keldysh approximation.

As discussed in Ref. 18, the noninteracting state is defined as an eigenstate of the unperturbed energy operator (2.6), not as a solution of the Schrödinger Eq. (2.4) in which the potentials $\vec{A}^g$ and U^g are set equal to zero. In the E-gauge this distinction does not play a role since the Hamiltonian H^E and the energy ε^E coincide. The noninteracting nuclear state is then a solution of Eq. (2.10) in which we drop the nucleus-field interaction term $-e_N\vec{E}\vec{r}$:

$$\Psi_N^E(\vec{r},t) = e^{-iE_N t/\hbar} \quad \Phi_N(\vec{r}) , \tag{2.11}$$

where Φ_N is an energy eigenstate

$$\left(\frac{\vec{p}^{\,2}}{2M} + V(\vec{r})\right) \Phi_N(\vec{r}) = E_N \; \Phi_N(\vec{r}) . \tag{2.12}$$

If we want to express the noninteracting nuclear state in the R-gauge we have to be cautious. In order to obtain an eigenstate of the energy operator ε^R (2.6) with the same constant energy eigenvalue E_N we must apply the unitary transformation (2.5) and (2.9). We then obtain a wavefunction which is different from Eq. (2.11)

$$\Psi_N^R(\vec{r},t) = e^{ie_N\vec{A}^R(t)\vec{r}/\hbar} \; e^{-iE_N t/\hbar} \quad \Phi_N(\vec{r}) . \tag{2.13}$$

From now on we shall drop the superscript in $\vec{A}^R$ and refer with the notation $\vec{A}$ to the vector potential in the R-gauge (2.8).

For the calculation of the electron wavefunction we use the second part of the Keldysh approximation and neglect the Coulomb interaction of the emitted electron with the residual nucleus. However, the interaction between the electron and the external field should be taken into account exactly. Hence, we need an exact solution of the Schrödinger equation for a free particle ($V = 0$) in the external field $\vec{A}(t)$.

We notice now that it is more convenient to solve this problem in the R-gauge

$$i\hbar \frac{\partial}{\partial t} \Psi_e^R(\vec{r},t) = \frac{1}{2m}(\vec{p} - e\,\vec{A}(t))^2 \; \Psi_e^R(\vec{r},t) \tag{2.14}$$

instead of the E-gauge

$$i\hbar \frac{\partial}{\partial t} \Psi_e^E(\vec{r},t) = \left(\frac{\vec{p}^{\,2}}{2m} - e\vec{r}\,\vec{E}(t)\right) \Psi_e^E(\vec{r},t) \tag{2.15}$$

since the canonical momentum $\vec{p}$ of the free particle is a constant of motion in the R-gauge, but not in the E-gauge. It is well known in classical mechanics [20] that the i-th component p_i of the canonical momentum $\vec{p}$ is conserved if neither the vector potential $\vec{A}^g$ nor the scalar potential U^g depend on the i-th component x_i of the space coordinate r. In particular, if $\vec{A}^g$ and U^g

are spatially uniform $\vec{p}$ is conserved. This also holds true in quantum mechanics: an operator is a constant of motion if it does not explicitly depend on time and if it commutes with the Hamiltonian H^g. Hence, if $\vec{A}^g$ and U^g do not depend on $\vec{r}$ the eigenvalue of the canonical momentum $\vec{p}$ is a constant of motion. This is true for the Hamiltonian H^R in the R-gauge (2.8) with the long wavelength approximation for the field. If we make the additional assumption that the vector potential $\vec{A}(t)$ is switched on and off initially and finally, i.e. $\vec{A}(t) = 0$ for $|t| > t_0$, we can identify the conserved canonical momentum with the momentum outside of the field, which would be measured by a spectrometer. Again, this holds only true in the R-gauge. On the other hand the canonical momentum is not a physical quantity in the sense of Ref. 19. Hence its eigenvalues are different in different gauges. Especially in the E-gauge (2.7), where the vector potential vanishes and the scalar potential does depend on $\vec{r}$, the canonical momentum $\vec{p}$ coincides with the mechanical momentum $\vec{\pi}(t) = \vec{p} - e\vec{A}(t)$ and its eigenvalue is no longer conserved.

Since it is always convenient to exploit the existence of conserved quantities, we like to solve Eq. (2.14) instead of Eq. (2.15). Only in the R-gauge can we make the following ansatz for the electron wavefunction: Ψ_e^R is characterized by the eigenvalue $\vec{p}$ of the canonical momentum and we assume that it factorizes into an unperturbed plane wave and a function f(t) which depends only on time, since also the field depends in the long wave length approximation only on time

$$\Psi_e^R(\vec{r},t) = f(t)\, e^{-i(Et - \vec{p}\vec{r})/\hbar} \quad .$$

In the nonrelativistic theory E is related to the canonical momentum by

$$E = \frac{\vec{p}^{\,2}}{2m} \tag{2.16}$$

Only in the absence of the field denotes E the kinetic energy. By inserting the ansatz into the Schrödinger equation (2.14) and using Eq. (2.16) one obtains the equation of motion of f(t)

$$i\hbar \frac{\partial}{\partial t} f(t) = \left(-\frac{e}{m}\vec{p}\,\vec{A}(t) + \frac{e^2}{2m}\vec{A}^2(t)\right) f(t) \quad .$$

Integrating this equation yields the exact solution of Eq. (2.14):

$$\Psi^{R}_{e}(\vec{r},t) = V^{-\frac{1}{2}} \times$$

$$\times \exp\{-\frac{i}{\hbar}\left[Et - \vec{p}\vec{r} - \frac{1}{2m}\int_{t_o}^{t} d\tau(2e\vec{p}\vec{A}(\tau) - e^2\vec{A}^2(\tau))\right]\} \qquad (2.17)$$

where V denotes the normalization volume. The lower limit of integration in Eq. (2.17) only contributes a constant phase to the wavefunction and is therefore insignificant. The analogous solution of the Dirac equation is known as the Volkov solution [21].

It is easy to check that the wavefunction (2.17) is an eigenstate of the operator of the canonical momentum with eigenvalue $\vec{p}$. Since the operators of the canonical momentum $\vec{p}$ and of the mechanical momentum $\vec{\pi}^R = \vec{p} - e\vec{A}$ commute in the long wavelength approximation, Ψ^R_e is also an eigenstate of $\vec{\pi}^R$ and of the operator of the kinetic energy in the R-gauge

$$\varepsilon^R \bar{\Psi}^R_e(\vec{r},t) = \frac{1}{2m}(\vec{p} - e\vec{A}(t))^2 \bar{\Psi}^R_e(\vec{r},t) \ . \qquad (2.18)$$

The nonrelativistic Volkov solution Ψ^R_e has a time dependent energy eigenvalue, in contrast to the non-interacting state Ψ^R_N (2.13) which has a constant energy eigenvalue E_N.

In order to obtain the solution of Eq. (2.15) in the E-gauge we have to carry out the gauge transformation (2.5) and (2.9) on Ψ^R_e with the result

$$\Psi^E_e(\vec{r},t) = e^{-ie\vec{A}(t)\vec{r}/\hbar} \ \Psi^R_e(\vec{r},t)$$

$$= V^{-\frac{1}{2}} \exp\{-\frac{i}{\hbar}\left[(\vec{p}-e\vec{A}(t))\vec{r} - \frac{1}{2m}\int_{t_o}^{t} d\tau(\vec{p}-eA(\tau))^2\right]\} \qquad (2.19)$$

where we used Eq. (2.16). We note that Ψ^E_e depends only on the eigenvalue of the mechanical momentum $\vec{\pi}(t) = \vec{p} - e\vec{A}(t)$. This is a physical quantity and has therefore the same value in any gauge. Here we expressed $\vec{\pi}$ by the gauge dependent canonical momentum and vector potential, both given in the R-gauge.

Throughout these lecture notes we shall describe the laser field by a circularly polarized monochromatic plane wave with frequency ω, propagating in $\hat{e}_3$-direction

$$\vec{E}(t) = E_0 (\hat{e}_1 \sin \omega t + \hat{e}_2 \cos \omega t) . \tag{2.20}$$

The corresponding vector potential in the R-gauge reads

$$\vec{A}(t) = A_0 (\hat{e}_1 \cos \omega t - \hat{e}_2 \sin \omega t), \; A_0 = \frac{E_0}{\omega} \tag{2.20b}$$

By inserting this vector potential into Ψ_e^R (2.17) and dropping the phase which results from the lower integration limit we obtain

$$\Psi_e^R (\vec{r},t) = V^{-\frac{1}{2}} \exp \left[- \frac{i}{\hbar} (E + \frac{\nu^2}{2} mc^2) t - \vec{p}\vec{r} \right] \times$$
$$\times \exp \left[iz \sin(\omega t + \phi) \right] \tag{2.21}$$

Here we introduced the azimuthal and polar angles ϕ and θ of the momentum $\vec{p}$

$$p_1 = p_T \cos \phi, \; p_2 = p_T \sin \phi, \; p_T = \sqrt{p_1^2 + p_2^2} = p \sin \theta . \tag{2.22}$$

The field dependence is expressed by the dimensionless parameter ν, which was introduced earlier

$$\nu = \frac{e A_0}{m c} = \frac{e E_0}{m \omega c} = \frac{m c^2}{\hbar \omega} \frac{E_0}{E_c} \tag{2.23}$$

E_c is again the critical field strength $E_c = m^2c^3/e\hbar = 1.3 \times 10^{16}$ V/cm. For practical calculations it can be convenient to express ν either by the intensity I and wavelength λ of the field

$$\nu^2 = \frac{e^2 \lambda I}{2\pi^2 \varepsilon_0 m^2 c^5} = 7.3 \times 10^{-19} \lambda^2 | \mu m^2 | I | W/cm^2 | ,$$

or by the photon density ρ and the wavelength λ

$$\nu^2 = \frac{e^2 \hbar \lambda \rho}{\pi \varepsilon_0 m^2 c^3} = 4.3 \times 10^{-27} \lambda | \mu m | \rho | cm^{-3} | .$$

Finally, the dimensionless amplitude z of the oscillating term in the exponent of Eq. (2.21) is given by

$$z = \frac{\nu c p_T}{\hbar \omega} = \frac{\nu c p}{\hbar \omega} \sin \theta \ . \tag{2.24}$$

For circular polarization the $\vec{A}^2$-term in Eq. (2.17) is constant, $\vec{A}^2(t) = A_0^2$, and gives rise to the field dependent energy shift $\nu^2 mc^2/2$ in Ψ_e^R (2.21). In a relativistic context, this term is usually interpreted as a contribution to an effective mass. This becomes apparent from the relativistic Hamiltonian

$$H = (m^2c^4 + c (\vec{p} - e \vec{A})^2)^{\frac{1}{2}}$$

$$= ((m^2 + e^2 A_0^2/c^2)c^4 + c^2 \vec{p}^2 - 2e \vec{p} \vec{A})^{\frac{1}{2}} \ .$$

Hence, in the following, we shall refer to this term as the effective mass correction. It will play a crucial role in the following Sections.

The $\vec{p} \vec{A}$ term of Eq. (2.17) is the origin of the oscillating exponent exp(iz sin) in Ψ_e^R (2.21) and responsible for the fact that the kinetic energy (2.18) is not a constant of motion. It can be interpreted in terms of the electron undergoing field induced multiphoton transitions. We see this with the help of the generating function of the Bessel function $J_n(z)$

$$e^{iz \sin \alpha} = \sum_{n=-\infty}^{\infty} J_n(z) \, e^{in\alpha} \ . \tag{2.25}$$

This enables us to rewrite the nonrelativistic Volkov wave-function in the form

$$\Psi_e^R(\vec{r},t) = V^{-\frac{1}{2}} \sum_{n=-\infty}^{\infty} J_n(z) \, e^{in\phi} \ \times$$
$$\times \exp \{- \frac{i}{\hbar} \left[(E + \frac{\nu^2}{2} mc^2 - n\hbar\omega)t - \vec{p}\,\vec{r}\right] \} \ . \tag{2.26}$$

Although we only deal with classical fields we can interpret the wavefunction (2.26) in terms of an infinite sum over n absorbed or emitted photons.

The time dependent part of Ψ_e^R includes an effective energy which changes in discrete steps of $\hbar\omega$. This represents multi-photon processes of any order n. We can estimate the most likely energy transfer from the field to the electron by considering the behavior of $J_n^2(z)$ as a function of the order n for fixed z. When the rapid oscillations are averaged over, $J_n^2(z)$ increases slowly with increasing $|n|$ until it reaches its maximum around $|n| = z$. For $|n| > z, J_n(z)$ as a function of n decreases rapidly to zero. Hence it is most likely that about z photons are absorbed by the electron, corresponding to an energy transfer $z\hbar\omega = \nu c p_T$. In an external field with ν of the order of unity the electron can therefore absorb energy from the field up to the order of its own kinetic energy, typically some MeV for nuclear beta decay. This strong energy absorption encourages us to expect a considerable laser effect on nuclear beta decay. However, we realize already that this will be counteracted by the increase of the electron's effective mass.

Finally, we need the neutrino wavefunction. Since the neutrino neither interacts with the laser field nor with the residual nucleus, it is well described by a plane wave solution of the Klein-Gordon equation

$$\Psi_\nu(\vec{r},t) = V^{-\frac{1}{2}} \exp\left[-\frac{i}{\hbar}(E_\nu t - \vec{q}\,\vec{r})\right] \tag{2.27}$$

with the relativistic dispersion

$$E_\nu = c\,|\,\vec{q}\,|\,. \tag{2.28}$$

TRANSITION RATE

We now intend to calculate in the framework of the model derived in the last Section the effects of a strong laser field on measurable quantities like the spectrum of the emitted electrons and the lifetime of the nucleus. The transition amplitude reads in first order of perturbation theory with respect to the weak interaction

$$F(T) = -\frac{i}{\hbar}\int_{-T/2}^{T/2} dt\ \langle \Psi_e\,\Psi_\nu\,\Psi_{N,f}\,|\,gV\,|\,\Psi_{N,i}\rangle\,. \tag{3.1}$$

The weak interaction is denoted by (gV) where V is a dimensionless operator and g is the coupling constant for beta decay, $g = 1.4\times10^{-49}$ erg cm^3. According to the Keldysh approximation the electron wavefunction Ψ_e is given by the exact solution (2.21) or (2.18) for a free particle in a circularly polarized field and the nuclear wavefunctions $\Psi_{N,i}$ and $\Psi_{N,f}$ are given by the noninteracting solution (2.11) or (2.13), respectively. The uncharged neutrino is described by the plane wave (2.27). By taking the wavefunctions of all charged particles consistently in the same gauge, either the E or the R-gauge, we obtain in any case

$$F(t) = -\frac{i}{\hbar V}\int_{-T/2}^{T/2} dt\, e^{i/\hbar(E + v^2/2\, mc^2 + E_\nu - Q)t} \times$$

$$\times\, e^{-iz\sin(\omega t+\phi)} < e^{i/\hbar[\vec{p} - e\vec{A}(t) + \vec{q}]\vec{r}}\, \Phi_f \,|gV\,|\, \Phi_i >. \tag{3.2}$$

If we choose the wavefunctions in the R-gauge we have to make use of the relation $e_{N,f} - e_{N,i} = -e$ in order to obtain Eq. (3.2). For the energy balance one has to take into consideration the mass of the electron which is created during the beta decay. Hence the value Q in Eq. (3.2) is given by

$$Q = E_{N,i} - E_{N,f} - mc^2 . \tag{3.3}$$

Since we are only interested in allowed nuclear beta decay we can drop the factor $\exp\{i(\vec{p}-e\vec{A}+\vec{q})\vec{r}/\hbar\}$ in the matrix element. The physical justification for this approximation is the fact the electron and the neutrino do not carry away any angular momentum in an allowed beta decay so that the spatial dependence of their wavefunctions can be neglected. In the nonrelativistic theory this is a fair approximation. The nonrelativistic limit implies $|\vec{p} - e\vec{A}(t)| \ll mc$, as discussed above. Since the spatial integraion in the matrix element extends over the radius R of the nucleus, which is of the order of some 10^{-13}cm, the exponential factor can be estimated for all nuclei by

$$|(\vec{p} - e\vec{A})\vec{r}/\hbar| < |\vec{p} - e\vec{A}|\, R/\hbar \ll Rmc^2/\hbar c \lesssim 0.02 \ll 1 .$$

Since we now dropped the factor $\exp i e\vec{A}\vec{r}/\hbar$, the discussion of the preceding Section on the question of when to take which wave function, viz. (2.11) or (2.13), appears obsolete. Nevertheless, we have emphasized this point because it is important in principle and of crucial significance in the case of forbidden beta-decay [16, 32].

By the same arguments the space dependent factor $\exp(i\vec{q}\vec{r}/\hbar)$ in the neutrino wavefunction can be replaced by unity. If the nuclear matrix element is abbreviated by

$$g V_{fi} = \langle \Phi_f | g V | \Phi_i \rangle$$

the transition amplitude for allowed beta decay takes the form

$$F(T) = - \frac{i}{\hbar V} gV_{fi} \int_{-T/2}^{T/2} dt \exp(- iz \sin(\omega t + \phi)) \times$$
$$\times \exp \left| \frac{i}{\hbar} (E + \frac{\nu^2}{2} mc^2 + E_\nu - Q)t \right| . \tag{3.4}$$

With the help of the expansion (2.25) Eq. (3.4) can be rewritten in the form

$$F(T) = - \frac{i}{\hbar V} gV_{fi} \sum_{n=-\infty}^{\infty} J_n(z) e^{-in\phi} \times$$
$$\times \int_{-T/2}^{T/2} dt \exp \left| \frac{i}{\hbar} (E + \frac{\nu^2}{2} mc^2 + E_\nu - Q - n\hbar\omega)t \right| . \tag{3.5}$$

This representation again allows for a simple interpretation in terms of n photons emitted or absorbed by the electron.

For the calculation of the decay rate we only need the transition amplitude (3.5) in the limit $T \to \infty$:

$$\lim_{T\to\infty} F(T) = - \frac{i}{\hbar V} gV_{fi} \sum_{n=-\infty}^{\infty} J_n(z) e^{-in\phi} \times$$
$$\times 2\pi \delta((E + \frac{\nu^2}{2} mc^2 + E_\nu - Q - n\hbar\omega)/\hbar) . \tag{3.6}$$

With the help of the standard relation [22]

$$(2\pi \delta(E))^2 \to 2\pi T \delta(E)$$

we obtain the transition rate per unit time

$$w = \lim_{T\to\infty} \frac{1}{T} \mid F(T) \mid^2$$

$$= \frac{2\pi}{\hbar} \frac{1}{V^2} |gV_{fi}|^2 \sum_{n=-\infty}^{\infty} J_n^2(z)\ \delta(E+\frac{\nu^2}{2} mc^2 + E_\nu - Q - n\hbar\omega) \ . \tag{3.7}$$

Only diagonal terms $n = n'$ contribute to $|F(T)|^2$ in the limit $T \to \infty$ since the off-diagonal terms yield products of two delta functions with different arguments. Due to reasons which will become clear below it is convenient to introduce the dimensionless quantity

$$n_0 = (E + \frac{\nu^2}{2} mc^2 - Q)/(\hbar\omega) \ .$$

Eq. (3.7) can then be rewritten in the form

$$w = (2\pi/\hbar^2 V^2 \omega) |gV_{fi}|^2 \sum_{n=-\infty}^{\infty} J_n^2(z)\ \delta(n-n_0 - E_\nu/\hbar\omega) . \tag{3.8}$$

In the next step we will simplify the sum over the Bessel functions. In high power laser fields the argument z of the Bessel function can achieve values of 10^3 to 10^6 as discussed above. Hence, up to 10^6 terms can contribute to the sum (3.8). An excellent approximation for Bessel functions of large order n and for $z > |n|$ is Debye's asymptotic expansion [23]:

$$J_n(z) = \sqrt{\frac{2}{\pi}}\ (z^2 - n^2)^{-\frac{1}{4}} \cos\left(\sqrt{z^2 - n^2} - n \arccos n/z - \pi/4\right) \ .$$

The rapidly decreasing exponential tail of $J_n(z)$ for $|n| > z$ will be neglected, i.e. we set $J_n(z) = 0$ for $|n| > z$. This approximation is only exact in the limit $z \to \infty$. According to Eq. (2.24) this corresponds to the classical limit $\hbar \to 0$. As long as z is a large number this classical limit is well justified. Furthermore, after squaring $J_n(z)$ we can average over the oscillations in $J_n^2(z)$ since it is hardly possible in any experiment to determine the intensity parameter ν and therefore also z to such an accuracy that the phase of the Bessel functions is precisely defined. Also these rapid oscillations of $J_n^2(z)$ as a function of n are a quantum feature [24]. This procedure then yields a powerful approximation for the square of Bessel functions

$$J_n^2(z) \cong 1/\pi \frac{\theta(z^2 - n^2)}{\sqrt{z^2 - n^2}} \tag{3.9}$$

The θ-function denotes the usual step function $\theta(x) = 1$ for $x > 0$ and $= 0$ for $x < 0$.

Since we are not interested in a particular term of Eq. (3.8) with a fixed number n of transferred photons, but only in the total effect of very many multiphoton terms, we can replace the sum in Eq. (3.8) by an integral. By inserting the approximation (3.9) into Eq. (3.8) and carrying out the n-integration we obtain the transition rate per unit time in the quasiclassical limit

$$w = \frac{2|gV_{fi}|^2}{h^2 v^2 \omega} \frac{\theta(z^2 - (n_0 + E_\nu/\hbar\omega)^2)}{\sqrt{z^2 - (n_0 + E_\nu/\hbar\omega)^2}} \tag{3.10}$$

If we recall that z and n are proportional to $(\hbar\omega)^{-1}$ we notice that in the quasiclassical approximation, which applies in the limit $\hbar\omega \ll mc^2$, the transition rate w becomes independent of the field frequency ω. It depends on the external field only via the intensity parameter ν (2.23).

ALTERNATIVE DERIVATIONS OF THE QUASICLASSICAL TRANSITION RATE

There are two further instructive and easy methods to derive Eq. (3.10) which illustrate its classical character. In the first part of the current Section we calculate the time integral for the transition amplitude (3.4) with the help of the stationary phase method [25] instead of expanding the integrand in terms of Bessel functions. The method of stationary phase approximates the integral over a rapidly oscillating function by the contributions from the regions around stationary points to the integrand [26]:

$$\int_a^b dt\; e^{ixf(t)} \cong \sum_n \sqrt{\frac{2\pi i}{x f''(t_n)}}\; e^{ix f(t_n)}$$

Here x is a large positive variable and f(t) a real function of the real variable t, so that the integrand is rapidly oscillating unless f(t) is stationary. Hence the major contribution to the value of the integral arises from the vicinity of the points t_n

at which $f'(t_n) = 0$. The sum over n extends over all stationary points in the interval [a,b]. The stationary phase method becomes exact in the limit $x \to \infty$, i.e. for infinitely large exponents. Hence the use of this approximation method for the time integration in Eq. (3.4) implies the transition to the classical limit $\hbar \to 0$ (i.e. $z \to \infty$) already in the transition amplitude F(t).

In order to apply the stationary phase method to Eq. (3.4) we first have to determine the stationary points t_n of the exponent of the integrand:

$$\cos(\omega t_n + \phi) = 1/z\ (n_0 + E_\nu/\hbar\omega)\ .$$

In order to obtain stationary points we find again the condition $z^2 \geqslant (n_0 + E_\nu/\hbar\omega)^2$. We furthermore note that the second derivative of the exponent has the same form

$$\pm\ \omega^2\sqrt{z^2 - (n_0 + E_\nu/h\omega)^2}$$

at all stationary points. Thus we obtain for the transition amplitude in the semiclassical limit

$$F(T) = -\frac{\sqrt{-2\pi i}}{V\hbar\omega}\, gV_{fi}\, \frac{\theta(z^2 - (n_0 + E_\nu/\hbar\omega)^2)}{(z^2 - (n_0 + E_\nu/\hbar\omega)^2)^{1/4}} \quad x$$

$$x \sum_{|-T/2,T/2|} \exp\left\{ i\left[(n_0 + E_\nu/\hbar\omega)\left(\arc\cos \frac{n_0 + E_\nu/\hbar\omega}{z} - \phi\right) \mp \right.\right. \tag{4.1}$$

$$\left.\left. \mp\sqrt{z^2 - (n_0 + E_\nu/\hbar\omega)^2} - 2\pi_n (n_0 + E_\nu/\hbar\omega)\ \right]\right\}.$$

Cross terms from different stationary points will once again not contribute to the transition rate $|F(T)|^2$ since the phases at the stationary points t_n are randomly distributed for all practical purposes so that the cross terms cancel each other in the limit $T \to \infty$. We have two stationary points per time interval $2\pi/\omega$ and hence $\omega T/\pi$ stationary points in the entire integration region [-T/2, T/2]. We thus find from counting all stationary points in $|F(T)|^2$ the transition rate per unit time

$$w = \lim_{T\to\infty} 1/T \; |F(T)|^2$$

$$= \frac{2|gV_{fi}|^2}{v^2\hbar^2\omega} \frac{\theta(z^2 - (n_0 + E_\nu/\hbar\omega)^2)}{\sqrt{z^2 - (n_0 + E_\nu/\hbar\omega)^2}} \tag{4.2}$$

This result is identical with the quasiclassical approximation (3.10) of the Bessel function approach.

There is an even simpler argument leading to Eq.(4.2) which is classical from the outset. We can write the transition rate per unit time for the decay in the absence of the laser field as

$$w_{free} = \frac{2\pi}{\hbar V^2} |gV_{fi}|^2 \, P_{free}(E) \, , \tag{4.3}$$

where

$$P_{free}(E) = \delta(Q - E_\nu - E) \, . \tag{4.4}$$

is the electron energy distribution function and $E = p^2/2m$ the kinetic energy. Of course, in the absence of the field the electron energy E is, for specified Q, a function of the neutrino energy E_ν only. In the presence of the field the kinetic energy of an electron is no longer conserved but time-dependent (cp. Eq. (2.61)), so that

$$E(t) = 1/2m \, (\vec{p} - e\vec{A}(t))^2$$

$$= \frac{\vec{p}^{\,2}}{2m} + \frac{\nu^2}{2} mc^2 - \hbar\omega z \cos(\omega t + \phi) \tag{4.5}$$

with $\vec{p}$ the conserved canonical momentum. We now assume that when the field is switched on, V_{fi} in Eq. (4.3) is unaffected but the electron energy distribution $P_{free}(E)$ is changed into

$$P(E) = \omega/2\pi \int_{t_0}^{t_0+2\pi/\omega} dt \quad \delta(Q - E_\nu - E(t)) \, . \tag{4.6}$$

This is the time average over the instantaneous electron energy distribution. If we recall that

$$\delta(f(x)) = \sum_i \frac{\delta(x - x_i)}{|f'(x_i)|}$$

where the sum is over all zeroes x_i of $f(x)$, and notice that $Q-E_\nu-E(t)$ has two zeroes within $t_0 \leqslant t \leqslant t_0 + 2\pi/\omega$, we find that

$$w = 2\pi/\hbar V^2 \, |g \, V_{fi}|^2 \, \bar{\rho}(E) \tag{4.7}$$

agrees with Eq. (4.2). This procedure illustrates once again that the approximation (3.9) for the Bessel functions as well as the stationary phase method correspond to a purely classical treatment of the electron-laser interaction.

By comparing Eqs. (4.4) and (4.6) we can already anticipate a general feature of the electron energy distribution which will be more extensively discussed in the next Section. Due to the field, the kinetic energy

$$E_{free} = \frac{\vec{p}^{\,2}_{free}}{2m} = Q - E_\nu$$

in Eq. (4.4) is replaced by E(t) of Eq. (4.5). Now, if $|e\vec{A}| >> |\vec{p}_{free}|$, $|\vec{p}|$ will be much larger than $|\vec{p}_{free}|$ in order that

$$\vec{p}^{\,2}_{free} = (\vec{p} - e\vec{A}(t))^2$$

can be satisfied as prescribed by Eqs. (4.4) and (4.6). Hence, for strong fields, we expect the electron energy spectrum to extend to much higher energies.

ELECTRON DISTRIBUTIONS

In order to obtain the angular distribution and the energy spectrum of the emitted electrons the transition rate w has to be integrated over the phase space of the neutrino and the electron. The total decay rate Γ is given by

$$\Gamma = \int \frac{V\, d^3p}{(2\pi\hbar)^3} \int \frac{V\, d^3q}{(2\pi\hbar)^3} \, w \quad . \tag{5.1}$$

It is convenient to use polar coordinates for the momenta $\vec{p}$ and $\vec{q}$ and to apply the dispersion relations (2.16) and (2.28) for the electron and the neutrino. This yields

$$d^3p = m\sqrt{2mE} \quad \sin\vartheta \quad dE\, d\vartheta\, d\phi \quad ,$$

$$d^3q = 1/c^3\, E_\nu^2 \quad \sin\vartheta_\nu \; dE_\nu\, d\vartheta_\nu\, d\phi_\nu \; .$$

For the transition rate w in Eq. (5.1) we use the semiclassical limit (3.10) which was shown to be an excellent approximation. Since w (3.10) depends only on E_ν and via z and n_0 on E and θ the integrals over θ_ν, ϕ_ν and ϕ are trivial in Eq. (5.1). After substituting E_ν by

$$x = n_o + E_\nu/\hbar\omega$$

Γ takes the form

$$\Gamma = \frac{\sqrt{2\, m^3}}{4\pi^4\, \hbar^7\, c^3}\, (\hbar\omega)^2 \left|q\, V_{fi}\right|^2 \int dE d\vartheta \sqrt{E} \quad \sin\vartheta \quad \times$$

$$\times \int_{n_o}^{\infty} dx\, (x - n_o)^2 \; \frac{\theta(z^2 - n^2)}{\sqrt{z^2 - n^2}}$$

We have two restrictions for the integration variable x which have to be satisfied simultaneously: (i) $n_0 \leqslant x$, corresponding to the condition of positive neutrino energy $E_\nu > 0$, and (ii) $-z \leqslant x \leqslant z$ originating from the θ-function. Therefore we have to distinguish between three cases: (i) if $n_0 > z$ the total decay rate Γ vanishes; (ii) if $-z \leqslant n_0 \leqslant z$ the x-integration extends from n_0 to z; (iii) if $n_0 < -z$ the x-integration extends from -z to +z. This leads to the elementary integrals

$$\int_{n_o}^{z} dx \, \frac{(x - n_o)^2}{\sqrt{z^2 - x^2}} = -\, 3/2\, n_o \sqrt{z^2 - n_o^2} + (n_o^2 + z^2/2)(\pi/2 - \arcsin n_o/z),$$

$$\int_{n_o}^{z} dx \, \frac{(x - n_o)^2}{\sqrt{z^2 - x^2}} = \pi(\, n_o^2 + z^2/2\,) \; .$$

The angle and energy distribution of the electrons then reads

$$\frac{d^2 \tau}{dE\, d\partial} = \frac{2m^3}{4\pi^4 h^7 c^3} \left| g\, V_{fi} \right|^2 (h\omega)^2 \; E \; \sin\partial \; \theta(z-n_o) \quad x$$

$$- 3/2\; n_o \quad z^2 - n_o^2 \; + (n_o^2 + \; z^2/2)(\pi/2 - \text{arc sin} \;\; n_o/z) \qquad \text{if } - z \leq n_o \leq z \;, \tag{5.2}$$

$$\pi \; (n_o^2 \; + \; z^2/2\;) \qquad \text{if} \;\; n_o < - z \;.$$

Since n_0 and z depend on E and θ, the condition $z \geqslant n_0$, i.e.

$$E + \frac{\nu^2}{2} mc^2 \; - \; Q \leq \; c\,\nu \quad 2\,m\,E \quad \text{Sin}\,\partial$$

yields the limits of allowed emission angles θ and of possible electron energies E. For $E \leqslant Q - \nu^2 mc^2/2$ this condition is always satisfied. But for higher energies electrons are only emitted into a certain angular range. The limits for the emission angle are for fixed energy E

$$\text{Sin}\,\partial \;\geqslant \begin{cases} 0 & \text{if} \quad E \leq Q - \dfrac{\nu^2}{2}\, m\,c^2 \\[2ex] \dfrac{E + \dfrac{\nu^2}{2} m\,c^2 - Q}{c\,\nu \quad 2m\,E} & \text{if} \quad E > Q - \dfrac{\nu^2}{2}\, mc^2 \end{cases} \tag{5.3}$$

The limits for the electron energy can be obtained from the condition $\sin\theta \leqslant 1$:

$$E_{min} = \begin{cases} 0 & \text{if} \quad Q \geq \dfrac{\nu^2}{2}\, m\,c^2 \\[2ex] (\; \dfrac{\nu^2}{2}\, m\,c^2 \; - \; Q)^2 & \text{if} \quad Q < \dfrac{\nu^2}{2}\, m\,c^2 \end{cases} \tag{5.4}$$

$$E_{max} = \quad (\; \frac{\nu^2}{2}\, m\,c^2 \; + \; Q\;)^2 \quad .$$

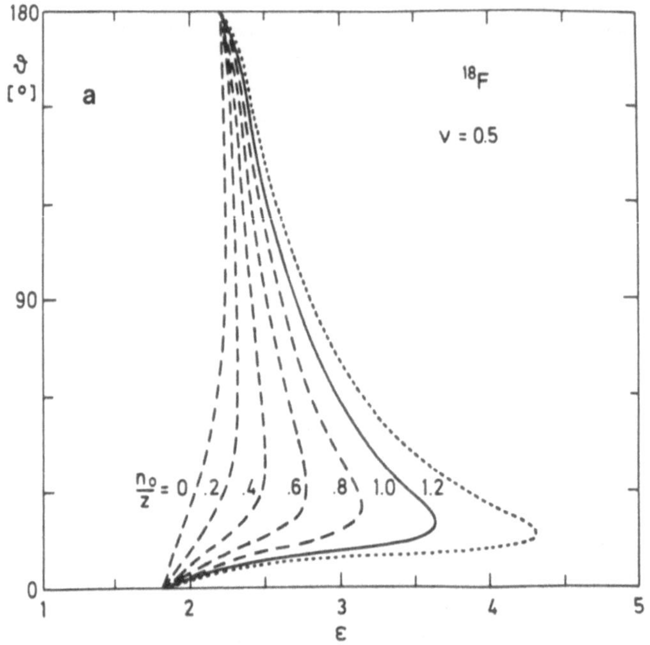

Figure 1 shows for the relativistic theory the possible (E,θ) values of the emitted electrons in a laser field with an intensity parameter $\nu = 0.5$ for two examples. ^{18}F decays via β^+ decay to ^{18}O with a moderate energy release Q = 634 keV, ^3H decays via β^- decay to ^3He with the very small energy release Q = 18.6 keV. (Note that we defined Q as the mass difference between nuclei, not between neutral atoms). The solid line in Fig. 1 shows the boundary of the integration area: on the left or inside the solid line, respectively, we have the allowed (E,θ) values. On the right or outside this curve $n_0 = z$, respectively, we have the nonclassical regime with extremely few events which are neglected in the present classical theory. Whereas Eq. (5.3) yields a symmetric electron distribution around $\theta = 90°$ with respect to the laser beam axis, the relativistic calculation yields an asymmetric distribution. Electrons at the high energy end of the spectrum are mainly emitted in forward direction. The dotted lines indicate a value of $n_0/z > 1$ up to which the Bessel functions are already so extremely decreased that this area cannot play a role in the nonclassical regime anymore. Nonclassical corrections to Eq. (5.2) can only be due to (E,θ) points extremely close to the solid line. Furthermore, the dashed lines in Fig. 1 show contours of constant ratios $n_0/z < 1$. The smaller this ratio is at some point (E,θ), the higher is the number of events that contribute to $d^2\Gamma/dEd\theta$ at this point.

For the example ^3H a laser field with $\nu = 0.5$ is already so strong that the effective mass exceeds the Q value. For all

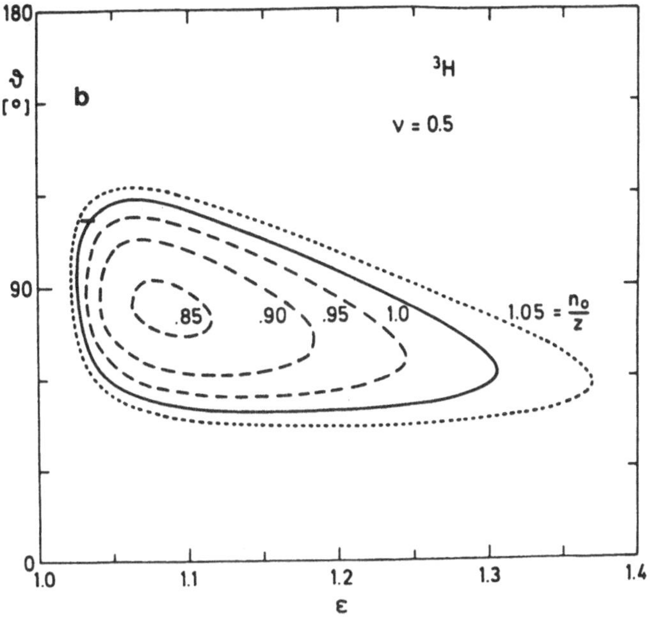

Fig. 1. Area of allowed emission angle and energy (solid curve) and contours of constant ratio n_0/z for a field intensity $\nu = 0.5$ and for the nuclei (a) ^{18}F and b) ^{3}H. The relativistic electron energy is scaled in units of mc^2: $\varepsilon = (m^2c^4 + p^2c^2)^{\frac{1}{2}}/mc^2$. These curves come from the complete relativistic theory.

energies n_0 is then positive. This is in contrast to the field-free situation where n_0 is always negative. The given combination of Q and ν allows only configurations with $n_0/z > 0.83$. The value n_0 denotes the minimum number of photons which the electron has to absorb from the field. In this example the electron must absorb energy from the field since the energy release of the nuclear decay alone is not sufficient to account for the effective mass of the electron in the field. Since n_0 for all (E,θ) values is rather close to z, only relatively few multiphoton terms contribute to $d^2\Gamma/dEd\theta$ at one particular (E,θ) point.

All these effects can also be seen in the spectrum of the electrons. We obtain the energy distribution of the electrons by integrating Eq. (5.2) numerically over θ between the limits (5.3). Figure 2 shows the results of the corresponding relativistic calculation, again for the two examples ^{13}F and ^{3}H. A similar plot can be found in Ref. 27. With increasing field intensity the electron distribution extends to much higher energies. On the other hand, the maximum of the spectrum

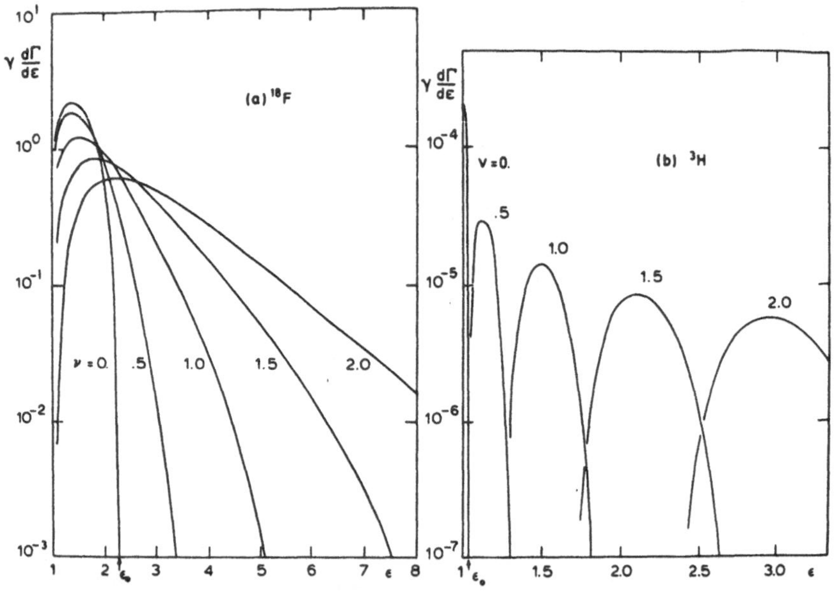

Fig. 2. Energy distribution of the electrons for different field intensities for the nuclei a) ^{18}F and b) ^{3}H. The relativistic electron energy is scaled in units of mc^2 as in Fig. 1, ε_0 is related to the Q value of the decay by $\varepsilon_0 = 1 + Q/mc^2$, and γ denotes a constant factor of dimension (energy x time). These curves are calculated from the complete relativistic theory.

decreases with increasing ν. The area under the curve, which represents the total number of emitted electrons and therefore the lifetime of the nucleus, seems to be more or less constant for the different field intensities. We shall investigate this question in the next Section.

In the ^3H spectrum in Fig. 2b we see again the effect of the strong effective mass $\nu^2 mc^2/2$ exceeding the Q-value of the reaction. For sufficiently high ν-values the spectrum does not begin at E = 0 anymore, but at some finite energy. This reflects again the fact that the electrons from the low energy end of the laser-free spectrum must absorb a considerable amount of photons from the field in order to raise their effective mass, since the available Q value is not sufficient.

One possibility to obtain the total decay rate Γ and the lifetime of the nucleus $\tau = 1/\Gamma$ is the numerical integration of Eq. (5.2) between the limits (5.3) for the angle and (5.4) for the energy. However, it is also possible to calculate Γ analytically from Eq. (3.10) by choosing another set of variables for the electron momentum and another order of integration. The dependence of z on $(p_1^2 + p_1^2)^{\frac{1}{2}}$ suggests, for example, the following set of variables

$$s = p_1^2 + p_2^2 , \qquad \phi = \arc\cos \frac{p_1}{\sqrt{p_1^2 + p_2^2}} , \; p_3$$

in terms of which

$$d^3p = 1/2 \quad ds\, d\phi d\, p_3$$

For the neutrino momentum $\vec{q}$ we use again polar coordinates. With this choice for $d^3\vec{p}$ and d^3q and with the semiclassical transition rate (3.10) the total decay rate Γ (5.1) reads after carrying out the three trivial angular integrations

$$\Gamma = \frac{|g\, V_{fi}|^2}{8\pi^4\hbar^8 c^3\omega} \int_0^\infty dE_\nu\, E_\nu^2 \int_{-\infty}^{\infty} dp_3 \int_0^\infty ds \qquad (6.1)$$

$$\times \quad \theta(z^2 - (n_0 + E_\nu/\hbar\omega)^2)\,(z^2 - (n_0 + E_\nu/\hbar\,)^2\,)^{-\frac{1}{2}}$$

After expressing z and n_0 in terms of s and p_3 the θ-function yields the following restrictions for the integration limits if the integrations are carried out in the indicated order (first ds, then dp_3 and finally dE_ν):

$$S\pm = m^2\,\nu^2c^2 + 2\,m\,(Q - E_\nu) - p_3^2 \pm 2m\nu c \sqrt{2m(Q-E_\nu) - p_3^2} \geq 0$$

$$(p_3)\pm = \pm\sqrt{2m\,(Q - E_\nu)} ,$$

$$(E_\nu) + = Q.$$

The total decay rate Γ can then be rewritten in the form

$$\Gamma = \left(m|gV_{fi}|^2/4\pi^4 h^7 c^3\right)\int_0^Q dE\ \nu E^2 \nu \int_{|2m(Q-E_\nu)|^{\frac{1}{2}}}^{-|2m(Q-E_\nu)|^{\frac{1}{2}}} dp_3 \qquad x$$

$$\int_{s_-}^{s_+} ds\ \left[(s_+ - s)(s - s_-)\right]^{-\frac{1}{2}} \tag{6.2}$$

The value of the integral

$$\int_{s_-}^{s_+} ds/\ (s_+ - s)(s - s_-) = \pi$$

is independent of the intensity parameter ν, and so are the limits of the remaining integrations over p_3 and E_ν. Hence, after integration over ds the dependence on the external field drops out of the expression (6.2) for the total decay rate. This means that in the quasiclassical limit the total decay rate Γ is not affected by the external field, nor are the partial decay rates $d\Gamma/dp_3$, $d\Gamma/d\phi$ and $d^3\Gamma/d^3q$. Field effects only show up in $d\Gamma/ds$ and in the related partial rates $d\Gamma/dE$, $d\Gamma/d\theta$, $d\Gamma/dp_1$, $d\Gamma/dp_2$, etc.

The last two integrations in Eq. (6.2) over dp_3 and dE_ν are easily carried out. We then obtain for Γ the nonrelativistic limit of the total decay rate of allowed nuclear beta decay in the absence of an external field:

$$\Gamma_{free} = |2m^3| \ /2\pi^3\hbar^7 c^3\ |g\ V_{fi}|^2 \int_0^Q dE\ \ |E|^{\frac{1}{2}}\ (Q - E\ )^2$$

$$= 8|2|^{\frac{1}{2}}\ /105\pi^3\ c^6\hbar^7\ (mc^2)^{3/2}\ \ Q^{7/2}\ |g\ V_{fi}|^2 \tag{6.3}$$

Again, there is a simpler way to realize the field independence of Γ in the classical limit: according to Eqs. (4.5-7)

$$\int wd\ p = 2\ \pi/\hbar v^2\ |g\ V_{fi}|^2 (\omega/2\pi)\quad x$$

$$x\int_{t_o}^{t_o+2\pi/\omega} dt \int d^3p\ \ \delta(Q - E_\nu - \ 1/2m\ (\vec{p} - e\ \vec{A}(t))^2) \quad , \tag{6.4}$$

where the integrations over t and $\vec{p}$ have been commuted. If we now transform the integration variable $\vec{p}$ to $\vec{p} - e\vec{A}(t)$, the field dependence is entirely eliminated. Obviously, in order to achieve this, it suffices already to integrate over the two components of $\vec{p}$ which lie in the plane of $\vec{A}(t)$.

The relativistic theory also leads to the result that the nuclear lifetime is unaffected by the external field. The physical importance of the effective mass term $\nu^2 mc^2/2$ for this result should be stressed again. We recall that the effective mass is due to the $\vec{A}^2$-term of the interaction Hamiltonian in the R-gauge. If the field is strong enough so that the effective mass exceeds the Q value of the decay emitted electrons <u>must</u> absorb the energy difference from the field. We thus have <u>two</u> competing processes: since the electrons absorb energy they have a larger phase space and the nucleus should decay faster. On the other hand, due to the effective mass only those electrons are emitted which absorb between n_0 and approximately z photons. The distribution of the photon absorption with respect to the photon number n is given in the classical limit by the approximation (3.9) for the Bessel functions or by the classical distribution (4.6). With increasing ν, i.e. with increasing phase space, the minimum of n_0/z in the entire (E,θ) plane approaches unity so that the portion $(z - n_0)z$ of photon absorptions that still leads to decay becomes very small. In the limit of classical electron field interaction these two effects, increasing the phase space and overcoming the effective mass, cancel out each other exactly.

In a different context, namely the hyperfine splitting of atomic levels in a strong radio frequency field, the just mentioned cancellation has been noticed previously: by merely replacing the electron mass m by the effective mass $(m^2+e^2\vec{A}^2/c^2)^{\frac{1}{2}}$ in the Hamiltonian, an effect was predicted which was not corroborated by experiment [33]. It was later shown [34], that in the long wave length limit this effect is cancelled by the $\vec{p}\vec{A}$-term which was initially neglected. This is another example for the interplay between the $\vec{p}\vec{A}$- and the $\vec{A}^2$- term. The argument following Eq. (6.3) shows clearly that this cancellation is just a consequence of the minimal coupling interaction.

However, one should keep in mind that the field-independence of the nuclear lifetime holds only in the classical limit $\hbar \to 0$ for the electron-field interaction. This limit is well justified for all beta decaying nuclei in the presence of laser fields with intensity ν in the order of unity and with optical or longer wavelengths. However, the quasiclassical limit breaks down for fields with shorter wavelengths (x-ray regime) and for (fictitious) nuclei with tiny Q values $Q \ll mc^2$. In these cases the parameter z which corresponds to the maximum number of absorbed photons cannot be considered a large number anymore. The classical approximation, however, holds only for multiphoton transfers of very high order. The classical limit furthermore breaks down for superintense fields when $\nu \gg 1$. Then the effective mass term becomes so large that the minimum value of the parameter n_0, which corresponds to the minimum number of absorbed photons, comes very close to z, i.e. there are no (E,θ) configurations with $n_0 \ll z$ anymore. In the Bessel function procedure this implies that (E,θ) configurations beyond the $n_0 = z$ border become important. For the stationary phase method the case $\nu \gg 1$ means that higher order terms in the expansion around the stationary points have to be taken into account. These three ways to get away from the validity of the semiclassical limit will be demonstrated again in the next Section.

THE NONCLASSICAL REGIME

In order to calculate the quantum effects of the electron-field interaction we have to attempt to calculate the total decay rate analytically as far as possible without any approximations. We therefore start from the exact expression (3.7) for the transition rate per time in terms of Bessel functions and choose again the variables (s,ϕ,p_3) for the electron momentum as in the last section. The sum over the Bessel functions in Eq. (3.7) cannot be carried out anymore. We choose the order of integrations so that we avoid an integration over the Bessel function. The three integrations over the angular variables are trivial, as well as the integrations over the neutrino energy E_ν and over p_3. We thus obtain an exact expression for the nuclear lifetime in a laser field

$$\Gamma = \frac{1}{30\,\pi^3 c^3 \hbar^7} \frac{|g\,V_{fi}|^2}{m^2} \sum_{n=-\infty}^{\infty} \int ds \; J_n^2\left(\frac{\nu c}{\hbar\omega} |s|^{\frac{1}{2}}\right) \times \tag{7.1}$$

$$\times \left(2m\left(Q + n\hbar\omega - \nu^2/2\,mc^2\right) - s\right)^{5/2} \theta\left(2m\left(Q+n\hbar\omega - \nu^2/2\,mc^2\right)-s\right)$$

The summation over n and the integration over s can be carried out analytically. For this purpose the sum of the squared Bessel functions has to be transformed in a proper way, and the calculation of the total decay rate Γ is somewhat lengthy [28].

However, there also exists a completely different approach for the exact calculation of the total decay rate Γ in the presence of an external field. It is possible to express the field influence on Γ by an operator acting on the decay rate Γ_{free} (6.3) of the beta decay in the absence of the field [29]. The exact expression for the nonrelativistic rate Γ of allowed beta decay reads

$$\Gamma = \exp\{-\frac{i}{\hbar} M(-i\hbar \frac{\partial}{\partial Q})\} \quad \Gamma_{free}. \tag{7.2}$$

For circularly polarized fields the operator M has the form

$$M(x) = \frac{\nu^2}{2} mc^2 x \{1 - (\frac{\sin\frac{\omega x}{2}}{\omega x/2})^2\} \tag{7.3}$$

As usual, these operators are defined by their power series expansion

$$\exp\{-\frac{i}{\hbar} M(-i\hbar \frac{\partial}{\partial Q})\} = \sum_{l=0}^{\infty} \frac{1}{l!}\} - \frac{i}{\hbar} M(-i\hbar \frac{\partial}{\partial Q})\} \quad .$$

and

$$-\frac{i}{\hbar} M(-i\hbar \frac{\partial}{\partial Q})\} = \frac{\nu^2}{2} \sum_{k=1}^{\infty} \frac{c_k}{4^k} (mc^2 \frac{\partial}{\partial Q})(h \frac{}{Q})^{2k} \quad .$$

The coefficients c_k are given by

$$c_k = \sum_{n=0}^{k} \frac{1}{(2n+1)!\,(2(k-n)+1)!} \quad .$$

Noticing that x occurs in Eq. (7.3) in combination with ω as the product ωx, we can simplify the operator considerably. The k-th term of the operator M applied to the free decay rate Γ_{free} differs by the order $(\hbar\omega/Q)^2$ from the (k-1)-th term. In the limit $\hbar\omega \ll Q$ it is therefore sufficient to take only the first term k = 1 in M into account. Consistently also $\exp\{-iM/\hbar\}$ can then be cut off at $\ell = 1$. To first order of $(\hbar\omega/Q)$ the total decay rate is thus given by

$$\begin{aligned} \Gamma &\cong \{1 + \frac{\nu^2}{24} (mc^2 \frac{\partial}{\partial Q})(\hbar\omega \frac{\partial}{\partial Q})^2\} \quad \Gamma_{free} \\ &= \{1 + \frac{35}{64} \frac{mc^2}{Q} (\frac{\hbar\omega}{Q})^2 \nu^2\} \quad \Gamma_{free} \quad . \end{aligned} \tag{7.4}$$

The enhancement of the total decay rate due to the external field is then to first order of $(\hbar\omega/Q)$

$$R = \frac{\Gamma}{\Gamma_{free}} = 1 + \frac{35}{64}\frac{mc^2}{Q}\left(\frac{\hbar\omega}{Q}\right)^2 \nu^2 . \quad (7.5)$$

Expressing the intensity parameter ν by the critical field strength E_c (2.23) the enhancement can be written in the form

$$R = 1 + \eta(Q)\left(\frac{E_o}{E_c}\right)^2 , \quad \eta(Q) = \frac{35}{64}\left(\frac{mc^2}{Q}\right)^3 . \quad (7.6)$$

This result agrees with the nonrelativistic limit in Ref. 28. As mentioned at the end of the preceding Section we can obtain a considerable enhancement either for nuclei with very small Q values $Q \ll mc^2$, or for x-ray fields or for high field intensities $\nu > 1$.

Figure 3 shows the enhancement R (7.5) as a function of ν for a Nd laser with $\hbar\omega = 1.17$ eV and for several Q-values. For $Q \ll mc^2$ the enhancement already becomes very large for relatively moderate field strengths. It should, however, be mentioned that the chosen Q values are considerably smaller than the Q values of any existing beta decaying nuclei. For ^3H Eq. (7.5) requires already $\nu = 4x10^3$ for a Nd laser in order to obtain R = 2, corresponding to a field strength E_0 of $1.2x10^{14}$ V/cm!

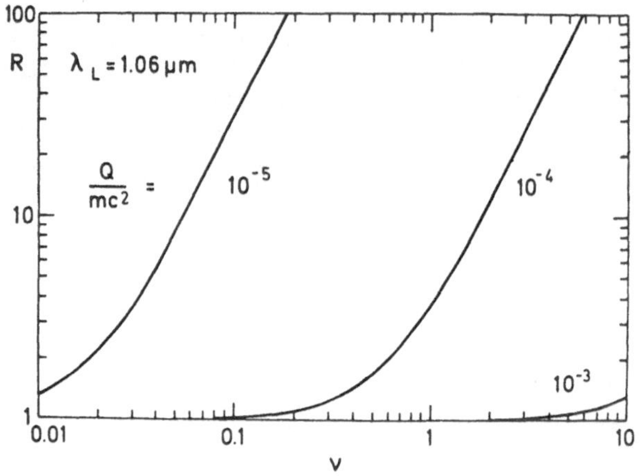

Fig. 3. Enhancement R of the total decay rate Γ due to an external field of wavelength $\lambda = 1.06$ μm as a function of the intensity parameter ν for different Q values.

It is interesting to note that in the relativistic theory in first order of $(\hbar\omega/mc^2)$ the enhancement R still has the form (7.6). Just the function η which depends only on the Q value of the decay has to be replaced by a relativistic expression [28]. This implies that the field dependence of the enhancement R is fairly well described by the nonrelativistic theory in Eqs. (7.5) or (7.6) for all field parameters. Hence the nonrelativistic theory yields also for $\nu >> 1$ a good estimate of the total decay rate which justifies its use for Fig. 3. Furthermore, the nonrelativistic total decay rate also gives for the entire range of Q-values of realistic nuclei approximately the correct answer, i.e. it also holds for electron energies $E >> mc^2$. We can see this from Fig. 4 which compares the Q-dependent factor of Eq. (7.6) in the relativistic and in the nonrelativistic theory. For $Q/mc^2 < 0.1$ both curves of η are identical and up to $Q/mc^2 \lessapprox 100$ they never differ by more than a factor of 4. This proof that the nonrelativistic theory yields reliable enhancements even for $\nu >> 1$ and $Q >> mc^2$ is the physical justification for its use throughout these lecture notes. However, there are, of course, measurable relativistic effects in the angular and in the energy distribution of the electrons and polarization effects are not at all accounted for by the nonrelativistic treatment.

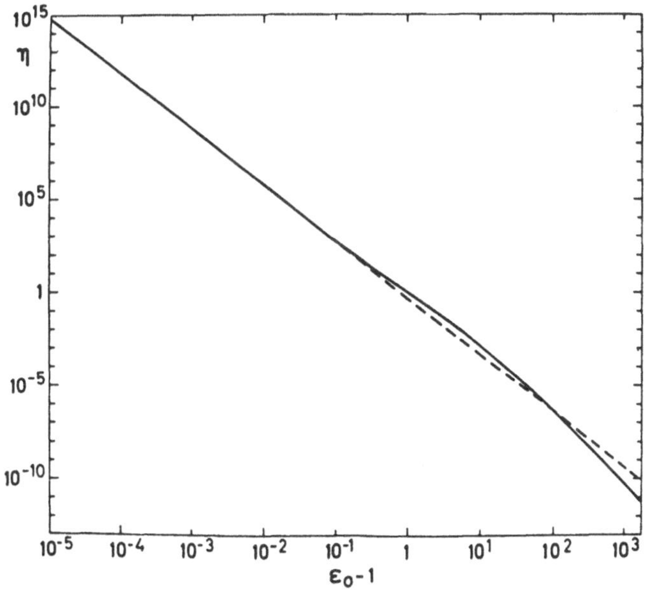

Fig. 4. Q-value factor η (7.6) in the Dirac theory (solid curve) and in the Schrödinger theory (dashed line) as a function of $Q/mc^2 = \varepsilon_0 - 1$.

SUMMARY

We have shown that the modification of the beta-decay lifetime due to an external electromagnetic field is a pure quantum effect. For presently feasible laser fields the interaction between the emitted electron and the field is well described by the classical limit $\hbar \to 0$, so that the nuclear lifetime is not affected. In order to obtain appreciable quantum effects in nuclear beta decays either (i) the photon energy has to be of the order of the Q value of the decay [30] or (ii) the field strength must be comparable with the critical field strength. We have furthermore seen in Fig. 4 that the modification of the lifetime is to a fair approximation a nonrelativistic quantum effect. Relativistic quantum corrections are small for all realistic nuclei.

Crucial for the result that the classical electron-field interaction does not give rise to a change in the nuclear lifetime, is the interplay between the $\vec{p}\vec{A}$-term of the interaction which is responsible for the energy exchange between the electron and the field, and the $\vec{A}^2$-term which contributes to the effective mass. For intense field problems the $\vec{A}^2$-term is very important and cannot be neglected as it is often done in quantum optics.

The result found here contradicts directly Refs. 10 and 14, which predicted a considerable enhancement of free neutron and nuclear beta decay rates with lasers available at present and casts doubt on Refs. 13 and 15 in which similar enhancement for γ decay and forbidden beta decay are obtained. We can state that the numerical results in all of these four publications are (due to completely different reasons) incorrect. In Refs. 10 and 14 the "enhancement" of the decay rate is due to a careless handling of sums over Bessel functions [25] as was discussed in these lecture notes. In Ref. 13 the underlying physical situation is not correctly modeled [31] and the calculations in Ref. 15 are misguided by a wrong interpretation of the non-interacting nuclear wave function (2.13) and include an algebraical error, after whose correction the proposed enhancement of forbidden decay rates disappears [32]. Obviously, Nature does not want her more elementary constituents to be tampered with by inadequate means.

REFERENCES

1. R. Daudel, Rev. Sci. 85:162 (1947); E. Segre, Phys. Rev. 71:274 (1947).
2. E. Segre and C. E. Wiegand, Phys. Rev. 75:39 (1949); 81:284 (1949); R. F. Leininger, E. Segre, and C. E. Wiegand, Phys. Rev. 76:897 (1949), 81:280 (1949).
3. K. T. Bainbridge, M. Goldhaber, and E. Wilson, Phys. Rev. 84:1260 (1951), 90:430 (1953).

4. For a review, see G. T. Emery, Ann. Rev. Nucl. Sci. 22:165 (1972).
5. A. G. W. Cameron, Astrophys. J. 130:452 (1950).
6. P. B. Shaw, D. D. Clayton, and F. C. Michel, Phys. Rev. 140:B1433 (1965).
7. R. F. O'Connell and J. J. Matese, Nature 222:649 (1969); Phys. Lett. 29A:533 (1969); J. J. Matese and R. F. O'Connell, Phys. Rev. 180:1289 (1969); L. Fassio-Canuto, Phys. Rev. 187:2141 (1969).
8. A. I. Nikishov and V. I. Ritus, Zh. Eksp. Teor. Fiz. 46:776 (1964) [Sov. Phys. JETP 19:529 (1964)].
9. See, for example, A. I. Nikishov and V. I. Ritus, Zh. Eksp. Teor. Fiz 46:1768 (1964) [Sov. Phys. JETP 19:1191 (1964)]; V. I. Ritus, Zh. Eksp. Teor. Fiz. 56:986 (1969) [Sov. Phys. JETP 29:532 (1969)]; V. A. Lyul'ka, Zh. Eksp. Teor. Fiz. 69:800 (1975) [Sov. Phys. JETP 42:408 (1975)].
10. I. G. Baranov, Iz. Vyssh. Uchebn. Zaved. Fiz. 17:115 (1974) [Sov. Phys. J. 17:533 (1974)].
11. I. M. Ternov, V. N. Rodionov, V. G. Zhulego, and A. I. Studenikin, Yad. Fiz. 28:1454 (1978) [Sov. J. Nucl. Phys. 28:747 (1978)]; Ann. der Phys. 37:406 (1980).
12. L. D. Landau and E. M. Lifschitz, The Classical Theory of Fields, (Pergamon, Oxford, 1975).
13. B. Arad, S. Eliezer, and Y. Paiss, Phys. Lett. 74A:395 (1979).
14. W. Becker, W. M. Louisell, J. D. McCullen, and M. O. Scully, Phys. Rev. Lett. 47:1262 (1981).
15. H. R. Reiss, Phys. Rev. C27:1199,1229 (1983).
16. L. V. Keldysh, Zh. Eksp. Teor. Fiz. 47:1945 (1964) [Sov. Phys. JETP 20:1307 (1965)].
17. See, for example, A. de Shalit, and H. Feshbach, Theoretical Nuclear Physics, Vol. I, (Wiley, New York, 1974).
18. R. R. Schlicher, W. Becker, J. Bergou, and M. O. Scully, Proc. of the NATO Advanced Study Institute on Quantum Electrodynamics and Quantum Optics in Boulder, Colorado, 1983, ed. by A. O. Barut.
19. C. Cohen-Tannoudji, B. Diu, and F. Laloe, Quantum Mechanics, (Hermann/Wiley, Paris, 1977)
20. M. Goldstein, Classical Mechanics (Addison-Wesley, Reading, Mass., 1965).
21. D. M. Volkov, Zs. Phys. 94:250 (1935).
22. J. D. Bjorken and S. D. Drell, Relativistic Quantum Mechanics, (McGraw-Hill, New York, 1964.)
23. G. Abramowitz and I. A. Stegun, Handbook of Mathematical Functions, (National Bureau of Standards, AMS 55, 1972)
24. N. G. Kroll and K. M. Watson, Phys. Rev. A8:804 (1973).
25. W. Becker, G. T. Moore, R. R. Schlicher, and M. O. Scully, Phys. Lett. 94A:131 (1983).
26. See for example, A. Erdelyi, Asymptotic Expansions, (Dover Publications, New York, 1956)

27. I. M. Ternov, V. N. Rodionov, A. E. Lobanov, and O. F. Dorofeev, Pis'ma Zh. Eksp. Teor. Fiz. 37:288 (1983) [JETP Lett. 37:342 (1983)].
28. I. M. Ternov, V. N. Rodionov, and O. F. Dorofeev, Zh. Eksp. Teor. Fiz. 84:1225 (1983) [Sov. Phys. JETP 57:710 (1983)].
29. W. Becker, R. R. Schlicher, and M. O. Scully, to be published.
30. W. Becker, R. R. Schlicher, M. O. Scully, M. S. Zubairy, and M. Goldhaber, Phys. Lett., to be published.
31. W. Becker, R. R. Schlicher, and M. O. Scully, to be published.
32. W. Becker, R. R. Schlicher, and M. O. Scully, Phys. Rev. C., to be published.
33. J. R. Movat, C. E. Johnson, H. A. Shugart, and V. J. Ehlers, Phys. Rev. D3:43 (1971).
34. T. W. B. Kibble, A. Salam, and J. Strathdee, Nucl. Phys. B96:255 (1975).

THERMOHYDRODYNAMIC INSTABILITIES : BUOYANCY-THERMOCAPILLARY CONVECTION

M.G. Velarde and J.L. Castillo

Departamento de Física Fundamental, U.N.E.D.

Apartado 50.487, Madrid, Spain

1. INTRODUCTION

Spontaneous, free or natural convection belongs to those physical phenomena that have fascinated people since the time of Archimedes (213 B.C.). Such convection usually arises when some interplay develops between different transport phenomena and external constraints, such as gravitational, electrical,or magnetic fields. Its many varieties provide, however, beautiful examples of problems among the most fertile and difficult in functional analysis, nonequilibrium thermodynamics, and hydrodynamics.

The term *convection* (*convectio*, carrying or converging) seems to have been coined by W. Prout (1834) to denote a mode of propagation of heat (by fluid motion) which had been described by Count Rumford forty years earlier to account for the heat transport in a hot apple pie. This concept actually covers a broader spectrum, in that mass, momentum, or energy in some form other than heat may be the object of transport.

C. Varley (1836), an English microscope maker, seems to have been the first to report "motions of extremely curious and wonderful character in fluids undergoing evaporation". These motions were made apparent with the aid of a microscope by incorporating finely divided coal into the drops and films studied. Weber (1855) also observed a tesselated structure and motions in a layer of alcohol and water on the slide of a microscope, but attributed the circulation observed to electric forces and thus gave a wrong interpretation of his observations. The purely thermal origin of the phenomenon and a correct interpretation was given later by Lehmann (1888). Many other subsequent studies by various authors described convection in cellular from. May it suffice to mention the remarkable and extensive work of Tomlinson (1864, 1873), Van der Mensbrugghe (1869, 1873), and above all James Thomson (1855, 1882), a lecturer in Civil Engineering at Belfast, and an older brother of Lord Kelvin (W. Thomson). Thomson gave a sound theoretical explanation of the phenomena that he and others had seen in a wine glass. Incidentally, these phenomena probably had been noticed since the invention of this universal beverage, for in Proverbs XXIII.31 it is said: "Look not thou upon the wine when it is red, when it giveth its colour in the cup, when it glideth down smoothly".

For this evaporative convection driven by unbalanced surface tension stresses | Thomson (1855, 1882)| a home experiment has been suggested by Brunt (1951): "The simplest illustration of such motion is obtained by pouring cheap gold paint into a small vessel to a depth of say, a quarter of an inch. In cheap gold paint the liquid medium is usually benzene or some other highly volatile liquid, while the "gold" is in the form of thin flakes. The evaporation of the liquid leads to such rapid cooling at the top of the layer that marked instability is immediately produced. As the thin flakes tend to set themselves parallel to the motion, they are not visible at the center or periphery of the cell, where they are seen edgewise from above. The cell will thus appear to have clear liquid at the center and at the outer boundary".

It is remarkable that Thomson advocated the two driving mechanisms known today as being operative in such cellular convection. For in his 1882 communication Thomson also referred to buoyancy forces to explain the tesselations in a tub of warm evaporating wash-water seen by chance at a roadside inn, on the occasion of a countryside excursion of the Belfast Naturalists' Field Club. Thomson went even further by observing that a similar tesselated formation is often to be noticed in a bowl of beef broth or other clear soup containing minute flocculent particles in suspension, when it is left to cool. It was the turn of the last century that Bénard (1900) reported on carefully controlled experiments of convective motions in thin horizontal liquid layers heated from below. After the publication (1900, 1901) of two lengthy papers based on his Ph.D. dissertation, Bénard and collaborators continued to work extensively on the same subject, seeking in the phenomena they studied a tentative explanation of a large number of apparently disparate problems.

Bénard worked with layers thinner than about a millimeter (aspect ratio of 1/100 and less with error to about 1 μm) lying on a metallic plate which was heated and maintained at a uniform temperature. The upper surface of the liquid (mostly whale spermaceti, with a melting temperature of 46 °C) was in free contact with the ambient air while the temperature of the bottom plate was considerably higher, being on occasion heated to 100 °C . Bénard observed two distinct phases in the convective phenomena produced under these conditions.

First, when the vertical temperature drop was large enough, a random motion of the fluid resulted. Shortly thereafter, the first phase -of relatively short duration (increasing with fluid viscosity from a few seconds up to several minutes) -appeared, in which the fluid formed cells of almost regular shapes. In this phase, the cellular cross sections showed nearly regular polygons of four to seven sides. During the second stage the cells became equal and regularly spaced hexagons filling up the plane. Thus the limit of the second phase was a steady regime of prisms with vertical boundaries and hexagonal cross sections. The liquid rose in the core of the cell, moved outward at the top, descended at the outer periphery and moved inward at the bottom. Bénard made the circulation visible by pouring into the fluid a few grains of lycopod of about 20 µm diameter, whose individual motion he was able to follow in detail. He correctly characterized the spatial periodicity of the phenomenon by defining its wavelength as the distance between centers of the hexagonal parallelepipeds. Bénard attributed an important role in the phenomenon to surface tension inhomogeneities, without however elaborating further on this point (Bénard, 1901, pp.92, 134,135). He carefully studied the free surface deflection, by sophisticated means originally developed by Foucault in the telescope-making industry, and gave quantitative estimates of the maximum values of depression and elevation from the surface level. He estimated the maximal surface deflection at 0.5 µm for a 1 m m deep layer under his best experimental circumstances.

Another result also described by Bénard is that as the thermal gradient is decreased the surface deflection is first drastically reduced, then disappears very slowly with the temperature drop. This was achieved by merely letting the spermaceti layer cool off from 100 °C down to solidification (46 °C). Bénard's thermal gradients were of about 1°/ m m. Incidentally, Dauzère (1908) produced solidified Bénard cells by quickly cooling a thin layer of melted beeswax undergoing convection. Bénard also estimated the angular and linear convective velocities and the mean periods of circulation of suspended particles in the fluid.

Bénard heuristically, and correctly, attributed the surface deflection to surface tension tractions:" *La tension superficielle à elle seule, provoque déjà une depression au centre des cellules et un excès de pression sur les lignes de faîte qui séparent les cuvettes concaves les unes des autres*" (Bénard, 1901, p. 92; see also p. 134). He also gave an estimate of minimum curvature radius at both the depression in the cells' centers (30-50 cm) and the elevation at the cells' edges (10-15 cm), for 100 C and 1 mm layer depth.

Bénard measured carefully the surface area s of the cells, and compared it with the depth of the fluid layer in a number of experiments. He noticed that s varied linearly with d, or, as he put it, "*les prismes cellularies seraient semblables, quelles que soient leurs dimensions lineaires.*" s/d and λ/d define a dimensionless wavelength. We have here a quantitative law giving a marked increase of the wavelength with increasing supercritical Marangoni number, since the Marangoni number depends on d^2. Unfortunately, it was not until the 1950's that scientists realized the necessity of incorporating surface--tension stresses in a dynamical model of Bénard convection. This became a necessity when Block (1956) found Bénard cells in horizontal layers of fluid when the higher temperature was on the *upper-side* . A straightforward and illuminating theoretical description of surface-tension-driven convection has been given in a paper by Pearson (1958), though Pearson's model did not adequately accomodate all of Block's findings. In retrospect, it is regrettable that Bénard did not explore the role of the surface tension inhomogeneities which, however, he recognized as important in his experiments . A persistent misinterpretation of Bénard's hexagonal cells causes writers even today to illustrate buoyancy--driven convection with some of the beautiful original pictures of Bénard.

Further experiments on Bénard surface-tension-driven convection have been conducted by Terada (1928), Volkovisky (1939), Block (1956), Koschmieder (1966, 1967), and others. Volkovisky and Terada worked with fluid layers flowing horizontally with a controlled preimposed horizontal velocity. Their results reinforced the picture already presented by Bénard and we shall not discuss them here. Block was the first to point out that Bénard's cells can be obtained by *cooling* a layer *from below* . Block was also able to suppress convection by adding tensioactive agents to the upper surface of the layer. These facts, together with the results of experiments conducted by astronauts on board spaceships, where gravity was $10^{-6}\, g$, unambiguously manifest that gravity played a minor role, if any, in most of Bénard's original experiments. Surface-tension-driven convection can be considered as a specific phenomenon qualitatively different from buoyancy-driven flows. Theories that support this view have been developed first by Pearson (1958), and later, and more realistically, by Sternling and Scriven (1959), Scriven and Sternling (1964), Smith (1966), and Bentwich (1971). There is also the work of Nield (1964), who described the convective flows when surface tension *and* buoyancy are operating in the fluid layer. More recently Davis and Homsy (1981) and Castillo and Velarde (1982) have extended previous theories by incorporating the surface deformation to the buoyancy -thermocapillary problem. In the remaining of these Notes we provide necessary or sufficient conditions for the onset of convection in Bénard layers when we consider a binary mixture heated from below and open to the ambient air. Thus we account for the eventual role played by an impurity also affecting the surface tension.

2. INTERFACIAL CONVECTION : HEURISTIC ARGUMENTS

Relevant parameters in a Rayleigh-Bénard geometry are the thermal Rayleigh number and the solutal Rayleigh number when we consider a fluid mixture

$$Ra = \alpha g d^3 \Delta T/\nu\kappa \qquad \text{and} \qquad Rs = \gamma g d^3 \Delta N/\nu D,$$

respectively, where α and γ are the thermal and solutal expansion coefficients, g is the gravitational acceleration, $\Delta T/d$ is the thermal gradient across the layer of thickness d, and ν is the kinematic viscosity of the mixture. κ and D are the heat and mass diffusivities. $\Delta N/d$ denotes the impurity gradient. With and open interface, surface tension tractions must be considered. This is accounted with the inclusion of the thermal and solutal Marangoni numbers (the latter is usually called the Elasticity, number),

$$M = -(\partial\sigma/\partial T)\, d\Delta T/\mu\kappa \qquad \text{and} \qquad E = -(\partial\sigma/\partial N_1)\, d\Delta N_1/\mu D,$$

respectively. Here σ is the surface tension (liquid-air, say) and N_1 is the mass-fraction of the impurity, which for convenience is considered the heavier component of the binary mixture. Consideration of surface tension tractions does not necessarily forces the consideration of the deformation of the interface for they may be operating even if the interface remains level. If, however the deformation is to be considered at least two more dimensionless groups appear, the Bond number (Bo or G) and the capillary (or crispation) number:

$$Bo = \rho g d^2/\sigma_0 \qquad \text{and} \qquad C = \mu\kappa/\sigma_0 d,$$

respectively, where ρ accounts for the density of the fluid mixture and σ_0 is some mean value of the surface tension σ, $\mu = \nu\rho$ is the dynamic viscosity. The Bond number estimates the strength of the gravitational forces with respect to the surface tension and thus large values of G correspond to rather flat, and level interfaces whereas low values of the Bond number exist when interfaces tend to be spherically shaped (at thermodynamic equilibrium this means minimization of free energy). Rather low Bond numbers appear in experiments aboard spacecrafts where gravity might decrease in four or six orders of magnitude the value on Earth. Then capillary lengths may reach the order of the meter and for this reason there is no need of a container for the handling of liquids at very low Bond numbers. The capillary number compares dissipation to surface tension forces. To a first approximation dissipation tends to damp out all inhomogeneities whereas, as before, surface tension tends to bend interfaces. For the most common experiments C varies between 10^{-2} and 10^{-7}. Large values of C appear, however, when the operation takes place near a critical point where the interfacial tension goes to zero or when we handle extremely thin films.

It is of some interest to read the definition of the capillary number from another perspective. Consider an interface where a capillary wave may develop. Let ξ be an estimate of its length. The actual value could eventually be d, the cell gap in a Rayleigh-Bénard (Marangoni) experiment. Then a quantity, the (mechanical) time constant of such a disturbance upon the interface (the interface may be likened to a stretched membrane) can be defined through the relation

$$\tau_\xi^2 = \rho\xi^3/\sigma \qquad \text{(from now on } \xi = d\text{)}.$$

With heat and mass diffusion the other two time constants are

$$\tau_V = d^2/\nu \qquad \text{and} \qquad \tau_T = d^2/\kappa \quad .$$

We have

$$C = \tau_d^2 / \tau_V \tau_T$$

and thus the large values of the capillary number correspond to the case of disturbances that decay so fast on the thermal and momentum dissipation scales that this happens before the mechanical disturbance, the "wave" decays, i.e., before the interface returns to the level position. Restoring mechanical forces (potential energy) make the interface overshoot the level position thus leading to interfacial oscillations. It is this potential energy that provides in a fluid layer the possibility of reversing an initially given fluid motion. Note that the above given argument can be extended to the case of an interface where some chemical reaction takes place . It suffices to replace the heat diffusivity by the appropriate chemical constant and abnormally large separation in the different time constants of the problem may result in oscillatory motions of the interface.

Still two more parameters are needed for a fluid mixture heated below or above: the Prandtl number, $P = \nu/\kappa$ and the Schmidt number, ν/D. Low Prandtl number fluids also tend to show oscillatory instabilities as there inertial terms tend to dominate dissipation. Finally the combination

$$A = (Ra + Le\ Rs)\ C/G$$

is a quantity that estimates the validity of the Boussinesquian approximation (for a single component $A = \alpha \Delta T$). In the following we shall consider values of A and C small with emphasis, however, on the role of C when the Rayleigh and Marangoni numbers compete or cooperate for the onset of instability. All results will show the dependence on P, Le and G and, for illustration, cases with vanishing Rayleigh numbers are also discussed.

3. BUOYANCY - THERMOCAPILLARY EQUATIONS

We use the following conventions and notations (Figure 1.): d is the mean distance between two infinitely extended surfaces; the lower is a rigid, heat conducting plate held at constantly controlled temperature. The upper surface is the one open to the ambient air. For simplicity is considered adiabatic (poor heat conductor). The fluid enclosed between these two surfaces in an incompressible binary mixture and we restrict consideration to a two-dimensional problem. Thus x and z denote, respectively, the horizontal and vertical coordinates. The ambient air is assumed to have negligible density and *dynamic* viscosity.

For universality in the description the following units (scales) are introduced: d for length, d^2/κ for time, κ/d for velocity, ΔT and ΔN_1 for temperature and mass fraction of the components, respectively. $\mu\kappa/d^2$ and σ_0 for pressure and surface tension, respectively. The open surface, S(t), is located at

$$z = 1 + \eta(x,t).$$

$\vec{n}$ denotes the outward unit normal vector to S,

$$\vec{n} = (-\partial\eta/\partial x,\ 1)/N$$

whereas $\vec{t}$ is the unit tangent vector

$$\vec{t} = (1, \partial\eta/\partial x)/N$$

and the curvature is $K(\eta)$,

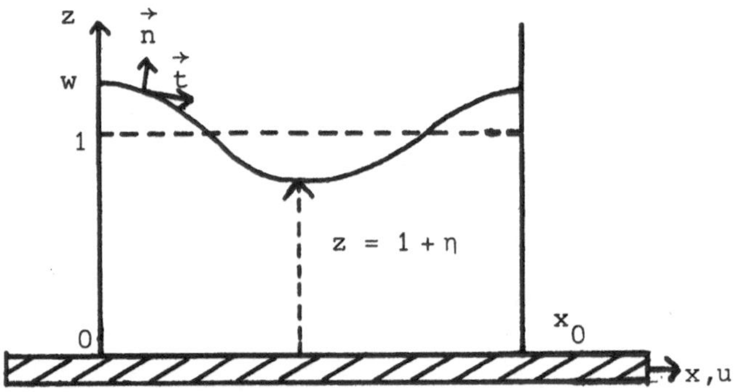

Figure 1. An exaggerated view of the open interface with bottom plate at z=0.

$$K(\eta) = (\partial^2\eta/\partial x^2)/N^3 \quad \text{with} \quad N = \{1 + (\partial\eta/\partial x)^2\}^{1/2}$$

The stress balance at the deformable interface is

$$\begin{aligned} \tau_{ij} n_j = & - (G/C)(\eta + A\eta^2/2) n_i \\ & + (K/C)\{1 - MC(\theta - \eta) - Ele(\Gamma - \eta)\} n_i \quad (1) \\ & + t_i (\vec{t}.\vec{\nabla})\{M(\theta - \eta) + Ele(\Gamma - \eta)\};\ i,j=1,2 \end{aligned}$$

where θ, Γ and $\vec{v}$ account for disturbances upon the temperature, mass fraction and velocity of the initially steady rest state. The stress tensor is

$$\tau_{ij} = -p\,\delta_{ij} + \varepsilon_{ij} \quad \text{with} \quad \varepsilon_{ij} = \partial v_i/\partial x_j + \partial v_j/\partial x_i \quad (2)$$

δ_{ij} is the Kronecker delta and the summation convention over repeated indices is assumed.

The kinematic boundary condition at the interface is

$$\partial\eta/\partial t = N v_i n_i \quad \text{on} \quad z = 1 + \eta \tag{3}$$

The condition that the heat flux is prescribed at the open interface leads to the condition

$$(\vec{n}.\vec{\nabla})\,\theta = (1 - N)/N \quad \text{on} \quad z = 1 + \eta \tag{4}$$

For the impurity we also prescribe at the interface its flux. This leads to a simpler analysis. A generalization is given later on. For a fluid layer bounded by a copper plate at the bottom we take there a heat conducting plate, impervious to matter transfer and mechanically rigid. Thus we have

$$v_i = \theta = \Gamma = 0 \qquad \text{on} \qquad Z = 0 \tag{5}$$

We assume that originally the fluid layer is in motionless state with steady linear distributions of temperature and solute. Thus the evolution equations for disturbances upon the motionless steady state are the Navier-Stokes, Fick, and Fourier equations for the region $0 \leqq z \leqq 1 + \eta\,(x,t)$, $-\infty < x < +\infty$

$$P^{-1}(\partial v_i/\partial t + \vec{v}.\vec{\nabla} v_i) = \partial\tau_{ij}/\partial x_j + Ra\theta\, k_i + LeRs\Gamma\, k_i \tag{6}$$

$$\partial\theta/\partial t + \vec{v}.\vec{\nabla}\theta = \nabla^2\theta + w \tag{7}$$

$$\partial\Gamma/\partial t + \vec{v}.\vec{\nabla}\Gamma = Le\,\nabla^2\Gamma + w \tag{8}$$

$$\vec{\nabla}.\vec{v} = 0 \tag{9}$$

where $k_i = (0,1)_i$.

4. NECESSARY CONDITIONS FOR THE ONSET OF CONVECTION

We define the integral over the free interface for a quantity f as

$$\int_{S(t)} f \; ds = \int_0^{S_0(t)} f \; ds = \int_0^{x_0} f(z=1+\eta)\; N dx \tag{10}$$

where ds is an element of arc length along S(t), and $S_0(t)$ is the length in one period along x. A two-dimensional volume integral of f over a period is

$$\langle f \rangle = \int_0^{x_0} \int_0^{1+\eta} f(x,z,t) dz \; dx \tag{11}$$

Then we define the energy

$$E = P^{-1} \langle v^2/2 \rangle + \lambda \langle \theta^2/2 \rangle + \Lambda \; Le \langle \Gamma^2/2 \rangle$$
$$+ (G/C) \int_{S(t)} ds \; \tfrac{1}{2} \; (\eta^2 + A\eta^3/3)/N \tag{12}$$

where $|\eta| < 3/A$ (Boussinesquian approximation). λ and Λ are the linking parameters whose choice is dictated by the convenience in obtaining the largest parameter region of stability of the initial motionless steady state of the fluid layer. Thus a variational condition is introduced $\delta(dE/dt) = 0$ where δ accounts for an arbitrary variation subjected, however, to the conditions earlier indicated (mass conservation, boundary conditions and all that). We have

$$\delta(dE/dt + \langle 2p \vec{\nabla} \cdot \vec{v} \rangle + \beta \int_x \eta N \; dx) = 0 \tag{13}$$

The consideration of arbitrary values of the capillary or crispation number produces a formidable problem and a reasonable approach is to consider its contribution, i.e., the role of the surface deformation to a first order approximation. Thus we set

$$\eta = \eta^{(o)} + \eta^{(1)} C + O(C^2) \ ; \quad \eta^{(o)} = 0 \tag{14}$$

(note that first it moves and then it gets deformed which in turn affects the motion) together with similar expansions for the remaining quantities then the evolution problem for disturbances upon the initial state is reduced to the following Euler-Lagrange equations

$$2\,\tau_{ij}^{(o)} + (Ra + \lambda)\,\theta^{(o)}\,k_i + (Rs + \Lambda)\,Le\,\Gamma^{(o)} k_i = 0 \tag{15}$$

$$(Ra + \lambda)\,w^{(o)} + 2\lambda\nabla^2\theta^{(o)} = 0 \tag{16}$$

$$(Rs + \lambda)\,w^{(o)} + 2\,\Lambda\,Le\,\nabla^2\Gamma^{(o)} = 0 \tag{17}$$

$$\vec{\nabla} \cdot \vec{v}^{(o)} = 0 \tag{18}$$

together with the conditions on $z = 0$:

$$v_i^{(o)} = \theta^{(o)} = \Gamma^{(o)} = 0 \tag{19}$$

On $z = 1$ we have

$$w^{(o)} = 0 \tag{20}$$

$$2\tau_{ij}^{(o)}\, n_j^{(o)}\, t_i^{(o)} + M\,\partial\theta^{(o)}/\partial x + Ele\,\partial\Gamma^{(o)}/\partial x = 0 \tag{21}$$

$$2\lambda\partial\theta^{(o)}/\partial x - M\,\partial u^{(o)}/\partial x = 0 \quad \text{and} \tag{22}$$

$$2\,\Lambda\,Le\,\partial\Gamma^{(o)}/\partial x - E\,\partial u^{(o)}/\partial x = 0. \tag{23}$$

Solutions of the above posed problem can be sought in the form of exponentials exp (iax) where $\underline{a}$ denotes a Fourier decomposition mode. We can set

$$\theta^{(o)} = \{ \sum_{i=1}^{6} \Omega_i \exp(q_i z) + \Omega_7 (\exp(az)-\exp(-az)) \} \exp(iax) \tag{24}$$

where the q_i are the roots of the polynomial equation

$$(q_i^2 - a^2) + (\lambda + \Lambda) a/4 = 0 \tag{25}$$

There are similar expressions for the remaining quantities. To the lowest order approximation, necessary conditions for the onset of instability are obtained in the form of Taylor expansion in the capillary number. For instance, using the Marangoni number the motionless steady state of the fluid layer is stable provided that its value remains below

$M^{(o)} + C\, M^{(1)}$, where

$$M^{(1)} = -2a^{-2}(M+Le\ E) \int_x dx\ \tau_{ij}^{(o)}\, n_j^{(o)}\, n_i^{(o)}\, (\partial w/\partial x)^{(o)} \left\{ \int_x dx\ \theta^{(o)}\, (\partial w/\partial z)^{(o)} \right\}^{-1} \tag{26}$$

Here the superscript (O) accounts for values at vanishing crispation number. Similar expressions have been found for the remaining control parameters (Rayleigh and Marangoni numbers). When, however, C vanishes there is no gravity and both Marangoni numbers are positive, energy theory yields $M + E = 56.7$ for a critical wavenumber $a_c = 2.22$. Below this line the fluid layer is absolutely stable.

5. SUFFICIENT CONDITIONS FOR THE ONSET OF CONVECTION

5.1. Boundary conditions

For the sake of generality in this Section we consider various possible heat transfer conditions at the boundaries. Heat is assumed to flow across the boundary following Newton's law of cooling (Robin or Biot condition). We have

$$\lambda\, \partial \delta T/\partial Z = q^o\ \delta T \quad \text{at} \quad Z = 0 \tag{27}$$

where q^o is a parameter that accounts for the transfer characteristics of the boundary. The limits q^o going to zero and to infinity, respectively, correspond to the cases of an adiabatic and perfectly conducting surface.

The mass transfer across the boundary follows a similar law. We have

$$\mathcal{D}\ \partial\delta N_1/\partial Z = S^o\ \delta N_1 \qquad \text{at} \quad Z = 0 \tag{28}$$

where now the vanishing value of S^o corresponds to an impervious boundary and $S^o \rightarrow \infty$ corresponds to a permeable membrane, a porous boundary say. At the open surface we assume similar b.c. However we must include the deformation of the interface. Thus we have:

$$\lambda\ \partial\delta T/\partial Z = -q^m\ (\delta T - \xi\ \Delta T/h) \quad \text{at} \quad Z = h \tag{29}$$

and

$$\mathcal{D}\partial\ \delta N_1/\partial Z = -\ S^m\ (\delta N_1 - \xi \Delta N_1/h) \quad \text{at} \quad Z=h \tag{30}$$

5.2 . Evolution equations for infinitessimal disturbances

When nonlinear terms in the equations given in Section 3 are disregarded we have the evolution equations for infinitessimal disturbances. Then solutions of the problem are exponentials like

$$f(x,z,t) = \phi(z)\ \exp\ (\omega t + iax) \tag{31}$$

where ω is a complex frequency. Re ω is the quantity whose value determines stability. When Re $\omega > 0$ and a nontrivial solution $\phi(z)$ exits, the reference motionless state is unstable (sufficient conditions), a denotes the Fourier mode that becomes unstable. Let W, Θ, and Γ be the ϕ (z) corresponding to velocity along the z-axis, temperature and mass-fraction of the solute, respectively. Then using Equations (11) - (14) a straightforward calculation yields

$$(D^2 - a^2)^2\ W = 0 \quad \text{(no buoyancy)} \tag{32}$$

$$(D^2 - a^2)\Theta = -W \qquad \text{(no Onsager cross-transport)} \qquad (33)$$

$$(D^2 - a^2)\Gamma = -W/Le \qquad \text{(no Onsager cross-transport)} \qquad (34)$$

Here $D = d/dz$. Note that the only difference between Equations (33) and (34) is the appearance of a scaling factor in the latter. This factor, the Lewis number, $Le = \mathcal{D}/\kappa$, accounts for the large disparity in the mass and heat diffusion in liquids. Note also that Equations (32)-(34) describe the evolution of non-oscillatory disturbances ($\mathrm{Im}\ \omega = 0$) at neutral (or marginal) stability ($\mathrm{Re}\ \omega = 0$). The b.c. are

$$W = DW = 0 \qquad \text{at } z = 0 \qquad (35)$$

$$D\Theta - A^o\Theta = 0 \qquad \text{at } z = 0 \qquad (36)$$

$$D\Gamma - B^o\Gamma = 0 \qquad \text{at } z = 0 \qquad (37)$$

At the open surface

$$W = 0 \qquad \text{at } z = 1 \qquad (38)$$

$$D\Theta + A^m(\Theta - \eta) = 0 \qquad \text{at } z = 1 \qquad (39)$$

$$D\Gamma + B^m(\Gamma - \eta) = 0 \qquad \text{at } z = 1 \qquad (40)$$

$$(D^2 + a^2)W + M a^2(\Theta - \eta) + Ele\, a^2(\Gamma - \eta) = 0 \quad \text{at } z=1 \qquad (41)$$

where the dimensionless surface deformation is given by

$$\eta = (C/a^4)D(D^2 - 3a^2)W \qquad (42)$$

Note that A^o, B^o, A^m and B^m are all non-negative.

5.3. Results: sufficient conditions for instability

The linear problem (32) - (34) together with the b.c. yields a homogeneous system of algebraic equations. Then the necessary and sufficient condition for the existence of a non-trivial solution is that the following relation holds

$$Mg_1(a, B^o, B^m)\left[g_2(a, A^o) + Cg_3(a, A^o)\right]$$

$$+ E g_1 (a, A^o, A^m) \left[g_2 (a, B^o) + C\, Le\, g_3 (a, B^o) \right]$$

$$+ g_1 (a, B^o, B^m) g_1 (a, A^o, A^m) g_4(a) = 0 \quad (43)$$

where (X = A or B)

$$g_1 (a, X^o, X^m) = \beta (a + X^m) + (X^o - a)(a - X^m)/(X^o + a) \quad (44)$$

$$g_2 (a, X^o) = - \beta^3 + \beta^2 \left[2 + 4a - 2a^2 + 4a^3 + (X^o - a)(1 - 4a + 2a^2)/(X^o + a) \right]$$

$$- \beta \left[1 + 4a + 2a^2 - 2 (X^o - a)(1 - 2a - a^2 - 2a^3)/(X^o + a) \right]$$

$$+ (X^o - a)/(X^o + a) \quad (45)$$

$$g_3 (a, X^o) = -32\, a^3 \beta \left[\beta + (X^o - a)/(X^o + a) \right] \quad (46)$$

$$g_4 (a) = 8a (\beta^2 - 4a\beta - 1) \quad (47)$$

$$\beta = \exp (2a) \quad (48)$$

Equation (43) defines the neutral stability curve, *i.e.*, the *locus* of all values of marginal non-oscillatory instability. The utility of Equation (43) can be seen by specifying some of the particular examples of application. One such case appears when $A^o = B^o$, $A^m = B^m$ and either $C = 0$ or $Le = 1$. Hence Equation (43) reduces to the simple relation

$$M/Mc + E/Ec = 1 \quad (49)$$

where the subscript "c" denotes a critical value for the onset of instability, *i.e.*, Mc (respectively, Ec) corresponds to the minimum of (43) at vanishing E (respectively, M) all other parameters given and *a* variable.

For a single component liquid film (E = 0) the ratio M/Mc is a measure of the available energy with respect to the minimal amount required for the onset of instability. This is also the case for E/Ec in an isothermal binary mixture ($M = 0$, $E \neq 0$). Thus a tight coupling like (49) is expected to hold, at least to a first-order approximation every time that two competing or cooperating instability mechanims operate simultaneously . As a matter of fact one can argue that the most general relationship is

$$M/Mc + E/Ec > 1 \tag{50}$$

with equality only when the space scale is identical, i.e., when the wave number at the onset of instability coincides in both cases: $a_c(M_c, E = 0) = a_c(M = 0, E_c)$ which is not, in general , the case.

Another interesting result that comes from Equation (43) is the correction to the critical Marangoni number due to the deformation of the interface. Let the actual Marangoni number at the onset of convection be $M = M^{(o)} + C\,M^{(1)}$ where $M^{(o)}$ is the value when there is no deformation $C = 0$. Then for a specific case, say A^o ; $B^o \to \infty$, and $A^m = B^m = 0$, using (43) a straightforward computation yields

$$M^{(1)} = -1.11 \times 10^3 \quad (M^{(o)} + Le\,E) < 0 \tag{51}$$

which shows that according to linear stability anlaysis the deformation of the interface rather destabilizes the liquid film.

6. COMMENTS

Buoyancy-thermocapillary convection has been used to introduce various aspects and methods used in the theoretical study of hydrodynamic instabilities. Interfacial (thermocapillary) convection, however, is far from being a subject with a state of development such at that attained by buoyancy-driven convection where theory and experiment have reached an extraordinary level of sophistication. However, interfacial phenomena are very important both academically and from the engineering viewpoint. Progress in the understanding of interfacial patterning, surface (interface) oscillations, etc. when there is interplay between convection, diffusion (transport of heat and matter) at and across the interface and reaction (including phase transition) is at the basis of a needed breakthrough in crystal growth, membrane biophysics , and many other problems in various realms of science and technology.

We have used the normal mode linear stability analysis which is a standard technique to obtain sufficient conditions for instability of steady states. For time-dependent (periodic or aperiodic) flows this technique needs further sophistication (Floquet theory for time-periodic states and either an initial value problem or some *ad hoc* approximations for the aperiodic case). Another technique also discussed here is the energy method which we have used for a study of the evolution of arbitrary disturbances (thus including infinitessimal ones) upon a steady state. For the phenomena discussed in the introduction of these Notes and in the Notes by T. Riste, I. Zúñiga, A. Castellanos and J. Grupp to a first- approximation these two techniques are good enough to account for almost all the results. However, the reader should note that we have not mentionned singular perturbation techniques, Galerkin methods and direct numerical integration of the equations.

ACKNOWLEDGMENTS

This research has been sponsored by the C.A.I.C.Y.T. (Spain).

REFERENCES

Bénard, H., Rev. Gen. Sci. Pures Appl., *11* (1900), 1261.
Bénard, H., Ann. Chim. Phys., *23* (1901), 62.
Bentwich, M., Appl. Sci Res., *24* (1971), 305.
Block, M.J. , Nature, *178* (1956), 650.
Brunt, D., in *Compendium of Meteorology*, edited by T.F. Malone, p. 1255 (American Meteorological Society, Boston), 1951.
Castillo, J.L., and M.G. Velarde, J. Non-Equilib. Thermodyn. *5* (1980) 111.
Castillo, J.L., and M.G. Velarde J. Fluid Mech., *125*(1982), 463.
Dauzère, C., J. Phys, (Paris), *7* (1908), 930.
Davis, S.H., and G.M. Homsy, J. Fluid Mech., *98*(1980), 527.
Koschmieder, E.L., Beitr. Phys. Atmos., *39* (1966), 1.
Koschmieder, E.L., J. Fluid Mech., *35* (1967), 9.
Lehmann, O., Molekular Physik, *1* (1988), 276
Nield, D.A. , J. Fluid Mech., *19* (1964) 341.
Normand, C., Y. Pomeau and M.G. Velarde, Rev. Mod. Phys. *49* (1977) 581 .
Pearson, J.R.A., J. Fluid Mech., (1958), 489.
Prout, W., *Bridgewater Treatises.*, chap. 8, pag. 65 (Pickering, Condon), 1834.
Schechter, R.S., M.G. Velarde and J.K. Platten, Adv. Chem. Phys. *26* (1974) 265.
Scriven, L.E., and C.V. Sternling, Nature *187*(1960) 186.
Scriven, L.E., and C.V. Sternling, J. Fluid Mech., *19* (1964), 321.
Smith, K.A., J. Fluid Mech., *24* (1966), 401.

Sterling, C.V., and L.E., Scriven, A.I.Ch.E. Journal, *5* (1959), 514.
Terada, T., and Second year students of Physics, Tokyo Imp. Univ. Aero. Res. Inst. Rept., *3* (1928), 3.
Thomson, J., Philos Mag., *10* (1985), 330.
Thomson, J., Proc. R. Philos. Soc. (Glasgow), *13* (1882), 464.
Tomlinson, C., Philos. Mag., *27* (1864), 528.
Tomlinson , C., Philos. Mag., *46* (1873), 376 .
Van der Mensbrugghe, G.L., Mem. Couronnées Acad. R. Belgique , *34* (1869).
Van der Mensbrugghe, G.L., Mem. Couronnées Acad. R. Belgique , *27* (1873).
Varley, C., Trans. R. Soc. Arts Sci. Mauritius, *50* (1836), 190.
Velarde, M.G., and C. Normand, Sci. Amer. *243* (1980) , 92.
Velarde, M.G., and J.L. Castillo, in *Convective transport and instability phenomena* , pp. 235-264 (Braun-Verlag, Karlsruhe) 1982.
Volkoviski, V., Pub. Sc. Tech. Ministère de l'Air, *151* (1939).
Weber, E.H., Ann. Phys. (Pogg. Ann., Leipzig), *94* (1855), 447.

THERMOHYDRODYNAMIC INSTABILITY IN NEMATIC LIQUID CRYSTALS: A SUMMARY OF ARGUMENTS AND CONDITIONS FOR SOME SIMPLE GEOMETRIES

I. Zuñiga and M. G. Velarde

Departamento de Física Fundamental, U.N.E.D.

Apartado 50.487, Madrid (Spain)

1. INTRODUCTION

Convection in nematic liquid crystals has been the object of intense research over the past few years. It was predicted and experimentally verified that the anisotropic tensor properties of the fluid, taken Newtonian in flow characteristics generates genuine and novel qualitative and quantitative features when flow is induced by thermal constraints or buoyancy forces. For instance, in a Rayleigh-Bénard geometry steady convection exists when the fluid layer is heated from below or from above. Threshold values are drastically lower than those typical of isotropic fluids under similar conditions. Oscillatory modes have also been studied. Furthermore, the interplay of thermal constraints and magnetic or electric fields leads to a unusual richness in the phenomena so far observed. We shall summarize here some of the conditions for thermal convection in Rayleigh-Bénard and cylindrical geometries with no pretension to completeness. We merely want to introduce the subject developed in the lectures by Prof. T. Riste. For an introduction to the physics and the hydrodynamics of liquid crystals the reader may refer to the monographs by de Gennes (1974) or Chandrasekhar (1977).

2. RAYLEIGH-BENARD CONVECTION

Nematic liquid crystals (L.C.) behave mechanically like other fluids. Yet they optically respond as crystalline solids due to always existing spontaneous ordering in the alignment of their elongated molecules. There is, however, no spatial ordering of molecular centres which is a difference with respect to the standard lattice structure of a crystal . The average orientation of the molecules in a L.C. can be described by a director vector field whose evolution may be obtained from the balance equation of angular momentum in the fluid. The director field couples to the eventual flow velocity and thus adds another nonlinear evolution equation to the standard Navier-Stokes equations of an otherwise Newtonian fluid.

In a Rayleigh-Bénard geometry (Normand *et al*, 1977; Velarde and Normand, 1980) the simplest boundary conditions (b.c) for the director are : either the vector is parallel to the horizontal plates (planar configuration) or it is orthogonal to them (homotropic case). The case of a tilted director would not be considered here although some papers have dealt with in the recent literature. Transmission of the above given b.c. to the whole continuum can be achieved with the action of a magnetic field, say. For both types of configuration years ago Pieransky *et al* (1973) reported thresholds for steady convection . For the equivalent case of Bénard convection (horizontal liquid layer heated below) the threshold, *i.e.*, the critical temperature gradient was about five hundred times lower than the corresponding value for an isotropic liquid, a silicone oil, say. They also found that although planar layers reacted very much like the silicone oils the equivalent homeotropic layer heated from below was not unstable at least in the range of thermal constraints studied by these authors. A homeotropic layer was unstable, however, when heated from above. The latter case came as a surprise although a simple explanation of the phenomenon was given based on the anisotropic character of the thermal diffusion tensor. For both director configurations, steady convection was reported in the form of roll patterns very much like the convective structures found with silicone oils or any other isotropic fluids.

An estimate of the thresholds for steady convective instability together with the pertinent physics of the phenomenon in L.C. was given by Dubois-Violette (1974). In a typical nematic L.C. the heat diffusivity along the greater axis of the molecules is about twice the diffusivity along the orthogonal and shorter axis, thus leading to a preferential transport along the director field. Thus for either director configurations once the heat transport is initiated, at the bottom, say, it rather tends to develop faster along the director itself due to the self-focusing action of the anisotropy . On the other hand when a layer is heated warmer regions tend to move upwards as a result of the buoyancy forces in the fluid. Besides, the flow velocity induces two viscous torques upon the director. The overall viscous torque brings rotation of this vector until the elastic torque of the L.C. or an external field balance the action of the velocity gradients.

For a planar configuration the two viscous torques are $\Gamma_1 = \alpha_2 \, \partial v_z / \partial x$ and $\Gamma_2 = \alpha_3 \, \partial v_x / \partial z$, respectively (see Fig 1 for further details). With almost circular horizontal rolls the two velocity gradients have about the same value. Thus $|\Gamma_1| \gg |\Gamma_2|$ since α_2 is about two orders of magnitude larger than α_3 ($-\alpha_2 \simeq 10^2 \alpha_3$). With the above indicated result on the torques the L.C. molecules tend on the average to rotate in a way that makes them to follow the direction of vertical heat transport thus leading to a drastic lowering of the threshold for Bénard convection (See Figure 2 for further details). Warmer regions in the fluid layer tend to become warmer not only because there is heating but because there is the additional (self-focusing) action of the anisotropic heat diffusivity. For a homeotropic configuration this anisotropy rather tends to suppress the convective instability when the layer is heated below. As a matter of fact this is a case of competing effects where the simultaneous action of two causes of conflict lead to overstability and eventual limit cycle motions. When, however, the homeotropic layer is heated above the anisotropy leads to steady instability with apparent negative bouyancy for such an instability is not expected for isotropic single component liquid layers. Homeotropic layers share, in fact, quite a number of common features with liquid mixtures in double-diffusive and Soret -driven convection.

Planar L.C. layers heated from above also exhibit steady convective instability at low enough wavenumbers and high enough thermal gradients (Figure 2). For it the convective pattern is not made of almost cylindrical cells but rather the cells tend to be elongated along the horizontal axis (small enough horizontal wavenumbers at the onset of convection) then $\partial v_z / \partial x$ is much greater than $\partial v_x / \partial z$ and the net resulting action of the viscous torques can be reversed with respect to the action described earlier. An instability was indeed found by Velarde and Zuñiga (1979) with threshold about three orders of magnitude higher than in a planar layer heated below. Clearly, this prediction would have great experimental difficulties due to the large temperature gradients required for convection unless buoyancy could be made strong enough due to an increase in the acceleration field., *i.e.,* with a force stronger that the fravitational one. Such a possibility exists in the cylindrical geometry and we shall come back to this problem further below.

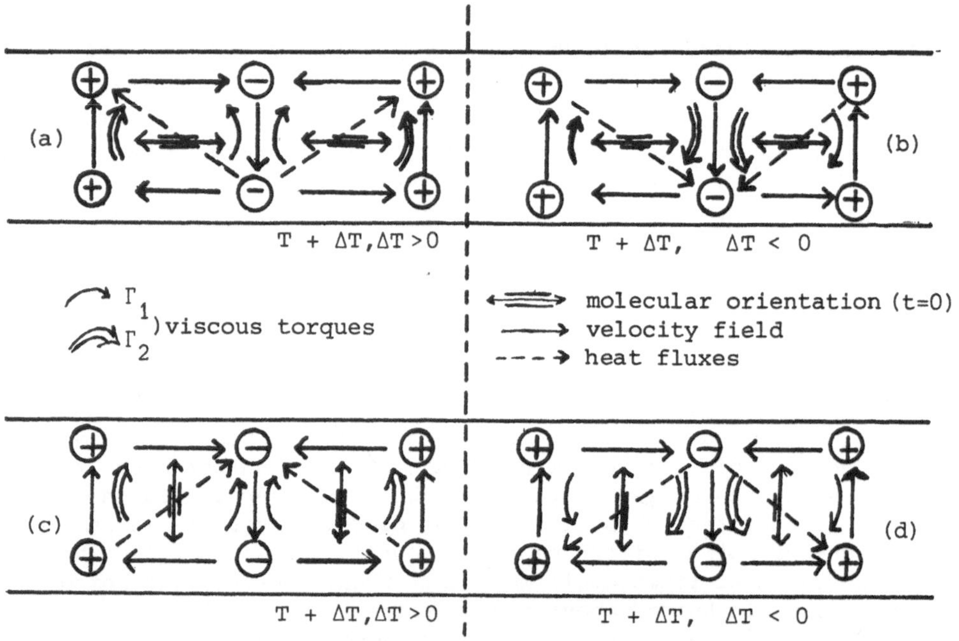

Fig.1. (a),(b): planar nematic layer heated from below, above (c),(d): homeotropic layer heated from below, above. In (a) and (d) the anisotropy helps destabilizing the system whereas in (b) and (c) the anisotropy plays the opposite role.

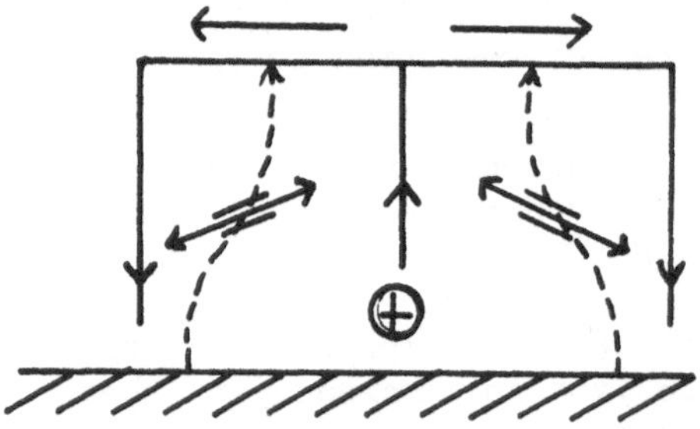

Fig.2. A planar nematic layer heated from below with an open surface. Here again the anisotropy helps destabilizing the state of rest thus leading to a drastic lowering of the threshold for Bénard-Marangoni convection.

Let us finish this Section with a discussion of oscillatory convection in homeotropic liquid layers heated below. By analogy with predictions given for the two-component Bénard problem (see, for instance, the Appendix in Velarde and Schechter, 1972, and the review paper by Schechter *et al*, 1974), Lekkerkerker (1977) suggested that an oscillatory instability is to be expected. The argument goes as follows. High enough values of the thermal gradient distort the director field which has a characteristic (relaxation) time scale $T_n \sim (\eta d^2/k)$, about ten thousand seconds in d = 5m m thick MBBA layers, say; k and η are the elastic constant and dynamic viscosity, respectively. The heat diffusivity operates on the time scale $T_{th} \sim (d^2/\kappa)$ which for the same MBBA layer is about ten seconds; κ is the heat diffusivity. Then if an oscillation is to exist its period, T, ought to be due to the large difference in the two time scales. Thus $T_{th} << T << T_n$ is the prediction. Velarde and Zuñiga (1979) did find the threshold conditions for such an oscillatory behavior with numerical predictions that agreed rather well with the observations made by Guyon *et al* (1979). Complete details, albeit not exemplified by any specific application of this limit cycle oscillation can be found in lecture notes by Velarde (1982,1983). In the latter reference the amplitude, period and stability of the limit cycle is obtained by means of a double-time, singular perturbation technique. Numerical estimates of the period together with further details can be found in Antoranz (1982). Oscillatory convection may appear either softly or as an inverted bifurcation depending on the Prandtl number and other transport or material constants of the liquid (Velarde and Antoranz, 1980). Dubois-Violette and Gabay (1982) have identified such inverted bifurcation for MBBA layers.

Another oscillatory convection was predicted by Velarde and Zuñiga (1979) when the action of an external magnetic field, which tends to align with it the director field, competes with buoyancy in a homeotropic homeotropic L.C. heated above (Fig.3). With a vertical magnetic field, a field parallel to the director, there is a contribution that adds to the elastic torque thus helping to maintain the stability of the liquid layer. With the increasing value of the external field there is a decrease of T_n, given earlier, and therefore T_n approaches T_{th}. At the eventually common value no oscillation is to be expected thus joining the predicted steady instability of the layer heated from below. Figure 3 provides a sketch of the theoretical result found. In the same figure we also see that when the layer is heated above, sufficiently high magnetic fields suppress steady convection as expected. In this case the anisotropy of the L.C. is suppressed by the action of the horizontal external field an the L.C. just

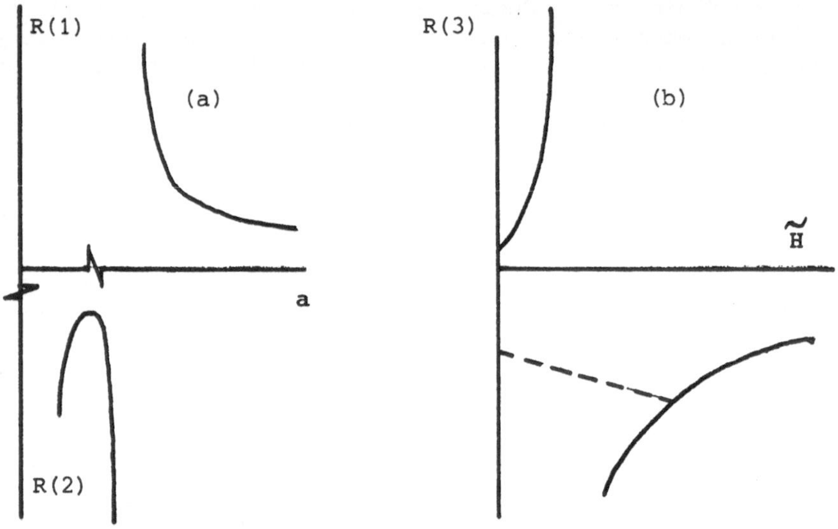

Fig.3. Neutral stability curves for (a) steady transitions in a planar layer heated from below $\{R(1)=R/2\pi^4$,positive$\}$and above$\{$ $R(2)=R/20\pi^4\}$,negative and (b) steady transitions (solid lines)and overstability(dashed line) in a homeotropic layer heated from above $\{R(3)=|R|/2\pi^4 .10^3\}$,positive) and below in the pressence of a magnetic field ($\tilde{H}$ denotes the field strength in dimensionless units). R is the Rayleigh number,i.e., the thermal gradient in dimensionless units;a is the wavenumber at the onset of instability. Depending on Prandtl and anisotropy numbers overstability may appear soft-or hardly induced.

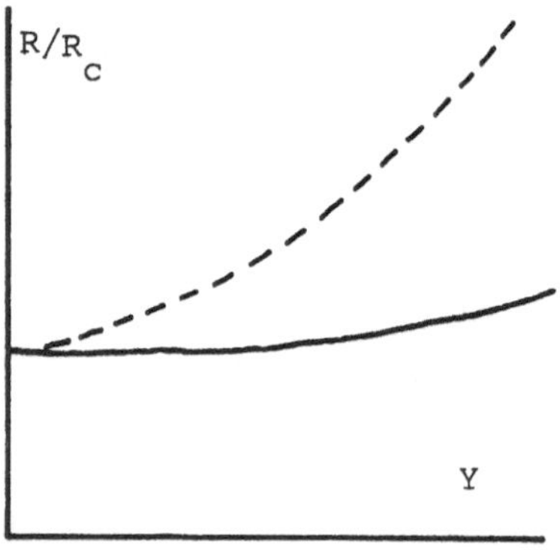

Fig.4. Variation of the Rayleigh number for the onset of instability in cylindrical containers with rotation. The dashed line denotes the threshold when the heating is from the outside and the solid line the opposite case. R_c denotes the Rayleigh number when Coriolis forces are irrelevant. Y denotes the angular rotation speed in dimensionless units.

behaves like a standard silicone oil, say. Furthermore for a layer heated below, high enough values of the field merely lead to the threshold values for an isotropic liquid layer heated below. Further details about Rayleigh-Bénard convection with and without magnetic fields can be seen in the Lecture Notes by T. Riste at this ASI (See also Salan *et al* (1983)).

Lastly, it should be mentionned how little work has been done with nematic L.C. with open and deformable interfaces. For the Bénard geometry there is to mention a prospective note by Guyon and Velarde (1979) where a heuristic approach permits a quick estimate of the thresholds for Marangoni-Bénard convection. Again, in a planar configuration heated from below, thanks to the more efficient transport along the director field heat is focused into the region of negative curvature of the director thus reducing the amount of energy needed for instability with respect to the equivalent isotropic liquid layer. A more quantitative analysis, including planar and homeotropic layers heated from below or above capable of exhibiting steady or overstable modes of convection, was provided by Velarde and Zuñiga (1979). Yet very little is known about interfacial phenomena in nematics and the evolution of the director field when approaching an open deformable surface.

3. ROTATING ANNULUS CONVECTION WITH RADIAL TEMPERATURE GRADIENTS

As indicated earlier, steady convection with too a large critical temperature gradient is not likely to be observed in a L.C. layer. For in most cases the L.C. would have no longer its anisotropic properties. Thus the Rayleigh-Bénard geometry is not suitable for studying convection in planar layers heated above. An alternative consists in choosing a set-up where large enough accelerations could compensate for moderate temperature gradients yet allowing high enough Rayleigh numbers. Barratt and Zuñiga (1983) have considered one such case by using the analogy, at least to the linear approximation between the Rayleigh -Bérnard problem and the convection in a rotating annulus. They studied the stability of a sample of nematic L.C. contained between two concentric, vertical cylinders rotating with constant uniform angular velocity about the vertical axis and subjected to a radial temperature gradient. The L.C. configuration was of the planar type with director placed parallel to the vertical axis at the initial time. They have shown that for small enough angular velocities the centrifugal force acts on the fluid very much like the buoyancy in the Rayleigh-Bénard geometry. They also considered the effect of Coriolis forces.

As expected by analogy , the neutral stability curves were not different from those obtained by Velarde and Zuñiga (1979) . Figure 4 accounts for some of the results found. Two cases of instability have been found. One corresponds to the case where the outer cylinder is heated and the other to the heated inner cylinder. In both cases the convective rolls appear with axis orthogonal to the initial orientation, the vertical axis, so that they define horizontal tori filling the gap one on top of the other like in the standard Taylor-Couette problem (Koschmieder 1981). When the outer cylinder is heated the critical gradient agrees well with the theoretical prediction obtained for planar L.C. with, however, values much lower than those reported by Carrigan and Guyon (1975). It must be noted that the latter authors were much more concerned with a qualitative rather than quantitative analysis of the role of Coriolis forces upon the convective instability threshold of the fluid layer. Their estimate was about four times the value in the corresponding Rayleigh -Bénard geometry. Moreover, if one considers their value at vanishingly small rotation it comes very close with the corresponding prediction by Barratt and Zuñiga (1983). Besides ,the rolls observed by Carrigan and Guyon are just vertical and orthogonal to those azimuthal predicted by Barratt and Zuñiga (1983). We wonder if the convective patterns observed by the other authors are not a secondary flow originated past the instability threshold of the rolls predicted by the linear theory. It would not be surprising if an hysteretic phenomenon exists for this second instability as the rolls found by Carrigan and Guyon rather tend to dissappear when the rotation vanishes.

When the inner cylinder is heated the rolls are elongated along the vertical axis as in the Bénard case discussed earlier. As the centrifugal force can be varied at will one can reduce the temperature gradient to a value such that the fluid keeps uniformly its L.C. anisotropy. Further details about this mode of convection in the cylindrical geometry can be found in the paper by Barratt and Zuñiga (1983)

ACKNOWLEDGMENTS

This research has been sponsored by the Stiftung Volkswagenwerk. The collaboration with Dr. P. J. Barratt was possible thanks to the U.K. - Spain bilateral scientific exchange agreement.

REFERENCES

Antoranz , J.C., Ph. D. Dissertation (1982), U.N.E.D. (Madrid)

Barratt, P.J., and I. Zuñiga, J. Physique, *44* (1983),311-321.

Carrigan, C.R. and E. Guyon, J. Physique Lett *36* (1975) L-145.

Chandrasekhar, S., *Liquid Crystals,* University Press, Cambridge, (1977).

Dubois-Violette, E., Solid State Commun. *14* (1974) 767.

Dubois-Violette, E. and M. Gabay, J. Physique *43* (1982) 1305-1317.

De Genes, P.G., *The Physics of Liquid Crystals,* Claredon Press (1974).

Guyon, E., P. Pieranski and J. Salan, J. Fluid Mech. *93* (1979) 65

Guyon, E., and M.G. Velarde, J. Physique Lett. *39* (1979) L-205

Koschmieder E.L. in *Order and Fluctuations in Equilibrium and Nonequilibrium Statistical Mechanics* J. Wiley and sons, N.Y. (1981).

Lekkerkerker, , J. Physique Lett. *38* (1977) L-277.

Normand, C., Y. Pomeau and M.G. Velarde, Rev. Mod. Phys. *49* (1977) 581- 624

Pieranski, P., E. Dubois-Violette, and E. Guyon, Phys. Lett. *30* (1973) 16.

Salan, J., and E. Guyon, J. Fluid Mech. *126* (1983) 13-26

Schechter, R.S., M.G. Velarde and J.K. Platten, Adv. Chem. Phys. *26* (1974) 265.

Velarde, M.G., and R.S. Schechter, Phys. Fluids *15* (1972) 1707.

Velarde, M.G., and I. Zúñiga, J. Physique *40* (1979), 725-731.

Velarde, M.G., and J.C. Antoranz, Phys. Lett *A80* (1980) 220

Velarde, M.G., and C. Normand , Sci. Am. *243* (1980) 92-108.

Velarde, M.G., in *Nonlinear Phenomena at Phase Transitions and Instabilities* (T. Riste, editor), Plenum Press, N.Y., (1981), 205-247.

Velarde, M.G., in *Evolution of Order and Chaos* (H.Haken, editor), Springer-Verlag, (1982) 132-145.

NEUTRON SCATTERING STUDIES OF PHASE TRANSITIONS IN

EQUILIBRIUM AND NONEQUILIBRIUM SYSTEMS

Tormod Riste

Institute for Energy Technology

2007 Kjeller, Norway

1. INTRODUCTION

Neutron scattering has proved to be a most valuable tool in studying condensed-matter phenomena. In my lectures I shall assume that you are unfamiliar with the technique, and have only little knowledge of the results obtained with it in the field of phase transitions. Before discussing the main theme, that of phase transitions in nonequilibrium systems, I shall give a tutorial review of some results on equilibrium phase transitions.

1a. Neutron scattering[1]

A conventional (steady state) reactor is still the most commonly used source of neutrons for scattering experiments. The neutrons provided from a reactor have a velocity distribution which corresponds to a Maxwell distribution. In order to be useful in a scattering experiment, the neutron beam has to be tailored to this purpose, i.e. to be collimated and monochromatized. A set of parallel slits in the neutron beam channel and a monochromator crystal usually fullfil this purpose. The wavelength (λ) of a neutron is related to its velocity (v) and energy (E) by

$$\lambda = \frac{h}{mv} = \frac{3.96}{v} = \frac{0.286}{\sqrt{E}} \ \text{Å} \qquad (1)$$

where v is in km sec^{-1} and E is in eV. h and m are the Planck constant and the neutron mass.

If the monochromator crystal is oriented to reflect from a set of lattice planes of spacing d, the glancing angle (θ) and λ are connected through the Bragg relation

$$\lambda = 2d \sin \theta \qquad (2)$$

By differentiation we see that the degree of monochromatization $d\lambda$ of the beam is determined by the angular spread $d\theta$, i.e. by the collimation and by the perfectness of the monochromator.

If we instead of λ introduce the wavenumber $k = 2\pi/\lambda$, the neutron momentum and energy may be written as $\hbar\vec{k}$ and $(\hbar^2/2m)k^2$, respectively. If $\vec{k}_o$ and $\vec{k}$ denote the incoming and outgoing wavevectors of the neutron and $\vec{\tau}$ a reciprocal lattice vector of the monochromator (a vector normal to the reflecting plane and of length $\tau = \frac{1}{2}d$), the Bragg relation may be written as equations expressing momentum and energy conservation:

$$\hbar\,(\vec{k} - \vec{k}_o) = \hbar 2\pi\vec{\tau} \qquad (3a)$$

$$\frac{\hbar^2}{2m}\,(k_o^2 - k^2) = 0 \qquad (3b)$$

$\hbar 2\pi\vec{\tau}$ is the momentum given to the neutron by the crystal as a whole in an elastic scattering process. Generalizing this relation to inelastic scattering, where the neutron and the crystal exchange momentum $\hbar\vec{q}$ and energy $\hbar\omega$, we get

$$\hbar(\vec{k} - \vec{k}_o) = \hbar 2\pi\vec{\tau} \pm \hbar\vec{q} \qquad (4a)$$

$$\frac{\hbar^2}{2m}\,(k_o^2 - k^2) = \hbar\omega \qquad (4b)$$

Inelastic neutron scattering may thus be looked at as a Doppler effect in which changes in direction of motion and energy of a plane neutron wave are induced by a condensed system of moving scattering centres. It follows from equations (1) and (4) that neutrons of thermal energies have wavelengths of atomic dimensions, and that thermal neutrons can show detectable shifts of both wavevector and energy when scattered by a solid or a liquid. This is in contrast to scattering of X-rays, for which the relative energy change is negligible and not detectable.

1b. Phase transitions in equilibrium systems[2]

Most phase transitions of everday life, like freezing of water, are discontinuous and involve large changes of density. The most studied, and best understood, transitions in physics are continuous, i.e. of second order. A very useful concept, when discussing phase transitions, is the order parameter, a quantity that is non-zero in one phase but vanishes in the other. In a ferromagnet the macroscopic magnetization (M) may be taken as an order parameter. If we normalize it by its value M_o at $T = 0$(K) as $m = M/M_o$ it varies continuously from 1 at $T = 0$ to 0 at $T = T_c$, the transition to the paramagnetic phase. This can be measured by classical, macroscopic methods and by neutron scattering. It is then found that m varies with the reduced temperature ε (= $T-T_c$)/ T_c as

$$m = \varepsilon^{\beta} \tag{5}$$

for temperatures relatively close to T_c. β is found to be in the range 0.3 - 0.5 for systems with 3-dimensional symmetry. A low (non-classical) value of β is a signature of the presence of fluctuations, to be discussed below.

The fact that a neutron couples to, or "sees", the order parameter is due to its own magnetic moment interacting with the atomic moments of the scatterer. The ordered ferromagnetic structure gives rise to Bragg reflections whose intensity is proportional to the order parameter squared[3]:

$$I_{Bragg} \underset{r \to \infty}{\sim} \langle m(0)\ m(r)\rangle_T = \langle m\rangle_T^2 \tag{6}$$

This formula first expresses the intensity by the correlation between local order parameters a distance r apart, of which the narrow Bragg peaks are responsible only for the asympotic($r \to \infty$) part.

In addition to the narrow Bragg peaks there is a more diffuse, but not uniform, intensity component centered at the Bragg positions. This diffuse intensity has its maximum at T_c and originates from order parameter fluctuations. In Fig. 1 is shown

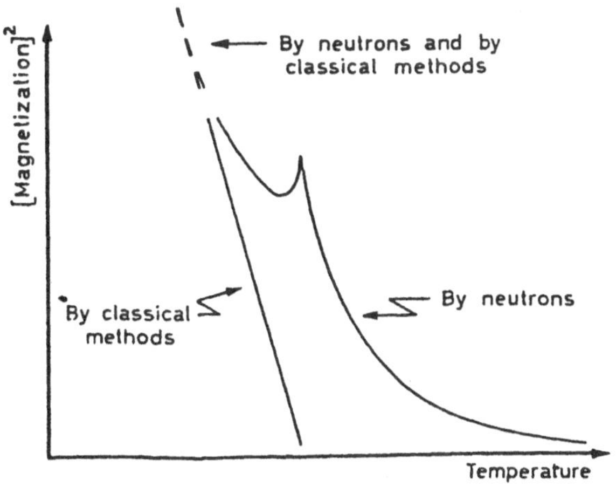

Fig. 1 Measurement by two methods of the disappearance of the magnetization at the critical temperature of Fe_3O_4

schematically the disappearance of the magnetization, as seen by classical methods. Also shown is the same phenomenon in a real neutron experiment[4] when the spectrometer is set for the Bragg condition of a magnetic reflection. The observation of the fluctuating component requires a probe that can sample phenomena on the scale of the magnetic subunits, hence only neutrons will do. Formula (6) will include the diffuse component if we drop the limitation $r \to \infty$. The diffuse component alone is given by

$$I_{diffuse} \sim \langle m(0) - \langle m \rangle_T \rangle_T \langle m(r) - \langle m \rangle_T \rangle_T \tag{7}$$

i.e. by the correlation between order-parameter fluctuations a distance r apart. In Fig. 2 is shown a schematic picture of a ferromagnet in its paramagnetic state,but near T_c. A fluctuation is now a ferromagnetic state that extends over a correlation range ξ. According to general principles of diffraction this will give a diffuse peak of angular width $\sim \xi^{-1}$. Measurements give

$$\xi = \xi_0 \varepsilon^{-\nu} \tag{8}$$

where ν is another critical exponent, normally in the range 0.5 - 0.7. $\xi \to \infty$ as $T \to T_c$, which means that long-range fluctuations

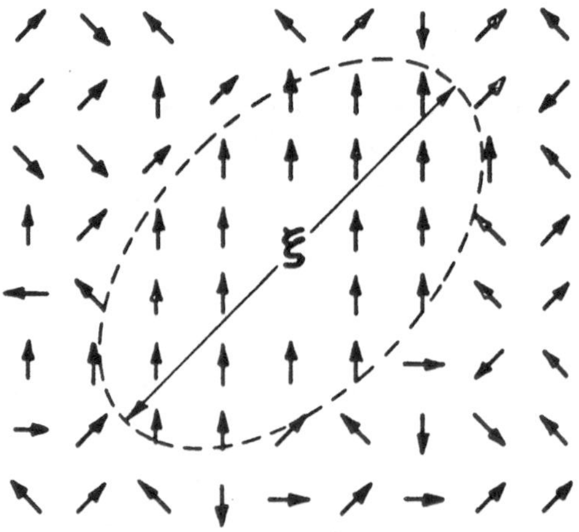

Fig. 2 Schematic picture of a local ferromagnetic region of extension (correlation range) ξ, created by fluctuations at $T \gtrsim T_c$.

easily form, since the difference in free energy of the two phases vanishes. This general behaviour implies that critical phenomena are insensitive to lattice distances and finer details of the type of interaction. Only the dimension of the lattice and the number of components of the order parameter are important, and define a set of universality classes in which the critical exponents (β,ν, etc.) fall. This is an implication of the scaling theory for critical phenomena.

Fig. 3 shows how this scattering from critical fluctuations peaks at T_c. The curves were measured at a very low-flux reactor (200 kW!), using a polyenergetic incident beam and turning the crystal a few degrees away from the setting for Bragg scattering. There exists also a method for getting rid of the fluctuation component, such as to allow measurement of the order parameter in isolation. This is through the polarized beam technique[6].

The configuration built up by an order-parameter fluctuation, as in Fig. 2, is time-dependent. Outside the critical temperature region a magnetic system has a strong preference for order ($T < T_c$) or disorder ($T > T_c$), and there are strong restoring

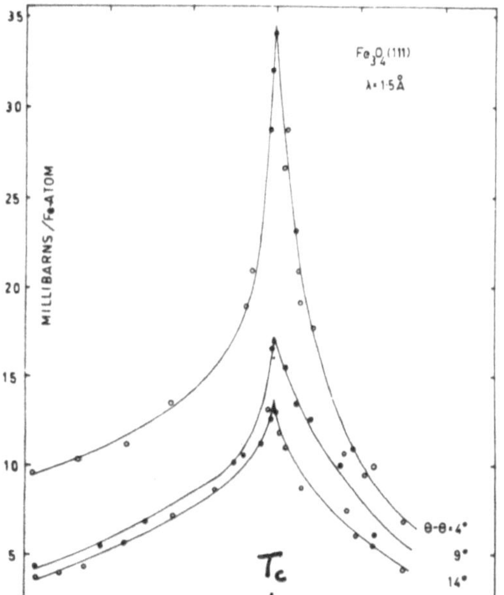

Fig. 3 Diffuse magnetic scattering as a function of temperature, as observed for different settings from the Bragg position of the crystal.

forces acting to reestablish the average state. This implies short relaxation times. In the critical region the ordered and the disordered phases are almost equally stable and the restoring forces are weak. Hence the relaxation time of the fluctuation is long and the conjugate quantity $\hbar\omega$, the inelasticity in the scattering, is weak. Experimentally and theoretically the thermodynamic relaxation time that we have just discussed is governed by

$$\tau \sim \varepsilon^{-1} \qquad (9)$$

To detect a possible inelasticity in the scattering we need an energy analyzer built into our neutron spectrometer, e.g. in the form of a third axis provided with an analyzer crystal. When studying structural phase transitions the possibility for doing energy analysis is an essential requirement. Around 1950 Anderson[7] and Cochran[8] proposed the socalled soft-mode mechanism for a displacive, structural phase transition: a phase transition occurs when the frequency of a vibrational mode tends towards zero. Approaching the transition from the high-temperature side, the "soft phonon" motion freezes into the lattice as a static displacement at the transition temperature T_c. The displacement gives rise to additional Bragg reflections for $T < T_c$, whose intensity again follows the order parameter squared. This is shown for $SrTiO_3$ in Fig. 4, which defines a $\beta = 0.34 \pm 0.01$. Again we can isolate more

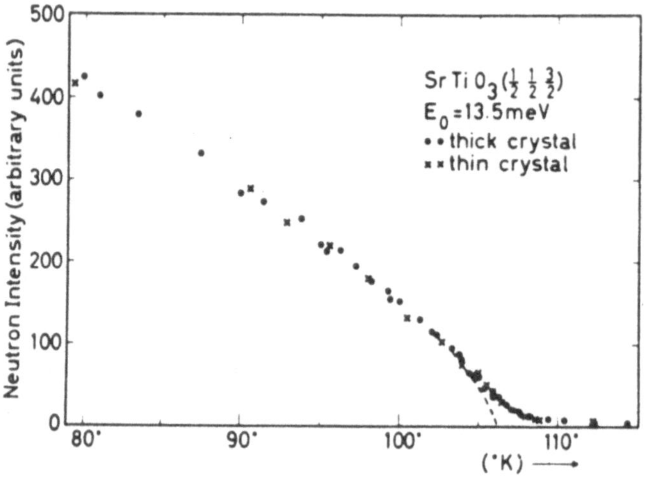

Fig. 4 Neutron intensity versus temperature for a superlattice reflection in $SrTiO_3$.

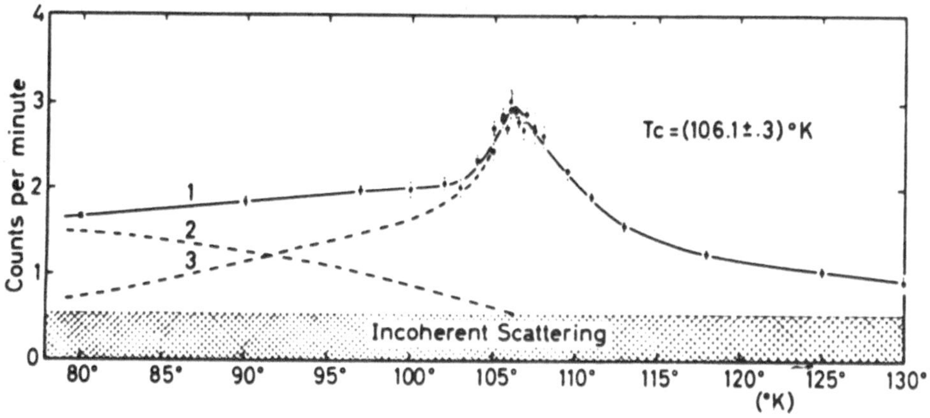

Fig. 5 Same as Fig. 4, but the crystal is rotated 0.9° away from its Bragg position. Below T_c the intensity is decomposed into residual Bragg scattering (curve 2) and diffuse scattering (curve 3).

or less the fluctuation component by rotation of the crystal, as seen in Fig. 5. This diffuse component may now in principle consist of two parts: a weakly inelastic part associated with critical fluctuations and an inelastic part due to the soft phonon. An energy analysis of the diffuse component[10] brings this out, see Fig. 6. Although there is some difficulty with such a model, in particular with the origin of the central intensity component, it illustrates our point.

In equilibrium phase transitions much of the interest is presently on multicritical phenomena[11]. A multicritical point is a point of sudden change of behaviour on an otherwise smooth phase boundary, and appears as a result of competing interactions and competing order parameters. The canonical example is that of a uniaxial antiferromagnet in a magnetic field. Depending on the strength of the field, the magnetic moments may like to be either parallel or normal to the field direction. At sufficiently high temperatures the moments are in a disordered, paramagnetic state. At one point of temperature and field the three phase boundaries meet and define a multicritical point.

2. NEUTRON SCATTERING STUDIES OF NONEQUILIBRIUM PHASE TRANSITIONS

Much of our knowledge about nonequilibrium phase transitions comes from studies of lasers[12]. If we want to use neutron scatter-

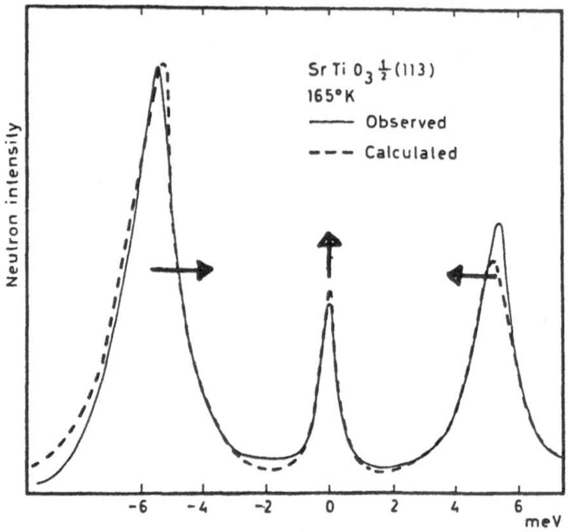

Fig. 6 Energy scan of diffuse scattering in $SrTiO_3$, showing a central component and soft phonon side bands at $T=T_c+60^0$. Arrows indicate softening of phonons and increasing central peak as $T \rightarrow T_c$.

ing we must look for other candidates, and the convective Rayleigh-Bénard instability seems to be the best choice. The order parameter here is the convective velosity, but does it couple to the neutron probe? In general not, but if we look to liquid crystals we are saved. In a nematic liquid crystal there is a coupling between molecular orientation and flow[13], and the intensity of neutron scattering depends on this orientation. Hence a liquid crystal provides us with a built-in pointer for the velocity of flow, and the neutron can read off the position of this pointer.

From our discussion above of equilibrium phase transitions we know what to search for and measure in the convective system:

the order parameter
    temperature variations, index β
    fluctuations, range ξ and relaxtion τ
scaling
soft mode

The diffraction pattern reflects the liquid structure, not the convective structure. The intensity of the diffraction maxima depends, however, on the orientation of the molecules and hence on the order parameter, as explained above.

In equilibrium systems we use neutron scattering to measure microscopic distances and times through their conjugates, i.e. momentum and energies. In a nonequilibrium convective system the relevant distances (cell structures, correlation ranges) and times (relaxation times) are macroscopic, typically ~ cm and ~ sec, respectively. These are much beyond the reach of conventional neutron scattering technique, which can be used for distances shorter than 10^{-5} cm and times shorter than 10^{-10} sec. Consequently we have to try a real-space and real-time method. The roll pattern has to be mapped through scanning with a narrow beam. The relaxation, or in general the time dependence, must be found by real-time counting of the scattered intensity.

Experimental setup

We have used three parallelepipedic sample cells with the following characteristics:

Cell no.	side walls	material bottom	top	dimension (mm)
1	Al	Al	Al	38x38x5
2	Al	Al	Al	25x25x3
3	st.steel	copper	Al	25x50x4

The first dimension given is vertical, the other two are horizontal. The cells were filled with fully deuterated para-azoxyanisole (PAA). The nematic liquid crystalline range is given as $118 < T < 135C$, but a pure specimen, as ours, may be supercooled to ~ 100C. A vertical temperature gradient is obtained by setting the difference of the power fed to electrical heating elements in the top and bottom plates of the vessel. The temperature of the bottom plate was kept constant throughout the experiment and the vertical temperature difference was monitored by thermistors or thermocouples. ΔT was found to stay constant within $\pm$ .01C for hours. A magnetic field in the range 0-1.2 kG could be applied to the sample.

The neutron wavelength used was 1.25Å, and the intensity was always recorded at $1.8Å^{-1}$, i.e. at the maximum of the first liquid diffraction peak[14]. When the molecules of nematic PAA are aligned by a magnetic field along the (horizontal) scattering vector, the bisector of the scattering angle 2θ, the intensity is a minimum. When molecules are aligned vertically, the intensity is a maximum.

In the first experiment[15a] we used sample cell No. 1 and alignment along the scattering vector. The onset of flow at the convection threshold (ΔT_c) is marked by a sudden increase of the neutron intensity. We found $\Delta T_c(H) = \Delta T_c(0)[1+(H/H_c)^2]$ with $H_c = 25G$. The family of curves obtained at different fields can be fitted fairly well to a mean-field curve with $\beta = 0.5$, as seen from Fig. 7, provided that we allow for a mixture of higher harmonics of the roll pattern at higher gradients[15b]. The critical index of the nth

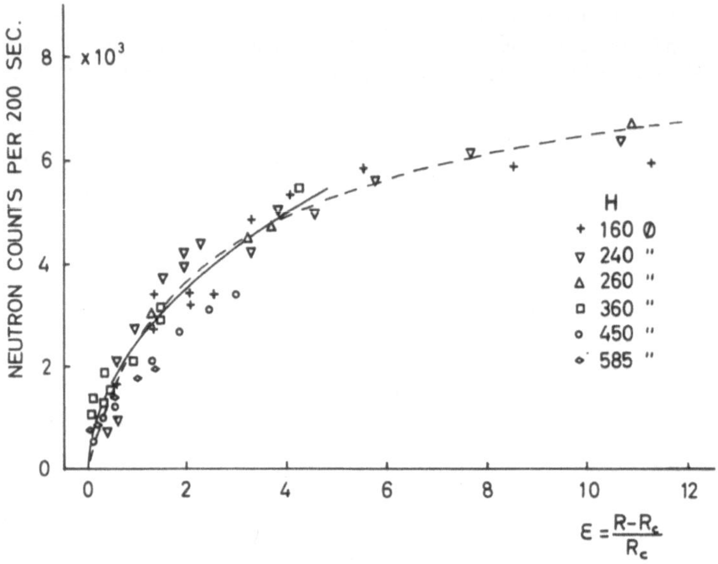

Fig. 7 Data collapse for the order parameter (i.e. neutron intensity) for the convective flow in PAA, when plotted against the reduced vertical temperature gradient. A constant high-field background has been subtracted. Broken curve calculated from mean-field theory with allowance for admixture of higher-harmonic rolls.

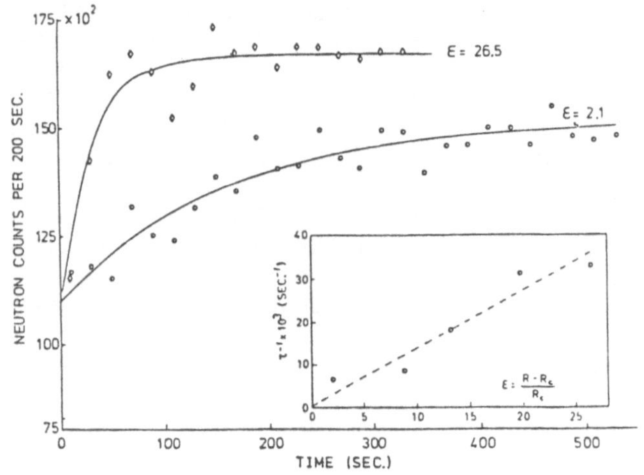

Fig. 8 After suppression of the convective flow by a strong field, the exponential build-up of the flow in a weak field (48G) is observed by neutron scattering. The inset shows the inverse time constant τ versus ε.

harmonic is n/2. This data collapse is a feature of scaling. By a sudden change of the applied field we could measure (see Fig. 8) the exponential build up of the flow, i.e. the order-parameter relaxation. The inset shows the expected slowing-down of the relaxation.

Another series of experiments[16-18] has been performed on sample cells 2 and 3, and are still in progress. The field is vertical, hence the onset of flow is marked by a sharp decrease of the intensity[14]. In this geometry one expects[19] the steady flow regime to be preceded by one of time-dependent flow. This is indeed borne out by neutron measurements of the order parameter on cell No. 3. Fig. 9 shows first a continuous transition to a phase extending from about 2.5 to 7.5 degrees in temperature difference and another phase above the latter value. There is hysteretic behaviour in the transition between the two phases. The upper stability limit of the lower phase is marked by the 1.2 kG-curve. The lower stability limit of the upper phase is marked by the broken 0.6 kG-curve. The 1.2 kG-curve is a common curve observed for H > 1.1 kG, for which the lower phase is found to be extinguished. For H < 1.1 kG a lower phase is observed and it is multistable: The order parameter may reside in a series of levels, as indicated in the figure. The number of levels is difficult to assess with certainty. The deepest levels are most stable, i.e. we may have to wait for days before a jump to higher levels is seen. The jumping frequency may be increased by additional variation of H.

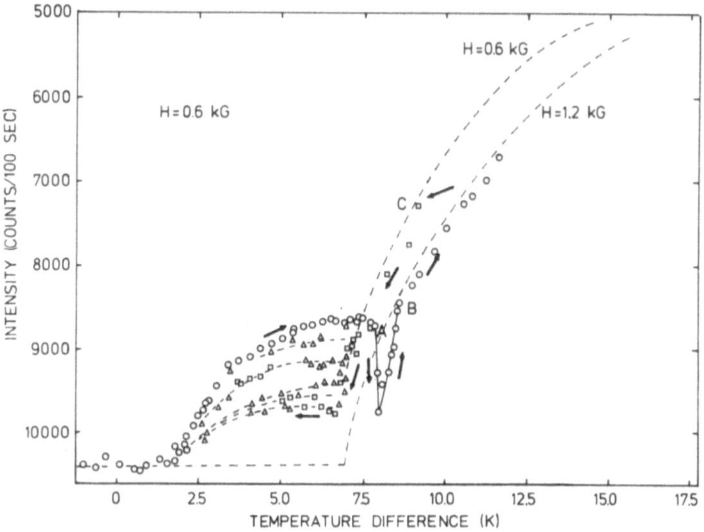

Fig. 9 Neutron intensity versus temprature difference (ΔT) for PAA in a vertical field H = 0.6 kG. Circular and square points are for a complete cycle of increasing and decreasing ΔT, respectively. Triangular points are measured after additional variation of H. Notice intensity dip between A and B.

Long-time observations at a fixed gradient setting in the upper part of the time-dependent regime indicate that the jumping sequence is from bottom to top through consecutive levels, and back through the reverse sequence. At each level faster intensity fluctuations may or may not be seen. They sometimes have a quasi-periodic character of period $\lesssim$ 1 hr. The whole time behaviour may thus be characterized by a slow, strange attractor (time constant ~ 10 hrs) on top of a faster strange attractor (time constant ~ 1 hr). A phase portrait of the "fast" attractor is shown in Fig. 10, which was observed for cell No. 2.

The multistability observed in the time-dependent regime possibly has some connection with problems that are currently being discussed: that of the number of modes governing two-dimensional viscous flow[20], and the distribution of wave vectors present in a roll pattern[21]. In the near future we shall trace out the roll structures corresponding to the different levels by using a pinhole collimation and scanning across the sample cell, as we once did for cell No. 1[15a].

The phase diagram deduced from measurements at a number of fields is shown in Fig. 11. The phase lines meet in a bicritical point.

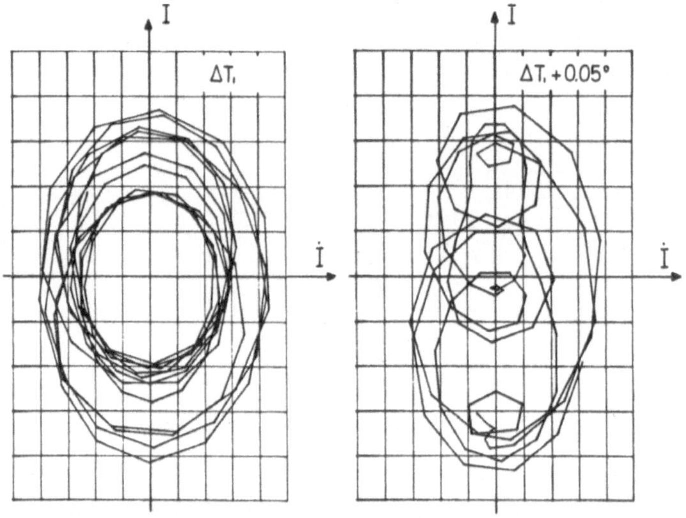

Fig. 10 Phase portrait ($\dot{I}$,I) deduced from data in the oscillatory regime, showing stronger nonlinearity for increasing ΔT.

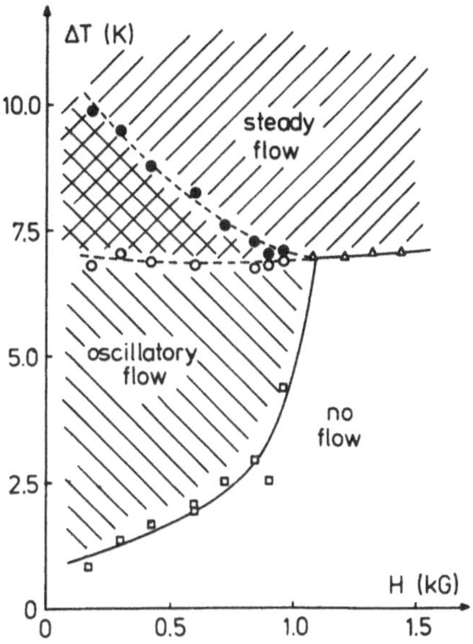

Fig. 11 Phase diagram showing confluence of first-order (broken) lines and second-order (full-drawn) lines at a bicritical point. First-order lines are upper and lower stability limits of oscillatory and steady flow regimes, respectively.

Theoretical work on nonequilibrium macroscopic systems invariably lend support to mean-field theory, implying classical exponents and negligible influence of fluctuations. The neighbourhood of the bicritical point of Fig. 11 is not sufficiently accurate that we can distinguish possible fluctuation effects. It is striking, however, that the curves of the static order parameter versus ΔT, in the vicinity of the first bifurcation, have increasing slope as H decreases. By computer fitting our experiments give for the critical exponent β for the oscillatory regime: $0.35 \pm .01$ at $H = 0.3$ kG and $0.52 \pm .02$ at $H = 0.6$ kG. For the steady regime we get $\beta = 0.55 \pm .02$ at $H = 1.2$ kG.

In some earlier work[16] we have tentatively ascribed an intensity maximum near the first bifurcation at low fields to critical fluctuations, and given heuristic arguments for its existence. We do not exclude, however, that these indications of nonclassical behaviour are artifacts of the experiment, connected with multistability of the convection pattern. In equilibrium phase transitions we associated the intensity maximum with an increasing correlation length (ξ) of the fluctuations as $T \to T_c$. One could in principle measure ξ by neutron scattering in a convection experiment, as has been done by light scattering[22]. The method consists of a real-space mapping of the spatial decay of the convective flow into a neighbouring, but shallower, portion of the sample cell.

In equilibrium, continuous phase transitions there is proportionality between the soft mode frequency and the order parameter. For cell No. 2 the quasiperiodic regime was sufficiently wide that we could make a plot[16] that possibly supports this relation.

To conclude then, we have shown that the convective instability in a nematic liquid crystal is a nonequilibrium phase transition with many of the same features observed in equilibrium phase transitions:

an order parameter with conventional values of the critical exponents, with thermodynamic slowing-down of its relaxation and with scaling properties. With proper choice of an applied magnetic field a multicritical phase diagram could be observed. The main difference from an equilibrium system is the existence of multi-stability and time dependence. In a quasiperiodic regime the fundamental mode possibly has a soft-mode character. On a more speculative side an intensity maximum for cell No. 2 is assigned to critical fluctuations.

REFERENCES

1. For a useful introduction, see G.E. Bacon: Neutron Diffraction, Oxford University Press (1975)
2. H.E. Stanley: Phase Transitions and Critical Phenomena, Oxford University Press (1971)
3. To simplify our formuale, we tacitly assume a system of 1D-Ising spins. For complete formulae, see W. Marshall and S.W. Lovesey: Theory of Thermal Neutron Scattering, Clarendon Press, Oxford (1971)
4. T. Riste (unpublished)
5. T. Riste, J. Phys. Chem. Solids 17 (1961) 308
6. O. Steinsvoll and T. Riste, J. Magn. & Magn. Mater. 14 (1979) 187
7. P.W. Anderson, Fizika dielektrikov (ed. G.I. Skanavi) Akad. Nauk SSR, Moscow (1960) p. 290
8. W. Cochran, Phys. Rev. Lett. 3 (1959) 412
9. T. Riste, E.J. Samuelsen, K. Otnes and J. Feder, Sol.St. Comm. 9 (1971) 1455
10. S.M. Shapiro, J.D. Axe, G. Shirane and T. Riste, Phys. Rev. B6 (1972) 4332
11. Multicritical Phenomena (ed. R. Pynn), Proc. of NATO ASI, Geilo, Norway, April 1983, Plenum (in press)
12. V. Degiorgio in Nonlinear Phenomena at Phase Transitions and Instabilities (ed. T. Riste) Plenum (1982) p. 181
13. P.G. de Gennes: The Physics of Liquid Crystals, Clarendon Press, Oxford (1974)
14. R. Pynn, K. Otnes and T. Riste, Sol. St. Comm. 11 (1972) 1365 M. Kohli, K. Otnes, R. Pynn and T. Riste, Z. Phys. B 24 (1976) 147

15a T. Riste, K. Otnes and H.B. Møller, Neutron Inelastic scattering 1977, International Atomic Energy AGency, Vienna (1978), Vol. I, p. 511
15b O. Kvernvold, Int. J. Heat & Mass Transfer 22 (1979) 395
16. T. Riste and K. Otnes, Proc. Yamada Conf. Sept. 1982, Physica 120B (1983) 376
17. K. Otnes and T. Riste, Proc. EPS General Conf. on Condensed Matter Physics, April 1983, Helv. Phys. Acta 56 (1983) 837
18. T. Riste and K. Otnes in ref. 11
19. H.N.W. Lekkerkerker, J. Phys. Lett. (1977) L 277. Se also ref. 18 for other references, and M.G. Velarde (this volume)
20. C. Foias, O.P. Manley, R. Temam and Y.M. Treve, Phys. Rev. Lett. 50 (1983) 1031
21. P. Manneville, Phys. Lett. 95A (1983) 463
22. J.C. Wesfreid, Y. Pomeau, M. Dubois, C. Normand and P. Bergé, J. Phys. (Paris) 39, (1978) 725

ELECTROTHERMAL INSTABILITIES IN DIELECTRIC LIQUIDS

A. Castellanos[(1)], P. Atten[(2)] and M.G. Velarde[(3)]

(1) Facultad de Físicas, Sevilla, Spain
(2) Labo Electrostatique, C.N.R.S., Grenoble, France
(3) Física Fundamental , U.N.E.D., Madrid , Spain

1. INTRODUCTION

Electrohydrodynamics (EHD) studies the interplay of mechanical and electrical forces in fluids |1-4|. In a first approximation it is assumed that electrical currents are very weak and therefore magnetic effects are negligible. Maxwell's equations are then reduced to Gauss' law and the charge conservation law. If a thermal gradient is added to the system interesting new phenomena appear, which may completely modify the previously existing EHD effects. In fact, the combined action of an electric field and a thermal gradient have long been studied with the aim of enhancing the heat transfer between metallic boundaries |5-7|.

The largely varied physical circumstances, where electrical forces with or without thermal gradients govern the dynamics of the system under consideration make for the growing importance of EHD in the physics and technological fields.

We shall consider the simplest possible EDH system, while still exhibiting all the above mentioned effects, i.e., a horizontal liquid layer of depth h , confined between two metallic electrodes at different electrical potentials and temperatures, and subjected to an injection of charges from either of them (unipolar injection) . Initially, the layer is at rest and under increasing thermal and electrical constraints it will exhibit transitions to more and more complex spatio-temporal patterns of convection . We shall concentrate here on i) the physical mechanisms responsible for the onset of convection and ii) the nature, supercritical or subcritical, of this transition (for related problems see References |8-11|).

2. ELECTROTHERMAL EQUATIONS

We consider a horizontal dielectric fluid layer, depth h, bounded by two plane-parallel metallic electrodes at electric potentials ϕ_0 and ϕ_1, respectively. θ_0 and θ_1 define their temperature.

The electrothermohydrodynamic equations of the problem are

$$\nabla \cdot (\varepsilon \vec{E}) = q \tag{1}$$

$$\frac{d}{dt} q + \nabla \cdot (q K \vec{E}) = 0 \tag{2}$$

$$\nabla \cdot \vec{v} = 0 \tag{3}$$

$$\rho \frac{d \vec{v}}{d t} = - \nabla p + \eta \nabla^2 \vec{v} + \rho g i_z + q \vec{E} - \tfrac{1}{2} E^2 \nabla \varepsilon + \nabla \{ \tfrac{1}{2} E^2 (\partial \varepsilon / \partial \rho) \} \tag{4}$$

$$\frac{d \theta}{d t} = \kappa \nabla^2 \theta \tag{5}$$

where ε is the dielectric constant, $\vec{E}$ the electric field, $\vec{v}$ the velocity field, p the pressure, ρ the mass density of the fluid, θ the temperature, K the ionic mobility, κ the thermal diffusivity (thermometric conductivity), i_z the unit vector along the vertical z-axis, η the dynamic viscosity and q the electric charge.

Equations (1)-(5) are closed using the following equations of state

$$\rho = \rho_0 \{ 1 - \alpha (T - T_0) \} \tag{6.a}$$

$$\varepsilon = \varepsilon_0 \{ 1 - e_1 (T - T_0) \} \tag{6.b}$$

$$K = K_0 \{ 1 + k_1 \ (T - T_0)\} \tag{6.c}$$

where the subscript "0" denotes a reference value generally taken at $z = 0$. α , e_1 and k_1 account for the corresponding first derivatives of ρ, ε and K with respect to temperature.

The expression for the so-called electric force, which acts on the dielectric liquid is

$$\vec{f}^e = q \vec{E} - \tfrac{1}{2} E^2 \nabla \varepsilon + \nabla [\tfrac{1}{2} \rho \ E^2 (\partial \varepsilon / \partial \rho)_T]$$

The first term is the ordinary Coulomb force per unit volume, where q is the free charge density and $\vec{E}$ is the electric field. The second term represents a force which will appear whenever an inhomogeneous dielectric constant ε is in an electric field. The third term, called the electrostictive force, is the gradient of a scalar and could be included in pressure, and therefore it does not seem to have any influence on the dynamics.

3. THE ONSET OF STEADY CONVECTION IN A SIMPLE APPROACH

An exact solution of the above posed differential problem is at present beyond our possibilities. The major difficulty rests on the nonlinear inertial and coupling terms in the various equations . Thus we shall try to account in an approximate way for all phenomena of interest . Firstly, we assume the existence of a convective cell where to a first-approximation the ascending portion of fluid, cross-section S_1, moves at an average speed V_1 . The descending flow, cross-section S_2 , has velocity V_2. Secondly, the two portions of the convective cell are interpreted as parts of a hydraulic circuit. The liquid flows as a consequence of a pressure gradient originated in the electrical and buoyant forces. We have

$$f = \rho g + q E - \tfrac{1}{2} E^2 \frac{d}{dz} \varepsilon = f \ (z, \vec{v}) \tag{7}$$

Thus the average pressure induced in a column of fluid of height d is

$$P_m = \int_0^d \{ f(z, + v) - f(z, - v)\} \, dz \tag{8}$$

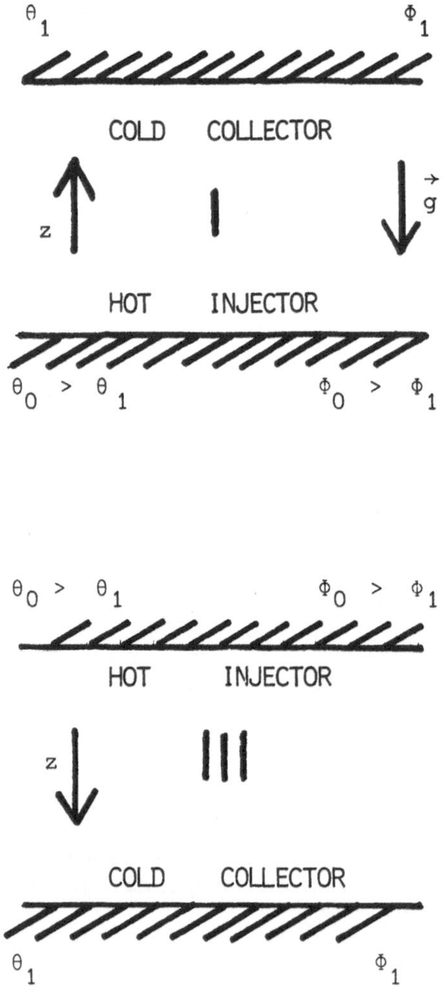

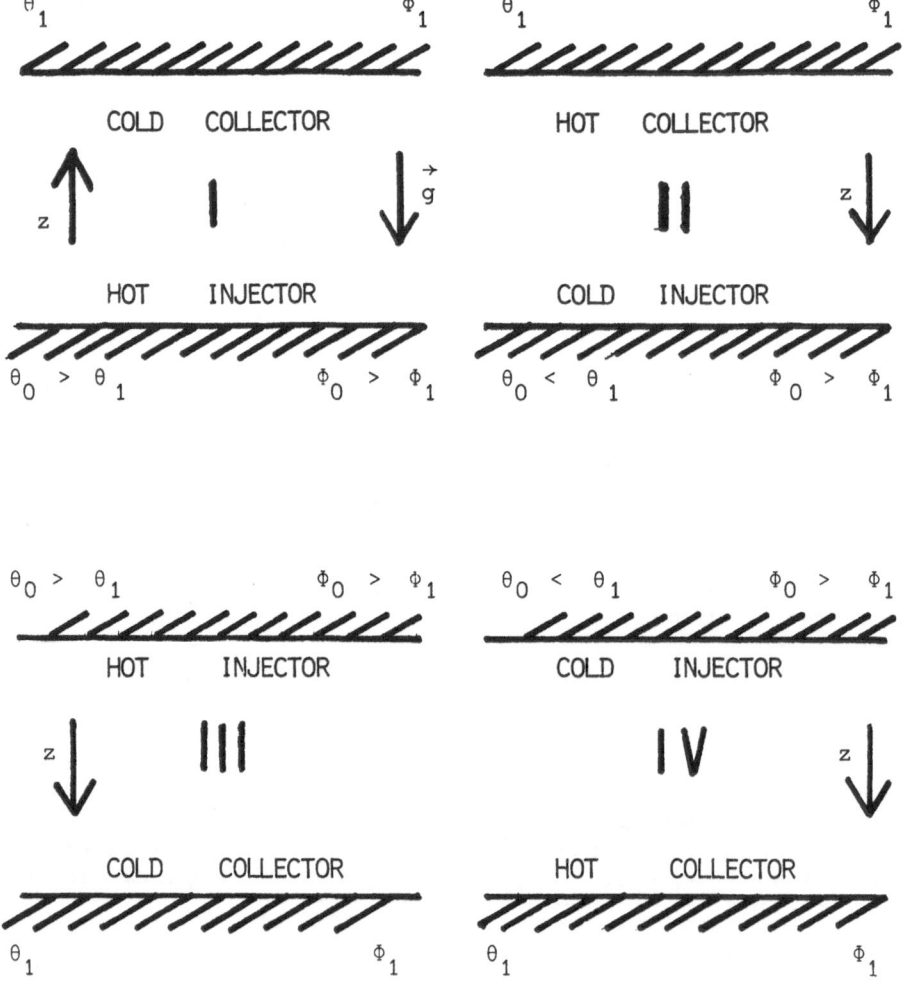

Fig.1. The four possible cases: (I) heating and injecting from below, (II) heating from above and injecting from below, (III) heating and injecting from above, and (IV) heating from below and injecting from above.

There is, however, a viscous resistance, P_r. For a cross-section of radius R, Poiseuille's law gives

$$P_r = \gamma \eta h/R^2 \tag{9}$$

where γ is a numerical constant that dependes on geometry (usually $\gamma \sim 10$). Then the net resulting pressure drop in the column is $(P_m - P_r)$. At the motionless steady state we have the equilibrium condition.

$$P_m = P_r \tag{10}$$

which yields the neutral stability locus. Convection sets in when $P_m > P_r$.

Let us now evaluate the two relevant quantities P_m and P_r for various specific cases.

4. CONVECTIVE FLOW AND THE PHASE TRANSITION PICTURE

i. Weak unipolar injection

To a first approximation the electric field, $\vec{E}$, can be obtained from Equations (1)-(2) with v = constant, $\varepsilon = \varepsilon_0$ and $K = K_0$. Thus we have

$$E(z,v) = E_0 \left\{ (1 + v/K_0 E_0)^{1/2} \left(1 + \frac{2 q_0 z}{\varepsilon_0 E_0} + \frac{v}{K_0 E_0}\right)^{1/2} - \frac{v}{K_0 E_0} \right\} \tag{11}$$

where $E_0 = E(0,v)$ is fixed by the condition $\phi_0 - \phi_1 = \int_0^h E(z,v)dz$ we can always set $E(z,v) \equiv E(z,0) + \delta E(z,v)$ which defines δE.

Then the Coulombian contribution of (11) to P_m is

$$P_m = \int_0^h E \{ \delta q(z, v) - \delta q(z, -v) \} dz \tag{12}$$

where for weak injection we have used the approximation $\delta f = E\,\delta q + q\,\delta E \simeq E\,\delta q$. Using now Equations (9),(10), (12), we have

$$\gamma\,\eta\, h/R^2 = \varepsilon_0 E_0^2 (q_0/\varepsilon_0 E_0)^2 h^2/K_0 E_0 + \qquad (13)$$

$$+ \varepsilon_0 E_0^2 (q_0/\varepsilon_0 E_0)^2 h^2 v^2/(K_0 E_0)^3$$

This equation can be written in a more suitable form by introducing the velocity scale (ν/h). We have

$$\gamma\,(h/R^2) = T\,C^2 + u^2 M^4 C^2/T \qquad (14)$$

where $u = V\,h/K$. $T = \varepsilon_0 (\phi_0 - \phi_1)/K_0\,\eta$ is a dimensionless group that accounts for the electrical constraint at the horizontal plates. The quantity $C = q_0 h^2/\varepsilon_0(\phi_0 - \phi_1)$ accounts for the strength of the unipolar injection . $M = (\varepsilon_0/\rho_0)^{1/2}/K_0$ relates the hydrodynamic and ionic mobilities. $\nu = \eta/\rho_0$ is the kinematic viscosity of the fluid and $Pr = \nu/\kappa$ is the Prandtl number.

The limit case $u = 0$ corresponds to the linear approximation. At vanishing u we have

$$T\,C^2 = \gamma\,(h/R)^2 \qquad (15)$$

Thus for $T\,C^2 > \gamma\,(h/R)^2$ the motionless steady state obtained with (10) is unstable. Actually, for a convective cell such that $R \sim h/4$ we find $T\,C^2 \simeq 160$ (the exact value is 220)

Another consequence of Eq (14) is that instability and thus transition from the motionless state to convective motion is possible for lower values of $T\,C^2$. Indeed the coefficient of u^2 in Equation (14) is positive definite and non-vanishing values of u exist below the threshold given by the linear theory. We have an inverse bifurcation with subcritical instability, metastability and hysteretical phenomena. This result also agrees well with a direct numerical integration of the Equations (1) - (6) and with the experimental results.

ii. Bénard convection

As a by-product or the hydraulic model calculation we obtain known results for thermal convection. The equilibrium condition (10) yields

$$Ra \ (1 - u^2 \, Pr^2 /60) = 6 \gamma (h/R)^2 \approx 10^3 \qquad (16)$$

This value is somewhat low due to the neglect of lateral diffusion effects. The exact result is 1708 for rigid boundaries with $Ra = \rho_0 \; g \alpha (\lambda g \beta) h^4 / \kappa \eta$.

iii. Weak unipolar injection coupled to Bénard convection

In this case the electrical and thermal effects are coupled together through the Navier-Stokes equations. A procedure similar to the previously developped scheme for E or T yields

$$(T/T_c + Ra/Ra_c) + (M^4/T - Ra \, Pr^2/60) u^2 = 1 \qquad (17)$$

Thus in the (T, Ra) plane we have instability whenever $(T/T_c + Ra/Ra_c) > 1$. Note that when heating from below $(Ra > 0)$ the thermal gradient is indeed destabilizing whereas the electrical constraint is always destabilizing. It also appears that when the layer is heated from above we can trigger subcritical (finite amplitude) instabilities, i.e., an inverted bifurcation with hysteretic phenomena. An inverted bifurcation is also possible when heating the layer from below provided $T \; C^2$ is large enough.

iv) $\varepsilon(T)$ effects in the absence of injection

In this case the electric field is given by $\frac{d}{dz}(\varepsilon E) = 0$. We get

$$E(z,v) = (\phi_0 - \phi_1)/ \varepsilon \int_0^h \varepsilon^{-1} \, dz \qquad (18)$$

with $\varepsilon(z,v) = \varepsilon_0 \{ 1 - e_1 [\theta(z,v) - \theta_0] \}$. Then to a first-order approximation the dielectric forces provide

$$P_m = - \tfrac{1}{2} \int_0^h E^2(z,0) \frac{d}{dz} [\delta \varepsilon (z,v) - \delta \varepsilon (z,-v)] \, dz \qquad (19)$$

If, however, we also have a temperature gradient a straightforward analysis gives the locus of neutral stability

$$Ra_c = (Ra + L^2 Ra^2 \; Pr \; Sa) \; [\; 1 - Pr \;\; u \, / \, 60 \;] \tag{20}$$

with $L = |e_1 \, \nu\kappa/\alpha \, g \, h^3|$ and $Sa = \rho_0 \, \varepsilon_0 \, (\phi_0 - \phi_1)^2 \, \eta^{-2}$. L accounts for the temperature-dependent permitivity effects and Sa is a measure of the electrical potential in the absence of injection. Thus the effect of a non-constant $\varepsilon(t)$ is always destabilizing. We have an unstable stratification with respect to the electrical forces: in the regions where ε grows the field intensity disminishes. Moreover, as the coefficient of U^2 in Equation (20) is always negative for a layer heated from below we expect a direct bifurcation, i . e ., a supercritical and continuous transition.

v) The two, $\varepsilon(T)$ and $K(T)$, effects with weak unipolar injection and thermal gradient.

The last case in our study corresponds to the competition or cooperation of the various effects discussed earlier. In this case the field is given by

$$(KE + v) \frac{d}{dz}(\varepsilon \; E) = (K_0 \, E_0 + v) q_0 \tag{21}$$

Define $E = E_0 \; E_r$, $\varepsilon = \varepsilon_0 \; \varepsilon_r$ and $K = K_0 \; K_r$. The solution of (21) is

$$E_r = \varepsilon_r^{-1} \{ 1 + 2 \, v/K_0E_0 + (1 + V_0/K_0E_0)(2q_0/\varepsilon_0E_0)(z + \int_0^h \varepsilon_r \, dz/K_r) +$$

$$+ (2v/K_0E_0)\varepsilon_r/r + (v/K_0E_0)(\varepsilon_r/K_r) |^{1/2} - (v/K_0E_0)\varepsilon_r/K_r \}$$

(22)

Then a straightforward, albeit lengthy calculation, yields the following neutral stability locus

$$[\,(1 + S)\, T/T_c + Ra/Ra_c\,] + \{\, (M^4/T)\,[\, C + \lambda \, g \, (L + N) Ra\,]^2$$

$$- Ra\ Pr^2/60\ Ra_c - S\ Pr^2/60\ T_c \} u^2 = 1 . \qquad (23)$$

where $S = [L^2 Ra + \lambda g\ (2L + N)]\ T\ Ra\ Pr/6\ M^2$, $N = k_1\ \nu\kappa/\alpha|g|\ h$ and $\lambda g = g/|g|$. Note that $\varepsilon(T)$ and $K(T)$ combine together through the factor $(2L + N)$ which agrees well with results obtained from a more accurate analysis. On the other hand at vanishing u we reobtain the linear stability predictions corresponding to the case of injection and heating from below, respectively, $\lambda g = -1$ and positive Ra.

5. OVERSTABILITY

In the preceding sections we discussed the onset of convective instability in the form of steady motion. The heuristic argument leading to the possible existence of overstable motion goes as follows. Consider a fluid parcel displaced upwards from its original equilibrium position with velocity, w. Because of the velocity w the parcel gets an increase in charge δq_w, that tends to push it further away from its original equilibrium position. As soon as the fluid parcel moves up it enters a region of lower relative temperature where the parcel stays hotter than its surrounding, due to the time scale set by the heat diffusion constant of the liquid. In this time scale the fluid parcel experiences a decrease in charge δq_θ that for $S > 1$ yields a net charge variation of $(\delta q_w + \delta q_\theta) < 0$. Such a negative charge fluctuation induces a restoring force that pushes down the fluid parcel, leading to an overshooting of its original equilibrium position. When the parcel moves down with a velocity $-w$ this velocity causes a decrease in charge $\delta q_{-w} = \delta q_w$ while at the same time the parcel enters now regions that have higher relative temperature. Once more the local temperature inhomogeneity causes a variation in the parcel's charge density which in this case is negative $\delta q_{-\theta} = - \delta q_\theta$. This again brings a restoring force that pushes the particle back to its original equilibrium position with another overshooting in the time scale of the heat diffusivity. It is clear that an oscillatory motion can be sustained provided the Prandtl number is high enough. Thus the predicted oscillation in the fluid is due to the electrical forces involved in the problem and cannot be linked to any inertial effects in the liquid. A rough estimate of the overstability frequency can be obtained as follows. Let us first consider the case $L = 0$ and $N \neq 0$, in the absence of buoyancy forces. Then the equation of motion of the fluid parcel is

$$\partial w/\partial t = (\eta/\rho)\partial^2 w/\partial z^2 + (\delta q_{\bar{w}} + \delta q_0)E_s/\rho \tag{24}$$

$$= -\nu a^2 w - (dq_s/dz)w/\rho a k_s - q_s k_1 E_s \delta\theta$$

where we have assumed that $E\delta q_w$ varies in phase with w. This assumption amounts to neglecting the ionic transit time, τ_i, relative to the period of the oscillatory motion. Indeed, we have $\tau_i/\tau_\nu = 1/R \simeq 10^{-3}$.

$$\partial w/\partial t = [-\nu a^2 - (dq_s/dz)/\rho a K_s] w + q_s k_1 E_s \beta z/\rho \tag{25}$$

Eq. (25) is a standard equation for a oscillator damped or amplified in accordance to the sign of the damping term in w. Note that the viscous effect $(-\nu a^2)$ competes with the direct velocity effect on charge fluctuation $\delta q_{\bar{w}}$. Thus for positive and high enough values of $(-dq_s/dz)$ an oscillation is expected to be amplified when dq_s/dz decreases and becomes smaller with respect to the viscous damping. Then for an oscillation to be sustained we must have a larger electric field. Conversely when dq_s/dz becomes positive ($NRa > C$ implies $dq_s/dz > 0$), we expect that the system cannot exhibit sustained oscillations and the static equilibrium would be asymptotically stable. On the other hand at the onset of overstability as the electric force component $E_s \delta q_{\bar{w}}$ must compensate for the viscous damping we expect that the threshold will be about the value for steady instability. The critical frequency of oscillation is determined by the magnitude of the indirect charge perturbation effect $E_s \delta q_\theta$ which gives

$$\omega = [q_s E_s N Ra/\rho d]^{1/2} \tag{26}$$

For the general case $N \neq 0$, $L \neq o$ the expected frequency is

$$\omega = [q_s E(N + 2L) Ra/\rho d]^{1/2} \tag{27}$$

ACKNOWLEDGMENTS

This research has been sponsored by the Stiftung Volkswagenwerk. The collaboration with Dr. P. Atten was possible thanks to the France-Spain scientific exchange agreement.

REFERENCES

1. N. Felici, Révue Gén. Electricité *78* (1969), 717-734.
2. W. V. R. Malkus and G. Veronis, Phys. Fluids *4* (1961) 13 and references therein.
3. P. Atten and J.C. Lacroix, J. Mécan. *18* (1979) 469.
4. W. J. Worraker and A.T. Richardson, J. Fluid Mech *93* (1979) 29, *109* (1981) 217
5. T. B. Jones, Adv. Heat Transfer *14* (1978) 107
6. A. Castellanos and M.G. Velarde, Phys. Fluids *24* (1981),
7. A. Castellanos, P. Atten and M.G. Velarde, Phys. Fluids, *27* (1984) 1606.
8. A. Castellanos, P. Atten and M.G. Velarde, J. Non-Equilib. Thermodyn., to be published.
9. C. Normand, Y. Pomeau and M.G. Velarde, Rev. Mod. Phys. *49* (1977) 581-624.
10. M. G. Velarde and C. Normand, Convection, Sci. American *243* (1980) 92-108.
11. M. G. Velarde in NONLINEAR PHENOMENA AT PHASE TRANSITIONS AND INSTABILITIES (T. Riste, Ed.) Plenum Press, New York (1982) 205-247.

ELECTROHYDRODYNAMIC INSTABILITIES IN NEMATICS WITH HOMEOTROPIC BOUNDARIES

J. Grupp

Physikalisches Institut
Westfälische Wilhelms-Universität
D-4400 Münster, West Germany

INTRODUCTION

In experiments on thin films of nematic liquid crystals (NLC) two typical classical alignments of the director field are often used: vertical (homeotropic) or parallel (planar) to the boundaries. In this brief report we deal with experiments that are performed on thin homeotropic films of the nematic MBBA [N-(p-methoxybenzylidene)-p-butylaniline] having a negative dielectric anisotropy ($\Delta\varepsilon = \varepsilon_{\parallel} - \varepsilon_{\perp} < 0$).

EHD CONVECTION FLOW IN SAMPLES WITH HOMEOTROPIC BOUNDARIES

When an applied voltage exceeds a threshold value U_{th} the Freederickzs /1/ transition occurs and the director field is deformed. In fig. 1a, this is schematically shown for an applied voltage of $U = 2U_{th}$. When the voltage is increased to a level $U \geqslant U_c$ the EHD convection flow appears in such a deformed director field. In fig. 2, this flow is presented by arrows and the director field resulting from the balance of the electric, elastic and viscous torques is qualitatively shown. Obviously the director field of both convection rolls is asymmetric in contrast to the case of planar boundaries. So the values of the $\hat{c}$ vectors (definition in fig. 1b) in the boundary zones of the left convection roll are smaller than in the boundary zones of the right one. This means a higher effective viscosity η in the left boundary zone: $\eta_l > \eta_r$. The forces on the boundaries exerted by the flow are $f_l = \eta_l \cdot v_{x,z}$ and $f_r = \eta_r \cdot v_{x,z}$ per unit area, where $v_{x,z}$ denotes the spatial derivative of v_x (fig. 3) with respect to the z-axis. η and $v_{x,z}$ are the values averaged over the corresponding boundary zones. Due

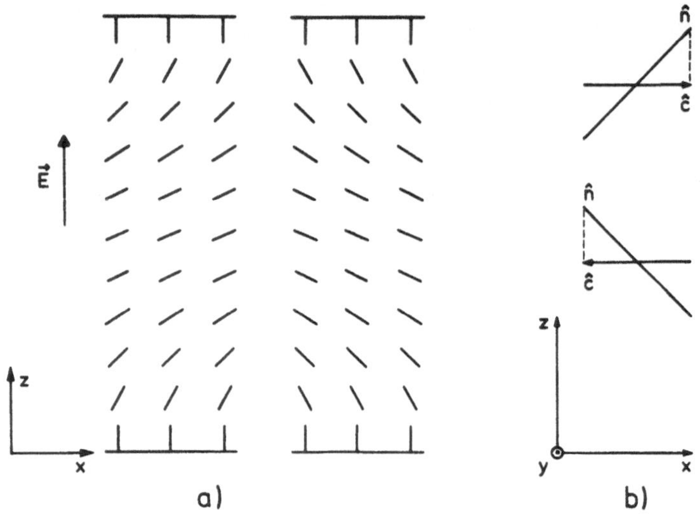

Fig. 1: a) Director fields for an applied voltage $U = 2\,U_{th}$. The external electric field $\vec{E}$ is parallel to the z-axis. Both the deformations can occur if the director is restricted to the z-x-plane.
b) Definition of the $\hat{c}$ vector: projection of the director $\hat{n}$ on to the x-y-plane as shown in two examples.

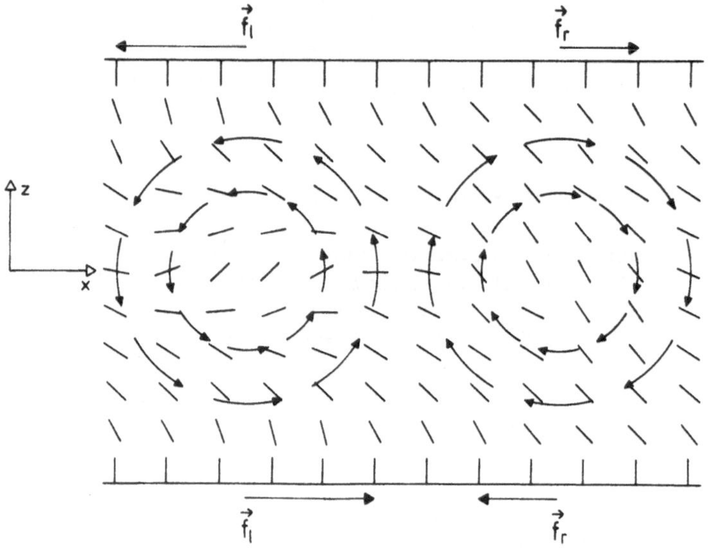

Fig. 2: Qualitative presentation of the director field in the presence of EHD convection flow. The surface forces per unit area that are exerted by the convection flow are shown for the left (l) and the right (r) convection roll.

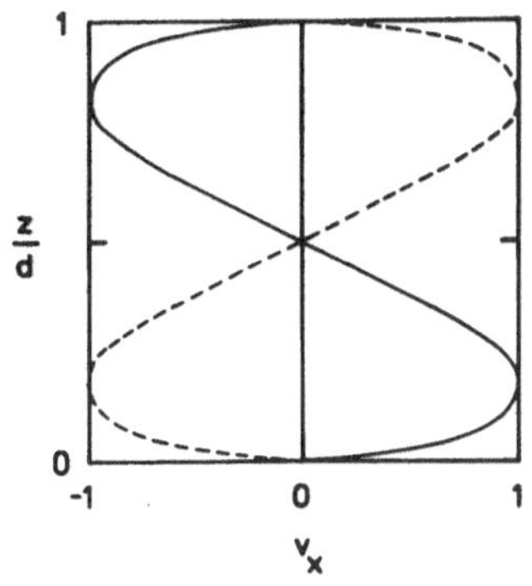

Fig. 3: v_x in arbitrary units is qualitatively shown as function of the reduced coordinate z/d (d ≡ thickness of the sample). The full curve refers to the x-position of the center of the left and the dotted curve to that of the right convection roll. Both the velocity fields are assumed to be symmetric in a first approximation.

to the differences in the effective viscosities the left convection cell exerts a larger surface force. If the top plate is borne without friction, this plate is shifted to the left.

In samples with homeotropic boundaries, domains with different directions of the $\hat{c}$ vector normally appear, when the Freederickzs transition occurs. Using external constraints (e.g. a magnetic field) a single direction of $\hat{c}$ can be obtained. In such uniformly deformed samples surface forces can be detected when EHD instabilities appear /2/.

EXPERIMENTS AND RESULTS

In our experiments the NLC [(d) in fig. 4a] is placed between two horizontal circular glass discs /3/. The upper disc (c) is suspended with rotational symmetry by a thin quartz thread (a). A uniform tangential direction of the $\hat{c}$ vector can be given in the sample e.g. by an appropriate torsion shear (fig. 4b). Applying a voltage $U \geqslant U_c$ to such a sample the convection rolls appear with a preferred radial orientation /2/. The surface forces exert a torque that causes a rotation of the suspended disc.

In the experiments, periodic torsion pendulum oscillations of the suspended disc are observed, when an a.c. voltage $U \geqslant U_c$ (conducting regime /4/) is applied /5/. These oscillations are presented in a two dimensional phase space: ϕ as function of $\dot{\phi}(\equiv d\phi/dt)$. The limit cycles shown are reached after some initial cycles, when the voltage is applied.

The angular motion of the suspended disc can be expressed by the following equation of torques, neglecting inertial effects:

$$k\dot{\phi} + D\phi = M, \qquad (1)$$

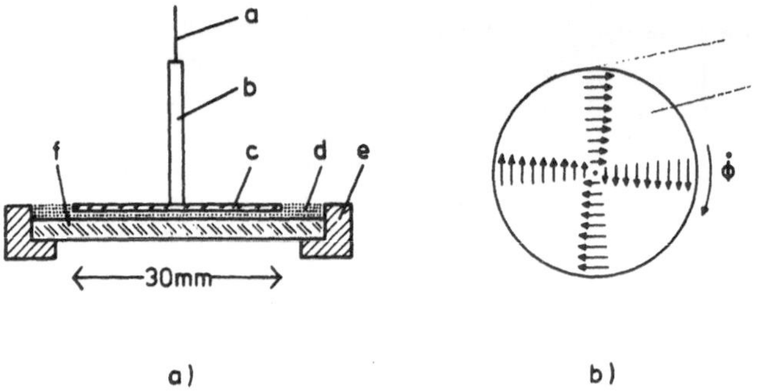

Fig. 4: a) Schematic presentation of the shear cell: (a) quartz thread, (b) pin, (c) suspended glass disc, (d) NLC and (e) bearing of (f) the lower glass disc.
b) ĉ vector field of the sheared sample schematically shown for the midplane of the sample (top view).

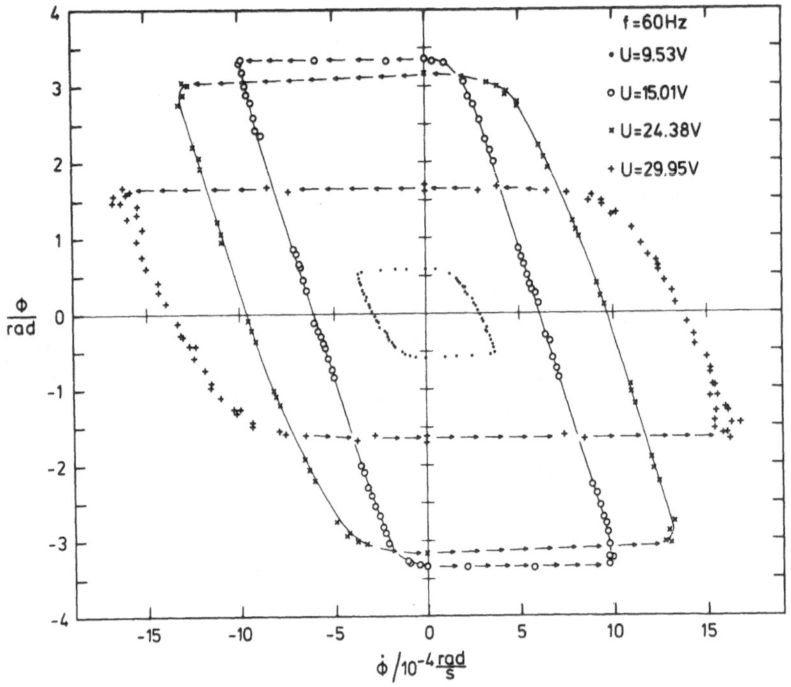

Fig. 5: Limit cycles in the two dimensional phase space for different applied a.c. voltages (f = 60 Hz, conducting regime): ϕ as function of $\dot{\phi}$.

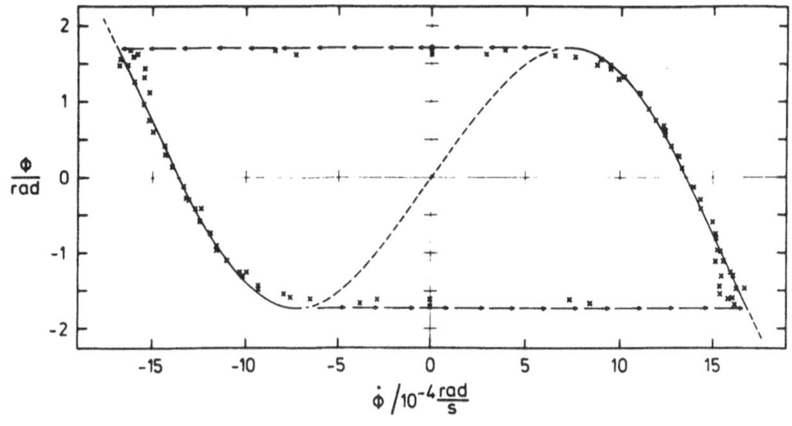

Fig. 6: Limit cycle (U = 29,95 V) fitted by a power series including terms up to fifth order. The dotted parts denote the instable branches. The jumps of $\dot{\phi}$ are indicated by arrows.

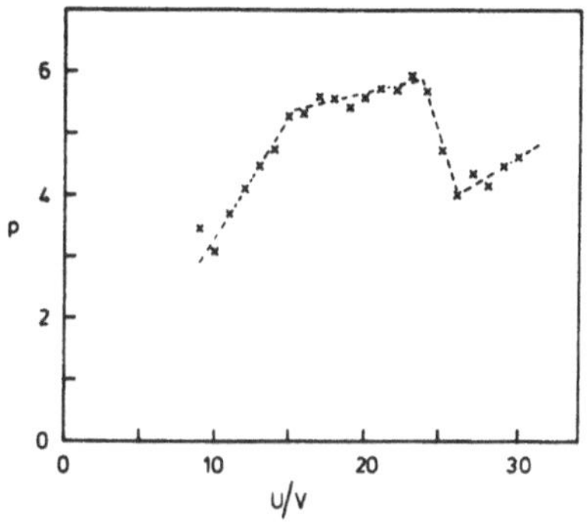

Fig. 7: The dimensionless quantity p as a function of the applied voltage U.

where k is the friction coefficient, D the torsion coefficient of the thread and M the torque due to the convection rolls. k and M are functions of $\dot{\phi}$: $k = k(\dot{\phi})$ and $M = M(\dot{\phi})$. These functions can be expressed as a power series /5/. Taking into account inversion symmetry of the limit cycles we obtain /5/:

$$M(\dot{\phi}) = A_1\dot{\phi} + A_3\dot{\phi}^3 + A_5\dot{\phi}^5 + \ldots \text{ and}$$
$$k(\dot{\phi})\cdot\dot{\phi} = B_1\dot{\phi} + B_3\dot{\phi}^3 + B_5\dot{\phi}^5 + \ldots \qquad (2)$$

Inserting into equation (1) ϕ is yielded as a power series of $\dot{\phi}$:

$$\phi = C_1\dot{\phi} + C_3\dot{\phi}^3 + C_5\dot{\phi}^5 + \ldots \qquad (3)$$

where $C_i = (A_i - B_i)/D$, $i = 1,3,5,\ldots$

Using such a power series including terms up to the fifth order the measured limit cycles can be well fitted, as is shown in fig. 6.

Characteristic quantities of the torsion pendulum oscillations are the period τ, the amplitude ϕ_{max} and the angular velocity $\dot{\phi}_o$ at the zero angle. These quantities can be related by a dimensionless quantity /6/:

$$p = \tau\cdot\dot{\phi}_o/\phi_{max}$$

In the case of harmonic oscillations p equals 2π. In fig. 7, the quantitiy p determined from our experimental results is shown as function of the applied voltage. Here, the investigated voltage range can be divided into a few intervals. In each of these, p is approximately a certain linear function of U. As the oscillations are driven by the EHD convection flow, these intervals seem to correspond to different regimes of the convection flow. This supposition could be checked by microscopic or light scattering investigations of the samples during the pendulum oscillations.

REFERENCES

1. V. Freederickzs, V. Tsvetkov, Phys. Z. d. Sowjetunion 6, 490 (1934)
2. J. Grupp, manuscript submitted to J. Physique (Paris)
3. J. Grupp, Rev. Sci. Instrum. 54, 754 (1983)
4. Orsay Liquid Crystal Group, Mol. Cryst. Liq. Cryst. 16, 229 (1972)
5. J. Grupp, Solid State Commun. 44, 627 (1982)
6. J. Grupp, Ph.D. dissertation, Münster 1981

GYROTACTIC BUOYANT CONVECTION AND SPONTANEOUS PATTERN FORMATION IN ALGAL CELL CULTURES

John O. Kessler

Department of Physics
University of Arizona
Tucson, Arizona 85721
U.S.A. *

Abstract

Regular convection patters may form spontaneously in isothermal liquids which contain swimming microorganisms. The energy for this dissipative process is supplied by the swimmers. Individual cell trajectories are guided by gravity and vorticity so that cells accumulate toward regions of the liquid where the downstreaming velocity is a maximum. This concentrative mechanism, named "gyrotaxis", has been proven by the demonstration that swimming cells focus at the axis of a downward cylindrical Poiseuille flow of the cell culture. Since the density of the cells exceeds that of the liquid in which they swim, gyrotaxis reinforces vorticity. This convection pattern producing system has been named "Gyrotactic Buoyant Convection (GBC). At sufficient average cell concentration, GBC can cause localised intermittent concentration pulses.

INTRODUCTION

The direction of locomotion of single algal cells is affected by gravity and the velocity field of the aqueous growth medium in which they swim. This modification of the orientation of individual cell trajectories can result in cell concentrative effects and

* Visiting at the Department of Applied Mathematics and Theoretical Physics, Cambridge University, Silver Street, Cambridge CB3 9EW, England until June 1984.

cooperative interaction which generates fluid convection patterns.

Cultures of *Dunaliella*, a single cell green alga, were generally used in the experiments reported. The cells are approximately ellipsoidal in shape. They contain a relatively dense posterior chloroplast and two thin anterior flagella which are used for forward locomotion. The cells usually vary in diameter from 5 to 15 μm and the flagella are of similar lengths. Swimming speeds as high as 100 $\mu m\ s^{-1}$ are observed. The average is somewhat less, perhaps 50 $\mu m\ s^{-1}$. Because of uneven flagellar motion, which may be purposeful or accidental, swimming trajectories may be intermittent, curved or irregular, and many cells rotate axially in the course of their forward progress. The culture medium used in experiments quoted here approximates seawater (1.03 $gm\ cm^{-3}$), and the average density of the cells is about 10% greater than that of the medium, as determined by observation of the sedimentation of killed cells.

It is observed that *Dunaliella* populations tend to swim upward. This behaviour is generally called "negative geotaxis". For concreteness, one may imagine the cells as rotationally symmetric around the swimming axis, but with a centre of mass located a distance L behind their geometric centre [1]. The cells then have a mass moment $\underline{M} = m\underline{L}$, where $\underline{M}$ points to the rear of the cell, and m is the cell mass. The gravitational torque $\underline{M} \times \underline{g}$ tends to orient the cells toward an upright position. If the cells swim forward with velocity $\underline{V}_c$, one may write $\underline{M} = -\gamma \underline{V}_c$. Generally γ will be a tensor which depends on various other externally supplied stimuli, as well as the fluctuating internal condition of the cell. Here, for simplicity, γ will be considered a time independent scalar. That assumption implies that the cells will swim strictly upward in a still fluid environment, which is, in fact, the average behaviour under uniform illumination.

GYROTAXIS AND FOCUSING

An object, such as an algal cell, which is placed in a fluid with vorticity will experience a torque $\beta \nabla \times \underline{v}$, where $\underline{v}$ is the fluid velocity. It will be assumed for simplicity that β is a scalar proportional to viscosity μ and the third power of an effective cell dimension which includes the effect of the ellipsoidal or irregular shape and the moving flagella. Combining the gravitational and fluid dynamic torques one obtains the total torque [2]

$$\underline{T} = \beta \nabla \times \underline{v} - \gamma \underline{V}_c \times \underline{g} \ . \qquad (1)$$

Under steady state conditions $\underline{T}$ vanishes, except for cell swimming fluctuation effects. The resultant trajectories produce a cell concentrative phenomenon that will be called "gyrotaxis", because of the crucial role of vorticity [3]. The average velocity of cells when $\underline{T}=0$ will be denoted by $\underline{V}$, without subscript c.

One may consider the simple special case of a cell culture slowly flowing through a cylindrical tube. In conventional coordinates (r,ϕ,z),

$$\underline{v}=-v_o(1-\frac{r^2}{R^2})\hat{\underline{z}} , \tag{2}$$

where v_o is the maximum velocity in this downward Poiseuille flow, R is the tube diameter, r the local radius, and $\hat{\underline{z}}$ the upward unit vector. Since $\underline{g}=-g\hat{\underline{z}}$ and $\nabla\times\underline{v}=-(2rv_o/R^2)\hat{\underline{\phi}}$, Eq.1 with $\underline{T}=0$ specifies the radial component of $\underline{V}$, $V_r=-(2rv_o\beta)/(\gamma gR^2)$. The orientation angle θ of the cell trajectory, relative to $\hat{\underline{z}}$, is given by $\sin\theta=(2rv_o\beta)/(mgLR^2)$, using $\gamma=mL/V$. It can be shown that the azimuthal component of $\underline{V}$ vanishes.

This derivation shows that a cell culture in a downward cylindrical Poiseuille flow will focus at the centre of the tube. The only assumptions needed are the simple fluid dynamical and gravitational torques and an absence of strong free will on the cells' part. The effect has been observed, Fig. 1. For the experimental conditions $v_o\simeq10^{-1}$ cm s^{-1}, $V\simeq10^{-2}$ cm s^{-1}, and travel distance 3 mm, a focus time of 100 s is predicted, and that is in the observed range. The parameter β was estimated by $3\pi\mu a^3/2$, where $\mu=10^{-2}$ gm cm^{-1} s^{-1}, $a=5\times10^{-4}$ cm. The cell density is $\simeq1$ gm cm^{-3} and L was estimated at 0.01a, as a result of observing the spontaneous rotation of killed cells.

It might be thought that the observed cell focusing phenomenon is due to some particle-flow-in-tube effect, i.e. the theory given is sufficient to yield focusing, but is it necessary? a simple vertical tall glass U-tube in which a quantity of algal culture is displaced to one leg and allowed to flow slowly into the other (by an air valve connected to the end of one leg of the apparatus) provides a simple proof of the necessity of the theory. Eq. (2) holds on the downflow side; on the upflow side the sign is changed, v_z is positive. Then the sign of V_r changes, which implies radially outward movement of the cells. This effect is indeed observed: initially there is visible focusing only on the downflow side. Eventually cells are seen to accumulate around the periphery of the ascending liquid column. The direction of $\underline{g}$ relative to the flow direction is crucial, as is the active swimming of the cells. Dead cells do not focus.

CONVECTION PATTERN FORMATION

Regions of the fluid where cell concentration differs from the average sink or rise, since concentration is directly proportional to density. This buoyant convection is reinforced by gyrotaxis: cells accumulate toward sinking fluid, thereby increasing vorticity. The process will be named Gyrotactic Buoyant Convection (GBC). The concentration variations which occur in GBC are reduced by random cell swimming and by cell-cell collisions. GBC generates convection/concentration patterns (Figs. 2 and 3) and streamers (Fig. 4) near the bottom of a cell culture. "Bioconvection" due to upswimming of a cell population, but neglecting gyrotaxis, has been described previously [4]; it cannot account for the type of streamers observed or for focusing.

To describe GBC from a continuum viewpoint, and in the Boussinesq approximation, the forcing term to be added to the Navier-Stokes equation is $\underline{g}nU\Delta\rho$, where n is the cell concentration, U the cell volume, and $\Delta\rho$ the difference between cell and aqueous medium densities. The volume fraction nU ranged from 10^{-5} to 10^{-3} in experiments where pattern formation was observed.

The cell flux $\underline{j}$ includes the gyrotactic term $n\underline{V}$ and a diffusive term to account for random behaviour and collisions then

$$\frac{Dn}{Dt} = -\nabla.\underline{j} = -\nabla\cdot(n\underline{V} - D\nabla n) \quad , \tag{3}$$

where $\underline{V}$ is obtained from Eq.1 with $\underline{T}=0$. The Navier-Stokes equation is

$$\frac{D\underline{v}}{Dt} = -\nabla p + \underline{g}(\rho_o + nU\Delta\rho) \quad , \tag{4}$$

where ρ_o is the growth medium density and p is pressure. An approximate solution can be obtained for steady state in a slot, at a horizontal plane of symmetry, with the assumptions $v_z = v_z(x)$, $n = n(x)$ and $j_x = 0$. Then

$$n = n_o \exp -\frac{\beta}{\gamma g D} v_z \quad . \tag{5}$$

Putting this result into Eq. (4), with $D/Dt = 0$, and using appropriate boundary and fluid conservation conditions, one obtains (numerically) a standing wave pattern with v_z nearly sinusoidal. The concentration n is therefore periodic and shows exponential maxima which account for the sharp appearance of the convection patterns (Fig. 2) which are visualized by the cell concentration. The length and velocity scales derived from these equations agree with experiment, for reasonable assumptions of the parameters. Further discussion is too lengthy for inclusion here.

COOPERATIVE EFFECTS IN THE TIME DOMAIN

If $|\underline{M} \times \underline{g}| >> \beta|\nabla \times \underline{v}|$, the cells simply swim upward. If the inequality is reversed, steady gyrotaxis ceases. The cells rotate with an unsteady angular velocity. Steady gyrotaxis requires $\sin\theta \leq 1$.

In the downward Poiseuille flow previously considered, if shear is increased focusing becomes ever less efficient as $\sin\theta \leq 1$ applies to fewer and fewer members of the cell population. For the case of efficient focusing, with $\sin\theta < 1$ initially, as cells stream to the tube axis, the cell concentration n, and hence the fluid mass density, increases there, while decreases occur toward the periphery. Thus the velocity profile changes from parabolic to centrally peaked. This effect increases $\nabla \times \underline{v}$ locally and can lead to $\sin\theta > 1$. The result is the appearance of cell concentration pulses, centred on the flow axis of symmetry, see Fig. 5. Pulse initiation is probably due to oscillations of cell concentration near the axis, as $\sin\theta$ oscillates around unity. A secondary effect arises from differences in sinking velocity of cell concentration pulses of different sizes. Large pulses catch up with smaller ones and swallow them. The cell pulses themselves have a vortex structure which appears similar to that of thermals. If conditions are well regulated and steady, cell concentration pulses occur at approximately regular intervals. Tall cell cultures containing spontaneously generated convection patterns include GBC streamers which contain cell concentration pulses very similar to those observed in the tubular flow.

OTHER ORGANISMS AND CONDITIONS

There are many macroscopic and microscopic fluid dynamical aspects of gyrotaxis and GBC which need extensive investigation. The types of patterns which can be obtained may be altered with weak directional illumination; they also vary from one organism to another. Eventually, it will be possible to infer microscopic information concerning swimming behaviour or cell geometry from observation of the macroscopic patterns. Experiments in the present series have also demonstrated that not only *Dunaliella* but *Euglena*, *Chlamydomonas*, and *Carteria* all exhibit GBC. This statement is made on the basis that cultures of each of these organisms exhibit spontaneous streamer generation near the bottom of the culture-containing vessel.

DISCUSSION

The spontaneous formation of geometric patterns by micro-organism containing fluids has features of both equilibrium and nonequilibrium phase transitions. The phenomenon may certainly be called "synergetic" in the sense that millions of independent cells work together coherently [5] coupled through the properties of the fluid and gravity.

Superficially, GBC seems analogous to Rayleigh-Bénard convection. Actually there is a big difference. In GBC, nothing is transported through the fluid layer, and the energy supplied to operate the dissipative structure is stored and transduced within entities that take part in the internal dynamics. This situation is typical for biological systems. It can definitely not be claimed that dissipative structure is associated with improving any macroscopic throughput.

A somewhat closer physical correspondence is found in ferromagnetism. One might consider vorticity as analogous to a mean magnetic field which exerts a torque on subentities which, acting cooperatively, produce more vorticity. The gravity field, in this conception, merely provides a directional framework for the action, much like the crystal lattice does in ferromagnetism. Certainly, the ferromagnetism analogy is "good" in the sense that there is no throughput, that there is a spontaneous appearance of domains in the absence of an external field ($\underline{H}$ or $\nabla \times \underline{v}$), and that an external field (e.g., a supplied Poiseuille flow-associated $\nabla \times \underline{v}$) can produce a single domain. The laser is another metaphor. The pumping is internal, $\underline{g}$ corresponds to cavity quality and vorticity is analogous to the radiation field.

APPLICATIONS

The work with *Dunaliella* was begun because this organism has remarkable environmental adaptability and potential commercial utility [6,7,8]. GBC allows cells to traverse vertical distances more quickly than they can swim individually. This effect is one aspect of cell harvesting and may also be of importance in algal ecology. Gyrotaxis could account for some of the plankton-concentrative effects in wind-driven Langmuir convection and will prove useful in cell separations based on motility and dynamic properties.

ACKNOWLEDGEMENTS

I should like to acknowledge the support of the University of Arizona Physics Department, the Office of Arid Lands, and the University of Arizona Foundation. I have greatly appreciated the help and advice of S.H. Davis, R.W. Hoshaw, N.R. Mauney, J.W. O'Leary and especially of Dick Kassander, without whom the work would not have happened.

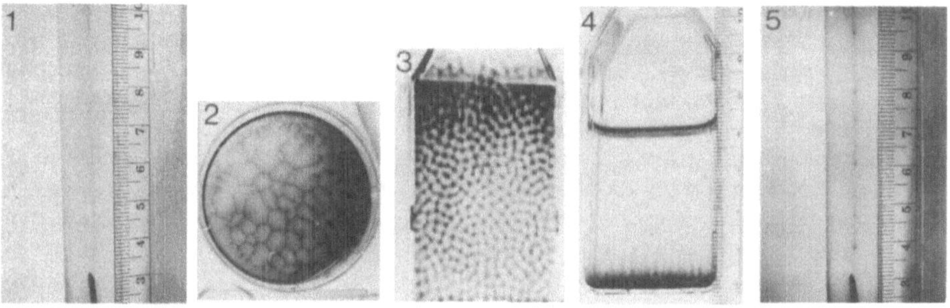

Figure 1. A culture of *Dunaliella tertiolecta* in artificial seawater is flowing through the vertical tube with an average flow rate of approximately 0.5 cm s^{-1}. The darker region on the axis is the column of focused cells.

Figure 2. Convection pattern in a 1 cm layer of a *Dunaliella* culture with cell concentration approximately 10^7 cells cm^{-3}. Cells accumulate in the dark regions which are downwelling. The cell concentration provides the flow visualization. Slowly varying pattern form.

Figure 3. As in Fig. 2, but near steady state (from a TV screen).

Figure 4. GBC streamers form near the bottom of the culture flask. *Carteria sp.*, approximately 10^6 cells cm^{-3}.

Figure 5. As in Fig. 1, at a flow rate of 0.1 cm s^{-1}. Concentration pulses which sink at approximately 0.3 cm s^{-1}.

REFERENCES

1. A.M. Roberts, Geotaxis in Motile Organisms, J. Exp. Biol. 53, 687 (1970). Also A.M. Roberts, Hydrodynamics of Protozoan Swimming, in Biochemistry and Physiology of Protozoa, Vol. 4, 2nd Ed., M. Levandowsky and S. Hutner, eds., Academic Press, New York, 1981, pp. 5-66.
2. A similar development is given in A.M. Roberts, the Biassed Random Walk and the Analysis of Microorganism Movement, in Swimming and Flying in Nature, Wu, Brokaw and Brennen, eds., Plenum, New York, 1975, pp. 377-393. A rudimentary version of the theory is given by Lord Rothschild, in Spermatozoan Motility, D.W. Bishop, ed., Am. Assoc. Adv. Sci., Washington, D.C., 1962, pp. 13-29.
3. "Rheotaxis", described by Roberts [3], is one component of gyrotaxis.
4. S. Childress, Mechanics of Swimming and Flying, Cambridge University Press, Cambridge, 1981, provides a list of references.
5. H. Haken, Introductory Remarks, in Evolution of Order and Chaos, H. Haken, ed., Springer Verlag, Berlin, 1982, pp. 1-4.
6. A. Ben-Amotz and M. Avron, Glycerol, β-Carotene and Dry Algal Meal Production by Commercial Cultivation of *Dunaliella*, in Algae Biomass, G. Shelef and C.J. Soeder, eds., Elsevier/North Holland, Amsterdam, 1980.
7. J.O. Kessler, M.D. Hurley, and B. Kingsolver, A Novel Harvest Technology for *Dunaliella* Phycoculture, in The Future of Small Energy Resources, R.F. Meyer and J.C. Olson, eds., UNITAR, McGraw Hill, New York, 1983, pp. 513-516.
8. J.O. Kessler, Algal Cell Harvesting, U.S. Patent 4,324,067, 1982.

EXPERIMENTAL INVESTIGATIONS OF PRECIPITATION PATTERNS

Stefan C. Müller

Max-Planck-Institut für Ernährungsphysiologie

Rheinlanddamm 201, D-4600 Dortmund, F.R.G.

1. INTRODUCTION

If a soluble electrolyte is placed in contact with a second electrolyte and, on interdiffusion, both react to form a weakly soluble salt, a metastable supersaturated solution can be produced. For many combinations of electrolytes at suitable concentrations the subsequent precipitation process does not occur continuously in space. Instead, well-separated bands of precipitate appear parallel to the front surface of the diffusing ions, sequentially in a period of hours to several days starting at the initial interface of the two reagents. Usually a gel-forming material is added to the solutions to prevent sedimentation and hydrodynamic convection. Such a periodic precipitation, commonly called the Liesegang phenomenon, was first observed in 1896 /1/ and since then investigated by many authors /2/.

In a typical "two-dimensional" Liesegang experiment diffusion occurs from the center of a flat circular container, such as a petri dish. The resulting pattern consists of a set of concentric rings as shown in Fig. 1A for precipitation of lead iodide formed by diffusion of iodide ions into a solution containing lead ions at substantially lower concentration. When the two reagents are placed in a test tube the direction of diffusion is confined to essentially one spatial coordinate and the precipitate is deposited in a "one-dimensional" set of parallel bands. Examples with lead iodide and magnesium hydroxide as the precipitating salts are shown in Fig. 1B and C, respectively. If the initial difference between the electrolyte concentrations as well as the concentrations themselves are sufficiently high numerous sharp bands appear at reproducible distances x_n from the origin of the imposed gradient. A spacing law x_{n+1}/x_n = constant is often fulfilled but deviations have been reported /3,4/.

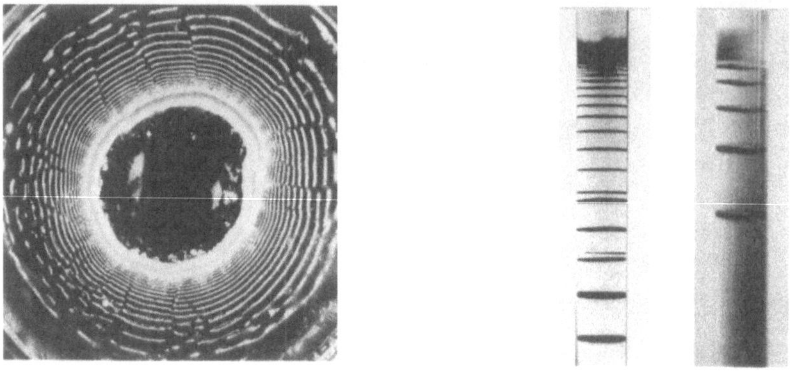

A B C

Fig. 1. Precipitation patterns in the presence of concentration gradients (Liesegang structures). (A) Concentric rings of PbI_2 in 1% agar (0.12 M KI, 0.012 M $Pb(NO_3)_2$). (B and C) Parallel bands of PbI_2 in 1% agar (0.17 M KI, 0.009 M $Pb(NO_3)_2$) and of $Mg(OH)_2$ in 9% gelatin (11.3 M NH_4OH, 0.4 M $MgSO_4$), resp. Scale bars: 1 cm.

Macroscopic structure also arises from a metastable system in the absence of concentration gradients. For instance, if a solution of appropriate concentration of lead iodide is prepared at elevated temperature, agar gel is added, and the solution is supercooled sufficiently to induce spatially uniform nucleation of colloidal lead iodide, precipitation occurs inhomogeneously in patterns which are more irregular but contain a dominating wavelength of 0.5 mm to 1 cm /5-8/. Examples of such structures are presented in Fig. 2.

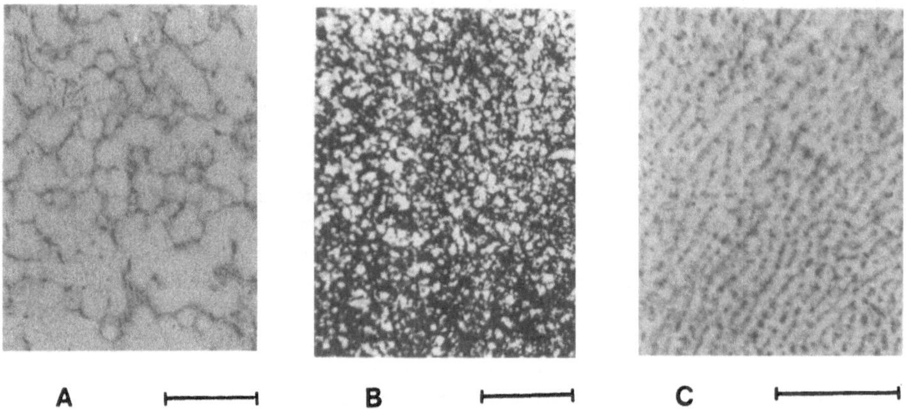

A B C

Fig. 2. Precipitation patterns without imposed concentration gradients. PbI_2 in agar with resp. concentrations: (A) 5.2 mM, 1%; (B) 6.4 mM, 1%; (C) 5.4 mM, 0.5%. Scale bars: 1 cm.

Various physico-chemical processes have been proposed for theories of Liesegang band formation /2/. Most approaches postulate that nuclation is discontinuous in space and that the spatial pattern of nucleation determines the band locations. Among these prenucleation theories only Ostwald's supersaturation hypothesis /9/ has been modelled mathematically /10/. Assuming that the growth of the nucleated particles at a band location inhibits further nucleation in the immediate vicinity recurrent banding can be produced. The significance of these models has been questioned, however, by further theoretical analysis which shows a continuously advancing nucleation front /11/. Experimental evidence to be described in the following sections also leads to the conclusion that patterned precipitation is a postnucleation phenomenon which occurs after a continuous distribution of colloid has been established /4,12/. This assumption is included in a theory which proposes that bands arise by means of coagulation of colloid /13/, but the existence of clear regions between bands has not been derived convincingly. A postnucleation theory which has been developed for the explanation of structure formation in "gradient-free" systems /6,14-16/ (see Fig. 2) is based on the hypothesis of a chemical instability due to the competitive growth of colloidal particles (Ostwald ripening), coupled with diffusion. It emphasizes that due to surface free energy contributions the solubility of colloidal particles is a decreasing function of size. Thus, large particles grow at the expense of smaller ones. For very slow particle growth kinetics competition occurs on a macroscopic scale /15/. There is evidence that this concept is also applicable to cases with concentration gradients (Liesegang structures) /8,12,17/.

This article summarizes results of recent experimental investigations concerned with the spatio-temporal evolution of Liesegang systems prior to visible band formation /12/, the relation between precipitating systems with small gradients and without gradients /4,8/, some structural details of pattern formation in initially uniform solutions /7/, and several structures of remarkable complexity /18/. The possible significance of these results for the competitive particle growth theory is briefly pointed out.

2. SPATIAL AND TEMPORAL SEQUENCE OF EVENTS IN A LIESEGANG SYSTEM

An experimental study of banded precipitation of $Mg(OH)_2$ in gelatin (see Fig. 1C) was performed by visual observations and by measurement of transmitted light, scattered light, deflection of a transmitted light beam (as an indication of changes in the refractive index), and by gravity effects as discussed in detail in Ref. /12/. A sequence of events was observed during the entire period from the time when the two electrolytes NH_4OH and $MgSO_4$ are placed in contact with each other to the completion of the final pattern. By mixing the $MgSO_4$ with an indicator the motion of a pH front due to the diffusion of OH^- ions was measured which corresponds to an ion product three times larger than the solubility product of the $Mg(OH)_2$ salt.

In association with the passage of that front, at any point along the axis of the system the onset of nucleation of colloidal particles was detected by a small increase in index of refraction. After a given interval of time following the nucleation front these particles have grown to give rise to distinct light scattering and turbidity appears with a moving front. Both the pH front and the turbidity front move continuously through the system. Hence, a major finding is the observation that the formation of colloidal particles is continuous in space. A smooth distribution of colloidal material is established a substantial interval of time prior to the next event, which is the onset of a substantial localized gradient of index of refraction at the prospective band positions. This event signals structure formation which occurs by a focusing mechanism in that a band is formed in a narrow space interval and the regions on either side of the band become depleted in colloidal particles, as shown below.

Two types of measurements in which optical techniques were applied to determine the sequence of events are presented in Figs. 3 and 4. The spatial distribution of the intensity of scattered light I_s plotted in Fig. 3 was obtained after three bands had already formed. The averaged intensity in front of the third band (dashed line) represents the essential characteristics of the spatial turbidity profile, while local peaks in I_s are not reproducible /12/. The next band to be formed appears within a relatively uniform turbidity region (at $x \gtrsim 9$ mm) a substantial time interval after the formation of colloidal particles. Note that there is a local minimum in the vicinity of the third band (dotted part of horizontal line), corresponding to a zone which is partially depleted in colloidal material.

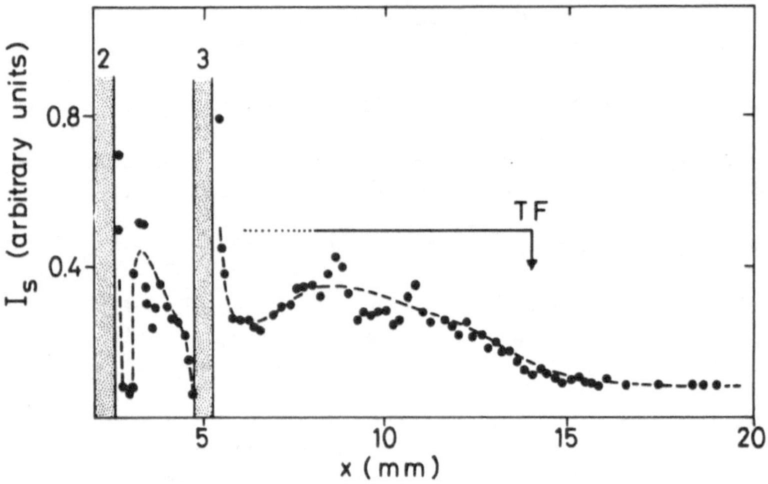

Fig. 3. Intensity of scattered light I_s vs coordinate x of a $Mg(OH)_2$ Liesegang system (5.5 M NH_4OH, 0.4 M $MgSO_4$ in 9% gelatin), after 320 min. Shaded areas represent visible bands. Horizontal line: turbidity region; TF: turbidity front.

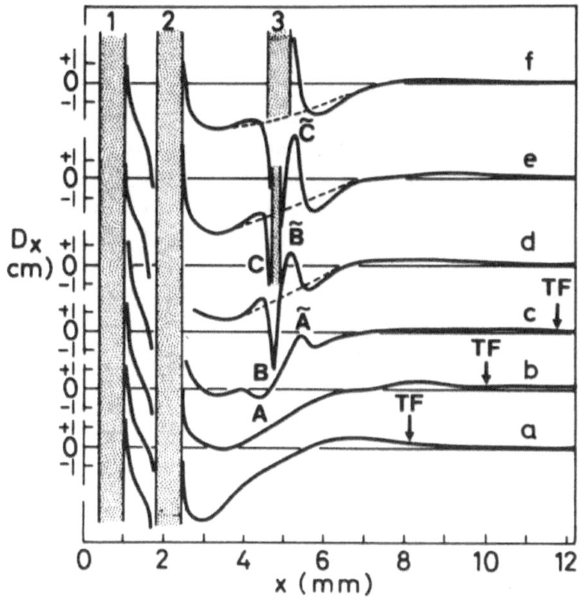

Fig. 4. Beam deflection D_x vs coordinate x of a $Mg(OH)_2$ Liesegang system as in Fig. 3. Times after start of experiment: (a) 110 min, (b) 165 min, (c) 180 min, (d) 200 min, (e) 220 min, (f) 260 min. TF: turbidity front.

A more detailed investigation of the evolution of structure was performed by recording the amount of deflection D_x of a transmitted narrow laser beam at various times prior to completion of the third band (Fig. 4). D_x is proportional to the refractive index gradient and thus determined by the local concentrations of the different solutes in the sample. As pointed out in Ref. /12/ some of the features of the measured curves can be interpreted in terms of the volume fraction of colloidal $Mg(OH)_2$. It follows from graphs (a) and (b) of Fig. 4 that during a long interval of time there exists a broad zone of continuously changing refractive index which roughly coincides with the turbidity region. The first indication of structure formation is a wave-like modulation around x $\gtrsim$ 5 mm [extremes A, Ã in graph (c)] which emerges from a previously smooth region within only a few minutes. It represents a localized peak in $Mg(OH)_2$ colloid. This pattern becomes more pronounced with sharper peaks and a smaller distance between both extremes [graph (d)] until a thin region of colloid turns into visible precipitate [shaded area in graph (e)]. In Fig. 4f the band has grown to its final width. In the immediate vicinity of the extremes labeled B, B̃ and C, C̃ there are small deviations between the measured D_x values and the smooth background curve from which the structure evolves (dashed lines). These deviations are consistent with the reduction in scattering intensity I_s on both sides of the band (Fig. 3) which indicates a reduction in colloid concentration.

3. PATTERN FORMATION IN THE PRESENCE OF LOW CONCENTRATION GRADIENTS

Tradionally Liesegang patterns are prepared with strong initial concentration gradients and sharp bands form at reproducible locations as shown in Fig. 1. An experimental study was performed with emphasis on the behavior of PbI_2 precipitation patterns when the concentrations of reactants are systematically decreased /4,8/. The determination of critical concentrations below which no patterns appear leads to the conclusion that the initial concentration difference $\Delta = \frac{1}{2}[I^-] - [Pb^{2+}]$ and concentration product $\sigma = [Pb^{2+}][I^-]^2$ are useful parameters to characterize the onset of structure formation. The deterministic spacing law specified in the Introduction is obeyed in systems with high values of Δ and σ. However, when the concentration gradients are lowered (small Δ or σ) there are significant deviations from this law and the structure is increasingly stochastic in spatial location. Measurements of the statistical distribution of band locations have been represented by histograms /8/, as illustrated in Fig. 5 for a PbI_2 system with four bands. Gaussian distributions are fitted to bands 1 to 3, while for the 4th band, in a region of very low gradients, the probability of band occurence is almost evenly distributed in a broad space interval. This observation of randomness is considered to be important for elucidating the connection to patterns in gradient-free systems (see below).

An interesting phenomenon shall be briefly mentioned: a "spatial bifurcation" occurs of a single Liesegang band into two clearly separated bands of precipitate, both located within an extended region of low-density colloid of PbI_2, when Δ is increased from almost zero to a positive but comperatively small value while σ is low and maintained constant. The main features are represented in the rough sketch of Fig. 6, which is derived from light transmission measurements discussed in detail elsewhere /8/. Recent calculations within the scope of the competitive particle growth theory lead to a process of band splitting which has a striking similarity to these experimental findings /17/.

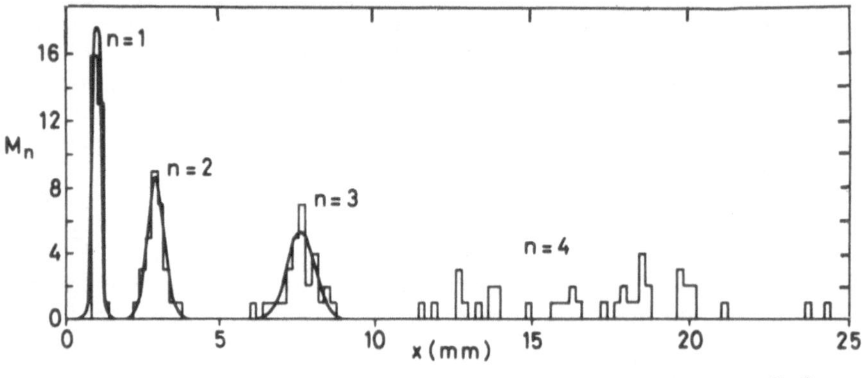

Fig. 5. Histograms of PbI_2 band locations ($\sigma = 2.1 \times 10^{-5} M^3$, $\Delta = 13.6$ mM) labeled with the respective band numbers n.

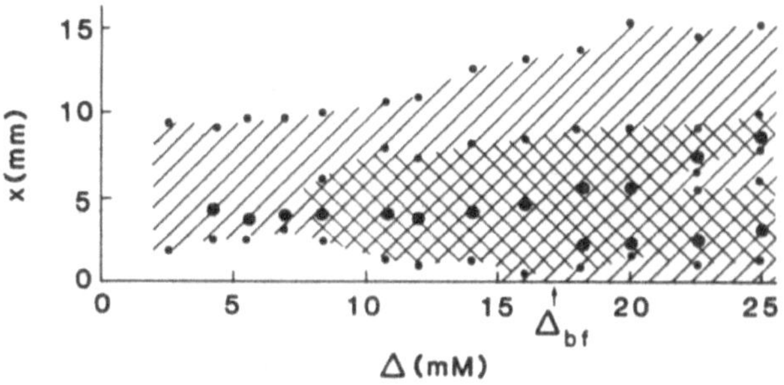

Fig. 6. Sketch of spatial bifurcation of PbI_2 precipitation bands. Location of bands (large circles) and extent of surrounding zones of visible colloid (shaded areas) as a function of at constant $\Delta = 1.8 \times 10^{-5} M^3$. Δ_{bf}: bifurcation point.

4. PATTERN FORMATION IN INITIALLY UNIFORM COLLOIDS

The appearance of spatially inhomogeneous distributions of precipitate in initially homogeneous supercooled solutions of a weakly soluble salt has been investigated for PbI_2 in an agar gel /5-8/. The macroscopic inhomogeneities formed on a length scale of the order of 0.5 mm to 1 cm are observed within a comparatively small range of initial salt concentrations (from 4.8 mM to 6.6 mM) while for concentrations outside this range, both lower and higher, the distribution of PbI_2 remains uniform /7/. Different varieties of structure have been reported which can be roughly characterized as two-dimensional networks (Fig. 2A), wavelike structures (Fig. 2C), or speckled patterns /8/. Generally, a stochastic element prevails. Inspection with a light microscope reveals that the visibly dense regions of PbI_2 consist of a large number of predominantly independent colloidal particles. Some coagulation occurs but the structures are not formed by contiguous crystallites. The particle size distribution is consistent with the main characteristics of reaction-limited colloidal growth kinetics /7,19/.

Further analysis of the mesoscopic features of the two network type patterns A and B of Fig. 2 leads to the following result: The small increase of the initial salt concentration from 5.2 mM to 6.4 mM, corresponding to a change in supersaturation by only a factor of 1.3, results in a pronounced decrease by a factor of 4 to 5 in the macroscopic length scale, the average radius of particles, and the average interparticle distance in the dense regions of the structures. The similarity of that change in all of the observed length scales points to a strong correlation between the macroscopic and mesoscopic structural properties of gradient-free precipitation patterns.

5. COMPLEX PATTERNS IN PERIODIC PRECIPITATION PROCESSES

In Liesegang systems prepared according to the standard procedure patterns may appear that are more complex than the concentric rings or parallel bands most frequently found. Some of these structures are ring systems in which radially aligned pattern formations are superposed as, for instance, in Fig. 1A, where the rings consist of a number of convex segments of PbI_2 separated from each other by small gaps devoid of visually detectable precipitate. The gaps extend out across the ring system in roughly radial direction. Similarly, radially aligned sets of pocket-like structures instead of convex segments may form /18/. Still another "radial" structure which appears repetitively in subsequent rings is the set of dislocations shown in Fig. 7A for a PbI_2 system. (Similar dislocations have been observed for $Ag_2Cr_2O_7$ rings /20/.) Irregular patterns within concentric rings have been reported which are reminiscent of inhomogeneous precipitation patterns in gradient-free systems /18/. A curiosity in a one-dimensional Liesegang experiment is presented in Fig. 7B. A spiral of $Mg(OH)_2$ precipitate occupies the region where the first three precipitation bands are usually located. Spiral formation has also been found in PbI_2 and $Ag_2Cr_2O_7$ systems /18/. The experimental procedure to produce such helicoidal structures in a reproducible manner is not known.

6. CONCLUSIONS

The experimental results presented in this article provide evidence that formation of macroscopic structure in precipitation is governed by postnucleation processes in systems both with and without concentration gradients. Recurrent banding comes about by a focusing mechanism: a sharpening of bands during their evolution is

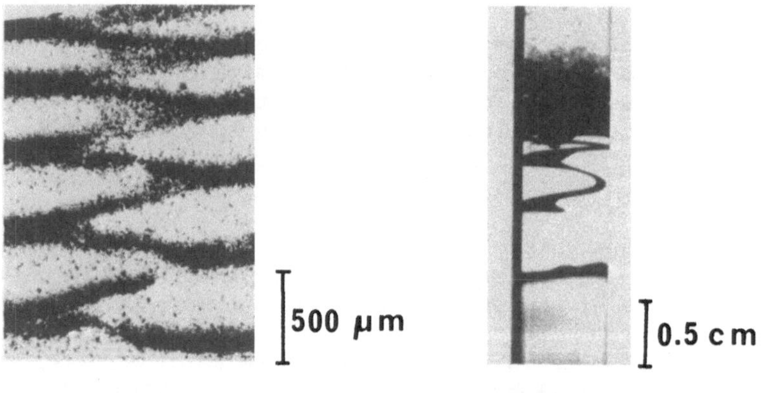

A B

Fig. 7. (A) Dislocation within a set of concentric PbI_2 rings in agar (0.12 M KI, 0.015 M $Pb(NO_3)_2$). (B) Helicoidal band of $Mg(OH)_2$ in gelatin (5.5 M NH_4OH, 0.37 M $MgSO_4$).

accompanied by partial depletion of colloidal material in their neighborhood. The location of bands is deterministic in the presence of large concentration gradients and becomes stochastic when these gradients are decreased. In the limit of no gradients various types of random structures are found. The results lead to the suggestion that the origin of pattern formation is the same in all cases and lend support to theories in which the structure formation is associated with the competitive growth of colloidal particles coupled with diffusion, as has been pointed out in Refs. /6,14-16/. Numerical simulations also show that the competitive particle growth mechanism can at least qualitatively account for some of the more complex structural features in systems with low gradients /17/.

REFERENCES

1. R.E. Liesegang, Naturwiss. Wochenschr. 11, 353 (1896).
2. E.S. Hedges, Liesegang Rings and Other Periodic Structures, Chapman and Hall, London, 1932; K.H. Stern, Chem. Rev. 54, 79 (1954).
3. E.S. Hedges and R.V. Henley, J. Chem. Soc. 1928, 2714; A. Packter, Nature (London) 175, 556 (1955).
4. S.C. Müller, S. Kai, and J. Ross, J. Phys. Chem. 86, 4078 (1982).
5. M. Flicker and J. Ross, J. Chem. Phys. 60, 3458 (1974).
6. D. Feinn, P. Ortoleva, W. Scalf, S. Schmidt, and M. Wolff, J. Chem. Phys. 69, 947 (1978).
7. S.C. Müller, S. Kai, and J. Ross, J. Phys. Chem. 86, 4294 (1982).
8. S. Kai, S.C. Müller, and J. Ross, J. Phys. Chem. 87, 806 (1983).
9. W. Ostwald, Lehrbuch der Allgemeinen Chemie, Engelmann, Leipzig (1897).
10. C. Wagner, J. Colloid Sci. 5, 85 (1950); S. Prager, J. Chem. Phys. 25, 279 (1956); J.B. Keller and S.I. Rubinow, J. Chem. Phys. 74, 5000 (1981).
11. G. Venzl and J. Ross, J. Chem. Phys. 77, 1302 (1982).
12. S. Kai, S.C. Müller, and J. Ross, J. Chem. Phys. 76, 1392 (1982).
13. N.R. Dhar and A.C. Chatterji, Kolloid-Z. 37, 2 (1927); ibid. 37, 89 (1927); S. Shinohara, J. Phys. Soc. Japan 29, 1073 (1970).
14. R. Lovett, P. Ortoleva, and J. Ross, J. Chem. Phys. 69, 947 (1978).
15. G. Venzl and J. Ross, J. Chem. Phys. 77, 1308 (1982).
16. R. Freeney, S.L. Schmidt, P. Strickholm, J. Chadam, and P. Ortoleva, J. Chem. Phys. 78, 1293 (1983).
17. S.C. Müller and G. Venzl, Lecture Notes in Biomathematics, (W. Jäger, Ed.), Springer, 1983, in press.
18. S.C. Müller, S. Kai, and J. Ross, Science 216, 635 (1982).
19. I.M. Lifshitz and V.V. Slyozov, J. Phys. Chem. Solids 19, 35 (1961); C. Wagner, Z. Elektrochem. 65, 581 (1961).
20. C.V. Raman and K.S. Ramaiah, Proc. Indian Acad. Sci. Sect. A9, 455 (1939).

OPTICAL DIAGNOSIS IN FLOWS - APPLICATION - EXPERIMENTS IN COMBUSTION

Louis Boyer

Department of Combustion - L.A. 72 CNRS
Université de Provence
13397 Marseille cedex 13, France

INTRODUCTION

The most important probe to study the systems and their dynamics is the electromagnetic radiation. For instance until the last two decades any knowledge of the universe had been obtained through processing of its visible or non visible light. In the labs, where the experiments can be prepared, numerous investigations are based on interaction of light with matter. Some of them give visualization,[1] others, such as absorption of ultraviolet, visible and infrared radiation provide detailed informations on energy levels. of molecules and atoms ,[2]. Here our aim will be to present how scattering of light provides many solutions for measuring physical quantities like velocity, temperature, concentration, their mean values as well as their fluctuations in flows with applications to an interesting reactive flow : combustion. Due to the importance of the subject we have been led to restrict the lectures to some aspects of optical diagnostics. This course concentrates on elastic and quasi elastic light scattering. Very important topics like Raman and coherent antistokes Raman scattering, Laser Induced fluorescence and a recent promising technique, the optogalvanic spectroscopy which give instantaneous and local informations on temperature and concentration will not deal with here. As a first approach of the latter subjects we suggest the reading of "Laser probes for Combustion Chemistry" edited by D.R. Grosley,[2]. Today combustion science is considered from two convergent points of view. On one hand, physicists are able to give sophisticated models where the mechanisms of combustion (i.e. diffusion processes and chemical reactions) are taken into account as well as the hydrodynamics,[3,4]. On the other hand, the evolution of aircraft gas turbine combustion technology over the past forty years has been extremely impressive and recent developments have caused significant shifts in development emphasis toward combustion technology. For example, in addition to the necessary improvements to increase the thrust/weight ratio, new concepts and technology improvements are necessary to satisfy recent exhaust pollutant regulations. Thus, to check the theories, to give numerical values for numerical simulations or to control the characteristics of an engine, accurate tools have been developed with the advent of the lasers. As a matter of fact most of the relevant information must be obtained by **non invasive, local** and **instantaneous** measurements. These needs can be satisfied by using optical techniques and laser sources. Of course most of optical techniques used in hydrodynamics such as flow visualization can be used but specific problems arise in combustion with the existence of large gradients of temperature or concentration. The well-known monograph of F.J. Weinberg "Optics of flames"[5] remains the reference on classical methods. We do also mention the book of R.M. Fristrom and A.A. Westenberg on "Flame structure",[6] which gives a complete view of the techniques applied up to 1965. In the last decade new techniques essentially based upon the laser light scattering have been used successfully. The aim of the present course is to give the general principles of some of these methods. The important literature on laser diagnostics can be found in the proceedings of the International Symposiums of the Combustion Institute, in Combustion and Flame, Applied optics, the proceedings of the AIAA meetings or of the ICOGERS meetings (Progress in astronautics and aeronautics. This list is non exhaustive).

As an introduction to combustion the reader will refer to the lectures of P. Clavin in this volume or in,[7]. In the following, without loss of generality we will consider only the problems of the premixed flame. In most cases it can be considerated as an hydrodynamic surface of discontinuity which separates the fresh mixture from the burned gases. This surface $\alpha(x,y,t)$ (the flame thickness is very small compared to the wavelength of wrinckles, fig.1) moves with a local velocity $\dot{\alpha}(x,y,t)$ in relation to the turbulent velocity field $u(x,y,t)$. The main measurements to be performed are velocity of gases and flame, temperature and concentrations of different species ; all these informations are necessary to build a model of the combustion process, for example in an internal combustion engine. Moreover the structure of the flame i.e. the temperature and concentration profiles are related to the dynamical properties and a knowledge of these characteristics is an important goal. To summurize, table 1 gives a survey of the quantities of interest and the measurement techniques generally used.

Table 1. Some applications of light scattering techniques (extended to fluorescence and multiphoton absorption) in flow diagnostics.

	Scattering	Technique	Ref.
- Fluid velocity V_f	Mie	L.D.V.	32 to 40
- Particle velocity V_p	Mie	L.D.V.	32 to 40
- grad $\vec{v}$	forced Rayleigh (diffraction)	two point L.D.V.	32 to 40
		phase grating	48
- Fluctuations :			
Velocity $\langle \Delta V(o) \Delta V(\tau) \rangle$	Mie	L.D.V.	32 to 40,24
Entropy $\langle \Delta S(o) \Delta S(\tau) \rangle$	Rayleigh	Photon beating	8
pressure $\langle \Delta P(o) \Delta P(\tau) \rangle$	Brillouin	Interferential spectrometry	15
concentration $\langle \Delta C(o) \Delta C(\tau) \rangle$	Rayleigh	Photon beating	8 -12
temperature $\langle \Delta T(o) \Delta T(\tau) \rangle$	Rayleigh	Scattered flux measurement	19 - 20
- Flame position α	Mie	Laser tomography	26
- Flame velocity $\dot{\alpha}$	Mie	Laser tomography	26
- Particle size d	Mie	Photon beating	14,42,43
		L.D.V.	42
		Diffraction	14
- Particles density n_p	Mie	Scattered flux measurment	14
- Temperature T	Raman	Grating spectrometry	2
	CARS	" "	9
- Concentration	Raman	" "	9
	Fluorescence	" "	49
	Multiphoton absorption	Opto galvano spectrometry	50

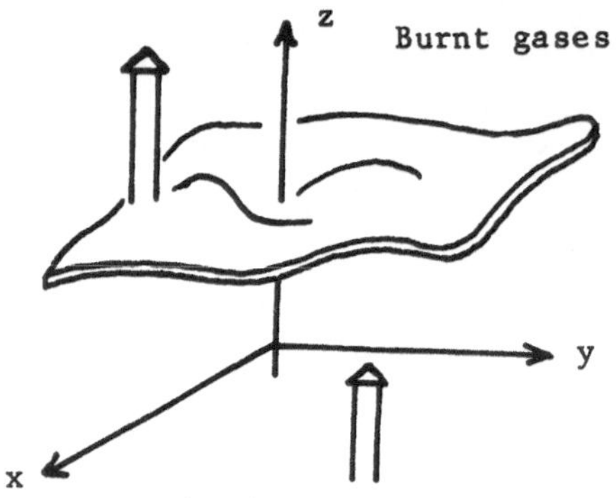

Fig.1 - The wrinkled flame : a hydrodynamical surface of discontinuity

A POWERFUL NON INVASIVE TOOL : THE INTERACTION OF LIGHT WITH MATTER

Properties of laser light

The theory of laser and the general characteristics of laser light have been given in details in Pr. Arechi's lectures. When using lasers we have simply to keep in mind some properties of practical interest.

Let us consider an argon-ion laser of common use. Several lines can be used:4880 Å , 5145 Å, etc... The intensity profile is very nearly an axisymetric gaussian :

$$I = I_o \exp(- \frac{8r_1^2}{D^2(x)}) \qquad (2.1)$$

where r_1 is the radial distance from the beam axis and D(x), the diameter defined for the $I_o e^{-2}$ intensivity, depends on the distance x along the beam.

In fact, in the far field the beam is a spherical wave diverging from a local waist located in the cavity and the divergence angle δ differs from the well-known result of diffraction theory only by a numerical factor :

$$\delta = \frac{4\lambda}{\pi D(o)} \qquad (2.2)$$

For the argon-ion laser D(o) = 1,5 mm and δ = 0.5 mrd. with a typical power from 10mW to 20W. The focusing properties of gaussian beams are somewhat different from the usual results of geometrical optics. Using a lens of focal length f, in the neighborhood of the focal point the focal waist is a region where the beam diameter is minimum and the light waves nearly planar with a gaussian intensity profile. This diameter is given by :

$$d = \delta f = \frac{4}{\pi} \frac{\lambda}{D(o)} f \qquad (2.3)$$

For example with f = 100 mm the spot size is about 30μm.

The second important property is the coherence. In laser Doppler velocimetry or more generally speaking in light beating spectrometry,[8] the coherence lenght L_c which is 10^2m for a Helium Neon laser is about 10 cm for a multimode argon-ion laser. If the geometric paths of two beams coming from the same laser differs from more than L_c, the relative phases are then at random and no interference (or beating) can occur. When operating with high power lasers the path lengths must be equal.

The last important property is the power. Consider for instance a 1W Ar-ion laser focused on a spot of 30μm diameter. The light intensity (energy for unit area and unit time) is $10^9 W/m^2$ i.e. 10^6 the light intensity of a sunny day. This is important in light scattering experiments and still more in non linear optics where the electric field must be very high. In such a case pulse lasers are used, and intensities of more than 10^{15} W/m^2 are easily obtained. So far, the laser light has been supposed quasimonochromatic with a linewidth $\Delta\nu$ related to L_c by

$$\Delta\nu = C/L_c \qquad (2.4)$$

If the frequency is not critical for L.D.A. or some light scattering experiments such as Rayleigh, Brillouin, Raman, Mie scattering it is necessary on the other hand to us tunable dye lasers in expements such as coherent antistokes Raman Scattering (CARS) or optogalvanospectroscopy. Finally, broadband lasers are used in certain CARS experiments,[9]. The statistical properties of laser light are presented elsewhere,[10,11]. The statistical properties of scattered light depends on the nature of the scattering medium,[11]. Nevertheless in most applications, where the number density of scatterers is high enough the scattered field is Gaussian to a good approximation,[12].

Elastic and quasielastic scattering

For elastic and quasielastic scatterings where the frequency shift of light is zero or very small compared to the frequency of the laser light, a classical description is adequate.
The incoming laser can be described by the corresponding incident field in the waist of a focused beam by :

$$\vec{E} = \vec{E}_o(\vec{r}) \exp i(\omega_o t - \vec{K}_o \vec{r}) \qquad (2.5)$$

$\vec{E}_o(\vec{r})$ is associated to the Gaussian intensity profile :

$$|\vec{E}_o(\vec{r})|^2 \simeq I_o \exp\left[-\frac{8(y^2+z^2)}{d^2}\right] \qquad (2.6)$$

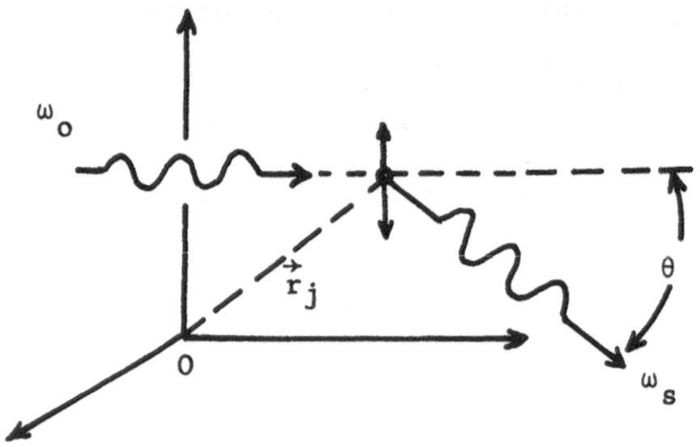

Fig.2 - Induced dipole scattering

ω_o is the angular frequency $\vec{K}_o$ the wave vector of the incident light with $\omega_o = CK_o$. We consider first a particle located in the scattering volume fig.2. Let ω_s and $\vec{K}_s$ the frequency and wave-vector of the scattered light respectively.

An important parameter is the scattering wave vector defined by :

$$\vec{K} = \vec{K}_s - \vec{K}_o \qquad (2.7)$$

For numerous experiments $\omega_s/\omega_o \simeq 1$ and $|\vec{K}_s| \simeq |\vec{K}_o|$, then (fig.3)

$$|\vec{K}| = 2|\vec{K}_o| \sin \theta/2 \qquad (2.8)$$

where θ is the scattering angle ($\theta = (\vec{K}_s, \vec{K}_o)$).

What is the scatterer ? any piece of matter, from the electron to the dust particle made of wood, chalk or aluminium oxide ; usually Rayleigh scattering stands for particle size much smaller than the wavelengh (i.e.<0.1μm) and Mie scattering for bigger particles. In this case scattering can be considered as a diffraction problem,[13,14].

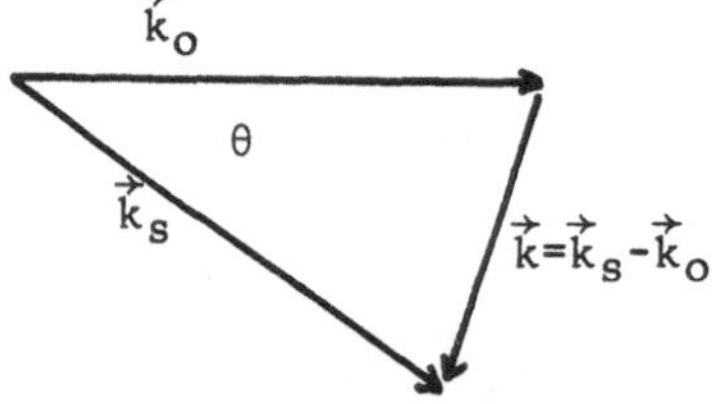

Fig.3 - The scattering wave vector $\vec{k}$

Now, let us consider a small particle an atom, a molecule or a small unit volume of matter. The mechanism of scattering is the radiation of the induced dipole under the action of the incident field $\vec{E}$ far from an electronic transition. The dipole polarization $\vec{P}$ is a function of $\vec{E}$ which can be developed to the first order :

$$P_i = \alpha_{ij} E_j \tag{2.9}$$

where α_{ij} is the polarizability.

In nonlinear optics, the extra terms must be taken into account ($\alpha_{ijkl}E_jE_kE_l$ for CARS experiment for instance).
$\vec{P}$ depends on space and time through the position of the particle which will give a phase modulation of the scattered field, the rotation and the deformation of the particle which appear as an amplitude modulation of the scattered field. The very important case of deformation or vibration is associated to the Raman scattering. Thus, it can be seen that the information is processed through amplitude , phase or frequency modulation just like in telecommunications.The theory of light scattering can be found in excellent monographs,[12,14,15,16] we give here the general result for anisotropic small rigid particle.

$$\vec{E}_s = \vec{K}_s \, x(\vec{K}_s \times \vec{E}_o)(\frac{\varepsilon_o}{\varepsilon})\frac{\exp[(i(\omega_o t-\vec{K}_s\vec{R})]}{4\pi R}\alpha \, e^{i\vec{K}.\vec{r}} \tag{2.10}$$

In case of scattering by a continuous isotropic medium $\alpha e^{i\vec{K}\vec{r}}$ must be replaced by

$$\alpha e^{i\vec{K}\vec{r}} \rightarrow \Delta X(\vec{r},t)e^{i\vec{K}\vec{r}}\, d\vec{r} \tag{2.11}$$

where $\Delta X\,(\vec{r},t)$ is the electric susceptibility ($\vec{P}=\varepsilon_o\,\Delta X\vec{E}$).

In these equations, $\vec{r}$ is the particle location, $R(>>\lambda)$ the distance from the observation point, ε_o, ε the dielectric constant of vacuum and of the medium respectively.
Eq.(2.10) can be written :

$$E_s = E_o\,\alpha\frac{\omega_o^2}{4\pi R\, C^2}(\frac{\varepsilon_o}{\varepsilon})\sin\phi \quad \exp\; i(\omega_o t+\vec{K}\vec{r}) \quad \exp(-\,i\vec{K}_s.\vec{R}) \tag{2.12}$$

where ϕ is the angle $(\vec{K}_s,\,\vec{E})$

On this simple equation several remarks must be made :
A/ The scattering vector K is the essential parameter in scatterings (not only for light scattering). The phase $\phi=\vec{K}\vec{r}$ which depends on the particle location is simply the sum of the phase $-\vec{K}_o\vec{r}$ of the incident beam and of the phase $\vec{K}_s\vec{r}$ of the scattered beam (fig.2). When $\vec{r}=\vec{r}(t)$ the product $\vec{K}\vec{r}$ appears like a phase modulation. When several particles j must be accounted for, the sum

$$\sum_j E_{sj} = \sum_j \exp\ i(\omega_0 t + \vec{K}\vec{r}_j) \qquad (2.13)$$

is an interference integral, so is eq.2.11 where it can be seen that scattering piks up the Fourier $\vec{K}$ component of the susceptibility $\Delta\chi(\vec{r},t)$

B/ The intensity of the scattered light, given by $|E_s|^2$ is proportional to R^{-2} like in an usual source.

C/ The intensity is also proportional to ω^4, the well-known property responsible for blue sky and red sunsets.

One scatterer with uniform motion : L.D.A.

We suppose a single scatterer running with a velocity $\vec{v}$ in the laboratory frame, it may be a dust particle in a flow :

$$\vec{r}(t) = \vec{r}(o) + \vec{v}t \qquad (2.14)$$

The phase is then :

$$\phi(t) = \vec{K}.\vec{r}(o) + \vec{K}.\vec{v}\ t$$

The frequency of the scattered light is thus shifted by the amount

$$\Omega_D = \vec{K}.\vec{v} \qquad (2.15)$$

which is the Doppler frequency.
It can also be written :

$$\Omega_D = 2\,\omega_0 \frac{v_{/\!/}}{C} \quad \sin\theta/2 \qquad (2.16)$$

where $v_{/\!/}$ is the $\vec{v}$ component along $\vec{K}$. We will come back to practical applications. We just give the magnitude for bacteria motion and gas velocity in a gas turbine combustion. In the first case $v \simeq 3\mu m/s$ in the second case $v \simeq 60 m/s$. The light frequency is $\nu_0 = \frac{\omega_0}{2\pi} \simeq 5.10^{14}$Hz, with $C = 3.10^8\ m.s^{-1}$ and $\theta = \pi$ we find for maximum frequencies $\nu_D = \frac{\Omega_D}{2\pi} = 10$ Hz for bacteria and 2 10^8Hz for the engine. In both cases the frequency shift is negligible compared to ν_0 and techniques which give the difference $\nu_s - \nu_0$ must be used both to get rid of inevitable drifts of the laser frequency and to obtain high accuracy measurements.

Several scatterers

The scattered field is the sum 2.13 :

$$\vec{E}_s = \sum_j \vec{E}_{sj} \qquad (2.17)$$

where $\vec{E}_{sj}$ is the scattered field radiated from the j particle. Information is contained in statistical properties of the random function $\vec{E}_s$. For Gaussian processes all correlation functions can be calculated from the second order correlation function,[11,12]

$$C_{E_s}(\tau) = \langle E_s(t)E_s^*(t+\tau)\rangle \qquad (2.18)$$

or from the power density function $S_E(\nu)$ the Fourier transform of $C_E(\tau)$:

$$S_E(\nu) \overset{T.F}{\Longleftrightarrow} C_E(\tau) \qquad (2.19)$$

A. Non correlated isotropic scatterers

It can be seen readily, using 2.12 that :

$$C_E(\tau) \sim N\langle e^{i\vec{K}.\,[\vec{r}_j(\tau)-\vec{r}_j(o)]}\rangle e^{-i\omega_0\tau} \qquad (2.20)$$

where N is the mean number of particles.

Two situations have to be considered. Let ℓ the mean free path of the particle, and suppose first that the inequality $K.\ell \gg 1$ holds. This is of course the case of the dust particle of L.D.A. but also the case for the molecules of a gas at low pressure. In this case :

$$C_E(\tau) \sim N\langle e^{i\vec{K}\vec{v}_j\tau}\rangle e^{-i\omega_0\tau} \qquad (2.21)$$

where the average must be weighted by the Maxwellian distribution. One finds readily :

$$C_E(\tau) \sim N\,\exp\left[-\frac{1}{2}\;K^2\langle V^2\rangle\tau^2\right]\;e^{-i\omega_0\tau} \qquad (2.22)$$

The correlation function and furthermore the spectrum centered on ν_0 are Gaussian functions.
The Gaussian profile is characterised either by the mean time

$$\tau_0 = \left[K\langle V^2\rangle^{\frac{1}{2}}\right]^{-1} \qquad (2.23)$$

or by the linewidth of the spectrum

$$\Gamma_{Hz} = K\langle V^2\rangle^{\frac{1}{2}} \qquad (2.24)$$

Suppose now $K\ell \ll 1$ which is the case for macromolecules diluted in a solvent, $r_j(t)$ is given by a random walk, the corresponding correlation function is then given by [12].

$$C_E(\tau) = C_E(o)\exp\left[-\frac{\tau}{\tau_0}\right]\;e^{-i\omega_0\tau} \qquad (2.25)$$

where the relaxation time τ_0 is

$$\tau_0 = \left[DK^2\right]^{-1}$$

D is the molecular diffusion coefficient given in first approximation by the classical Einstein's formula

$$D = \frac{K_BT}{6\pi\eta a} \qquad (2.26)$$

a is the particle radius, and η the viscosity of the fluid. By taking the Fourier transform, the spectrum is a Lorentzian profile centered on ν_0 with a line width at half maximum equal to DK^2 :

$$S(\omega) = \frac{S(o)}{(\omega - \omega_0)^2 + [DK^2]^2} \quad (2.27)$$

These examples show a general property of the spectrum of scattered light : the spectrum associated with the variable responsible for the fluctuations of E_s is identical to the spectrum of the variable itself but translated of ν_0.

B. Correlated scatterers (see for instance P. Lallemand in[8])

We will restrict to the hydrodynamic limit : $K\ell \ll 1$. Then,

$$E_s \sim \sum_j e^{i\vec{k}\vec{r}_j(t)} = \sum_j \int \delta(\vec{r}-\vec{r}_j)e^{i\vec{k}\vec{r}}\, d\vec{r} \quad (2.28)$$

using the Dirac distribution or $E_s \sim \int \sum_j \delta(\vec{r}-\vec{r}_j)e^{i\vec{k}\vec{r}} dr$

The sum $\sum_j$ is then interpreted as the number density :

$$\sum_j \delta(r-r_j) = \rho_0 + \delta\rho(r,t) \quad (2.29)$$

For $K \neq 0$ the correlation function is then proportional to :

$$C_E(\tau) \sim \int \langle \delta\rho(\vec{r},\tau)\delta\rho(\vec{r}_0)\rangle e^{i\vec{k}.\vec{r}}\, d\vec{r} \quad (2.30)$$

$$C_E(\tau) \sim \langle \delta\rho(K,\tau)\delta\rho(-K,o)\rangle \quad \text{for } K \neq 0$$

Except for the translation ν_0 that we did not mention, this result means that the E_s spectrum is identical to the spectrum of the K-component Fourier transform of the density.

The calculation of the correlation function of density is directly related to the equation of hydrodynamic. For a set of statistically independent variables such as entropy, pressure and concentration the spectrum of scattered light is easily interpreted (fig.4). The Rayleigh spectrum is composed of a line of HWHM equal to $D_T K^2 = \Gamma_s$ associated to entropy fluctuations, D_T is the coefficient of thermal diffusion. A second line of HWHM equal to $D_m K^2 = \Gamma_c$ is associated to concentration fluctuations, D_m is the coefficient of molecular diffusion. Usually - in liquids for instance - Γ_c and Γ_s are quite different. We have added to these spectra a depolarized line associated to orientational motions which is easily observed for anisotropic particles by using a polarizer which transmits only the scattered field perpendicular to the incident field, in this case the others components are eliminated. Finally, the fluctuations of pressure are propagating sound waves which give the Brillouin line shifted by a Doppler frequency : $\Omega_B = KV_s$.
Where V_s is the sound velocity (not the velocity of particles !). For the wave vectors K easily accessible the Brillouin frequency $\nu_B = \Omega_B/2\pi$ belongs to the hypersonic domain from 100MHz to 10GHz. In addition to measurement of velocity of sound, attenuation of

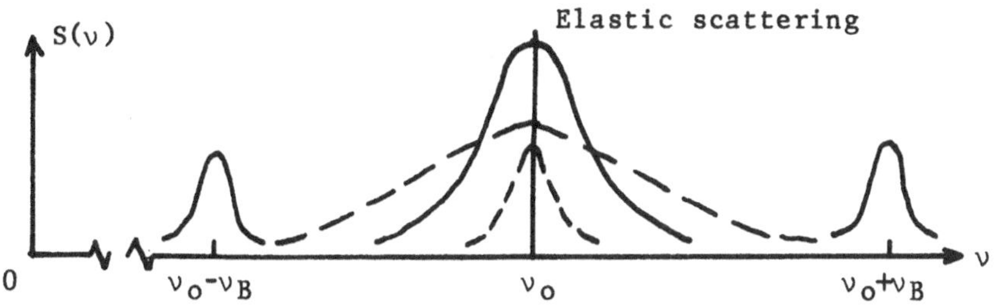

Fig. 4 - The spectrum of elastic and quasielastic scattered light : the central line is the sum of — entropy line ---- concentration line, — — orientational motions line (depolarized)and elastic scattering : a δ -- function convolved with the apparatus function .

hypersons is also obtained from linewidth measurements. This technique which provides a good way to access to viscoelastic properties of media at any temperature,[17] has been used successfully in flows near a transition,[18].

Mie scattering

General theory may be found in[13] but a recent and more extensive survey on the problem is given in[14]. When the size as well as the complex refractive index of a particle must be taken into account i.e. when the phase varies appreciably on the surface of the scatterer the calculations are generally difficult. Rigorous treatments have been done only for sphere, infinite cylinder and under certain conditions for oblate or prolate ellipsoids. The problem is to know both amplitude and phase of the scattered field far from the scatterer. The phase is very important when interference occurs,particularly in L.D.A. applications. Moreover, the calculation of intensity as a function of size and refractive index is of great interest for industry, where the need for measuring particle characteristics in situ appears in a wide range of applications.

It is convenient to use non dimensional sizes $\alpha_i = \frac{\pi a_i}{\lambda}$ where a_i is one of the characteristic dimensions. The curves of figure 5 represent the intensity of scattered light as a function of θ by 0.5μm and 2.0μm diameter polystyrene spheres suspended in water. They show a number of qualitative universal characteristics There are approximately α scattering lobes over the range $0<\theta<180°$. There is always a primary scattering lobe with a maximum at $\theta=0°$

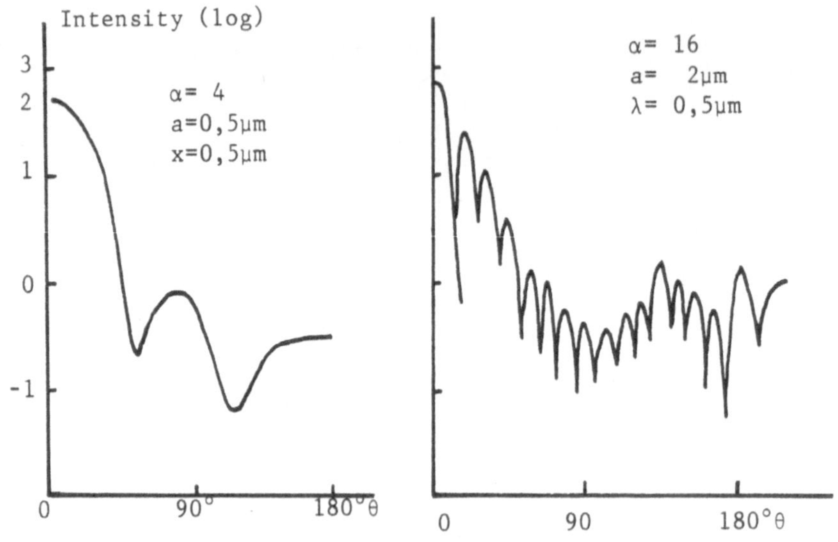

Fig.5 - Typical Mie scattering functions

very strong relatively to other lobes. The forward (θ=0°) and backward (θ=180°) scattering intensities which are equal for small particles are quite different for $\alpha>1$. For instance for 3μm diameter particles often used in LDA the forward scattered intensity is about 10^3 times greater than the back scattered intensity. Fig.6 shows a typical polar diagram of intensity which gives at first sight the scattering properties of the particle.

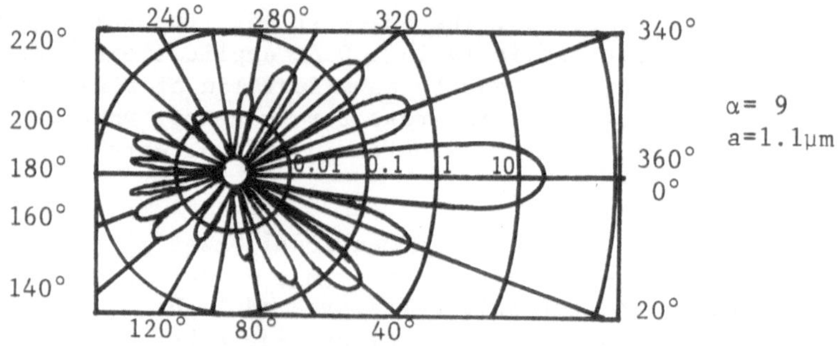

Fig.6 - Polar-log diagram of a typical Mie scattering function.

SOME APPLICATIONS OF RAYLEIGH AND MIE SCATTERINGS IN FLOW STUDIES

In the following we propose several examples chosen among numerous applications. These present techniques are widely used, others will be found with further details in references.

Temperature and velocity fluctuations measurements from Rayleigh scattering

From 2.30 it can be seen that the intensity of scattered light is proportional to

$$C_E(o) \sim \langle|\delta\rho(K,o)|^2\rangle \qquad (3.1)$$

where $\delta\rho(k,o)$ is the K-fourier component of the density of the medium - a gas for instance - where we suppose the local equilibrium to be reached. In other words the intensity is proportional to the mean square fluctuation in number of particles which is given by :

$$\langle\delta N^2\rangle = V\rho^2 K_B T X_T \qquad (3.2)$$

which follows from statistical fluctuation theory (In the hydrodynamic limit $k\to 0$, it is possible to ignore the K dependence and

$$\lim_{k\to o} \delta\rho(K) = \lim_{k\to o} \int d\vec{r}\, e^{i\vec{k}.\vec{r}}\, \delta\rho(\vec{r}) = \delta N).$$

In 3.2 ρ is the mean number density, V the scattering volume, X_T is the isothermal compressibility. The principle of the temperature T measurement is contained in 3.2. Of course the intensity depends also on the nature of species and on their variations through the polarizability which has not been explicited for sake of simplicity. However, in many combustion problems for instance where the major species is nitrogen in both products and reactants the intensity depends essentially on T. In fact, absolute measurements are very difficult and the main application of the method is to measure the variations of temperature in jet diffusion or premixed flame,[19]. The experiment is very simple. The scattered light is collected by a large aperture lens and the scattering volume imaged onto a photomultiplier (fig.7). An interference filter is necessary to reduce background from flame fluorescence. From the signal it is possible not only to obtain probability density functions as a function of the distances in the burner but also power density spectra Fig.8. Time-space correlation function has been also reported,[20]. In our opinion such a technique is very useful in laboratory experiments where the gases can be carefully filtered but in hostile environments the stray light due to dust particles or soot (in rich flames) may render the measurement illusive. The Rayleigh scattering method has also been used successfully in the determination of the flame structure i.e. the temperature profile,in a turbulent flow,[21].

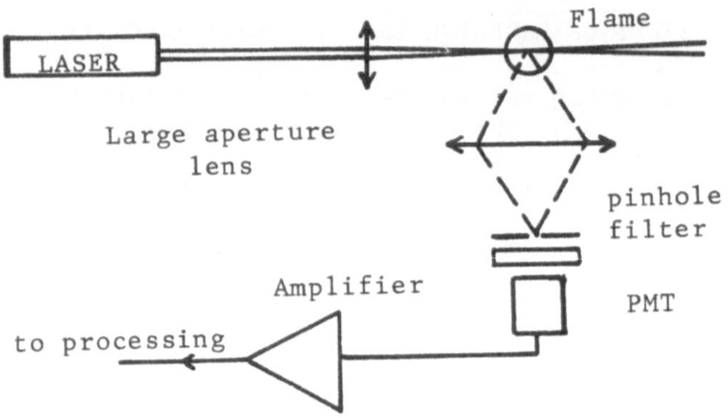

Fig.7 - Collection and detection of scattered light in intensity measurements.

Finally the same technique can also be used to obtain the mean square velocity $\langle \Delta V^2 \rangle$,[22,23,24] in turbulent flows where $\langle \Delta V^2 \rangle$ is directly related to pressure variations responsible for variations in scattered intensity.

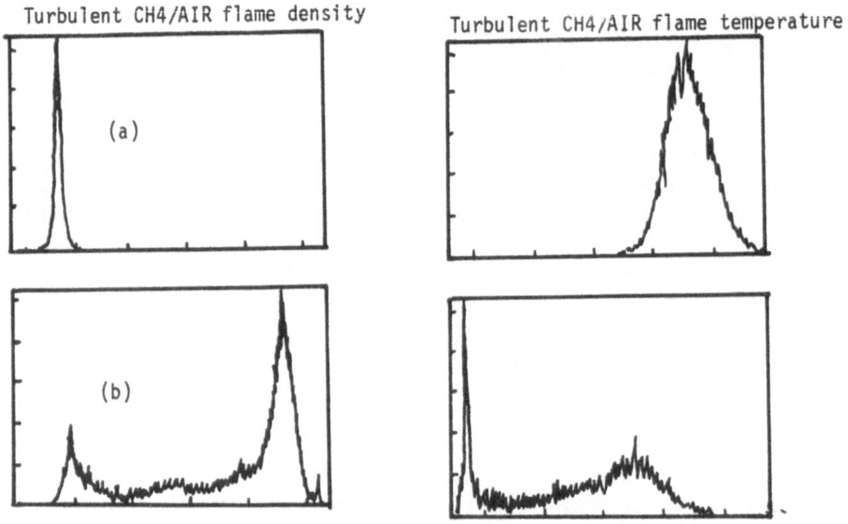

Fig.8 - Probability distribution of relative number density and temperature in a turbulent premixed flame a/ top b/ just ahead (ref. 19)

Concentration mapping in a mixing layer

This is an interesting application of Mie scattering : a study of the turbulent mixing layer in an axisymetric free jet,[25]. The jet is seeded with submicron particles (fig.9) and illuminated with a argon ion laser focused into a thin sheet passing through the jet axis. This plane is imaged onto a TV camera and the irradiance of each point, proportional to the light scattered by the particles, is then proportional to their number density. It is then possible to get local information on turbulent mixing or in other words on turbulent diffusion. To obtain an instantaneous concentration distribution the TV camera is gated on for 10μs. The reading time (with analog-digital conversion) for 10^4 points is equal to about 1s.

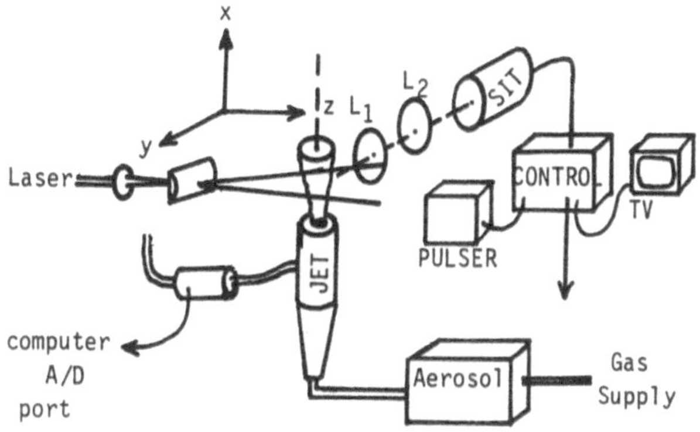

Fig.9 - Experimental arrangement for concentration mapping ref.25

It is thus impossible to get any information concerning the time evolution. Nevertheless, the profiles of mean concentration (fig.10a) and r.m.s. concentration fluctuations (fig.10b) have been measured with accuracy and excellent resolution.

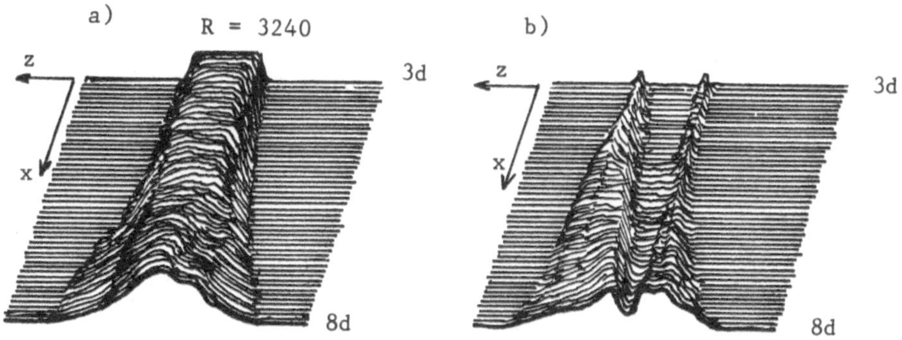

Fig.10 - Mean concentration a) and rms concentration fluctuation profiles (drawing from ref.25)

Laser tomography in combustion

Measurements of the flame position $\alpha(x,y,t)$ and of the flame velocity $\dot{\alpha}(x,y,t)$ are of prime interest in turbulent combustion studies. One of the relevant questions is the relation between the turbulent flow and the dynamics of the flame front.

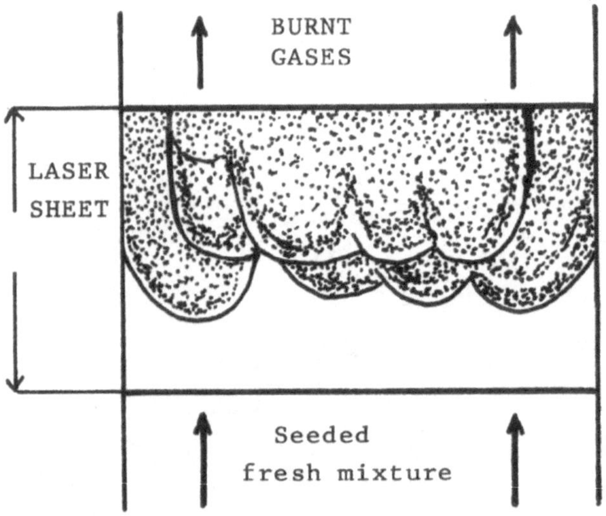

Fig.11 - Tomography of a turbulent flame front

We briefly describe the principle of the method, further details are given in,[26]. When a fresh (unburnt) gas mixture propagating along z (fig.11) is seeded with small oil droplets (1μm in diameter) such as used in laser anemometry, a laser beam passing through the mixture is visualized because the light is strongly scattered in all directions (regardless the anisotropy of scattering which has no influence). On some surface $\alpha(x,y,t)$ the gases burn in a thin zone of about 0.1mm in which the oil droplets vaporize (and then burn). A striking feature of the absence of droplets in burnt gases is the vanishing of light scattering. The laser beam cannot be visualized in the burnt mixture and the position of the frontier between the bright and dark regions is $\alpha(x,y,t)$(fig.11). The fig.12a shows a device where a vertical sheet of laser light crosses the burner. This sheet is imaged onto one or two slits (in case of space-correlation). The light flux passing through a slit is detected by a photomultiplier which gives a signal proportional to the flame position. Taking the derivative of this signal, we easily obtain $\dot{\alpha}$, the instantaneous fluctuating part of the velocity of the flame front. Several applications of this technique can be made :

a/ Stabilization of a flame. (Fig.12b)
In some cases, such as the study of flame instabilities,[27] or the study of the dynamical properties of flame fronts,[28], it is necessary to obtain an unattached flame whose average position is constant with respect to the burner. This situation can be achieved with a slightly divergent duct in which the flame is dynamicaly stabilized. The average position in the duct is then the position where the mean flame velocity and the mean flow velocity are equal. Nevertheless in a divergent duct the longitudinal and transverse velocity gradients stabilise the flame against local perturbations and these effects must be taken into account in the evolution equation of the flame front. In some circonstances they can be of the same magnitude as the intrinsic terms and thus can affect and even dominate the dynamical properties of the flame front. Moreover any velocity gradient must be avoided in experiments on cellular structures and therefore a cylindrical burner must be used.
Starting from a flat velocity profile in the cylinder, the evolution of this profile with z is very small on distances small with respect to the burner diameter and the gradient is negligible. Thus the flame front has a neutral stability position. At a first approximation,[3] we can write :

$$\dot{\alpha} = \bar{U} - U_F + u(t) \tag{3.3}$$

Where $\bar{U}$ is the mean flow velocity, U_F is the mean flame velocity and $u(t)$ is the fluctuating (turbulent) part of the flow velocity. The flame position is thus given by :

$$\alpha(t) = (\bar{U} - U_F)t + \int_0^t u(t')dt' \tag{3.4}$$

and it is evident that in practice the first term is never strictly zero and drifts will always be observed. By controlling the flow with a servo-loop it is possible to stabilize the flame front position from the tomography signal. Of course the time constant must be carefully choosen so as not to modify the intrinsic dynamics of the flame.

b/ Dynamical behavior of flame front in turbulent flow.
This subject is presented in detail further.

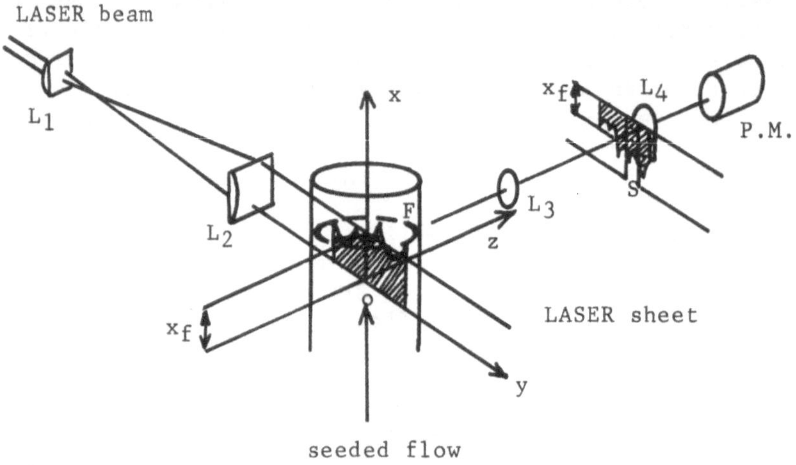

Fig.12a - Schematic arrangement for obtaining the instantaneous flame front position and velocity

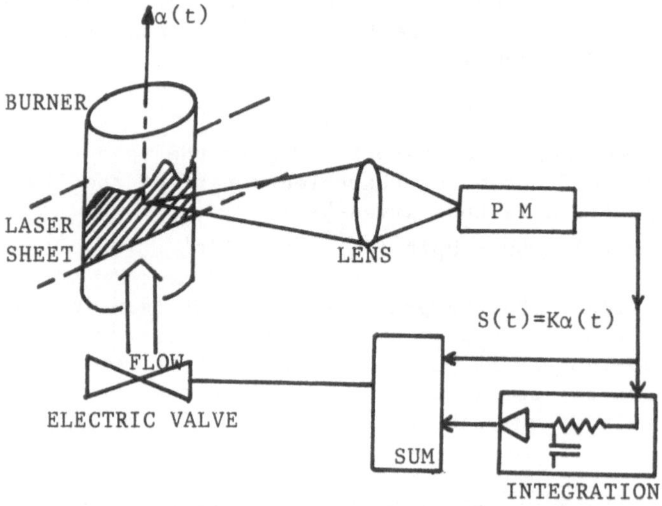

Fig.12b - A flame stabilization device

c/ Application in combustion engines

The behavior of a spark ignited flame has been studied and reported in,[29]. In this experiment, the gas mixture propagates vertically with a velocity higher than the flame velocity. Then the small expanding flame is carried upward with the gas flow. The laser sheet is horizontal and can be displaced along the vertical axis z. Like in the above devices the gas is seeded with oil droplets. Inside the front which is spherical at least just after ignition, there are no droplets, thus the laser sheet shows a dark circle with a diameter depending on time through the evolution of the flame. The light flux is a linear function of the area of the circle and the flame velocity can be deduced. It is thus possible to study the dependence of velocity as a function of different parameters such as the spark gap or the spark energy. In these experiments where a quantity associated to the scattered light is measured, the seeding must be uniform and independent of time. These requirements can be easily achieved at constant pressure. However inside a combustion chamber for instance where the pressure varies strongly with time, the scattered flux depends on time with the density of droplets. Nevertheless the frontier between unburnt and burnt gases is always visible and the position of the discontinuity in brightness of the laser ribbon gives the flame position. A more sophisticated image processing will be necessary in such investigation.

d/ Scanning laser tomography

In this configuration a non-expanded laser beam is swept along two vertical planes zx and zy. A horizontal square section of the burner of about $10 \times 10 cm^2$ is thus analyzed into n lines, the number of resolved points per line is given by 10/d(cm)where d is the diameter of the laser beam in the neighborhood of the flame. The displacement of the beam is obtained using either a set of n mirrors fixed on a rotating wheel or two opto-electronic deflectors. The "video-signal" collected by a photomultiplier gives an analogical picture of the surface $\alpha(x,y,t)$. An instantaneous map is obtained if the image-frequency is at least two times the highest frequency in the spectrum of $\alpha(x,y,t)$.

The signal/noise ratio in laser tomography

Two fundamental sources of noise must be considered, the fluctuation in the density of the seeding particles and the photon shot noise.

The probability distribution of seeding particles in the laser beam is Poissonian to a good approximation. The r.m.s. fluctuation in the number of seeding particles is thus $\sqrt{N}$ where N is the number of different particles present in the scattering volume summed over a time τ_m

$$N = \rho v \left[1 + \frac{\tau_m}{\tau_t}\right] \qquad (3.5)$$

Where ρ is the seeding density, v is the scattering volume, τ_m is the measurement time and τ_t is the transit time of the particles swept through this volume. The seeding limited signal to noise ratio is thus

$$S/N_s = [\rho v(1 + \tau_m/\tau_t)]^{\frac{1}{2}} \tag{3.6}$$

and is optimised by increasing the seeding density and the size of the scattering volume.

In the absence of seeding noise the signal-to-noise ratio for a shot noise limited measurement is well known and is given simply by

$$S/N_p = (n\ \tau_m)^{\frac{1}{2}}$$

where n is the mean number of photons detected per second. Finally, the ratio of photon noise to seeding noise is :

$$R = \left[\frac{\rho v}{n(\frac{1}{\tau_m} + \frac{1}{\tau_t})} - 1\right]^{\frac{1}{2}}$$

and thus the condition that photon noise be negligible is that the number of photons detected in the measurement time (whichever be the smaller) be much greater than the average number of particles in the scattering volume. To conclude, it is worthwhile coming back to the vaporization time τ_v of the particle. This is the only limitation of the method, it concerns the frequency domain. This time calculated for a 1μm droplet is about 10^{-6}s[30]. It must be smaller than the residence time τ_r of the droplet in the flame :

$$\tau_r = \ell/U_F$$

where ℓ is the flame thickness and U_F the flame velocity. At atmospheric pressure ℓ is about 0.3mm and the condition $\tau_v < \tau_r$ gives $U_F < 300\text{ms}^{-1}$. Thus laser tomography in combustion can provide a visualization of flame fronts even in a strongly turbulent medium such as the combustion chamber of jet propulsion engines.

Laser Doppler velocimetry

The basic formula of L.D.V. has been given by 2.15 and 2.16, the angular Doppler frequency Ω_D is simply given by :

$$\Omega_D = \vec{K}.\vec{v} \tag{3.7}$$

or using the magnitude of K , the Doppler frequency is

$$\nu_D = \frac{\Omega_D}{2\pi} = 2\nu_0 \frac{v_{/\!/}}{c} \sin\theta/2 \tag{3.8}$$

Whatever the technique of detection, the measurement will give the particle velocity. Then, to measure the flow velocity in turbulent medium it is necessary to choose carefully the size and density ρ_p (if possible) of the particles. The general motion equation for a sphere is given by (Landau, fluids dynamics) :

$$\frac{4}{3}\pi a^3 \rho_p \frac{d\vec{v}_p}{dt} = -6\pi\eta a(\vec{v}_r) + \frac{4}{3}\pi a^3 \rho_f \frac{dv_f}{dt}$$

$$- \frac{2}{3} \pi a^3 \rho_f \frac{d\vec{v}_r}{dt} - 6a^2 \sqrt{\pi \eta \rho_f} \int_{t_o}^{t} \frac{dv_r}{dt'} \frac{d'}{\sqrt{t-t'}} \qquad (3.9)$$

where index p and f refer to the particle and to the fluid respectively, η is the viscosity. This equation has been studied extensively,[31] but to a good approximation, in laser velocimetry only the Stokes drag term can be considered. The time response of a particle to variations of velocity is equal to

$$\tau_p = \frac{2}{9} \frac{a^2 \rho_p}{\eta} \qquad (3.10)$$

For example, assuming a 1μm oil dropplet in air at 1 atmosphere, then $\tau_p \simeq 3\mu s$..and the frequency forwhich it follows the flow within 3db is 50KHz. In some circonstances the settling velocity will be also taken into account, for the above example this velocity is about $30\mu ms^{-1}$. Finally it is whorwhile noticing that LDV can give both velocities of big particles and flow velocity when properly seeded. Except in the case of very high velocities where classical interferential spectrometry (spherical and plane Fabry-Pérot interferometer) is applied, the measurement of ν_D is obtained by optical heterodyning (or optical mixing) of the scattered field with a reference field like in radiotechniques. Because all photodetectors-eye, photographic plate, photodiode, photomultiplier-have a quadradic response the superposition of two fields with different frequencies will give a time dependent signal with a frequency equal to the difference of frequency of the two incoming beams. As an example (fig.13.a) consider the light intensity when the laser and scattered light are added :

$$I = \{E_0 \cos(\omega_0 t + \phi_1) + E_s \cos [(\omega_0 + \Omega_D)t + \phi_2]\}^2$$

The signal is proportional to the mean value of I integrated on the response time of the detector (10^{-8} s for instance for a photomultiplier). Then :

$$I = E_o^2 \cos (\omega_0 t + \phi_1) E_s^2 \cos^2 [(\omega_0 + \Omega_D)t + \phi_2\}$$

$$+ 2E_o E_s \cos(\omega_0 t + \phi_1) \cos [(\omega_0 + \Omega_D t + \phi_2)]$$

and

$$\langle I \rangle = \frac{1}{2} (E_o^2 + E_s^2) + E_o E_s \cos(\Omega_D t + \phi) \qquad (3.11)$$

The alternative current has just the Doppler frequency. Two requirements must be satisfied, the signal E_s must be high enough compared to the shot noise and the phase difference $\phi = \phi_2 - \phi_1$ must be less than the phase associated to the coherence length L_c of the laser :

$$\phi_c = \frac{2\pi L_c}{\lambda}$$

The best adjustment is obtained when the optical paths are equal. This is easily performed in the configuration of fig.13.b which is generally called the "real fringe" system. Two equivalent models can be used. The more convenient is to describe the modulation of scattered light when a particle moves in the interference zone of the two laser beams. Then the modulation frequency is :

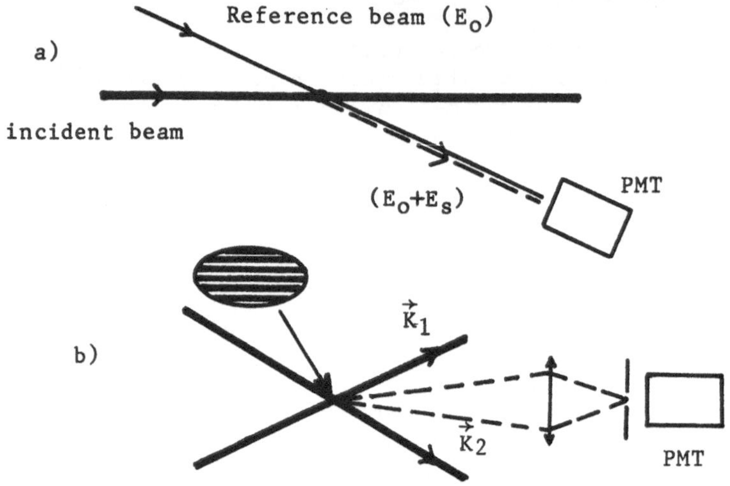

Fig.13 - Typical configurations in L.D.V.

$$\nu = \left(\frac{i}{v_{/\!/}}\right)^{-1}$$

where i is the fringe spacing and $v_{/\!/}$ the component of velocity parallel to $\vec{K}_f = \vec{K}_2 - \vec{K}_1$. If we denote here θ the angle $(\vec{K}_2, \vec{K}_1)$ then

$$\nu = 2\nu_o \frac{v_{/\!/}}{c} \sin\frac{\theta}{2} \qquad (3.12)$$

This result is similar to formula 3.8 where θ is the scattering angle. It is not surprising : as a matter of fact the scattered light impinging on the photomultiplier comes from particles irradiated with two different laser beams. So a simple calculation of the sum of the two scattered waves with two different Doppler frequencies will give the result 3.12. Both "real fringes" or light beating descriptions are adequate. This configuration is the most often used not only because it is convenient for practicing but also, as it can be seen, the signal frequency does not depend on the observation direction and large collecting apertures can be used. Other optical configurations are reserved to specific problems. Nevertheless, in all cases the above requirements must be satisfied.

The optical systems, particularly the beam splitters will not be described here and will be found in references. Theory and practice are well exposed in T.S.I. Manual,[32] or in the book of F. DURST et al,[33]. More details will be found in the Durrani and

Greated's,[34] book or in the Drain's,[35]. Moreover the reader is invited to refer to,[36,37,38,39]. L.D.V. is essentially limited by photon noise. The signal/noise ratio is given by,[40] :

$$S/N = \left(\frac{\text{amplitude of a.c. Doppler current}}{\text{shot noise}}\right)^2$$

$$= \frac{I_o}{4\hbar\omega}\ \frac{\eta}{K^2}\ \frac{1}{\Delta f}\ \frac{D^2}{P}$$

where I_o is the laser intensity, η the quantum efficiency of the detector, Δf the filter bandwidth (it is fixed by the Doppler frequency range associated to turbulence), D^2 is a parameter associated to the particle and collecting optics characteristics and finally P is a parameter associated both to the visibility of fringes and to the size of the particle compared to the fringe spacing. Generally, the intensity scattered by the particles seeded in the flow is roughly proportional to a^2,so that it is better to choose first the proper particle size instead of buying an expensive high power laser. However in gases, small particles of about 1μm diameter must be used with a one to ten watts laser power. To give an example, consider a typical foward scattering device using 1mm diameter beams from a 5m W He-Ne laser, focused by a lens of focal length 200mm and an angle θ equal to 20°. With a collecting angle of 10° and an oil droplet of 1μm in diameter one finds S/N=500 if Δf is 10MHz and η=0,1.

Diffusion broademing spectroscopy : an in situ measurement of particle size

This technique is L.D.V. on Brownian motion of particles diluted in a fluid. We suppose first that the fluid is at rest. The spectrum of scattered light by such particles given above (eq.2.27) is then :

$$S(\omega) = \frac{S(o)}{(\omega-\omega_o)^2+[DK^2]^2}$$

It is characterized by a Lorenzian profile with half-width at half-maximum equal to DK^2. D is the diffusion coefficient given either by 2.26 as a first approximation or by a modified Stokes-Einstein relationship :

$$D = \frac{KT}{6\pi\eta a}\ C(\ell/a)$$

where the slip conection factor $C(\ell/d)$ is given by [41] and ℓ/a is the mean free path/ particle size ratio. For monodisperse suspensions it is then easy to get the size from linewidth measurements (note the K^2 dependence, which isadecisive test). The detection technique is as well the light beating spectroscopy,[8] with harmonic analysis of the photocurrent spectrum or with photon correlation spectroscopy,[8,38] a technique which is utilized when the scattered light intensity is very low. In this case, the photocurrent is a series of pulses associated to photoelectrons which can be counted and digital correlators are then used. Photon correlation spectroscopy is also applied in L.D.V. experiments,[38] The applications of diffusion broadening spectroscopy are numerous. Examples can be found in biology (size of ribosomes of Escheridia Coli), medicine (size of spermatozoa), in solid state physics (crystal growth) etc... In combustion studies, this non-invasive technique is extensively applied to sprays and soot,[42,43]. Its limitation lies in the difficulty to come back to a size-distribution when the suspension is polydisperse. Generally a type of distribution is assumed and the calculated spectrum is compared to the recorded one. In my opinion, due to the limitations associated with the signal/noise ratio such a determination is often illusory. When the particles are suspended in a flow, the line is shifted by the Doppler frequency associated with the fluid velocity, this implies the selfbeating technique (without reference beam). However, the finite transit time of the particles carried by the flow in the laser beam must be much larger than the diffusion time $[DK^2]^{-1}$.

Application of Mie scattering to particle size measurement

An important literature is devoted to the application of Mie scattering to particle sizing,[14]. However, no one at the present time is able to compute rigorously the properties of the light scattered by an arbitrary shaped particle, nevertheless by assuming a spherical shape good results can be obtained. In addition to improvements in computational techniques, ameliorations in optical devices allow for better confidence in measurements. For example consider one small particle passing through the center of the scattering volume and a big one passing one edge. It is then very difficult, if not impossible to distinguish them only from the scattered intensities which may be equal. This shortcoming has been extensively studied and partially overcome by inversion mathematical scheme but it is completely eliminated by using two laser beams of distinct colors simultaneously. The first one with a top hat intensity profile obtained with a proper holographic filter is used for Mie scattering intensity measurements. On the other hand, the waist of the second laser beam focused on the same point is smaller than the first one. So the intensity is only measured when two scattering signals are obtained simultaneously, since then the particle is necesseraly within the control volume of the first laser. Furthermore, two crossing beams - instead of one - are used to locate the particle and give the L.D.V. information. Such device which gives simultaneously velocity and size has been recently described,[44].

EXPERIMENTAL STUDY OF THE FLOW-FLAME INTERACTION

We consider a wrinkled flame in a vertical burner where the

fresh mixture is propagating upward. The mean flame position is at rest in the laboratory frame. Two of the relevant questions are : how is the dynamic of flame under the action of a turbulent flow and how does it modify the flow up- and downstream ? Three mechanisms must be taken into account.

a/ The transverse diffusions of mass and heat associated to the wrinkles which change the local temperature and the local concentration of species. It can be shown that except for some light gases (H_2) these processes have stabilizing properties. In other words a wrinkled flame initially plane in a laminar flow tends to restore a plane shape. The parameter which describes these properties is specific for a given flame. The Markstein's length , is a function of the Lewis number Le (ratio of thermal to molecular diffusivities) and of the gas expansion γ:

$$\gamma = 1 - \frac{\text{burnt gas density}}{\text{fresh gas density}} \simeq 0.8$$

b/ The gas expansion associated with the variation of temperature of about 5 leads to the classical configuration of two non-miscible fluids of different densities. The surface of discontinuity between the fluids is an horizontal plane with the lighter fluid above, this is just the case of our flame when the fresh mixture is going upward. To conclude, gravity is a stabilizing or non-stabilizing effect according to the downward or upward propagation of the flame.

c/ The last effect which is purely hydrodynamic is the modification of the velocity field by wrinkling.

At a first approximation the local structure of the wrinkled flame is that of a planar laminar flame and thus the normal velocity of the front relative to the fresh gases u_L may be considered as constant. Behind the flame, the normal velocity u_B is increased, at constant pressure it is easy to show :

$$u_B/u_L = T_B/T_o$$

This ratio is greater than 5. So on the flame front there is a discontinuity of the normal gas velocity with a continuity of the tangential component. Now, a stream line which impinges on the flame front with a non zero incidence angle is deviated in a way similar to geometrical optics (fig.14)*. However this affects all

*Note however that the effect of "refraction" of stream lines (in flame) is somewhat different, the Desçartes's law must be replaced by $tg_i = \frac{T_B}{T_o} tg_r$, where i and r are the incidence and refraction angles respectively.

the flow field with a characteristic length of about the wavelength of the wrinkles. If only this effect is considered the flame is always unstable (fig.14). But, when all mechanisms are taken into account it can be shown that for a given mixture, there is a velocity threshold u_c under which the flame is stable for all wavelengths,[45]. The experiments presented in the following are performed in the stable regime of a propane-air flame diluted in nitrogen. Consider a modulation of the upstream velocity field, the flame front undergoes a wrinkling and induces a modification of the flow. This feedback effect has been recently analysed,[46] and transfer functions have been calculated. Far upstream the velocity field is not influenced by the flame and behaves as an exciter. From its spatio-temporal spectrum the transfer functions allow to calculate the velocity spectrum at any distance of the flame front and moreover the flame velocity spectrum. Let $x = \alpha(y,z,t)$ be the surface which describes the flame front. The evolution equation is

$$\dot{\alpha} = U_{x=\alpha} + C(\Delta\alpha - \frac{1}{u_L}\frac{\partial U}{\partial x}\Big|_{x=\alpha}) \qquad (4.1)$$

In this expression $\dot{\alpha}$ is the local fluctuation of flame velocity, $u_{x=\alpha}$ is the actual gas velocity at the flame front, C a constant function of the Markstein length and the last term is a curvature effect which can be neglected in a first approximation. The feedback effect comes from two relations :

a/ For small perturbations the flow velocity field is :

$$u = u_e + u_i \qquad (4.2)$$

where u_e is the exciter and u_i the induced velocity field.

b/ This velocity field is given in Fourier space by :

$$u_i = e^{kx} \frac{-\gamma g k + \frac{\gamma}{1-\gamma} k^2 u_L^2 + O(k^3)}{(2-\gamma)i\omega + 2ku_L} \alpha \qquad (4.3)$$

where k is the transverse wave vector (perpendicular to the mean flow velocity), x is the distance to the flame front. It can be seen on 4.3 that the gravity and hydrodynamic effects are strongly coupled. The classical velocity field associated to gravity waves is obtained when $u_L=0$; if the gravity effect is neglected then the k^2 term is the purely hydrodynamic one. Suppose k and ω very small, and neglect in 4.1 the diffusive term, then on the flame front (x=o) :

$$u_i \simeq - \frac{\gamma g}{2u_L} \alpha$$

and

$$\dot{\alpha} = - \frac{\alpha}{\tau g} + u_e \qquad (4.4)$$

where $\tau_g = \frac{2u_L}{\gamma g}$ is the relaxation time associated to the stabilizing mechanism of gravity. Equation 4.4 shows that a negative feedback occurs for $\omega\tau_g<1$ to give $u_i= -u_e$ in the low frequency limit. In the general case the diffusive term in 4.1 will give rise to a small difference. The general result is obtained from computational calculations. The predicted spectral density of the flame speed $S_{\dot{\alpha}}(\omega)$ and of the total velocity field $S_u(\omega)$ at different

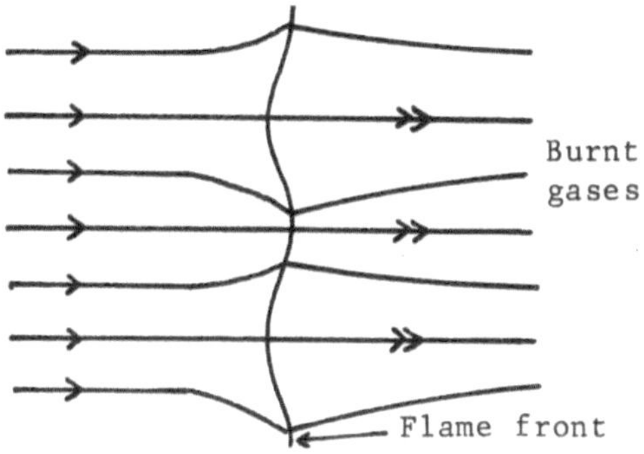

Fig.14 - The hydrodynamic Landau's instability and modification of the flow field

distances from the flame front are plotted fig.15. In order to check these theoretical results the flames were produced in a

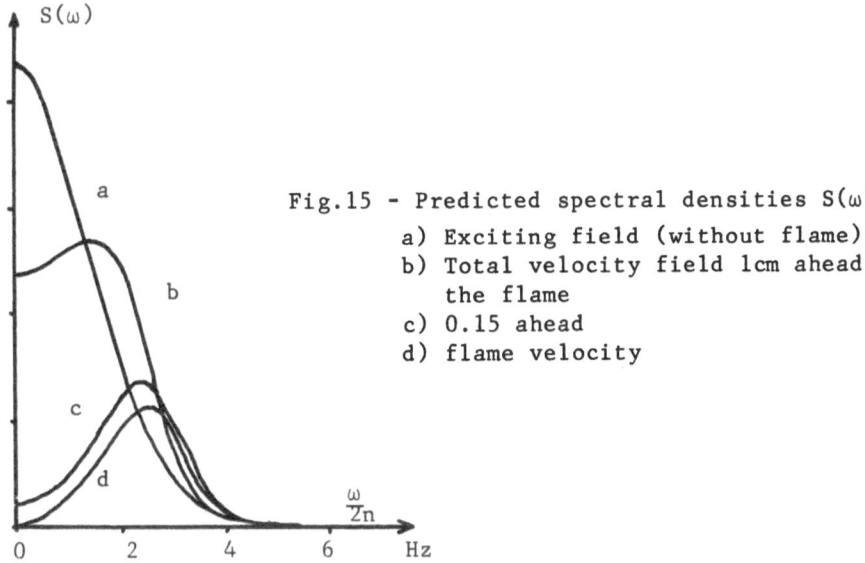

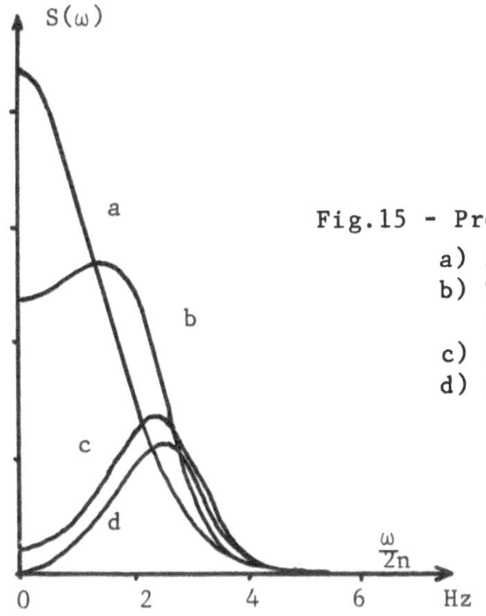

Fig.15 - Predicted spectral densities S(ω)
a) Exciting field (without flame)
b) Total velocity field 1cm ahead the flame
c) 0.15 ahead
d) flame velocity

cylindral glass tube 8cm in diameter. The turbulence of the gas stream was generated in a narrow pipe and then expanded to produce the required low flow velocity. An upstream grid was used to homogenize the turbulence. A downstream cooled grid permitted to decouple the dynamics of the flame from instabilities occurring downstream. The gas velocity was measured by L.D.V. The position of the front $\alpha(t)$ was measured by laser tomography and the flame speed obtained by taking the analogical time derivative of $\alpha(t)$. Further details are given elsewhere,[47]. The experimental results are plotted on figure 16 and show a good agreement with the predictions. It is worthwhile noticing that the two spectra $S_u(\omega)$ and $S_{\dot\alpha}(\omega)$ are obtained and calibrated by two independent techniques and their similitude in form and intensity are quite

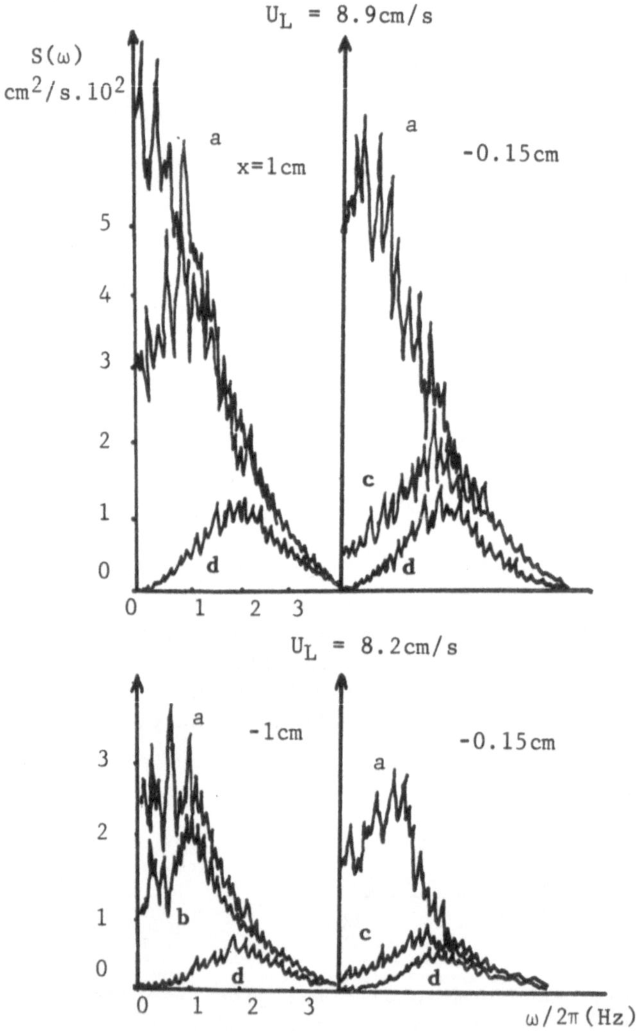

Fig.16 - Spectral densities at 1cm and 0.15cm from the flame front for flame speeds of 8.9 and 8.2cm/sec. a) exciting velocity field
b) Total velocity field 1cm ahead the flame
c) Total velocity field 0.15cm ahead
d) Flame velocity field (ref.47)

remarkable. To our knowledge these results are the first evidence of the feedback effect of a wrinkled flame on the upstream gas flow.

Acknowledgements

I am grateful to Mrs. M. Gillino and to Mr. J.P. Pahin for their technical assistance.

REFERENCES

1. W. Merzkirch, Flow visualization, Acad. Press, N.Y. (1974)
2. D. R. Crosley, Laser probes in combustion chemistry, American Chemical society, Washington DC (1980)
3. P. Clavin and F. A. Willimas, Theory of premixed-flame propagation in large -scale turbulence, J. Fluid Mech. 90, 589 (1979)
4. P. Clavin and F.A. Williams, Effects of molecular diffusion and of thermal expansion..., J. Fluid Mech. 116, 251 (1982)
5. F. J. Weinberg, Optics of flames, Butterworths, London (1963)
6. R. M. Fristrom and Westenberg, Flame structure, Mc Graw Hill, NY (1965)
7. P. Clavin, Dynamical behavior of premixed flame fronts in laminar and turbulent flows, in Prog. Energy Combust. Sci., Pergamon Press, Oxford to appear (1984)
8. H. Z. Cummins and F. R. Pike, Photon correlation and light leaking spectrometry, Plenum Press, NY (1973)
9. P. R. Regnier and J.P. Taran, Gas concentration measurement by cars, in Laser Raman gas diagnostic, M. Lapp and C.M. Penney, ed. Plenum Press, NY (1974)
10. F. T. Arecchi, this ASI proceeding
11. B. Crosignani, P. Di Porto and M. Bertolotti, Statistical properties of scattered light, Acad. Press., NY (1975)
12. B. J. Berne and R. Pecora, Dynamic light scattering, John Wiley, NY (1976)
13. M. Born and E. Wolf, Principles of Optics, Pergamon Press, Oxford (1975)
14. L. P. Bayvel and A.R. Jones, Electromagnetic scattering and its applications, Appl. Sci. Publishers, London (1981)
15. B. Chu, Laser light scattering, Acad. Press, NY (1974)
16. I. L. Fabelinskii, Molecular scattering of light, Plenum Press, NY (1968)
17. R. Vacher and L. Boyer, Brillouin scattering : a tool for the measurement of elastic and photoelastic constants, Phys. Rev. B6, 639 (1972)
18. M. Hareng and J. Leblond, Brillouin scatterinng in superheated water, J. Chem. Phys. 73(2), 622 (1980)
19. R. W. Dibble and R. E. Hollenbach, Laser Rayleigh thermometry in turbulent flames, Proceedings of the 18th Symposium (Inter-

national) on combustion, The Combustion Institute, 1489(1981)
20. M. Mamazian, L. Talbot, F. Robben and R.K. Cheng, Two-point Rayleigh scattering measurements in a V-shaped turbulent flame, Proceedings of the 19th Symposium (International) on Combustion, The Combustion Institute, 487 (1982)
21. J. R. Smith, The influence of turbulence on flame structure in an engine, Proceedings of the ASME winter meeting Phoenix (1982)
22. P. J. de Gennes, The observation of Pressure and velocity correlations in a turbulent flow, C.R. Acad. Sc, Paris, 262, 74 (1966).
23. H. L. Frish, Study of turbulence by spectral fine structure of scattered light, Phys. Rev. Let. 19, 1278 (1967)
24. J. C. Lelievre and J. Picard, Observation of laser light scattering by a turbulent gas without seeding, Phys. Letters A, Netherland 38A, 267 (1981)
25. M. B. Long and B.T. Chu, On the mixing and structure of an axisymetric turbulent mixing layer, Proceedings of the AIAA 13th fluid and plasma dynamics conference, Snowmass (1980)
26. L. Boyer, Laser tomographic method for flame front movement studies, Combust. and flame, 39, 321 (1980)
27. J. Quinard, G. Searby and L. Boyer, The stability limits and critical size of structures in premixed flames, Proceedings of the 9th International Colloquium on dynamics and explosions and reactive systems, Poitiers (1983)
28. G. Searby, F. Sabathier, J. Monreal, P. Clavin and L. Boyer, The feedback of a flame front on turbulent flows, ibid.
29. G. G. de Soete, Measurement of initial flame speed by laser tomography, Colloque International Berthelot-Vieille-Malard Le Chatelier, The Combustion Institute, Bordeaux 49, (1981)
30. G. A.E. Godsave, Studies of the combustion of drops in a fuel spray. The burning of single drops of fuel, Proceedings of the fourth Symposium International on Combustion, Cambridge (USA) 818 (1952)
31. A. T. Hjelmfelt and L.F. Mockros, Motion of discrete particles in a turbulent fluid, Appl. Sci. Res., 16, 149 (1966)
32. L. M. Fingerson and R. J. Adrian, "Laser velocimetry, theory, applications and techniques" TSI LDV Short course St Paul (USA) (1978)
33. F. Durst, A. Melling and J.H. Whitelaw, Principles and practices of L.D.A. Academic Press, NY (1976)
34. T. S. Durrani and C.A. Greated, Lasers systems in flow measurements, Plenum Press (1977)
35. L. E. Drain, The laser Doppler technique, John Wiley, Chichester (1980)
36. P. Buchhave et al., The accuracy of flow measurements by laser Doppler methods, Proceedings of the LDA. Symposium Copenhagen (1975)

37. H. D. Thompson and W.H. Stevenson, Laser velocimetry and particle sizing, Proceedings of the 3rd international workshop on LDV Purdue, Hemisphere Pub. NY (1979)
38. E. O. Schulz-Dubois, Photon correlation techniques in fluid mechanics, Proceedings of the 5th International Conference Kiel (Germany), Springer-Verlag, Berlin (1983)
39. D.F.G. Durao et al., Proceeding of the International Symposium on applications of LDA to fluid mechanics, Lisbonne (Portugal)(1982)
40. R. J. Adrian and W.L. Early, Evaluation of LDV performance using Mie scattering theory, in "Proc. of the Minnesota Symposium on laser anemometry, Minneapolis, 426 (1976)
41. N. A. Fuchs and A.G. Sutugin, Highly dispersed aerosols, Ann Arbor Science Pub. Ann. Arbor (1970)
42. M. Weill, P. Flament and G. Gouesbet, The status of the art in soots diagnostics by means of diffusion broadening spectroscopy, Proceedings of the NATO workshop on soot in combustion system, Le Bischenberg France, Plenum Pub. (1981)
43. S. S. Penner and P.H.B. Chang, Particle sizing in flames, in "Gas dynamics of detonations and explosions", Progress in Astronautics and aeronautics, vol.75, AIAA pub. NY (1981)
44. P. Flament, M.E. Weill and G. Gouesbet, Measurement of soot diameters by means of diffusion broadening spectroscopy. In ref.39 proceedings.
45. P. Pelcé and P. Clavin, Influence of hydrodynamics and diffusion upon the stability limits of laminar premixed flames, J. Fluid Mech., 124, 219 (1982)
46. G. Searby and P. Clavin, The induced velocity field by turbulent flames, Combust. Sci. Tech. submitted
47. G. Searby, F. Sabathier, P. Clavin and L. Boyer Phys. Rev. Let., to appear (issue of 17 oct. 83).
48. M. Fermigier, M. Cloitre, E. Guyon and P. Jeuffer, Application of forced Rayleigh scattering to studies of laminar and turbulent flows, J. Mec. Theor. Appl., 1, 123 (1982)
49. J. W. Daily, Laser induced fluorescence spectroscopy in flames, in "Laser probes for combustion chemistry", American Chemical Society, Washington DC (1984)
50. J.E.M. Goldsmith, Resonant multiphoton optogalvanic detection of atomic hydrogen in flames, Opt. Lett. 7, 437 (1982)

THEORY OF GASEOUS COMBUSTION

Paul Clavin[(1)] and Amable Liñán[(2)]

(1) LA-72, Université de Provence
13397-Marseille, France
(2) ETSI Aeronáuticos, Universidad Politécnica
Madrid-3, Spain

I - BASIC CONSIDERATIONS

I.1. Two feed-back mechanisms

The combustion processes are characterised by two non linear feedback mechanisms producing self-acceleration. One is of a thermal nature and the other is purely chemical.

The first one results from the fact that the overall chemical reaction involved in combustion is exothermal with a rate that is a strongly increasing function of the temperature. This produces a self-acceleration of the combustion process that is saturated by the reactants consumption

$$\nu'_F F + \nu'_O O \longrightarrow P + Q \nearrow \tag{1.1}$$

Fuel Oxidant Products Heat release
Reactants

ν' are the stoichiometric coefficients.
For simplicity the reverse reaction is neglected in (1.1).
Let's define W_F, W_O and W_P as the mass consumption (per unit time and per unit volume of reactive mixture) of fuel, oxidant and products respectively. From (1.1) one has :

$$W_F = W_O/\nu = -W_P/(1+\nu), \tag{1.1'}$$

with $\nu = \nu'_O M_O/\nu'_F M_F$ where M_i are the molar mass.

The thermal feedback is described by the nonlinear Arrhenius law :

$$W_F = \rho B\, Y_F^{n_F}\, Y_O^{n_O}\, e^{-E/RT} \tag{1.2}$$

Where Y_i, n_i and E are the mass fractions ($Y_i=\rho_i/\rho$) the order of reaction and the activation energy respectively. B is the frequency factor whose temperature dependance can be neglected in front of exp(-E/RT). $n_i \neq \nu'_i$ results from the fact that (1.1) is not an elementary reaction governed by collisions. For ordinary hydrocarbons flames the number of elementary reactions required to transform the fresh mixture into burnt products is of the order of 300 with 50 intermediate species ! This precludes a complete quantitative description of the combustion. In this course we will focus the attention on the effects that can be described by (1.1) and (1.2). Only few words will be said concerning the effects produced by the complex chemistry and by the diffusion of the intermediate species. The density dependance of B can be easily calculated only in the case of an elementary reaction where $\nu'_i=n'_i=1$ by noticing that, according to the elementary kinetic theory of gases the number of reactive collisions per unit time and unit volume is given by :

$$\frac{W_F}{\nu'_F M_F} = N_F\, N_O\, \underbrace{K\, e^{-E/RT}}_{\text{reaction constant}} \qquad (1.3)$$

with the molar concentration N_i defined by $N_i = \rho Y_i/M_i$ and where the prefactor K is predicted to be a constant independent of ρ and weakly dependent on T.

The second feedback mechanism is produced by the <u>chain branching reactions.</u> These reactions are <u>autocatalytic</u> reactions that produce more active intermediate species than they consume leading to a <u>self acceleration</u> of chemical process. Such a mechanism can be represented schematically by :

R (Reactants)	$\longrightarrow$	X (Active species)	chain generation	(1.4a)
R + X	$\longrightarrow$	2X + P_1 (Products)	chain branching	(1.4b)
m + X + X (m: Third body)	$\longrightarrow$	P_2 + m (Products)	chain termination	(1.4c)

Because of the strong non-linearity in temperature of the Arrhenius law (1.2), the thermal feedback dominates the kinetic one in the ordinary hot flames. And a good insight into the problem can be obtained by ignoring in a first step the details of the chemical kinetics aspects.

I.2. The conservation equations

When the combustion is assumed to be controlled by the overall chemical reaction (1.1) with the rate (1.2), the equations controlling the reacting flow involve the fuel and oxidant mass balance and the conservation of the energy :

SPECIES $\{\rho \frac{\partial}{\partial t} Y_i + \rho \underline{V} \cdot \underline{\nabla} Y_i - \underline{\nabla} \cdot (\rho D_i \underline{\nabla} Y_i)\} = -W_i$, $i = F, O$

transient convection diffusion production (1.5)

ENERGY $\{\rho \frac{\partial}{\partial t}(C_p T) + \rho \underline{V} \cdot \underline{\nabla}(C_p T) - \underline{\nabla}(\lambda \underline{\nabla} T)\} = + q W_F$

Fourier law

where the energy released in the reaction per unit mass fuel q is given by :

$$q = Q/\nu'_F M_F \tag{1.6}$$

C_p is the specific heat of the reactive mixture. D_i and λ are the molecular diffusivities of the species i and the heat conductivity of the reactive mixture respectively.

Two basic assumptions have been made in writing (1.5). The first concerns the transport properties of the reactive mixture where it has been assumed that the binary Fick law holds. This is well verified when the mixture is diluted in inert gases (as for the example the Nitrogen of the air). The second assumption concerns the equation for the energy conservation where the compressible effects have been neglected. These last effects are of a relative order of magnitude of Mach squared, and it is legitimate to neglect them for subsonic combustion but it is worthwhile to mention that they have determinant effects in detonations (supersonic-waves) that will not be considered here. This approximation is called the "isobaric approximation" where it is assumed that, according to the perfect gas law,

$\rho T = C^t$ (for simplicity the change in the molecular mass in also neglected) (1.7)

But it is clear that because of the presence of the flow field $\underline{V}$, the system (1.5) is not closed. In fact because of the expansion of the gas (described by (1.7)), the flow field is influenced by the combustion. Thus, the mass and momentum equations have to be added (in the general case) to the system (1.5)(1.7) and general combustion problems appear as phenomena where hydrodynamics is coupled with diffusion-reaction process. But in some simple cases (as the 1-d and steady case) this coupling disappears.

I.3. The adiabatic temperature of combustion

Consider the case of an homogeneous and adiabatic combustion. In this case, eq.(1.5) reduces to

$$-\frac{\partial Y_F}{\partial t} = +\frac{W_F}{\rho} = \frac{\partial (C_p T/q)}{\partial t} = -\frac{\partial (Y_O/\nu)}{\partial t} \tag{1.8}$$

$$Y_F - Y_O/\nu = Y_{Fu} - Y_{Ou}/\nu$$

$$Y_F + C_pT/q = Y_{Fu} - C_pT_u/q \tag{1.8'}$$

where the subscript u is for the initial mixture (unburnt gases)

$$\frac{\partial Y_F}{\partial t} = -Y_O^{n_O}\, Y_F^{n_F}\, Be^{-E/kT} \tag{1.9}$$

When $t\to\infty$ the reaction is completed by consumption of the limiting component. Let assume, for example, that F is the limiting component, in that case one has

$$t\to\infty\ ,\ Y_F = 0\ ,\ Y_O = Y_{Ou} - \nu Y_{Fu}$$

thus, according to (1.8'), the temperature attains a maximum value given by :

$$t\to\infty \qquad T = T_b$$

$$T_b = T_u + \frac{qY_{Fu}}{C_p} \tag{1.10}$$

eq.(1.10) expresses the conservation of the energy between the initial and the final time.

I.4. The two different kinds of combustion process

One can rewrite the species conservation equations (1.5) in the following symbolic manner

$$\mathbb{L}_F(Y_F) = -\frac{W_F}{\rho} \tag{1.11}$$

where the linear differential operator $\mathbb{L}_F$ is given by :

$$\mathbb{L}_F = \frac{\partial}{\partial t} + \underline{V}\cdot\underline{\nabla} - D_F\nabla^2 \tag{1.11'}$$

where ρD_F has been assumed to be constant. A similar operator $\mathbb{L}_O$ and $\mathbb{L}_T$ can be defined for Y_O and T.Let's define the characteristic mechanical time τ_m by the relation :

$$\mathbb{L}_F(Y_F) \sim \frac{Y_{Fu}}{\tau_m} \tag{1.12}$$

In steady cases, τ_m can be considered as the shortest mechanical time (convection, diffusion).

The characteristic reaction time $\tau_r(T)$ is defined by :

$$\tau_r^{-1}(T) = B\ e^{-E/RT}\ ,\ \frac{W_F}{\rho} = \frac{Y_F^{n_F}\ Y_O^{n_O}}{\tau_r(T)} \tag{1.13}$$

The equation (1.11) leads to :

$$\mathbb{L}_F(Y_F) = -\frac{Y_F^{n_F}\ Y_O^{n_O}}{\tau_r(T)} \tag{1.14}$$

Two extreme cases can be considered from (1.12) and (1.14) :

(1) $\tau_r \gg \tau_m$ **Frozen-flow**

the chemical reaction can be ignored in (1.14) that reduces to

$$\mathbb{L}_F(Y_F) = 0 \tag{1.15a}$$

(2) $\tau_r \ll \tau_m$ **Equilibrium flow**

in that case the chemical equilibrium must be realized

$$W_F = 0 \tag{1.15b}$$

and according to the irreversible reaction (1.1) and (1.2) this is possible only in two cases

$$Y_O = 0 \quad \text{or} \quad Y_F = 0 \tag{1.15b'}$$

One introduces the Damköhler number Da by

$$Da = \frac{\tau_m}{\tau_r}$$

in such a way that the frozen flow corresponds to the small Damköhler nb limit ($Da \to 0$) and the equilibrium flow to large Damköhler nb limit ($Da \to \infty$), in regions of small T and high T, resp.

Let us consider the instructive example of the combustion developed at the leading edge of a mixing layer of fuel and oxidant :

In the early stage of the mixing layer the mixing time is shorter than the reaction time ($Da \to 0$) and a frozen reactive premixed mixture is obtained with a rich composition on the side of the fuel flow and with a lean composition on the side of the oxidant flow. Then a combustion proceeds in this premixed mixture to transform the frozen flow in an equilibrium flow ($Da \to \infty$). This called a premixed combustion. But because of the two different types of composition (lean and rich) in the frozen flow, the burnt gases in the equilibrium flow present two different regions of equilibrium composition. One is defined by $Y_O = 0$ but with $Y_F \neq 0$ (burnt gases of a premixed combustion in a rich mixture where there is an excess of fuel) and the other characterized by $Y_F = 0$

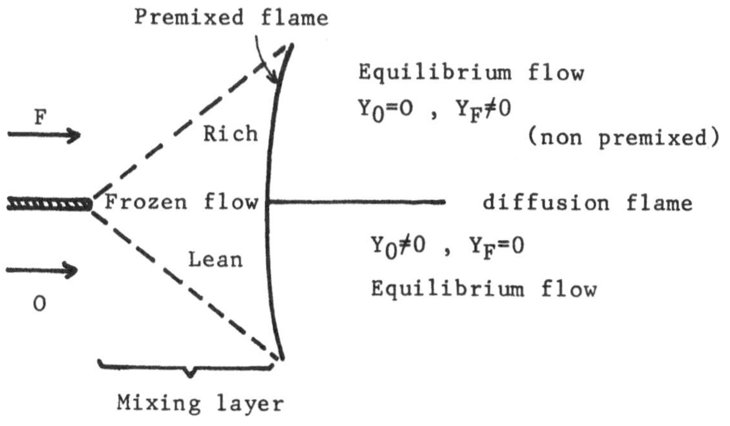

Fig.1.1

(burnt gases of a lean mixture) but with $Y_O \neq 0$. Then, the combustion takes place at the boundary of these two regions of equilibrium flow in the form of what it is called a diffusion flame to burn the fuel of one equilibrium flow with the oxidant of the other equilibrium flow.

1.5. The large activation energy

The very existence of frozen premixed mixtures as well as the thin transition layers shown in Figure 1.1 are due in fact, mainly to the high sensitivity of the reaction rate (1.2) to the temperature, associated with the large values of the activation energy encountered in combustion, $E \gg RT$, and to the exothermicity of the reaction.

Introducing the reduced activation energy β defined by :

$$\beta = \frac{E}{RT_b} \frac{(T_b - T_u)}{T_b} \quad \text{(Zeldovich number)} \tag{1.16}$$

ordinary values of β are close to 10;

$$\varpi^{-1} = \frac{\tau_r(T)}{\tau_r(T_b)} = \frac{e^{+E/RT}}{e^{+E/RT_b}} = \exp\left\{ + \frac{\beta(1-\theta)}{1+\alpha(\theta-1)} \right\} \tag{1.17}$$

where θ is the reduced temperature $\theta = \frac{T-T_u}{T_b-T_u}$, $(0<\theta<1)$, and α is the gas expansion parameter $\alpha=(T_b-T_u)/T_b$ (in ordinary flames $0.8<\alpha<1$). Thus, for $\beta \simeq 10$ and $\alpha \simeq 0.8$, eq.(1.17) leads to $\tau_r(T_u) \simeq 2.10^{22}\tau_r(T_b)$. As $\tau_r(T_b)$ is known to be of the order of 10^{-4}s, the reaction time at the ordinary temperature $T_u \sim 300°K$ is of the order of

$$\tau_r(T_u) \simeq 2.10^{18}\text{s} \tag{1.17'}$$

that can be considered as an infinite time at the human scale!

In fact eq.(1.17) shows clearly that the relative reaction rate (compared to the one at the adiabatic flame temperature) is transcendentally $O(e^{-\beta})$ small everywhere except when the temperature is close enough to the adiabatic flame temperature that is precisely when $1-\theta = O(1/\beta)$. See fig.1.2.

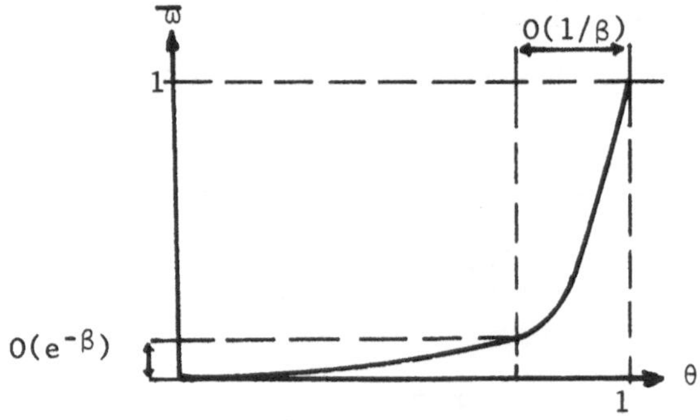

Fig.1.2

Thus it is interesting to notice that, in the limit of $\beta\to\infty$, $\omega = \tau_r^{-1}(T)/\tau_r^{-1}(T_b)$ goes to the singular limit $\omega = 0$ for $0<\theta<1$ and $\omega = 1$ for $\theta=1$. This behavior illustrates the high non-linearity of the sensitivity of the reaction rate to the temperature.

II - PREMIXED FLAMES

II.1. Position of the problem

A premixed flame is a subsonic chemical wave propagating in a premixed frozen mixture under the diffusive transport mechanism of mass and energy. In fact it appears as a thin transition layer between the frozen mixture ($Da\to 0$) and the equilibrium mixture ($Da\to\infty$). It turns out experimentally that the mass flux of frozen mixture crossing this transition surface is a characteristic not only of the thermodynamics conditions (temperature and pressure) and of the chemical composition of the frozen mixture but also of the geometrical configuration of the flame front and of the flow. The simplest case is the planar front moving with a constant velocity. In this case, the equasions (1.5) written in the moving frame of the front reduces to much simpler one dimensional and steady equations. $\underline{V} = (u,0,0)$ and the total mass conservation implies that the mass flux $m=\rho u$ is constant across the front.
A further simplification occurs when the mixture is far from the stoichiometric composition. In this case, one can neglect the change in the mass fraction of the abundant species and the combustion is controlled only by the limiting species. Then the system of equation (1.5) reduce to two coupled equations. In fact, in these 1-d and steady solutions, there is no direct coupling with the hydrodynamics and the deflagration waves are described by ordinary equations of diffusion-reaction :

$$
\begin{aligned}
m\frac{d}{dx} Y - \rho D \frac{d^2}{dx^2} Y &= - W \\
m \frac{d}{dx} (C_p T) - \lambda \frac{d^2 T}{dx^2} &= + qW
\end{aligned}
\qquad (2.1)
$$

with, according to (1.2), $W = \rho K\, Y^n\, e^{-E/RT}$

where $K = BY^{n_j}$ (the subscript j referring to the abundant species) and with the boundaries conditions :

(2.1')

Unburnt gases $x = -\infty$: $Y = Y_u$, $T = T_u$ (Frozen flow)

Burnt gases $x = +\infty$: $Y = Y_b = 0$ (Equilibrium flow)

See the shape of the corresponding profiles in fig. 2.0

The unknowns of the problem are the concentration and temperature profiles (that determines the flame structure) as well as the mass flux m appearing as an eigenvalue of the problem and defining the laminar flame velocity u_L (defined in the fresh mixture)

$$m = \rho_u\, u_L = \rho_b\, u_b \qquad (2.2)$$

In the moving frame of the flame front, u_L and u_b represent the flow velocity in the upstream fresh mixture and in the downstream burnt gases respectively, u_L and u_b can also be considered as the flame velocity relative to the fresh and burnt mixture respectively. Notice that because of the gas expansion ($\rho_b/\rho_u = T_u/T_b$), the flame velocity defined relative to the burnt gas, u_b, differs from the flame velocity relative to the fresh mixture, u_L. A direct integration of (2.1) from $x=-\infty$ to $x=+\infty$ shows that $T=T_b$ at $x=+\infty$ where T_b is given by (1.10). Thus, introducing the reduced quantity

$$\psi = Y/Y_u \quad \text{and} \quad \theta = (T-T_u)/(T_b-T_u) \qquad (2.3)$$

the boundaries conditions (2.1') for θ and $1-\psi$ appear to be identical

$$x=-\infty \;:\; \theta = 1-\psi = 0 \quad \text{and} \quad x=+\infty \;:\; \theta = 1-\psi = 1 \qquad (2.3')$$

Furthermore the equations of θ and $1-\psi$ differ only by the value of the diffusion coefficient :

$$m\frac{d}{dx}\psi - \rho D\frac{d^2}{dx^2}\psi = -\frac{1}{Y_u}W \qquad (2.4)$$

$$m\frac{d}{dx}\theta - \rho D_{th}\frac{d^2}{dx^2}\theta = \frac{1}{Y_u}W$$

where the thermal diffusivity is defined by $D_{th} = \lambda/\rho C_p$. Thus, when the Lewis number $Le = D_{th}/D$ is unity, the two equations of (2.4) are identical with the same boundary conditions, and it turns out that

$$\theta = 1-\psi \qquad x \in (-\infty, +\infty) \qquad (2.5)$$

and the problem is reduced to solve only one non linear thermal equation of reaction-diffusion type :

$$m\frac{d}{dx}\theta - \rho D_{th}\frac{d^2}{dx^2}\theta = \frac{\rho}{\tau_r(T_b)}\omega(\theta) \qquad (2.6)$$

with

$$\tau_r^{-1}(T_b) = \frac{K}{Y_u}\exp(-E/RT_b) \quad \text{and} \quad \omega(\theta) = (1-\theta)^n \exp\left\{-\frac{\beta(1-\theta)}{1+\alpha(\theta-1)}\right\} \qquad (2.6')$$

$\omega(\theta) = (1-\theta)^n \varpi(\theta)$ where ϖ has been defined in § I.5 .

With the boundary conditions

$$x=-\infty \;:\; \theta=0 \;;\; x=+\infty \;:\; \theta=1 \qquad (2.6'')$$

From an historical point of view, Mallard and Le Chatelier (1883) were the first not only to consider a premixed flame as a progressive chemical wave but also to give the basic mechanism of propagation and to provide the first experimental data on flame speed. They introduce the notion of inflammation temperature T_i under which the reaction is quenched and they consider the thermal propagation of a flame as successive inflammations of slices of frozen flow. A part of the heat released in a reacting slice is used to warm up the frozen slice just ahead where the combustion will start when the temperature will reach T_i. Thus, the combustion process was explained to possibly propogate in a frozen flow.

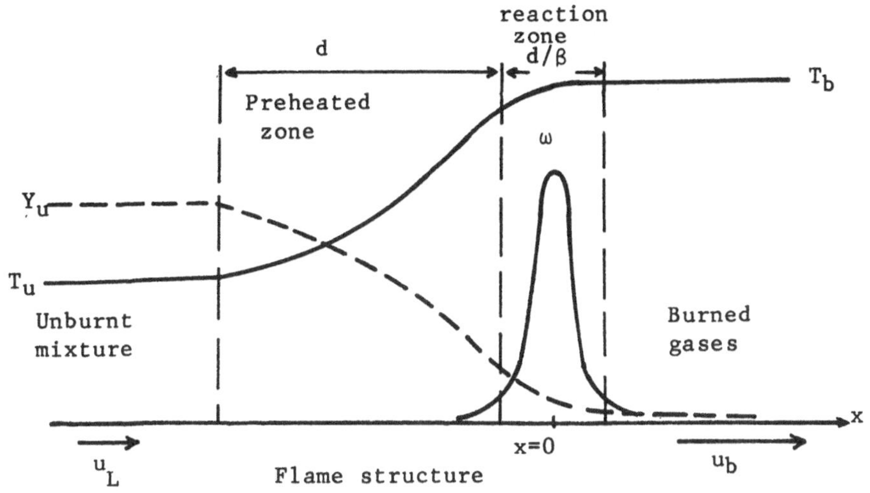

Fig.2.0

Taffanel (1913) and Jouguet (1913) wrote the corresponding differential equation similar to (2.6) and Taffanel obtained the correct expression of the laminar flame speed in terms of D_{th} and of the characteristic chemical time τ_r :

$$u_L \propto \sqrt{D_{th}/\tau_r(T_b)} \qquad (2.7)$$

Such an expression can be directly obtained by a dimensional analysis. With $D_{th} \sim 0.3\ cm^2/s$ and $\tau_r(T_b) \sim 10^{-4}s$ one obtains $u_L \sim 55 cm/s$.

II.2. Existence and unity

II.2.1. The cold boundary difficulty

From a mathematical point of view, the problem (2.6,6',6") is hill posed. A necessary condition for the existence of the solution is that the production term be zero in the upstream condition (frozen flow) ;

$$\omega(\theta=0) = 0 \qquad (2.8)$$

This condition is not exactly satisfied by the Arrhenius law (2.6'). But from a physical point of view, as shown by (1.17') and the discussion under fig.1.2, this problem is rather academic for the usual combustion problem. From a mathematical point of view, the problem has to be treated as an unsteady problem($\rho \frac{\partial}{\partial t}\theta$ must be added on the left hand side of 2.6) and according to Aldushin, Khudyaev and Zeldovich (1981) one has to look for travelling waves in a non steady, but homogeneous medium that evolves "in bulk" with the characteristic chemical time of the frozen flow :

$$\frac{\partial}{\partial t}\theta = \frac{1}{\tau_r(\theta)} \quad , \theta \text{ close to } 0 \Rightarrow \tau_r \text{ close to } \tau_r(T_u) \qquad (2.9)$$

This problem can be accurately solved by a two time scales method when the two time scales $\tau_r(T_u)$ and $\tau_r(T_b)$ are largely different ($\varpi_u \ll 1$). But, according to (1.17'), ϖ_u is so small in combustion problems ($\varpi_u \sim 10^{-22}$) that $\omega(\theta=0)$ can be physically considered as zero. In the past many different tentatives have been developed to modify the Arrhenius law in the "unburnt" side but, clearly, this modification has not to appear in the final result. It will be seen in the following that the introduction of such modifications is not necessary and that the problem is completely cured by considering the asymptotic limit $\beta \to \infty$.

II.2.2. The unicity

But, even when (2.8) is satisfied (cold boundary difficulty cured), it turns out that the uniqueness of solutions depends critically on the behavior of $\omega(\theta)$ near $\theta=0$. Since the work of Fisher (1937) and of Kolmogorov, Petrovsky and Piskonov (1937), it is known that (2.6,6") may admit a continuous set of solutions corresponding to a continuous spectrum of travelling speeds, m, as soon as $\omega'_0 = \frac{d}{d\theta}\omega(\theta)|_{\theta=0} > 0$. The problem is to determine what are the solutions selected by the physical situations. In the above mentioned work of (K.P.P.)(concerned with a biological problem where $\omega(\theta) \alpha\ \theta(1-\theta)$), it has been shown that the asymptotically stable solution (that is the only one relevant from a physical point of view) corresponds to the lower bounds m_{KPP} of the spectrum. Moreover m_{KPP} is found to be directly proportional to $\sqrt{\omega'_0}$. But on the other hand in the case of an inflammation temperature where :

$$\omega(\theta) = 0 \quad \text{for } 0<\theta<\theta_i<1 \quad \text{and} \quad \omega(\theta)>0 \quad \text{for } \theta_i<\theta<1 \qquad (2.10)$$

and where the inflamation temperature θ_i is a given positive constant ($0<\theta_i<1$), it has been proven that the system (2.6.6") admits only a unique solution (see Zeldovich (1948), Johnson-Nachbar (1963) and Gel'fand (1959)). As this unique solution corresponds to $m \neq 0$, it is clear that this solution is not related to the K.P.P. solution .The problem has also been proved to have a unique solution when $\omega_0<0$ and when $\omega(\theta)$ has only one zero for $0<\theta<1$. An intensive literature has been devoted to the general problem of the propagation of plane wave fronts controlled by equations of reaction and diffusion (see f.e. P. Fife (1978))and also Murray (1977)), but few comments are usually found concerning the transition from the K.P.P. solution to the unique solution of the θ_i-model.. The most pertinent ones have been developed by authors concerned by combustion phenomena namely Adulshin, Kudyaef & Zeldovich (1981). It is also worthwhile to consult the books of Frank-Kamenetskii (1969) and the one of Zeldovich & co-authors (1980). Interesting comments can also be found in the monography of P. Fife (1979).

After having cured the cold boundary difficulty as in the paper of Aldushin & Co (1981), the production term (2.6') reduces to

$$\omega(\theta) = (1-\theta)^n \left\{\exp(\beta(\theta-1))-\exp(-\beta)\right\} \qquad (2.11)$$

For simplicity one has neglect the term $\alpha(\theta-1)$ in (2.6') that will be proved to not be important. The shape of the production term (2.11) appears as an intermediate case that goes from a "mild" to a "sharp" non linear form as β increases from 0 to infinity (see fig. 2.1). For very small values of β $\omega(\theta) \sim \beta\ \theta(1-\theta)$ and for large values, when the transcendentally small terms $O(e^{-\beta})$ can be neglected, $\omega(\theta)$ can be considered as a θ_i model but, in addition, with $1-\theta_i = O(1/\beta)$. This last property allows to neglect the term $\alpha(\theta-1)$ in eq.(2.6') as well as the density change in the r.h.s. of eq(2.6) where ρ can thus be replaced by ρ_b. These approximations are used here for simplicity but they can be removed without difficulties
A simple change of space variable put the eq(2.6) in the following simpler form

$$\xi = x/d \ , \quad d = \sqrt{\rho D_{th}\tau_r(T_b)/\rho_b} \quad \text{and} \quad \rho D_{th} = C^t \qquad (2.12)$$

$$M\frac{d\theta}{d\xi} - \frac{d^2}{d\xi^2}\theta = \omega(\theta) > 0 \qquad (2.13)$$

With the boundaries conditions $\xi = -\infty$: $\theta=0$ Fresh mixture
$\xi = +\infty$: $\theta=1$ Burnt gases

and with $0<\theta<1$ for $\omega<\xi<+\infty$

the reduced front velocity M is related to m by :

$$M = m\ \tau_r(T_b)/\rho_b d = m\sqrt{\tau_r(T_b)/\rho_b\rho D_{th}}.$$

In the following, the attention is restricted to the cases $\omega(\theta)>0$ and $(0<\theta<1)$.

II.2.3. Orbits in the phase space

In order to get a better insight into the problem, it is worthwhile to consider the phase space (θ,P) with P defined by :

$$P = M\, d\theta/d\xi \ , \qquad (2.14)$$

and eq(2.13) takes the form :

$$\frac{1}{M^2}\frac{dP}{d\theta} = \frac{P-\omega}{P} \qquad (2.15)$$

The singular points $(\omega=0)$ are (1,0) and (0,0). The first one, corresponding to the burnt gases, is a saddle (see fig.II.2a) with two principal directions $Q^b_{\pm}$ $(Q^b_-<0,\ Q^b_+>0)$. Only the negative one Q^b_- is relevant because when $\omega>0$ one must have $P>0$ as well as $\theta<1$.

$$Q^b_{\pm} = \frac{M^2}{2}\left(1 \pm \sqrt{1-4\omega'_1/M^2}\right) \qquad (2.16)$$

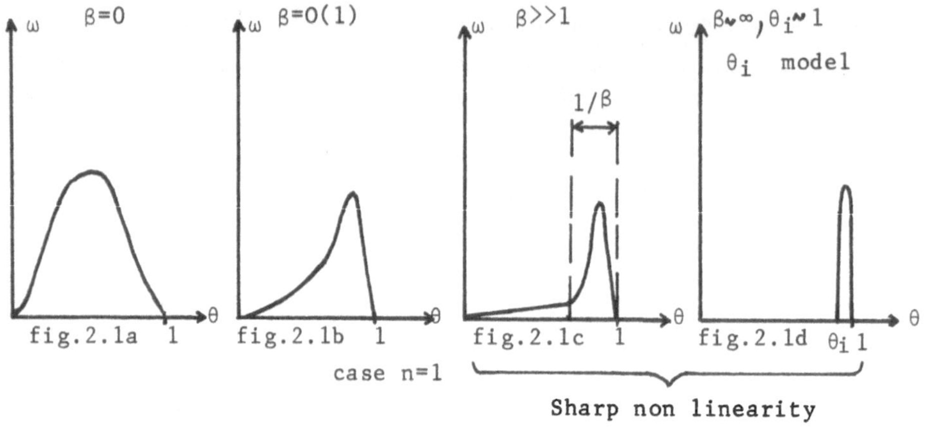

Fig.2.1

where $\omega_1' = \frac{d\omega}{d\theta}\Big|_{\theta=1} < 0$ for $n=1.(\omega_1' = 0,\ n=2)$

The second singular point (0,0) is related to the unburnt mixture (fresh gases), it is a spiral point when $M<2\sqrt{\omega_1'}$ and a node in the opposite case (see fig.II.2b)

$$M > 2\sqrt{\omega_0'} \tag{2.17}$$

When $\omega>0$, the solutions are associated with non negative values of θ and thus one has to consider only the case (2.17). No travelling waves can be obtained from (2.13) when the condition (2.17) is not satisfied. The two principal directions at the mode have a positive slope :

$$Q^u_{\pm} = \frac{M^2}{2}\left(1 \pm \sqrt{1-4\omega_0'/M^2}\right) > 0 \tag{2.18}$$

where

$$\omega_0' = \frac{d\omega}{d\theta}\Big|_{\theta=0} > 0$$

The solutions (2.13) are represented in the phase space by orbits leaving the saddle point (1,0) with the slope Q^b_- to reach the mode (0,0) with one of the two possible slopes Q^u_- Q^u_+ . The differential equation (2.13) being of first order, for each value of M one cannot have more than one trajectory linking the two singular points. The question is to determine the set of values of M (spectrum)associated with a solution. It is worthwhile to notice the following points :

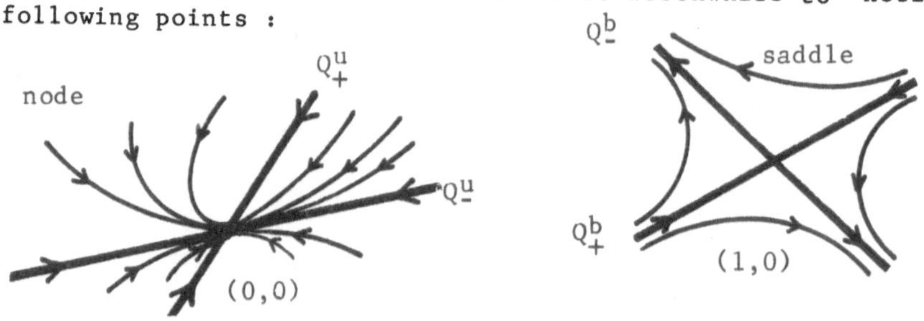

Fig.2.2

- a) For a given value of M, only one orbit can leave the saddle point tangent to Q_-^b and also only one can reach the node with a tangent equal to Q_+^u. But this is not the case for the other principal direction Q_-^u where a family of orbits $P_\kappa(\theta)$ tangent to Q_-^u can reach the node with the following behavior

$$P_\kappa(\theta) \quad Q_-^u \theta + \kappa\theta^{(M^2-Q_-^u)/Q_-^u} + \dots \qquad (2.19)$$

where κ is an arbitrary constant (Notice that according to (2.18) $\frac{M^2}{Q_-^u} -1))1)$

- b) In the limit $M\to\infty$, $P=\omega(\theta)$ is, according to (2.15), the trajectory as soon as $dP/d\theta$ remains bounded everywhere.

- c) $0<|Q_-^b|<|\omega_1'|$ and $M^2>Q_+^u>2\omega_0'>Q_-^u>\omega_0'$. Thus the trajectory $P(\theta)$ has to cross $\omega(\theta)$ where according to (2.15) $\frac{dP}{d\theta} = 0$.

- d) For the minimum value $M=2\sqrt{\omega_0'}$, $Q_+^u = Q_-^u = 2\omega_0$. For $M\to\infty$: $Q_+^u \to M^2$ and $Q_-^u \to \omega_0$

Before presenting the detailed analysis of a simplified model, let us present some general results.

II.2.4. General results

- i) When the curve $\omega(\theta)$ is concave as in fig. 2.1a, it has been shown by Kolmogorov & Co (1937) that a solution satisfying $0<\theta<1$ exists for all values of M larger than a lower bound called M_{KPP}. The KPP solution is governed by the behavior at the unburnt boundary condition $\theta=0$:

$$M_{KPP} = 2\sqrt{\omega_0'} \quad \text{with} \quad \omega_0' = \frac{d\omega}{d\theta}\Big|_{\theta=0} \qquad (2.19)$$

It has also been proved that the KPP solution is the physically relevant one in the sense that this solution is reached asymptotically in time ($t\to\infty$) from ordinary initial conditions ($t=0$)

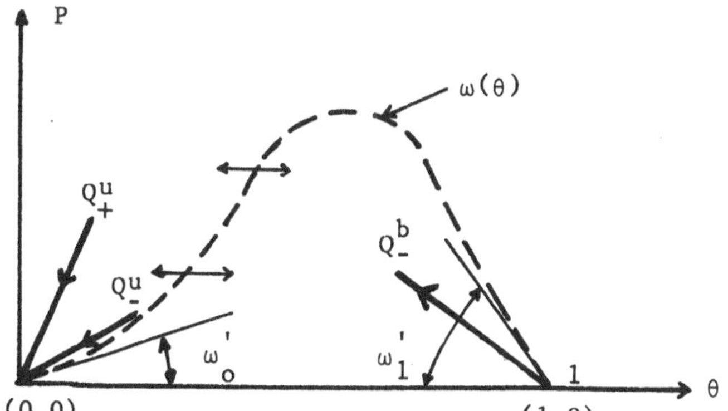

Fig.2.3

including a heavyside step function (see Fife 1979 and also Aronson & Weinberger 1978 for recent reviews). All these solutions $M>M_{KPP}$, including the KPP one, are tangent to the "ordinary" principal direction Q_-^u at the node of the cold boundary condition.

ii) When the production curve is no more concave everywhere in the interval $\theta\in[0,1]$, the lower bound can be higher than M_{KPP} defined by (2.19)(see Fife 1979). Let us call M_L the lower bound in this case, $M_L>M_{KPP}$. The corresponding L-solution is still the relevant one from a physical point of view but presents a different structure from all the other in the spectrum $M>M_L$. The L solution is the only one reaching the node (0,0) at the cold boundary tangent to the "extraordinary" principal direction Q_+^u. To the best of our knowledge, this property that has not been stressed enough before Aldushin & Co (1981) is of primary importance in premixed flames. In fact the relevant travelling wave solution that is always associated with the lower bound of the spectrum, changes of nature when the non linear term $\omega(\theta)$ becomes sharper and sharper. Such a sharpening can be obtained with the production term (2.11) by increasing the reduced activation energy β of the Arrhenius law. Notice that large values of β correspond to a large difference between the characteristic time scales τ_b and τ_u of the chemical production rate at its maximum value $\tau_b = \beta^n\tau_r(T_b)$ and close to the boundary controlled by $\omega_o' = O(\beta e^{-\beta})$, $\tau_u = \frac{e^\beta}{\beta}\tau_r(T_b)$:

$$\beta\to\infty \ , \ \tau_b/\tau_u \to 0 \ , \quad \tau_b/\tau_u = O(\beta^{n+1}e^{-\beta}) \tag{2.20}$$

As we will see in the next section, for large values of $\beta(\beta\to\infty)$, the L-solution goes to a limiting value that was obtained in 1983 by Zeldovich and Frank Kamenestkii

$$\lim_{\beta\to\infty} M_L = M_{ZFK}$$

The important point is that this ZFK solution is proved to not depend on τ_u but only on τ_b. But on the contrary, for $\beta\to 0$ the KPP solution holds and, as shown by eq(2.19), the solution is completely controlled by τ_u. Thus, the L-solution must insure a transition between these two extreme cases. In fact the L-solution, if it exists, is the only one that is not tangent to Q_-^u at the cold boundary and thus it is the only one in the spectrum that is not controlled by τ_u when $\omega_o'\to 0$. All the other solutions corresponding to $M>M_L$ contain in the upstream part of their temperature profile a long tail associated with τ_u which is determined by ω_o' . But according to the order of magnitude (1.17') the corresponding length scale is prohibitively long to be meaningful in ordinary experimental conditions. The L-solution is the only one that has a thin flame thickness controlled (as u_L see eq.2.7) by D_{th} and τ_b. Contrarly to the other solutions ($M>M_L$), the chemical reaction is not necessary for matching the cold boundary condition $T=T_u$ in the upstream part of the L-solution where the heat conduction is the dominant phenomenon. This peculiarity of the L-solution is related to the principal direction Q_+^u .

Furthermore the unique solution of the ϑ_i -model goes to the Z.F.K. solution in the limit $\theta_i \to 1$:

$$\lim_{\theta_i \to 1} M_{\theta_i} = M_{ZFK} \tag{2.22}$$

II.2.5. The exact solution for a particular model

As shown by Aldushin et al. (1981), the KPP solution (2.19) holds for (2.11) in a finite domain of β around zero. There exists a critical value β^* at which the L-solution ($M_L \geq M_{KPP}$) appears. In order to better understand this transition let us study the following model that can be exactly solved (another model is presented by Aldushin et al. (1981)) ;

$$\omega(\theta) = \begin{cases} \omega_o' \cdot \theta & ; \quad 0<\theta<1-\varepsilon \\ \frac{1}{\varepsilon^2} h(1-\theta) & ; \quad 1-\varepsilon<\theta<1 \quad \text{(see fig.2.4)} \end{cases} \tag{2.22'}$$

with $0<\varepsilon<1$ and where one of the two parameters h and ω_o' can be removed through adequate scaling. For convenience, let's keep ω_o' fixed (M_{KPP} fixed) and let vary h and ε.

In this model, the orbit leaving the saddle point (1,0) tangent to Q_-^b as well as the one reaching the node (0,0) tangent to Q_+^u are straight lines for $(1-\varepsilon)<\theta<1$ and $0<\theta<1-\varepsilon$ respectively. For h and ε given, let us consider the modification of the orbits (solution of the problem 2.13) when M decreases. For $M=\infty$, the orbit is $P=\omega(\theta)$ (see -b) and when M decreases two scenario are possible depending on the values of h and ε :

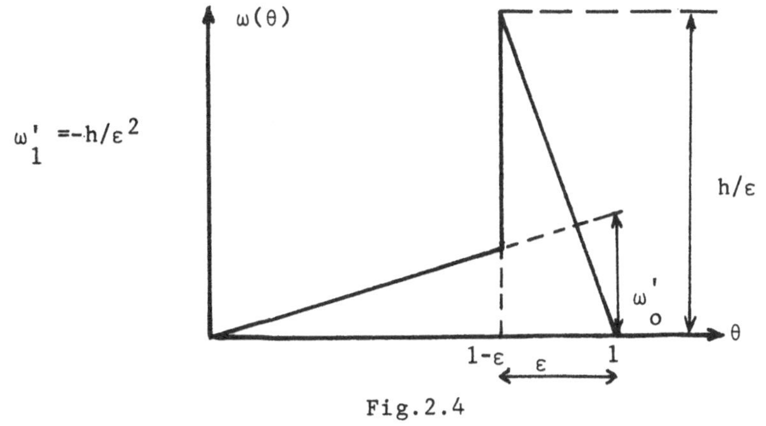

Fig.2.4

K.P.P. Scenario : $h<(1-\varepsilon^2)\omega_o$

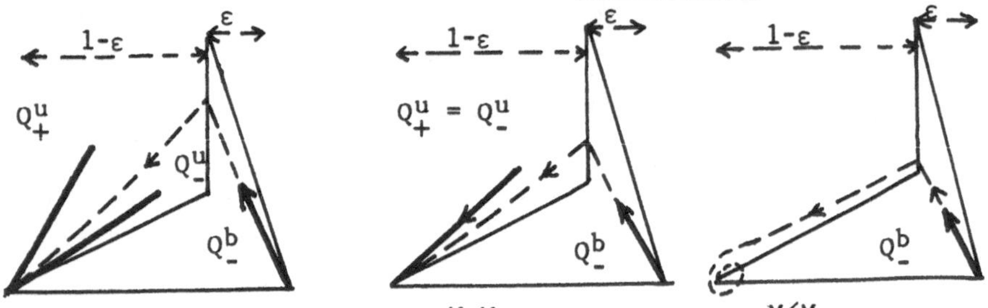

$M>M_{KPP}$ fig.2.5a $M=M_{KPP}$ fig.2.5b $M<M_{KPP}$ fig.2.5c

Fig.2.5

$$Q^u_{\mp}\cdot(1-\varepsilon)>|Q^b_-|\varepsilon \quad \text{for all} \quad M>M_{KPP} = 2\sqrt{\omega_o} \quad \text{(see fig.2.5)} \qquad (2.23)$$

The intersection of the extraordinary principal direction $Q^u_{\mp}$ with $\theta=1-\varepsilon$ above the intersection of the orbit with $\theta=1-\varepsilon$ for all $M>M_{KPP}$. The extraordinary principal direction cannot be used and all the possible trajectories reach the node (0,0) with the slope Q^u_-. Notice that in the fig.2.5b corresponding to the limiting case $M=M_{KPP}$, the orbit does not correspond to the straight line $P=Q^u_+\theta$ for $0<\theta<1-\varepsilon$.

L scenario : $h>(1-\varepsilon^2)\omega_o$

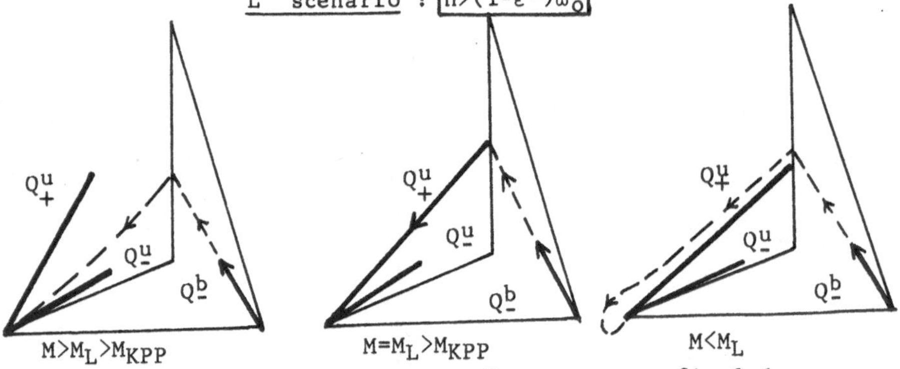

$M>M_L>M_{KPP}$ fig.2.6a $M=M_L>M_{KPP}$ fig.2.6b $M<M_L$ fig.2.6c

Fig.2.6

L-scenario : $h > (1-\varepsilon^2)\omega_o$

There exists a value M_L of M (with $M_L > M_{KPP}$) such that

$$Q_+^u \cdot (1-\varepsilon) = |Q_-|\varepsilon \tag{2.24}$$

corresponding to $$M_L = \left\{ \frac{h}{1-\varepsilon} + (1-\varepsilon)\omega_o \right\} / \sqrt{\frac{h}{1-\varepsilon} - \omega_o} \tag{2.24'}$$

For $M > M_L$ the situation is similar to the fig.(2 .5a), $Q_+^u(1-\varepsilon) > (Q_-^b)\varepsilon$

For $M = M_L$ the orbit consists of the two straight lines Q_+^u and Q_-^b

For $M < M_L$ the orbit has to cross $\theta=0$ before reaching the node (0,0) and the corresponding solution cannot be retained.

Transition : $h = (1-\varepsilon^2)\omega_o$

$$M_L = M_{KPP} = 2\sqrt{\omega_o} \tag{2.25}$$

The cases $M > M_{KPP}$ and $M < M_{KPP}$ are represented by figures similar to fig.(2 .5a) and fig.(2 .5c) respectively. When $M = M_{KPP}$ the orbit consists of the two straight lines Q_+^u and Q_-^b but contrarly to the fig.(2 .6b), $Q_+^u = Q_-^u$ in the fig.2 .7. When $h=(1-\varepsilon^2)\omega_o$, the KPP solution is the L solution because $Q_+^u = Q_-^u$.

From the eq(2.23) and (2.24), the propagating wave speed can be plotted in terms of h for different given values of ε. The characteristic shape is plotted in fig. 2.8.

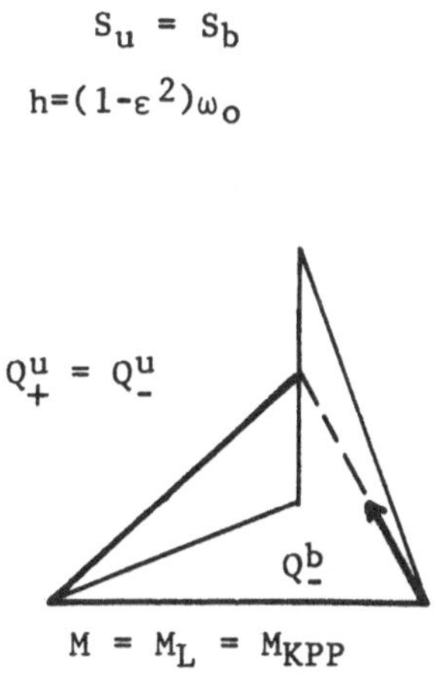

Fig.2.7

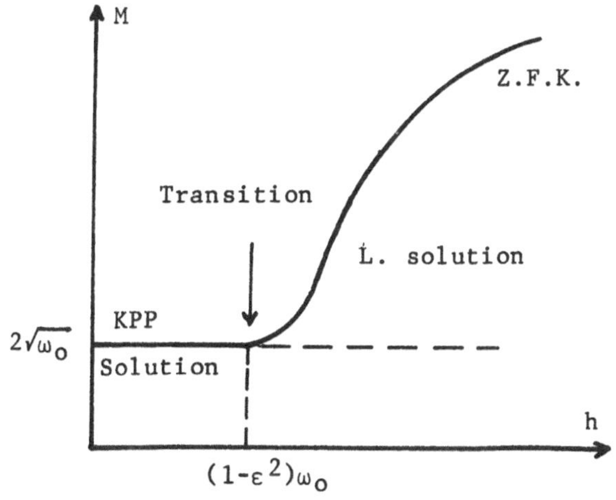

Fig.2.8

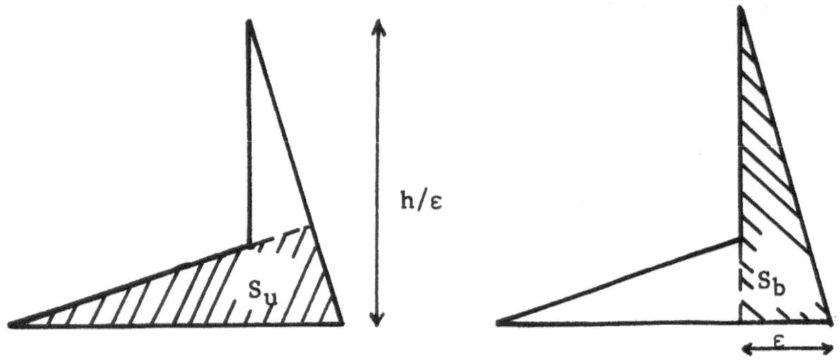

Fig.2.9

By noticing that $\frac{h}{\omega_0} + \varepsilon^2$ is the ratio of two surfaces S_b and S_u plotted in fig. 2.9, the critical condition for the transition between the K.P.P.-solution and the L-solution takes a very simple form :

$$S_u > S_b \quad : \quad \text{KPP solution}$$
$$S_u < S_b \quad : \quad \text{L solution} \qquad (2.26)$$

Such a criterium cannot be straight forward extended to general production terms as (2.11) to give precisely the corresponding critical value β^* of the Zeldovich number β.

$$\beta<\beta^* \quad : \quad \text{KPP solution}$$
$$\beta>\beta^* \quad : \quad \text{L solution} \qquad (2.26')$$

An estimate of β^* can be obtained by an approximate version of (2.26) involving the total aera $S_t = \int_o^1 d\theta\omega(\theta)$

$$S_t = 2S_u \qquad (2.27)$$

where S_u is the area under the slope at the origin+. For the production term (2.11), the criterium (2.27) gives $\beta^* \sim 6$ and $\beta^* \sim 8$ for n=1 and n=2 respectively. The numerical determination carried out by Higuera (1983) yields $\beta^*= 3.04$ and $\beta^*= 5.11$ for n=1 and n=2. These numerical results are in good agreements with the time dependent numerical analysis of Aldushin et al. (1981). Furthermore the critical condition has been obtained by Higuera (1983) for other models as the following one :

+ The criterium (2.27) was suggested by Zeldovich - private communication, July 1983.

$$\omega(\theta) = \theta(1-\theta)e^{\beta\theta} \quad , \quad \beta^* = 1.64 \qquad (2.28)$$

used by Zeldovich (1948) for an exothermic chain branching reaction (1.4b) with unity Lewis number for both the limiting component R and the intermediate X. The corresponding threshold for the onset of the L-solution, $\beta^*=1.64$, is found to be much more lower than for the model (2.11). These numerical values of β^* can be compared with the values of the Zeldovich number associated with the overall activation energy and the flame temperature occurring in ordinary flames (see fig. 2.10).

E kcal/mole T°K	20	30	40	50
300	0	0	0	0
350	4.10	6.16	8.22	10.27
1500	5.37	8.05	10.74	13.42
2000	4.28	6.42	8.56	10.7
2500	3.54	5.32	7.08	8.86

Values of the Zeldovich number (1.16) for T_0=300°K

Fig.2.10

It turns out that, even for cool flames observed in the low temperature range of the hydrocarbon oxidation ($T_b \sim 600°K$), the Zeldovich number β seems to be too large for the K.P.P. solution. But in such cool flames the multiple step-chemistry is expected to be a fundamental process and definitive conclusions cannot be obtained by the simplified model (2.6). For ordinary flames, $E \gtrsim 30$Kcal and $1500°K<T_b<2500°K$, the KPP solution is found to not be relevant and the solution appears to be more likely controlled by the L-solution. In addition, the corresponding values of β are large enough ($\beta \sim 10$) to make the exact solution accurately approximated by the dominant orders of an asymptotic expansion in large values of β ($\beta \to \infty$). This approach based on the early works of Zeldovich and Frank Kamenetskii (1938) is presented in the next section.

Let us finally recall that the study of the model (2.22') shows that the transition "K.P.P. solution→L solution" can be produced only by changing the "hot part" ($\theta=1$) of $\omega(\theta)$. Thus, the selection criterium that has been proved to be relevant for selecting the K.P.P. solution (see Zeldovich 1948) cannot be uniformly valid because it is based only on the linearized form of $\omega(\theta)$ around the cold boundary ($\theta=0$). This comment emphasizes the limits of validity of selection criteria such as the one proposed recently by Die and Langer (1983) in the context of propagating patterns.

II.3. The asymptotic expansion

In the limit of an infinitely large Zeldovich number β, the ratio of the characteristic times of reaction at T and T_b becomes singular (see fig.I.2) :

$$\lim \beta\to\infty \quad , \quad \frac{\tau_r^{-1}(T)}{\tau_r^{-1}(T_b)} = \begin{cases} 0 \, , & T_0<T<T_b \\ 1 \, , & T = T_b \end{cases}$$

For large values of β, the reaction can be considered as thermally quenched everywhere inside the flame except for temperature sufficiently close to the maximum value T_b , $(T_b-T)/(T_b-T_u)=1-\theta= O(1/\beta)$. And according to eq(2.6'), the reduced reaction rate $\omega(\theta)$ is negligible except in a thin reaction zone of thickness d/β separating the preheated zone where the flow is frozen(ω is trascendentally small) from the burned gases at equilibrium, $\psi=0$ $\omega=0$ (see fig. 2 .0). Because of the quasi-similarity of the profiles of temperature and limiting component, the consumption of the reactant stops the reaction soon after its initialisation.

In 1938 Zeldovich and Frank Kamenestkii developed an approximate solution of the flame propogation described by the L-solution of eq(2.6). This approximation valid for $\beta>>1$ has now been proved to be the dominant order of the asymptotic expansion first developed independently by Bush and Fendell (1970) for the model (2.6) and by Liñan (1971) for exothermic chain branching reaction. From eq(2.6) it appears clearly that the heat released by the chemical reaction is partly consumed by heating the reacting gas through the convective term $m\frac{d}{dx}\theta$ and partly carried away by heat conduction. When the reaction zone is thin, its temperature change is negligible and, at the dominant order, the chemical production is balanced by the heat conduction ;

$$\underline{\text{reaction zone}} : \quad - \rho D_{th} \frac{d^2}{dx^2}\theta = \frac{\rho}{\tau_r(T_b)} \quad \omega(\theta) \qquad (2.29)$$

After multiplication by $\frac{d}{dx}\theta$ the eq(2.29) can be easily integrated to compute the energy flux that leaves the reaction zone by heat conduction for preheating the frozen flow :

$$\rho D_{th} \frac{d\theta}{dx} = \sqrt{2 \frac{\rho D_{th}}{\tau_r(T_b)} \rho_b \ I_\theta} \qquad (2.30)$$

where according to (2.6) and (2.6')

$$I_\theta = \int_\theta^1 d\theta'\omega(\theta') = \frac{1}{\beta^{n+1}}\int_{\beta(\theta-1)}^{0} d\Theta\ \Theta^n \exp\left\{\frac{\Theta}{1+\alpha\beta^{-1}\Theta}\right\} \quad (2.30')$$

The main difficulty is to prescribe the lower bound appearing in I_θ and corresponding to the boundary between the reaction zone and the preheated zone. Because of its marked exponential temperature, the reaction rate falls so sharply with a decrease in temperature that, when β is large, the lower bound can be taken to be $-\infty$ ($\beta(\theta-1)\rightarrow-\infty$) without modifying notably the value of I_θ :

$$I_\theta = \frac{1}{\beta^{n+1}}\ \Gamma(n+1) + O\left(\frac{1}{\beta^{n+2}}\right) \quad (2.31)$$

with $\Gamma(n+1) = n!$ when $n \in N$

The heat conduction flux (2.30) entering the preheated zone balances the convection flux of energy leaving this zone only for a well defined value of the mass flux m corresponding to the unknown flame velocity. The solution of eq(2.6) in the preheated zone (where ω can be neglected) shows that the value of the convection flux is simply m :

preheated zone : $\theta = e^{x/d}$ (2.32)

with $d = \frac{\rho D_{th}}{m}$ and where the origin of the x-axis has been chosen such that θ (x=0)=1. Eq(2.32) yields :

$$\rho D_{th} \left.\frac{d\theta}{dx}\right|_{x=0_-} = m = \rho_u u_L \quad (2.33)$$

And eq(2.30) and (2.33) provide the expression of the laminar flame speed u_L. This approach can be systemized by an asymptotic expansion whose the first and second order are presented below. The main purpose of such a systematic expansion is not only to obtain a more accurate analytical solution of the planar flame but also to provide a framework that can be used to solve more complex cases as wrinkled fronts in non homogeneous flow. The corresponding works are briefly outlined in the following section and a more detailed presentation can be found in a recent review article (Clavin 1984). Let us consider the model (2.6) but where ρD_{th} is not necessarily assumed to be constant. When the independent variable $\xi = \int_0^x \frac{m}{\rho D_{th}} dx$ is introduced, eq(2.6) takes the form :

$$\frac{d\theta}{d\xi} - \frac{d^2\theta}{d\xi^2} = \Lambda\ A(\theta)\omega(\theta) \quad (2.34)$$

where

$$\Lambda = \frac{(\rho D_{th})_b \rho_b K_b}{m^2 \tau_r(T_b)} \quad (2.35')$$

and where $A(\theta)$ corresponds to the weak temperature dependence of the prefactor in the r.h.s. of (2.34)

$$A(\theta) = \frac{(\rho D_{th})(\rho K)}{(\rho D_{th})_b(\rho K)_b} \qquad (2.35'')$$

According to the result of Zeldovich and Frank Kamenestkii presented above, the unknown quantity Λ/β^{n+1} is expected to remain finite in the limit $\beta \to \infty$ and one must look for a solution expanded as follows :

$$\Lambda/\beta^{n+1} = \Lambda_o + \frac{1}{\beta}\Lambda_1 + 0(\frac{1}{\beta^2}). \qquad (2.36)$$

In the preheated zone $\omega(\theta)$ is transcendentally small and the solution of eq(2.34) yields :

Outer solution : $\xi<0$, $\theta(\xi) = e^{\xi} + t.s.t.$ (2.37)

where t.s.t. means transcendentally small terms.
To solve the problem in the reaction zone of thickness d/β located around $\xi=0$, the stretched variable η is introduced and considered as the independent variable in this thin zone :

$$\eta = \beta\xi \quad , \quad \frac{d}{d\xi} = \beta \frac{d}{d\eta}$$

Then, the solution is expanded in the following manner

$$(1-\theta) = \frac{\Xi_o(\eta)}{\beta} + \frac{\Xi_1(\eta)}{\beta^2} + 0(\frac{1}{\beta^3}) \qquad (2.38)$$

and using the corresponding expansion of $\Lambda A(\theta)\omega(\theta)$, eq(2.34) provides the equations for $\Xi_1, \Xi_2 \ldots$.

Inner equations

$$\frac{d^2\Xi_o}{d\eta^2} = \Lambda_o \Xi_o^n e^{-\Xi_o} ,$$

$$-\frac{d\Xi_o}{d\eta} + \frac{d^2\Xi_1}{d\eta^2} = \begin{cases} \Lambda_1\Xi_o^n e^{-\Xi_o} - \Lambda_o\gamma b_1 \Xi_o^{n+1} e^{-\Xi_o} - \Lambda_o\gamma\Xi_o^{n+2} e^{-\Xi_o} \\ \Xi_1\Lambda_o(n\Xi_o^{n-1} e^{-\Xi_o} - \Xi_o^n e^{-\Xi_o}), \end{cases} \qquad (2.39)$$

where

$$\gamma = \alpha \; ; \; b_1 = \frac{\partial Log(\rho D_{th}\rho K)}{\partial Log\ T}\Big|_{T=T_b} \quad ; \quad A = 1 - \gamma b_1 \frac{\Xi_o}{\beta} + 0(\frac{1}{\beta^2}) \qquad (2.40)$$

The boundary condition in the burnt gases yields :

$$\eta \to +\infty \quad : \quad \Xi_o \to 0 \quad , \quad \Xi_1 \to 0 \qquad (2.41)$$

According to the equations (2.39), the solutions Ξ_o and Ξ_1 present the following limiting behavior in the direction toward the preheated zone :

$$\eta \to \infty : \frac{d\Xi_o}{d\eta} \to Ct \neq 0 \qquad \frac{d^2\Xi_1}{d\eta^2} \to Ct \neq 0 \quad . \qquad (2.42)$$

The matching conditions (Van Dyke 1964, Cole 1968) of the inner solution (2.38) with the outer solution $\theta(\xi) = \theta_o(\xi) + \frac{1}{\beta}\theta_1(\xi) + 0(\frac{1}{\beta^2})$ of the preheated zone yields :

$\eta \to -\infty$:

$$: \; \Xi_o(\eta) = (\frac{\partial\theta_o}{\partial\xi}\Big|_{\xi=0_-})\eta + \theta_1(\xi=0_-) + t.s.t. \qquad (2.43)$$

$$\Xi_1(\eta) = \left(\frac{\partial^2\theta_0}{\partial\xi}\Big|_{\xi=0_-}\right)\frac{\eta^2}{2} + \left(\frac{\partial\theta_1}{\partial\xi}\Big|_{\xi=0_-}\right)\eta + \theta_2(\xi=0) + \text{t.s.t.}$$

According to the result (2.37), this yields

$\eta\to-\infty$:

$$\Xi_0(\eta) = \eta + \text{t.s.t.}$$
$$\Xi_1(\eta) = \frac{\eta^2}{2} + \text{t.s.t.} \tag{2.44}$$

The eigenvalues Λ_0 and Λ_1 are determined by prescribing the boundary conditions (2.41) and (2.42) (see Joulin & Clavin 1976) :

$$\Lambda_0 = \frac{1}{2\Gamma_{n+1}}$$

$$\Lambda_1 = \Lambda_0 \left\{b_1 \gamma \frac{\Gamma_{n+2}}{\Gamma_{n+1}} + \gamma \frac{\Gamma_{n+3}}{\Gamma_{n+1}} - \frac{1}{\Lambda\, \Gamma_{n+1}} \mathfrak{J}_n\right\} \tag{2.45}$$

where $\mathfrak{J}_n = \int_0^\infty d\Xi \left(1 - \sqrt{\frac{1}{\Gamma_{n+1}}} \int_0^\Xi dx\, x^n e^{-x}\right)$

As already mentioned, the dominant order Λ_0 is identical to the solution of Zeldovich and Frank Kamenestkii u_{ZFK}. The numerical evaluation of $\mathfrak{J}_{n=1}$ yields : $\mathfrak{J}_{n=1} = 1.344$,(see Fendell 1972). For n=1 and γ=0.85, the flame velocity u_L is, according to (2.35) and (2.45), given by :

$$u_L = u_{ZFK}\left(1 - \frac{0.57}{\beta} + 0(\frac{1}{\beta^2})\right) \quad \text{for} \quad b_1 = -\frac{3}{2} ,$$

and

$$u_L = u_{ZFK}\left(1 - \frac{1.2}{\beta} + 0(\frac{1}{\beta^2})\right) \quad \text{for} \quad b_1 = 0. \tag{2.46}$$

According to the values of β given in fig. 2.10, the results (2.46) shows that the first correction is not always negligible. By comparison with numerical solutions the results (2.46) are found to be accurate with less than 10% error for $\beta>3$ (Bush & Fendell 1971). For $\beta>5$, the error is less than 2%. In fact the very limitation of the asymptotic method toward the small value of the Zeldovich number β is the critical value β^* corresponding to the onset of the KPP solution. The coefficient γ appearing in the first term in the bracket of eq(2.45) is not present in the result of Fendell (1972) concerning n=1 but $L_e\neq1$. In order to get rid of the cold boundary difficulty, $A(\theta)$ was approximated by θ^{b_1} in the model used by Fendell. Such a modification of the Arrhenius law is not necessary in the framework of the asymptotic expansion $\beta\to\infty$ and can lead to irrelevant corrective terms when the hot boundary is modified.

This asymptotic method has been successfully applied to other one dimensional and steady cases such as : non unity Lewis number (Bush and Fendell 1970), volumetric heat losses for describing the thermal extinction (Buckmaster 1976, Joulin Clavin 1976), monopropellant droplet burning (Fendell 1972, Liñan 1976), stretch effect of a planar front stabilised in a stagnation point flow (Buckmaster 1982, Libby and Williams 1982), influence of flame holders and of the spherical geometry (Deshaies and Co 1981, Clarke and McIntosh 1980), in this last case the temperature in the burnt gas can decrease toward a room temperature small compared to the flame temperature and the corresponding inner zone structure is more complex (than the one presented above) and has been first studied by Linan (1974) in the context of diffusion flames. The study of the effects of two limiting components for reactive mixtures close to the stoichiometric composition has been studied by Joulin & Mitani 1981). More recently, the asymptotic method was successfully applied by Liñan and Clavin (1984) to a multiple step, kinetic scheme introduced by Adams and Stock (1953) to model the reactions with Hydrogen and Halogens. These approaches are very promising but, presently, they are limited to the cases where the complete chemical scheme can be reduced to few steps. For example, such a drastic reduction has not yet been performed successfully for the simplest hydrocarbons. Performance numerical methods are now available to solve the structure of flames sustained by the most complex kinetic scheme. But because of the huge amount of physico-chemical parameters necessary to describe the structure of flames sustained by complex kinetic scheme, the analysis of the corresponding numerical results are difficult. A promising avenue is to combine numerics and analytical treatments.

II.4. Dynamics of flame front

Let us consider the motion of flame fronts in flows that can be unsteady and non-uniform. When the length and time scales of the initial flow are larger than those associated with the planar flame (the flame thickness d and the transit time d/u_L where u_L is the laminar flame speed), the flame front can be considered, in first approximation, as a surface of discontinuity whose motion is controlled by two distinct factors : the normal burning velocity u_n associated with the mass flux of fresh mixture crossing the front and the value at the front of the flow velocity field. Each point of the front moves with a velocity equal to the difference between the values (at this point) of the upstream flow velocity and the normal burning velocity oriented toward the fresh mixture in the direction normal to the front.

As soon as one is concerned with wrinkled fronts and/or inhomogeneous flows, the flame **cannot be described by a pure reaction-diffusive model.** Because of the gas expansion produced by the temperature increase in the preheated zone, the streamlines are deflected across the tilted front and a strong coupling with hydrodynamics is developed. When the size D of the wrinkles of the front is large compared to the flame thickness d, the correponding fluid mechanical effects can be split in two distinct parts :

-i) The flame structure is locally influenced by the convective transfer produced in the preheated zone by the deflection of the streamlines.

-ii) Outside the flame where the gas density ρ and the temperature T are uniform, the flow field is also modified (from its initial value without combustion) upstream and downstream the flame on a distance D from the front.

The first effects i) produce a change of the normal burning velocity $u_n \neq u_L$. It must be noticed that, in addition to the convection transfer produced by the gas expansion, the diffusion fluxes of heat and mass play also a great part in this modification to flame structure of wrinkled fronts.

The second effects ii) results in a modification to the gas flow at the front.

Both of these effects i) & ii) influences the motion of the front, but the second is the stronger one at long wave lengths.

When the modification to flame structure is neglected, $u_n = u_L$ the flame could be considered as a passive surface in the sense that the motion is completely prescribed by the value of the flow at the front. Even in this case, the second effect ii) produces a strong hydrodynamical feedback in the motion of the front. This effect was first described in the pioneering works of Darrieus (1938) and Landau (1944) who computed the flow field induced by the front wrinkling when $u_n = u_L$ and d=0. The analysis was carried out at the linear approximation in the amplitude of the front corrugations and the induced flow velocity was found to be in phase with the front wrinkles. The resulting motion of the front reveals a strong instability mechanism of planar flames in uniform flows. The least fluctuation around the planar steady state solution (described in the preceding sections) is amplified under the hydrodynamical effect ii). As the wavelength becomes shorter, the front is more unstable and Darrieus and Landau concluded that planar fronts freely propagating in uniform mixture cannot exist. The instability mechanism appeared to be so strong that they conclude also that the combustion must be a self-turbulizing phenomena. In fact, only two parameters being involved in this theory, the gas expansion parameter γ and the laminar flame speed u_0, the growth rate σ of the instability is found to be proportional to the modulus k of the wave vector of the front wrinkles

$$\sigma = \sigma_1(\gamma) u_L k \tag{2.47}$$

where $\sigma_1(\gamma)$ is a positive adimentional quantity vanishing only in the unrealistic limit $\gamma \to 0$ (zero gas expansion, $\rho_u = \rho_b$).

The first attempt to take into account the effect of the modification to flame structure was performed in the fiftyth by Markstein (see his review paper 1965) who assumed the following phenomenological relation between u_n and the mean radius of curvature of the front R (R>0 when the front is concave toward the unburnt gases) :

$$\frac{u_n - u_L}{u_L} = \frac{\mathcal{L}}{R} \tag{2.48}$$

where $\mathcal{L}$ is a phenomenological length (called Markstein length) that was assumed to be proportional to the flame thickness d and whose the expression is a characteristic of the reactive mixture. It can easily be anticiped from eq(2.48) that the effects associated with the modification to flame structure can only change the dispersion relation (2.47) through a k^2 -term. And thus the large wavelength can never be stabilized by such a mechanism. Another stabilizing mechanism must be present in order to explain the large planar front observed experimentally.

The first analysis of the wrinkled flame structure was carried out by Barenblatt, Zeldovich and Istratov (1962) but in the diffusive-thermal model where the gas expansion effects i) and ii) are neglected. This model was extensively used these ten last years to culminate in the derivation by G. Sivashinsky (1977) of a non linear differential equation for the flame motion describing a self turbulizing behavior of the cellular structures (Michelson & Sivashinsky 1977). The asymptotic technique applied to solve this model in the limit of large values of the Zeldovich number $\beta \to \infty$, is presented in the paper of Joulin & Clavin (1979) that is devoted to the dynamical properties in the presence of heat losses that can produce the thermal extinction. Travelling and spinning waves as well as oscillatory fronts have been predicted by this model (for a review see the book of Buckmaster and Ludford 1982 and the review article of Sivashinsky 1983). But even as modified by Sivashinsky to take into account a weak gas expansion, this model underestimates the hydrodynamical effects that has to play a dominant role (see eq(2.47)).

Recently, the coupling between diffusion and hydrodynamics has been properly taken into account for describing the wrinkled flame structure in an analytical work by Clavin & Williams (1982). The asymptotic expansion $\beta \to \infty$ is used together with a multiscale method based on the assumption $\varepsilon = d/D$ smaller than unity. The corresponding result was used by Pelcé & Clavin (1982) to study the stability limits of planar fronts propagating downward. The results can be summarized as follow :

a) - In the approximation of a one step overall chemical reaction, the modification to flame structure by wrinkling is predicted to be a stabilizing mechanism for most of the ordinary hydrocarbon-mixtures; whatever the equivalence ratio may be; the only exception could be a mixture of very light reactive components (as hydrogen) diluted in nitrogen. This conclusion contradicts the result obtained by the diffusive-thermal model and appears as a typical effect of the mechanism i).

b) - The gas viscosity has a neutral effect on the stability properties of planar fronts.

c) - The acceleration of gravity g associated with the effects of the modification to flame structure by wrinkling can counterbalance the hydrodynamical instability for all the wave numbers when the flame velocity is low enough. The cellular threshold is predicted to be observable for flame velocity u_L varying between 5cm/s and 17cm/s with rich mixture of ordinary fuel for low flame velocities (u_L<12cm/s) and with lean composition for higher velocity.

d) - The cell size at the threshold can be expressed in terms of only the variable g, u_L and γ ; the detailed properties (chemical kinetics, transport processes...) do not enter into the final expression.

These predictions are in good agreements with the recent experiments of Quinard et al. (1983). Furthermore the induced velocity field has been recently recorded in the unburnt mixture by Searby et al. (1983) in the case of stable fronts stabilized in weakly "turbulent" flows. As predicted by the theory for planar stable flames, the induced velocity field is found to be out of phase with the front corrugations leading to a blocking of the low frequencies in the turbulence approaching the front (see the lecture of L. Boyer at the present summer school).

The analysis of Clavin & Williams (1982) concerning the flame structure of wrinkled fronts in a non homogeneous flows has been extended independently by Matalon & Matkowsky (1982) and by Clavin & Joulin (1983) to the nonlinear case of finite amplitudes of the front corrugations. As anticipated by the early phenomenological analysis of Karlowitz et al. (1953), the modification to the normal burning velocity u_n produced by the front curvature and by the flow inhomogeneities can be expressed in terms of only **one geometrical scalar** i.e. the total flame stretch experienced by the front.

$$u_n - u_L = - \mathcal{L} \left(\frac{1}{\sigma} \frac{d\sigma}{dt}\right) \qquad (2.49)$$

where σ is the surface element of the front and $\frac{d}{dt}$ its time derivative when each point of the flame moves as described at the beginning of this section. $\mathcal{L}$ is the Markstein length that depends, as the laminar flame velocity u_L, on the physico-chemical properties of the reactive mixture (transfer properties, chemical kinetics, etc...). The corresponding expression of $\mathcal{L}$ has been obtained for different cases (see the review article by Clavin 1984).

Except its limitation to the weak stretch, this surprisingly simple result is general and can be used in any flow configuration : stagnation point flow, spherical flames, turbulent flames... etc... The effects of the strong stretch has been recently investigated by Libby, Linan and Williams (1983) in the particular case of a planar front in a stagnation point flow. Once again, the effect of the gas expansion is proved to be important ; for example the flame extinction under strong stretch predicted by the "diffusive-thermal model" ($\gamma=0$) for one overall chemical reaction is no more accessible when the effects i) are properly taken into account).

It is worthwhile to express the result (2.49) in terms of the mean radius of curvature of the front R. At the same order, the modification to normal burning velocity can be expressed as (see Clavin & Joulin 1983) :

$$u_n - u_o = \mathcal{L}\, u_L \left\{\frac{1}{R} + \frac{1}{u_L}\, \underline{n} \bullet \underline{\underline{\nabla u}} \bullet \underline{n}\right\} \qquad (2.50)$$

where $\underline{n}$ is the unit vector normal to the front and $\underline{\underline{\nabla u}}$ is the "rate of strain tensor" of the upstream flow evaluated at the flame position. Each of the two terms in r.h.s. represents a contribution to the total flame stretch. The term $-u_L/R$ represents the stretch of the front moving in a uniform flow with a constant normal velocity u_L. This term is the effect of the non planar geometry of the front. The term $\nabla \bullet \underline{u} - \underline{n} \bullet \underline{\underline{\nabla u}} \bullet \underline{n}$ is known to represent the stretch of a surface convected by the flow field $\underline{u}$. (Here $\nabla \bullet \underline{u} = 0$ outside the flame). This second term is the effect of the non homogeneity of the flow.

Finally let us mention that the eq(2.49) and (2.50) provide a non linear equation for the front in a non uniform and/or unsteady flow. But, as already mentioned, the value of the flow field appearing in this equation is not a given quantity and, because of the hydrodynamics effects ii), this flow is, in fact, a functional of the flame surface. This aspect of the problem of the flame dynamics has been solved only in the linear approximation (see L. Boyer 1983). Nevertheless, for turbulent wrinkled flames, and when the corresponding random process is assumed to be stationary and homogeneous, the local equation for the front evolution provide an expression for the turbulent flame speed. The time average of the modification to normal burning velocity is found to be zero and the turbulent flame speed is proved to be given by simply the laminar flame speed times the mean area increase of the front.

III. DIFFUSION FLAMES

III.1. Position of the problems

In the broadest sense a diffusion flame is defined, see

Williams 1965, as any flame in which the fuel and the oxidizer are initially separated ; the term diffusion flame is synonymous with nonpremixed combustion. In diffusion flames mixing and chemical reaction take place simultaneously.

In a restricted sense, a diffusion flame is defined as a non-premixed flame in which most of the reaction occurs in a thin zone that separates the fuel from the oxidiser, and that, following Burke and Schumann 1928, it can be approximated by a surface.

Two types of problems may be encountered in the analysis of chemical reactions in unpremixed systems : (a) Problems of the evolution type, like unsteady mixing and combustion in mixing layers and boundary layers without a stagnation point, and (b) quasy steady problems like mixing and reaction in the stagnation region and quasi-steady droplet combustion.

For large activation energies, multiple solutions exist for the conservation equations in problems of the elliptic, quasi-steady type within a range of Damköhler numbers, bounded by an extinction Damköhler numbers, and by an ignition Damköhler number ; these solutions correspond to different combustion regimes,(Linan 1974). In problems of the evolution type the conservations equations are parabolic, because the terms representing the diffusive effects have second order spatial derivatives, while those terms representing the local heat accumulation or convective effects are first order with respect to time or with respect to a spatial derivative, that does not appear in the diffusive terms. Then the solution is uniquely determined in terms of the initial and boundary conditions With increasing values of the time-like variable the flow changes from nearly-frozen, with incipient effects of the chemical reaction, to a near-equilibrium, diffusion controlled mode of combustion (see fig.1.1); the transition occurs through premixed flames, of the deflagration or of the detonation type (see Liñan and Crespo 1976).

In the following, we shall present a detailed analysis of the diffusion flame structure, using as an example of quasi-steady problems, the diffusion flame at the stagnation region between to opposed jets of fuel and oxidiser. The results are, however, applicable to any general configuration. The analysis will be carried out for reactions that can be modelled by an Arrhenius overall irreversible reaction, but the analysis can be generalised for more general reactions.

The qualitative structure of diffusion flames is not strongly dependent on the value of the Lewis number of the species. The analysis simplifies considerably when we write $\rho D_\alpha = \lambda / C_p = \rho D_{th}$, corresponding to the equidiffusional approximation. In this case the transport operators on eq(1.5) coincide for all species, and when we take into acount (1.1'), the coupling functions, or Schwalb-Zeldovich variables,

$$Y_b - \nu Y_F \quad \text{and} \quad T+Y_F q/C_p$$

are found to diffuse as passive scalars.
Thus the systems of eqs(1.5,1',2) can be written in the form

$$\mathbb{L}(Y_F)= -BY_F^{n_F}\, Y_0^{n_0}\, e^{-E/RT} \tag{3.1}$$

$$\mathbb{L}(Y_0-\nu Y_F)= 0 \;, \quad \mathbb{L}(T+Y_F q/C_p)= 0 \tag{3.2}$$

and also $L(Y_N)=0$ for any inert species. Here $\mathbb{L}$ is defined by :

$$\mathbb{L}(Z) = \frac{\partial Z}{\partial t} + \underline{v} \bullet \underline{\nabla}(Z)-\frac{1}{\rho}\, \underline{\nabla}\bullet(\rho D_{th}\, \underline{\nabla}\, Z) \tag{3.3}$$

If we assume that we have, as in the example of fig. 1.1, two independent feed streams, one with fuel, with mass fraction Y_{F0} and temperature T_{F0}, and the other with oxidiser with mass fraction Y_{00} and temperature T_{00}, and no heat or mass flux on all the other boundary surfaces, then it is possible to write the solution of eq(3.2) in terms of the solution Z (mixture fraction) of the problem of eq(3.3) for an inert species, of unit concentration, Z=1, in the fuel feed stream, and zero concentration at the oxidiser stream (see for example Bilger, 1975, and Peters 1983).
Thus we can write the following Schvab-Zeldovich relations

$$Y_0-\nu Y_F= Y_{00} -(\nu Y_{F0} + Y_{00})Z \tag{3.4}$$

and

$$T + Y_F q/C_p = T_{00} +(T_{F0}-T_{00}+ Y_{F0}\, q/C_p)Z \tag{3.5}$$

independent of the kinetics, that appears only in eq(3.1).
In the non-reacting limit $Be^{-E/RT}\to 0$, when the chemical reaction is frozen, eq(3.1) simplifies to

$$\mathbb{L}(Y_F) = 0 \tag{3.6}$$

so that

$$\begin{aligned} Y_F &= Y_{Ff} = Y_{F0}\, Z \\ Y_0 &= Y_{0f} = Y_{00}\, (1-Z) \\ T &= T_f = T_{00} +(T_{F0}- T_{00})Z \end{aligned} \tag{3.7}$$

In the opposite limiting case (the Burke-Schumann limit), of very fast chemical reactions, $Be^{-E/RT}\to\infty$, we have chemical equilibrium, with

$$\begin{aligned} &Y_0= 0 \;, \quad Y_F = -Y_{00}/\nu +(Y_{F0}+ Y_{00}/\nu)Z \\ &T = T_{F0}+(T_{00}- T_{F0}+ Y_{00}q/\nu C_p)(1-Z) \end{aligned} \tag{3.8a}$$

for Z>Ze, on the fuel side, and

$$Y_F= 0 \;, \quad Y_0=Y_{00} -(\nu Y_{F0}+ Y_{00})Z$$

$$T = T_{O0} + (T_{F0} - T_{O0} + Y_{F0}\, q/C_p) Z \qquad (3.8b)$$

for Z<Ze, on the oxidizer side

A thin flame, located at Z=Ze given by

$$Ze = Y_{O0}/(Y_{O0} + \nu Y_{F0}) \qquad (3.9)$$

There $Y_O = Y_F = 0$ and $T=T_e$, given by the adiabatic flame temperature value

$$T_e = T_{O0} + (T_{F0} - T_{O0} + Y_{F0}\, q/C_p) Ze \qquad (3.10)$$

The flame sheet separates the fuel from the oxidiser.

To calculate the flame sheet position, Z=Ze, and the temperature and reactant concentration distributions, given by (3.8), we must solve the transport problem (3.3) for the mixture fraction. These distributions in the Burke-Schumann limit of infinitely fast reactions are diffusion controlled. It is easy to show, taking into account (3.8), that m''_F, the mass burning rate of fuel per unit flame surface, and the corresponding value m''_O for the oxidizer, given by

$$m''_F = \rho D_F \left(\frac{\partial Y_F}{\partial n}\right)_{e^+} \quad , \quad m''_O = -\rho D_O \left(\frac{\partial Y_O}{\partial n}\right)_{e^-} \qquad (3.11)$$

are related by

$$\nu m''_F = m''_O \qquad (3.12)$$

and the heat release per unit flame surface $q'' = \lambda \left[\frac{\partial T}{\partial n}\right]_{e^+}^{e^-}$ is related to m''_F by

$$q'' = \frac{q}{C_p}\, m''_F \qquad (3.13)$$

That is fuel and oxidiser reach the flame in stoichiometric proportions, and the heat release by the chemical reaction is conducted away from the flame. The jump relations (3.12,13) are also valid for non-unity Lewis numbers, when the relations (3.4) and (3.5), and therefore (3.9) and (3.10) do not hold. For non-unity Lewis numbers, and diffusion controlled combustion, we must solve the equations $\mathbb{L}_T(T)=0$, $\mathbb{L}_O(Y_O)=0$ on the oxidiser side of the flame where $Y_F=0$; and the equations $\mathbb{L}_F(Y_F)=0$, $\mathbb{L}_F(T)=0$, on the fuel side where $Y_O=0$. The concentration of both reactants must be zero at the thin flame, and the temperature continuous ; there are jumps however of the concentration and temperature gradients normal to the flame sheet, that satisfy the jump relations (3.12) and (3.13).

In order to describe with some details the transition from the frozen flow distributions of the form (3.7) to the equilibrium distributions of the form (3.8) we shall analyse the structure of counterflow diffusion flames. We shall make this description in the realistic limit of large values of the non-dimensional activation energy E/RT_e, when multiple solutions of the conservation equasions may exist, with associated bifurcation points, where jumps from a nearly frozen state to a near-equilibrium state, and vice versa, occurs.

In counterflow diffusion flames two streams, one of fuel and the other of oxidiser flow in opposite directions, setting up a diffusion flame between them where the chemical reaction takes place. The qualitative structure and the main characteristics of diffusion flames do not change when the effects, associated with density changes, of the heat release on the flow field are neglect. We shall, in addition assume for simplicity in the presentation of the structure, assume that the two jet velocities are equal. The flow field shown in fig.III.1 is axis-symmetric. Close to the stagnation point the flow field is given by :

$$u = -Ax \quad , \quad v = Ar/2$$

as a result of the constant density approximation, and for two identical jets. The factor A is the inverse of a residence time m the stagnation of the initial diameter D and velocity V of the jets.

It is well known that in the solutions of the conservation equations in the stagnation region the temperature and concentrations are functions only x, the distance normal to the mixing layer and, in unsteady problems, of time. We shall look for steady solutions, so that only x is involved as independent variable. The mixture fraction Z and the fuel mass fraction Y_F are given by the

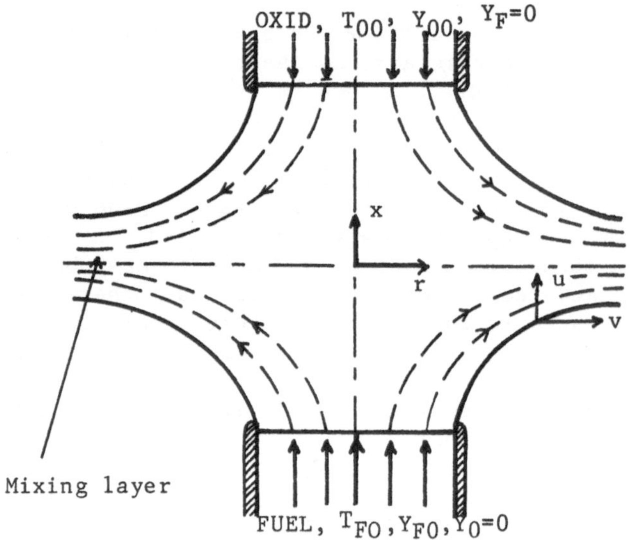

Fig.3.1

conservation equations

$$- Ax \frac{dZ}{dx} - D_{th} \frac{d^2Z}{dx^2} = 0 \qquad (3.14)$$

$$- Ax \frac{dY_F}{dx} - D_{th} \frac{d^2Y_F}{dx^2} = -Be^{-E/RT} Y_F^{n_F} Y_0^{n_0} \qquad (3.15)$$

where we have considered, for simplicity, that both ρ and the thermal diffusivity D_{th} are constant. We shall write these equations in non-dimensional form in terms of the variable

$$\eta = x/\sqrt{D_{th}/A}$$

where D_{th}/A is the characteristic thickness of the mixing layer. Thus we obtain

$$\eta Z_\eta + Z_{\eta\eta} = 0 \qquad (3.16)$$

$$\eta Y_{F\eta} + Y_{F\eta\eta} = D_\infty e^{-E/RT} Y_F^{n_F} Y_0^{n_0} \qquad (3.17)$$

to be solved with the boundary conditions

$$Z=1 \ , \ Y_F = Y_{F0} \quad \text{for} \quad \eta \to -\infty$$

$$Z=0 \ , \ Y_F = 0 \quad \text{for} \quad \eta \to \infty$$

if the mixing layer thickness is small compared with the size of the jets. Y_0 and T are given by eq(3.4,5) in terms of Y_F and Z. D_∞ is a Damköhler number $D_\infty = B/A$, the product of the frequency factor B and the characteristic diffusion time, or residence time, 1/A across the mixing layer. The solution of (3.16) can be written as :

$$Z = (1/2)\text{erfc}(\eta/\sqrt{2}) \qquad (3.18)$$

which together with eqs(3.4 and 5) and eq(3.17) determines $Y_F(\eta)$. In the non-reactive limit $D_\infty e^{-E/RT} \to 0$ the solution was given before as the frozen solutions eq(3.7). In the limit $D_\infty e^{-E/RT} \to \infty$, of infinitely large Damköhler numbers, we obtain the Burke-Schumann (B-S), chemical equilibrium, solution, given by eq(3.8). These limiting solutions are shown schematically in figs(3.2a, 2b) in terms of the mixture fraction Z as independent variable.
Notice that in the B-S limit there are jumps in the concentration and temperature gradients at the flame sheet, $Z=Z_e$, that acts as a sink for the fuel and oxidiser, and as a source for the products and thermal energy. The jumps are rounded off, as shown by the dashed lines in fig.III.2b due to finite rate effects.
The linear form of the profiles, in both limiting cases, when the mixture fraction Z is used as independent variable can be understood, when we notice that the transport operator appearing in eq(3.17) can be written in the form

$$\eta Y_{F\eta} + Y_{F\eta\eta} = Z_\eta^2 \, Y_{FZZ} \qquad (3.19)$$

where $Z_\eta^2 = (1/2\pi)\exp(-\eta^2/2)$, the local gradient of the mixture fraction , is a function of $2Z=\text{erfc}(\eta/\sqrt{2})$, whose variation across the mixing layer accounts for convective effects.

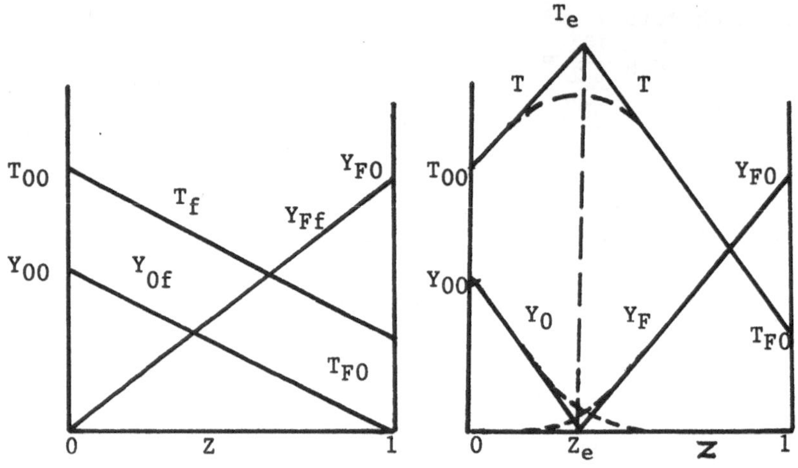

Fig.3.2a   Fig.3.2b

III.2. Ignition regime

For zero Damköhler numbers the temperature and concentration distributions take their frozen values, $Y_{Ff}=Y_{F0}\,Z$, $Y_{0f}=Y_{00}(1-Z)$ and $T=T_f=T_{00}+(T_{F0}-T_{00})Z$. The deviations from these values, for non zero small values of D_∞, can be calculated in first approximation, by using eq(3.17) with the right hand side evaluated in terms of the frozen solution, i.e. by using the equation

$$Z_\eta^2 \, Y_{FZZ} = D_\infty e^{-E/RT_f} \, Y_{Ff}^{n_F} \, Y_{0f}^{n_o} \tag{3.20}$$

with $Y_F(0)= 0$, $Y_F(1)= 1$

The solution gives the first two terms of an expansion of the solution of eq(3.17) in powers of the Damköhler number D_∞. It represents well the temperature and concentration distributions as long as the deviations of the chemical production term from its frozen flow value, appearing in the right hand side of eq(3.20) are small.

For large activation energies, $E/RT \gg 1$, the changes in the Arrhenius factor, $\exp(-E/RT)$, become significant, even with small changes in the temperature from its frozen flow value T_f. That is suffices to have variations of T from T_f small, $(T-T_f)/T_f= O(RT_f/E)$ in order to change the Arrhenius exponential by a factor of order unity, so that it can no longer be approximated by $\exp(-E/RT_f)$, although we can use, because $(T-T_f)/T_f \ll 1$, the Frank-Kamenetskii (1969), linearisation

$$\exp(-E/RT) \simeq \exp(-E/RT_f)\exp E(T-T_f)/RT_f^2 \tag{3.21}$$

of the Arrhenius exponent, when analysing the diffusion flame in the ignition regime.

Due to the much weaker sensitivity of the reaction rates to concentration changes, the factor $Y_F^{n_F} Y_O^{n_O}$ can often be approximated by its frozen flow value $Y_{Ff}^{n_F} Y_{Of}^{n_O}$ because in the ignition regime the relative changes in concentration are small of the order of RT_f/E. Although this is not the case when, for example, $T_{00}-T_{F0}>>RT_{00}^2/E$, when in the ignition regime the chemical reaction occurs on the oxidizer edge of the mixing layer, where the temperature is higher, and where we can write $Y_O=Y_{O0}$ but not $Y_F=Y_{Ff}$ in evaluating the reaction rate. See Liñán 1974.

In the particular case $T_{00}=T_{T0}=T_0$, when $T_f=T_o$, we can except from the solution of (3.20) that $(T-T_o)/T_o$ will be of the order RT_o/E, and therefore we shall be in the ignition regime for values of the Damköhler number such that

$$\frac{q}{C_pT_o} D_\infty e^{-E/RT_o} Y_{F0}^{n_F} Y_{00}^{n_o} = \frac{RT_o}{E}\mathcal{D} \tag{3.22}$$

with $\mathcal{D}$ a number of order unity.
In this distinguished limiting case, $E/RT_o\to\infty$ with $D_\infty\to\infty$ so that $\mathcal{D}$ is fixed of order unity, the deviations of Y_O, Y_F and T from their frozen values $Y_{00}(1-Z)$, $Y_{F0}Z$ and T_o are small of order RT_o/E; and are given in first approximation in terms of the solution, for $\phi= E(T-T_o)/RT_o^2$, of

$$Z_\eta^2 \phi_{ZZ} = -\mathcal{D} e^{\phi} Z^{n_F} (1-Z)^{n_o} \tag{3.23}$$

$$\phi(o) = \phi(1)= 0$$

and

$$Y_F/Y_{F0} = Z-\varepsilon_o\phi \quad , \quad Y_O/Y_{00}= 1-Z-(\nu Y_{F0}/Y_{00})\varepsilon_o\phi$$

where $\varepsilon_o=(C_pT_o/qY_{F0})(RT_o/E)$ is small for large values of the activation energy.

The limiting form eq(3.23) of the conservation equation for T, for large values of E/RT_o, results from linearisation eq(3.21) of the Arrhenius exponent and from the neglect of the reactant consumption in evaluating the reaction rate.

The reduced Damköhler number $\mathcal{D}$, defined by eq(3.22), is the only parameter, aside from the reaction orders η_F and η_o, entering in eq(3.23). When this equation is solved numerically, two solutions are found for all values of the Damköhler number smaller than an ignition value $\mathcal{D}_I$, of order unity, for which the two solutions coincide. No solution of the problem (3.23) exists for values of the Damköhler number $\mathcal{D}$ larger than $\mathcal{D}_I$. Thus, when we plot a certain norm of $\phi(Z)$, such as $\phi_{max}=E(T_{max}-T_o)/RT_o^2$ in terms of $\mathcal{D}$, for fixed values of the reaction orders, the resulting curve is of the form shown in fig. 3 .3 that corresponds to $\eta_F=\eta_o=1$, with a $\mathcal{D}_I= 2.59$.

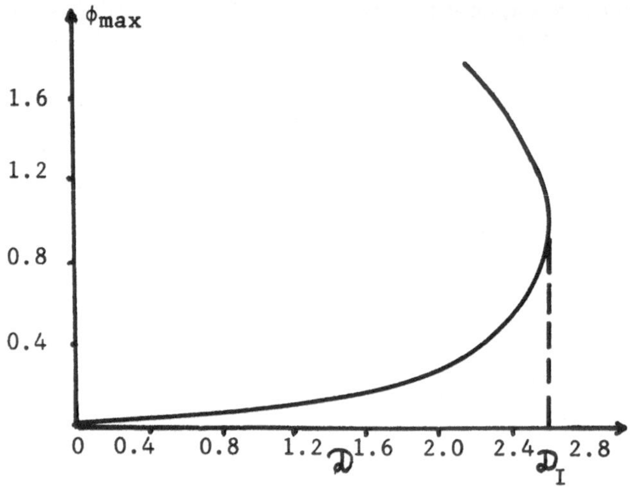

Fig.3.3

The existence of the two solutions is associated with the strong dependence of the reaction rate with temperature. The solution corresponding to the lower branch of Fig. 3.3 shows small temperature increments, associated with small reaction rates ; the solution, for the same Damköhler number, corresponding to the upper branch shows a much larger value of ϕ_{max} associated with larges values of $\exp\phi$. For values of $\mathcal{D} > \mathcal{D}_I$ the increments in temperature, and the corresponding values of $\exp\phi$, are two large to sustain a nearly frozen solution. A stability analysis of the two solutions represented in Fig. 3.3 would show the upper branch to be unstable ; and only the lower branch can be reached for large times in an unsteady process if we begin with initial conditions in an appropriate domain of attraction. When $\mathcal{D}$ is increased slowly by increasing, for example, the reactant or oxidizer concentration in the free streams the solution will follow the lower branch of Fig. 3 .3, until a sudden jump from the nearly frozen mode of combustion to a near equilibrium mode occurs if $\mathcal{D}$ is increased slightly above $\mathcal{D}_I$.

III.3. Diffusion controlled regime

We shall now describe the structure of the flame in the limiting case of large Damköhler numbers, $D_\infty \to \infty$, with E/RT_{00} fixed. This is the Burke-Schumann limit. As we indicated before, in this limit an infinitely thin reaction zone, or flame sheet, located at $Z=Z_e$, where the temperature is the adiabatic flame temperature T_e, given by eqs(3.9 and 10), separates a region $Z>Z_e$ where

$Y_F=0$. The temperature and concentration distributions are given in this limit by eq(3.8), and are shown schematically in fig. 3 .2b. For large, but finite values of D_∞, the distributions are still given by the B-S solution (3.8), outside a thin reaction zone located, as shown in fig. 3 .4, around Z_e. Let $\ell \ll 1$ be the thickness of the reaction zone relative to that of the mixing layer, then the mass fractions of the reactants Y_F/Y_{F0} and Y_0/Y_{00}, that coexist in the reaction layer are there of order ℓ ; and this is also the order of $C_p(T_e-T)qY_{F0}$, measuring the deviation of T from T_e. We require that all the terms appearing in the form of Schwabb-Zeldovich relations (3.4 and 5) given below

$$C_p(T-T_e)/qY_{F0}= -Y_F/Y_{F0}+ \{1-C_p(T_{00}-T_{F0})/qY_{F0}\}(Z-Z_e)$$
$$Y_0/\nu Y_{F0} = Y_F/Y_{F0}-(1+Y_{00}/\nu Y_{F0})(Z-Z_e) \tag{3.24}$$

are of the same order $\ell \ll 1$.
If we want to relate ℓ to D_∞, we use eq(3.17), noticing that, because Z_η^2 is of order unity (as long as Z_e is not close to 0 or 1), its terms are of order

$$\ell/\ell \quad , \quad \ell/\ell^2 \quad , \ D_\infty e^{-E/RT_e} \quad \ell^{n_F+n_o}$$

respectively, if, for simplicity, we assume Y_{00} and Y_{F0} to be of order unity. We thus see that the convective term is, of order unity, negligible when compared with the diffusion term, of order $1/\ell \gg 1$, that must be balanced by the reaction term. So that the value of ℓ given by

$$\ell = \{D_\infty e^{-E/RT_e}\}^{-1/(1+n_F+n_o)} \tag{3.25}$$

must be small compared with unity, if we want the reaction zone to be thin, and thereby the convective term to be negligible compared with the diffusion term. Notice that this approximation is equivalent to replacing the transport operator $\eta Y_{F\eta}+ Y_{F\eta\eta}$ by $(Z_\eta^2)_e Y_{FZZ}$, justified because the changes of Z_η^2 accross the reaction zone are small. An additional simplification, when solving eq. (3.17) in the thin reaction zone, results from replacing the Arrhenius exponential by $\exp(-E/RT_e)$, when we notice that for large Damköhler numbers

$$(T_e-T)/T_e \sim (qY_{F0}/c_pT_e)\ell$$

is small, and then

$$\exp(-E/RT) \simeq \exp(-E/RT_e)\exp\{E(T-T_e)/RT_e^2\} \tag{3.26}$$

and the last factor can be replaced by 1, if

$$(E/RT_e)(qY_{F0}/C_pT_e)\ell \ll 1$$

Therefore, when this last relation holds, for large values, according to eq(3.25), of the Damköhler numbers, the fuel concentration in the thin flame is given by the equation :

$$(Z_\eta^2)_e Y_{FZZ} = D_\infty e^{-E/RT_e} Y_F^{n_F} Y_0^{n_o} \tag{3.27}$$

to be solved, together with second relation (3.24) and the boundary conditions :

$$Y_F \to 0 \quad \text{for} \quad (Z_e - Z) \to \infty$$
$$Y_O \to 0 \quad \text{for} \quad (Z - Z_e) \to \infty \tag{3.28}$$

We should however point out that if, for example $0<n_F<1$, Y_F comes down to zero at a finite value of (Z_e-Z), to be determined as part of the solution. The decay to zero is exponential for $n_F=1$, and of the form $(Z_e-Z)^{-(2+n_0)/(n_F-1)}$ for $n_F>1$.

The problem (3.27-28) can be recast in a universal form, Liñan 1961, of the type

$$y_{XX} = y^{n_F}(\tilde{y}-X)^{n_o}$$
$$y_{-\infty} = 0 \ , \quad (\tilde{y}-X)_\infty = 0 \tag{3.29}$$

in terms of the variables

$$\tilde{y} = Y_F/Y_r \ , \quad X=(Z-Z_e)(Y_{FO}+Y_{OO}/\nu)/Y_r \tag{3.30}$$

where Y_r, the characteristic value of Y_F in the thin reaction zone, is given by the relation :

$$Y_r^{n_F+n_0+1} = (Z_\eta^2)_e (Y_{FO}+Y_{OO}/\nu)^2 \ \nu^{-n_0} \ D_\infty^{-1} \exp(E/RT_e)$$

and

$$Y_O/\nu Y_r = \tilde{y} - X, (T-T_e) = qY_r/C_p\{-\tilde{y}+mX\} \tag{3.31}$$

where

$$m = \{1-C_p(T_{OO}-T_{FO})/qY_{FO}\}/(1+Y_{OO}/\nu Y_{FO}) \tag{3.32}$$

In the particular case when the reaction order is zero for one of the reactants, the boundary conditions of eq(3.29) must be modified. For example if $n_0=0$, the oxidant concentration will come down to zero at a finite value X_0 of X , and the reaction term will be zero for $X>X_0$, where $\tilde{y}-X = 0$. Then the boundary condition $(\tilde{y}-X)_\infty=0$ must be replaced by the conditions $y-X=0$ and $y_X=1$ at $X=X_0$ to be determined, as part of the solution, as a function of n_F. If $n_F=1$ the solution of eq(3.29) is of the form $y=A\exp X$ and $A=1/e$ results from the conditions $A\exp X_0=X_0=1$. If $n_F\neq 1$, $\tilde{y}_X = \{2/(n_F+1)\}^{\frac{1}{2}}\tilde{y}^{(n_F+1)/2}$ and

$$\tilde{y}^{(1-n_F)/2} - 1 = (X_0-X)(n_F-1)/\sqrt{2(n_F+1)}$$

with X_0 given by $X_0 = \{(n_F+1)/2\}^{1/(n_F+1)}$, to insure that $\tilde{y}_X=1$ at $X=X_0$. Notice that for $n_F<1$, y will be identically equal to zero for $X<X_1$ when X_1 is given by $X_0-X_1 = \sqrt{2(n_F+1)}/(1-n_F)$.

The solution of (3.29), then describes the concentration and temperature profiles within the reaction zone if the Damköhler number is sufficiently large so that $Y_r/Y_{FO}\ll 1$. However, due to the large values that the nondimensional activation energy

$$(E/RT_e)(qY_{FO}/C_pT_e) = \beta_e \tag{3.33}$$

takes in practical cases, the nondimensional temperature drop $E(T_e-T)/RT_e^2$ in the reaction zone, of the order of $(Y_r/Y_{FO})\beta_e$, can become of order unity, and then the Arrhenius exponent can no longer be evaluated at the temperature T_e, but it should be calculated using the Frank-Kamenetskii approximation (3.26), when describing the structure of the thin reaction zone.

III.4. Extinction regime

The extinction regime corresponds to large activation energies $\beta_e \gg 1$, and precisely those Damdöhler numbers that lead to values of Y_r/Y_{F0}, calculated by (3.31), of the order $1/\beta_e$. This is also the order of the thickness of the reaction zone relative to that of the mixing layer.

Outside the reaction zone the reaction can be frozen due to the rapidly decreasing values of the Arrhenius exponential when the temperature drops. However, the temperature and concentrations are still given there by the Burke-Schumann solution in first approximation. In the thin reaction zone, in first approximation, Y_F, Y_0 and T are given by the solution of

$$(Z_\eta^2)_e Y_{FZZ} = D_\infty e^{-E/RT_e} e^{E(T-T_e)/RT^2_e} Y_F^{n_F} Y_0^{n_0} \tag{3.34}$$

together with eq(3.24), and the boundary conditions

$$Y_{FZ} = 0 \quad \text{for} \quad (Z_e - Z) \to \infty \quad ; \; Y_{0Z} = 0 \quad \text{for} \quad (Z - Z_e) \to \infty \tag{3.35}$$

obtained from the matching conditions with the Burke-Schumann solution, or first approximation of the outer solution.
The problem (3.34-35) can be recast in a simpler form, similar to (3.29), using variables of the type (3.30) with $Y_r = Y_{F0}/\beta_e$. That is, if

$$y = (Y_F/Y_{F0})\beta_e \quad , \; \xi = (Z - Z_e)(1 + Y_{00}/\nu Y_{F0})\beta_e \tag{3.36}$$

Then the problem (3.34-35) becomes

$$y_{\xi\xi} = \delta e^{-y+m\xi} \, y^{n_F} (y-\xi)^{n_0} \tag{3.37}$$

$$y_\xi = 0 \quad \text{for } \xi \to -\infty \, , \qquad y_\xi \to 1 \quad \text{for } \xi \to \infty \tag{3.38}$$

involving the parameters m, defined in (3.32), n_f, n_o and the reduced Damköhler number δ, defined by

$$\delta = (Z_\eta^{-2})_e D_\infty e^{-E/RT_e} \nu^{n_0} (Y_{F0}/\beta_e)^{n_F+n_0-1} (Y_{F0} + Y_{00}/\nu)^{-2} \tag{3.39}$$

of order unity in the extinction regime.

Notice that due to the presence of the Arrhenius variable factor $\exp(-y+m\xi)$ in eq(3.37), the chemical reaction can be frozen on one or both sides of the thin reaction zone, and thus we must use for eq(3.37) the boundary conditions (3.38), weaker than those used for eq(3.29). We thus allow for leakage $y_{-\infty}$ of the fuel, or $(y-\xi)_\infty$ of the oxidizer through the reaction zone, due to quenching of the chemical reaction in the extinction regime. Associated with this leakage, we shall find concentrations of order $1/\beta_e$ of the fuel on the oxidant side of the flame, and similar concentrations of oxidizer on the fuel side of the flame. The amount of reactants

$y_{-\infty}$ and $(y-\xi)_\infty$ leaking through the reaction zone must be calculated from the numerical solution of the problem (3.37-38) ; it is, for given reaction orders n_O and n_F, a function of δ and m.

In order to understand the reasons for finding reactant leakage, notice that $E(T-T_e)/RT_e^2 = -y+m\xi$, so that the temperature decreases with a slope m on the oxidant side of the flame, and with a slope (1-m) on the fuel side. It is then clear that we will not find leakage of fuel if m is zero or negative, when no quenching of the reaction takes place on the oxidant side ; no leakage of oxidant will occur when m>1. The numerical solution of the problem (3.37-38) can be found in Liñan 1974, for the case $n_O=n_F=1$, where $(4\delta)^{1/3}(y-\xi/2)$ was used as dependent variable, $(4\delta)^{1/3}\xi/2$ as independent variable ; the solution was found in terms of $\delta_0= 4\delta$ for several values of $\gamma=2m-1$. When m belongs to the interval (0,1), two solutions of the problem (3.37-38) exist for values of δ larger than an extinction value δ_E, a function of m for fixed values of n_O and n_F, and no solutions exist for $\delta<\delta_E$. This is shown in fig.III.4 where $(y-\xi)_\infty$ is plotted in terms of δ for several values of m, with $n_F=n_O=1$. Fig.III.4 can also be used to calculate the leakage of oxidant, for $n_F=n_O=1$ for the symmetrical case when m lies in the interval $(0.5<m<1)$ if m is replaced by 1-m and $y_{-\infty}$ by $(y-\xi)_\infty$:

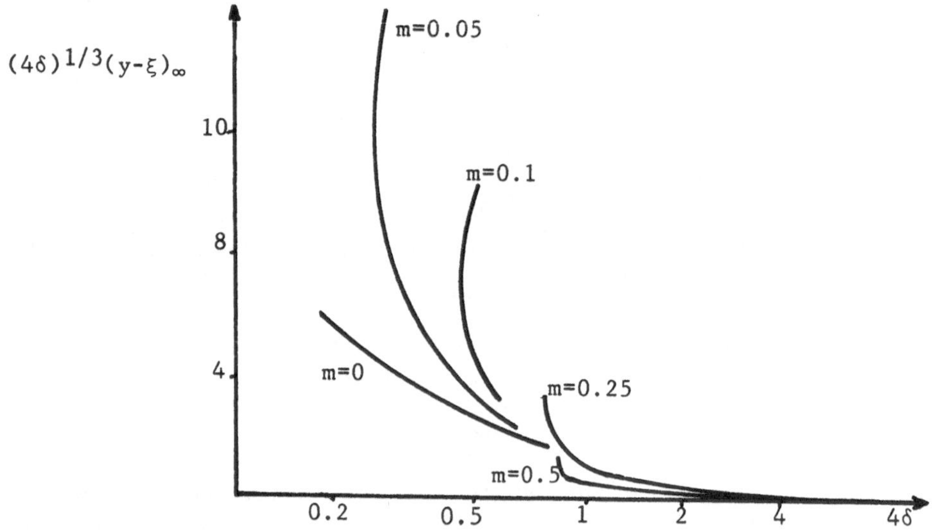

Figure 3.4

A good correlation of the extinction Damköhler number in this case was obtained by Liñan 1974, yielding

$$\delta_E=(e/2)\{m-2m^2+1.04m^3+0.44m^4\} \tag{3.40}$$

valid in the range $0<m<0.5$; and also in the range $0.5<m<1$ if m is replace by 1-m.

Of the two solution branches appearing in fig. 3 .4 only the lower branch, showing smaller deviations from the B-S solution, can be expected to be stable. The structure of this solution for large values of δ takes a universal form given by eq(3.29), where $\tilde{y}=\delta^{1/3}y$, $X=\delta^{1/3}\xi$; the Arrhenius factor $\exp(-y+m\xi)$ = $\exp\{\delta^{-1/3}(-\tilde{y}+mX)\}$ does not deviate significantly from unity in the reaction zone if $\delta\gg1$. The upper branch shows stronger non-equilibrium effects. For large values of δ the solution merges if m takes values around 0.5, with the solution corresponding to an unstable partial burning regime, when both reactants cross the thin reaction zone and coexist on both sides of the flame sheet. It merges with a premixed flame regime when m or 1-m are negative or small (see Liñan 1974).

For small values of m, or of (1-m), extinction conditions of the diffusion flame occur with a premixed flame regime. This can be shown, using the eq(3.37-38) that describe the reaction zone structure, as follows. We pose our problem as that of finding $\delta(b,m)$ that leads to an oxidant leakage $(y-\xi)_\infty=b/m$ with b of order unity and $m\ll1$. Then we anticipate the following structure of the reaction zone, shown schematically in fig. 3.5

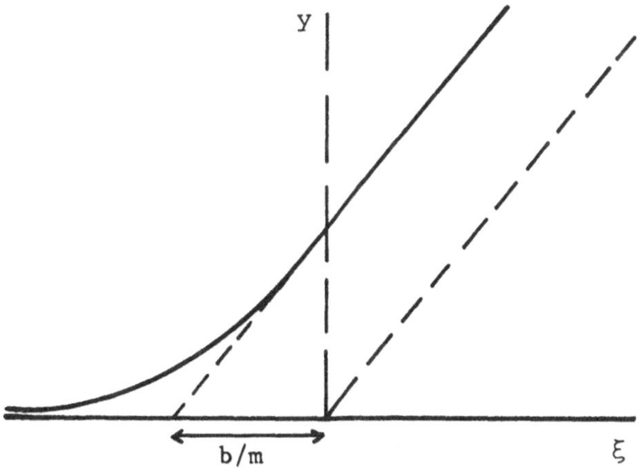

Fig.3.5

There is a reaction zone, of thickness of order unity in the ξ variable, located around $\xi= -b/m$, where b of order unity, is negative if m is negative. In this reaction zone y is of order unity, to the right the reaction is frozen, and y tends to zero for large values of $-\xi$. The solution will be written in the form of an expansion in powers of m, using $\zeta=\xi +b/m$ as independent variable. In terms of ζ eq(3.37) takes the form

$$y_{\zeta\zeta} = \delta e^{b}(b/m)^{n_0} e^{-y+m\zeta}\, y^{n_F}\, \{1+m(y-\zeta)/b\}^{n_0} \tag{3.41}$$

to be solved with the boundary conditions

$$y_\zeta = 0 \quad \text{at } \zeta \to -\infty \quad ; \quad y-\zeta = 0 \quad \text{at } \zeta\to\infty \tag{3.42}$$

When we introduce in these equations the expansions

$$y = y_0(\zeta)+ m\, y_1(\zeta)+\dots$$
$$\delta e^{b}(b/m)^{n_0} = \delta_0 [\, 1+md+\dots\,] \tag{3.43}$$

where y_0, y_1, as well as δ_0 and d, are assumed to be of order unity for ζ of order unity, we obtain the equation

$$y_{0\zeta\zeta} = \delta_0 y_0^{n_F}\, e^{-y_0} \tag{3.44}$$

and boundary conditions

$$y_0 \to 0 \text{ for } \zeta\to -\infty,\ (y_0-\zeta)\to 0 \quad \text{for } \zeta\to\infty \tag{3.45}$$

identical to those that describe the reaction zone structure in a premixed flame. Notice that $y_{0\zeta}\to 0$ for $\zeta\to-\infty$ implies that also $y_0\to 0$. A first integral of (3.44) yields, when using (3.45):

$$y^2_{o\zeta} = 2\delta_0 \int_0^{y_0} x^{n_F}\, e^{-x}\, dx$$

and then $\delta_0^{-1} = 2\Gamma(n_F+1)$, that provides in first approximation, for small values of m, the relation

$$2\Gamma(n_F+1)\ \delta e^{b}(b/m)^{n_0} = 0 \tag{3.46}$$

between δ and b. The relation $\delta(b)$ is single-valued, but the inverse relation $b(\delta)$, determining the oxidant leakage, $b=m(y-\xi)_\infty$, in terms of δ, although single-valued for $m<0$, is doubled valued for $m>0$ if δ is larger than an extinction value δ_E, as shown in fig. 3.4 for small values of m. At the extinction value, the Damköhler number is given by

$$\delta_E = (e\, m/n_0)^{n_0}/2\Gamma_{n_F+1} \tag{3.47}$$

where Γ_{n_F+1} is the gamma function of order n_F+1.
The equation giving the second term of the expansion (3.43) is

$$y_{1\zeta\zeta}=\delta_0 y_0^{n_F} e^{-y_0}\{y_1(n_F/y_0-1)+d+\zeta+n_0(y_0-\zeta)/b\} \tag{3.48}$$

to be solved with the conditions $y_{1\zeta}=0$ for $\zeta\to -\infty$ and $y_1=0$ for $\zeta\to\infty$. We notice that $y_{0\zeta}$ satisfies the homogeneous part of this equation. If we multiply both members of (3.48) by $y_{0\zeta}$ we obtain :

$$(d/d\zeta)(y_{0\zeta}y_{1\zeta}-y_{0\zeta\zeta}y_1)= y_{0\zeta}\delta_0 y_0^{n_F} e^{-y_0}\{d+\zeta+n_0(y_0-\zeta)/b\}$$

that we can integrate from $-\infty$ to ∞ to obtain

$$d\Gamma_{n_F+1} + (n_o/b)\Gamma_{n_F+2} + \alpha(1-n_o/b)=0 \qquad (3.49)$$

a relation determining the perturbation d in the Damköhler number in terms of b. Here α is given by the integral

$$2\alpha = \int_{-\infty}^{+\infty} x \frac{d}{dx}(y_{ox}^2)dx = \int_0^\infty (1-y_{ox})dy_o = \gimel_{n_0}$$

is a function of n_0 that can be evaluated numerically without difficulties. If we are simply interested in calculating the new extinction value of δ, we do not need to evaluate this integral, because d can be calculated using the first approximation for the extinction value of b, namely $b=n_o$, in (3.49). Then $d= -(n_F+1)$ and a two term expansion for the extinction value of the Damköhler number is

$$\delta_E = \{2\Gamma_{n_F+1}\}^{-1}(em/n_o)^{n_o}\{1-(n_F+1)m+\ldots\} \qquad (3.50)$$

Generalisation to other configurations and kinetic schemes

The analysis presented here was generalized to account for variable gas density and realistic variations of transport coefficients by Krishnamurthy et al.(1976), who consider a jet of oxidizer flowing normal to a condensed fuel surface ; the vaporisation rate of the fuel is controlled in this case by the balance between the heat feedback from the flame and the heat required for gasification. Extinction experiments carried out with this configguration have been used to obtain kinetic data for a variety of fuels. See Williams 1981. The quasisteady spherically symmetrical burning of a droplet was analysed by Law, 1975.

For a review of the use of the conserved scalar approach in turbulent combustion see Bilger 1976. When the assumption of infinitely fast chemistry is used together with the equi-diffusional approximation, the concentration of the main reactants and radicals as well as the temperature can be calculated in terms of the local instantaneous mixture fraction, that can be described in terms of its probability density function. When non-equilibrium effects become important, in addition to the mixture fraction, at least another "progress" variable must be included in the formulation. A presumed joint pdf for the conserved scalar and for the temperature as a progress variable was used by Peters et al.(1981) to derive a closed form expression for the mean turbulent reaction rate in the limit of large activation energies, showing low quenching would reduce the turbulent reaction rate.

Williams (1975) introduced the concept of laminar diffusion flamelet in turbulent combustion showing how mixing and chemical reaction occurs in thin regions that are wrinkled and strained by the turbulent fluctuations, but show a structure similar to that of the counter flow diffusion flames. For recent reviews on these laminar flamelets in diffusion flames, see Peters and Williams 1981 and 1983 and Peters 1983.

For a review of the numerical analysis of diffusion flame structure with multiple kinetics see Peters 1983, who also reviews the work that has been carried out to analyse these flames using asymptotic techniques. The use of these techniques is justified by the fact that the characteristic reaction times associated with the different reactions that enter into a reaction mechanism, differ very often by many orders of magnitude. Thus it is possible to justify by means of asymptotic techniques the use of the "partial equilibrium" assumption for some reactions, and the use of the quasi-steady approximations for some of the radical species.

The first attempts to describe by means of asymptotic techniques the diffusion flame structure with multiple kinetics were those of Clarke 1968 and 1969, see also Clarke 1975 and Allison and Clarke 1980, in connection with the Hydrogen-Air reaction. Melvin and Moss 1970 analysed the structure of Methane-Air flames for large Damköhler numbers. An analysis of diffusion flames involving a one-step forward and the backward reaction has been carried out by Peters 1979. For a review of additional asymptotic analysis of the structure and ignition processes of diffusion flames see Buckmaster and Ludford 1982.

Acknowledgements

The authors are grateful to Mrs M. Gillino and Mr. J.P. Pahin for their technical assistance.

REFERENCES

1. G. K. Adams and G. V. Stocks (1953). The combustion of hydrazine. Fourth (Int.) Symp. on Combustion, 239-248.
2. A. P. Aldushin, S. I. Khudyaev and Y. B. Zeldovich (1981). Flame propagation in the reacting gaseous mixture. Archivum Combustionis, 1 n°1/2, 9-21.
3. R. A. Allison, J. F. Clarke (1980).Theory of a hydro-oxyg.diffu. Combust. Sci Techn. **23**,113-123
4. D. G. Aronson and H. F. Weinberger (1978). Advances in Mathematics, vol.30 n°1, 33 and (1975) Non linear diffusion in population genetics Lecture Notes in Mathematics n°446, 5-40 Springer.
5. G. I. Barenblat, Y. B. Zeldovich and A. G. Istratov (1962). On diffusional thermal instability of laminar flame. In russian. Prikl. Mekh. Tekh. 2, 21-26.
6. R. W. Bilger (1976). Turbulent jet diffusion flames Prog. Energy Combust. Sci. 1, 87
7. L. Boyer (1983). Optical Diagnostics in flows. These Proceedings.

8. J. D. Buckmaster (1976). The quenching of deflagration waves. Combust. Flame 26, 151-162.
9. J D. Buckmaster and G.S.S. Ludford (1982). Theory of laminar flames, Cambridge Univ. Press.
10. J. D. Buckmaster and D. Mikolaitis (1982). The premixed flame in counterflow, Combust. Flame, 47, 191.
11. S. P. Burke and T.E.W. Schumann (1928). Industrial and Engineering Chemistry, 20, 998.
12 W. B. Bush and F. E. Fendell (1970). Asymptotic analysis of laminar flame propagation. Combust. Sci. Tech. 1, 421-428.
13. W. B. Bush and F. E. Fendell (1971). Asymptotic analysis of the structure of steady planar detonation. Combust. Sci. Tech. 2, 271-285.
14. J. F. Clarke and A. C. McIntosh (1980). The influence of a flame holder on a plane flame, including its static stability. Proc; R. Soc. London A372, 367-392.
15. J. F. Clarke (1968). On the structure of a Hydro. Oxyg.Dif.Flame Proc. Roy. Soc. A. 307, 283-302
16. J. F. Clarke (1969).Reaction broadening in H_2-O_2 diffusion flames Proc. Roy. Soc. A.312, 65-83
17. J. F. Clarke (1975).Parameter perturbations in flame theory. Progr. Aerospace Sci. 16, 3.29
18. P. Clavin and F. A. Williams (1982). Effects of molecular diffusion and of thermal expansion on the structure and dynamics of premixed flames. J. Fluid Mech. 116, 251-282.
19. P. Clavin and G. Joulin (1983). Premixed flames in large scale and high intensity turbulent flow. J. Physique Lettres, 44 L1-L2.
20. P. Clavin (1984). Dynamical behavior of premixed fronts in laminar and turbulent flows. Progr. Energy Combust. Sci. to appear.
21. J. D. Cole (1968). Perturbation methods in applied mathematics. Blaisdell, Waltham, Mass.
22. G. Darrieus (1983. Propagation d'un front de flamme. Unpublished works presented at "La Technique Moderne" in 1938 and at the "Congrès de Mécanique Appliquée, Paris, 1945".
23. B. Deshaies, G. Joulin and P. Clavin (1981). Etude asymptotique des flammes sphériques non adiatatiques. J. Mécanique 20, 691-735.
24. F. E. Fendell (1972). Asymptotic analysis of premixed burning with large activation energy. J. Fluid Mech. 56, 81-95.
25. P. C. Fife (1979). Mathematical aspects of reacting and diffusing systems. Lecture Notes in Biomathematics n°28, Springer Verlag.
26. R. A. Fisher (1937). The wave of advance of advantageous genes. Annals of Eugenics 7, 355-369.
27. D. A. Frank Kaminetskii (1969). Diffusion and heat transfert in chemical kinetics, Plenum Press.

28. I. M. Gelfand (1959). Uspeki Mat. Nauk (N.S. 14, n°2 (86), 87-158. Some problems in the theory of quasilinear equations. American Mathematical translations (1963) Series é vol.29, 295-381.
29. P. Higuera (1983). Private communication, Madrid, August 1983.
30. E. Jouguet (1913). Sur la propagation des déflagrations dans les mélanges gazeux, Compte rend. acad. sci. Paris,T.156, n°11, 872-876, see also "La mécanique des explosifs", Douin 1917.
31. G. Joulin and P. Clavin (1976). Analyse asymptotique des conditions d'extinction des flammes laminaires, Acata Astronautica 3, 223-240.
32. G. Joulin and P. Clavin (1979). Linear stability analysis of non adiabatic flames. Combust. Flame 35, 139-153.
33. G. Joulin and T. Mitani (1981). Linear stability analysis of two reactant flames. Combust. Flame 40, 235-246.
34. B. Karlowitz, J. R. Denniston, D. H. Knapschaefer, F.E. Wells (1953). Studies on turbulent flames, Fourth (Int.) Symp. on Combustion, 613-620.
35. A. N. Kolmogorov, I.G. Petrovskil, N.S. Piskunov (1937). A study of the equation of diffusion with increase in the quantity of matter and its application to a biological problem. Bjul. Moskovskovo Gos. Univ. 1 n°7, 1-72.
36. L. Krishnamunthy, F.A. Williams, K. Seshadri (1976).Asymptotic theory of diffusion flames extinctions in the stagnation Combust. and Flame, 26, 363.377
37. Y. Kuramoto (1978). Diffusion-Induced chaos in reaction systems. Supplement of the Progress of theoretical Physics 64 346-367. See also "Synergetics"(Bielefeld 1979)
38. L. Landau (1944). On the theory of slow combustion. Acta Physicochimica URSS vo.XIX n°1, 77-85.
39. C. K. Law (1975).Asymptotic theory for ignition and extinction Combust. and Flame, 24, 89.98
40. P. A. Libby, and F.A. Williams (1982). Structure of laminar flamelets in premixed turbulent flames, Combust. Flame, 44 287-303.
41. P. A. Libby, A. Liñan, F.A. Williams (1983). Strained premixed flames with non unity Lewis numbers. Combust. Sci. Tech. to appear.
42. A. Liñan (1971). A theoretical analysis of premixed flame propagation with an isothermal chain reaction. Contrat n°EOOAR 68-0031 Technical report 1 INTA. Madrid.
43. A. Liñan (1974). The asymptotic structure of counter flow diffusion flames for large activation energies. Acta Astronautica 1, 1007-1039.
44. A. Liñan (1976). Monopropellant droplet decomposition for large activation energies. Acta Astronautica 2, 1009-1029.
45. A. Liñan and P. Clavin (1984). Premixed flames with non branching chaiin reaction. J. Chem. Phys. submitted.

46. A. Liñan and A. Crespo (1976). An asymptotic analysis of unsteady diffusion flames. Combust. Sci. Technol. 14, 95-117.
47. A. Liñan (1961). On the internal structure of laminar diffusion flames. TN n°3 for INTA-AFOSR contract AF 61(052) - 221.(AD n°273069)
48. L. Mallard and H.L. Le Chatelier (1883). Recherches expérimentales et théoriques sur la combustion des mélanges gazeux. Annales des mines 4, 274-553.
49. G. H. Markstein (1964). Non steady flame propagation, Pergamon Press.
50. M. Matalon and B.J. Matkowsky (1982). Flames as gasdynamics discontinuities. J. Fluid Mech. 124, 239-259.
51. A. Melvin and J.B. Moss (1970). Fifteenth International symposium on Combustion, 625. The Combustion Institute. Pittsburgh.
52. D. M. Michelson and G.I. Sivashinsky (1977). Non linear analysis of hydrodynamic instability in laminar flames II. Acta Astronautica 4, 1207-1221.
53. J. D. Murray (1977). Lectures on non linear differential equations model in biology. Clarendon Press Oxford.
54. P. Pelcé and P. Clavin (1982). Influence of hydrodynamics and diffusion upon the stability limits of laminar premixed flames. J. Fluid Mech. 124, 219-237.
55. N. Peters and J. Warnatz (1982). Numerical methods in laminar flame propagation. Notes on numerical fluid mechanics 6 Viewieg.
56. N. Peters (1983).Laminar diffusion flamelet models in non pre. To appear in Prog. Energy Combustion Sci.
57. N. Peters, W. Hocks and G. Mohinddin (1981).Turbulent mean reaction rates in the .J. Fluid Mechanics 110, 411.432
58. N. Peters and F.A. Williams (1981). in : The role of coherent structures in modelling turbulence and mixing, J. Jimenez, Ed. Lectures Notes in Physics 136, 364, Springer.
59. N. Peters and F.A. Williams (1983).Lift off characteristics of turbu. jet diffusion flames. AIAA J. 21, 423.429
60. N. Peters (1979).Premixed burning in diffusion flames. Int. J. Heat and Mass Transfer 22, 691-703
61. J. Quinard, G. Searby and L. Boyer (1983). The stability limits and critical size of structures in premixed flames. These proceedings.
62. G. Searby, F. Sabathier, P. Clavin and L. Boyer (1983). The hydrodynamical coupling between the motion of a flame front and the upstream gas flow. Phys. Rev. Letters, to appear (issue of 17 oct.83).
63. G. I. Sivashinsky (1977). Non linear analysis of hydrodynamic instability in laminar flames I, Acta Astronautica, 4, 1177-1206.
64. G. I. Sivashinsky (1983). Instabilities, pattern formation, and turbulence in flames. Ann. Rev. Fluid Mech.15, 179-199.

65. M. Taffanel (1913). Sur la combustion des mélanges gazeux et les vitesses de réaction. Compt. rend. acad. sci. Paris 27 X 1913, 714-717 and 29 XII 1913, 42-45.
66. M. Van Dyke (1964). Perturbation Methods in fluid mechanics, Academic
67. F. A. Williams (1965). Combustion theory. Addison Wesley.
68. F. A. Williams (1981). Review of flame extinction. Fire Safety Journal 3, 163 -175
69. F. A. Williams (1975). Turbulent Mixing in Non-reactive and Reactive Flows, p.189, S.N.B. Murthy, Ed. Plenum
70. J. B. Zeldovich and D.A. Frank Kamenetskii (1938). A theory of thermal propagation of flame, Acta physicochimica U.R.S.S. IX n°2, 341-350.
71. J. B. Zeldovich (1948). Theory of flame propagation. Zhur. Fiz. Khim. USSR, 22, 27-49.
72. J. B. Zeldovich, G.I. Barenblatt, V. Librovich, G.M.Makviladze (1980). Mathematihecky Teoria Gorenya i Vzriva. Nauka.

BIFURCATION IN HETEROGENEOUS COMBUSTION

L. De Luca, G. Riva and C. Bruno

Dipartimento di Energetica-CNPM, Politecnico di Milano
32 Piazza Leonardo da Vinci
20133 Milano, Italy

1. INTRODUCTION

Heterogeneous combustion takes place at the interface separating two different physical phases. The burning of solid propellants for rocket propulsion represents typically this type of combustion. So-called composite solid propellants are mixtures of finely ground salts, such as NH_4ClO_4 (ammonium perchlorate = AP), and a polymer binder (for instance polybutadiene) in a ratio of order 4:1. At operational pressures combustion of AP-based solid propellants takes place in a thin zone close to the interface solid/gas. On the solid surface itself the polymer pyrolyzes forming gaseous products, and the AP decomposes into NH_3 and $HClO_4$ that react exothermically between themselves and with the polymer vapor in a diffusion flame. Details of the physics and chemistry of this type of combustion may be found elsewhere[1-2]. The purpose of this study is to present a mathematical model of the solid phase-gas system reacting at the interface, and to show that depending on the external variables (pressure, heat flux by radiation) and on the physico-chemical characteristics of the propellant (energy content, activation energies of the condensed phase decomposition reaction, etc.) burning occurs in different regimes. In mathematical terms, the response of the system (e.g., the interface temperature) bifurcates when some particular parameters are varied. Experimental evidence of this behavior has already been obtained and is presented elsewhere[3].

2. MATHEMATICAL MODEL

The one-dimensional deterministic energy conservation equation for the condensed phase, written in a coordinate system x fixed to the burning interface regressing with velocity r_b, is a reaction/ diffusion (Fourier) equation. Together with its boundary and initial conditions, its dimensionless form is[4]:

$$C\left(\frac{\partial\Theta}{\partial\tau} + R\,\frac{\partial\Theta}{\partial X}\right) = \frac{\partial}{\partial X}\left(K\,\frac{\partial\Theta}{\partial X}\right) + F_o \int_{\lambda_1}^{\lambda_2} f_1(X,\lambda)\,d\lambda + \varepsilon_c H_c - \dot{q}_{out}\,\frac{S\,d_{ref}}{A} \qquad (2.1)$$

$$X \leq 0, \quad \tau \geq 0$$

$$\Theta(X, \tau = 0) = \text{assigned}$$

$$\Theta(X \to -\infty, \tau) = \Theta_{-\infty}$$

$$\left(K\,\frac{\partial\Theta}{\partial X}\right)_{c,s} = \left(\frac{k_g}{k_{ref}}\,\frac{\partial\Theta}{\partial X}\right)_{g,s} + R\,H_s - \dot{q}_{out}(\Theta_s) \qquad (2.2)$$

The radiative flux term in (2.1) may be ≥ 0 and assumes a condensed phase absorbing between the wavelengths λ_1, λ_2. The dimensionless burning, or regression, rate R is given by

$$R = \exp\left(-E_s\left(\frac{1}{T_s} - 1\right)\right) p^{n_s} \qquad T_s \geq T_k$$

$$R = \left(\frac{T_s - T_m}{1 - T_m}\right)^w p^{n_s} \qquad T_k \geq T_s \geq T_m$$

and H_s by

$$H_s = \frac{Q_{s,ref} + T_{s,ref}(c_g - c_c(T_{s,ref})) - T_s(c_g - c_c(T_s))}{Q_{ref}}$$

Coupling with the gas phase occurs via (2.2) at the interface. The gas phase equations (continuity, momentum and energy equations) simplify if, as it is the case with AP-based propellants, the characteristic times of the gas are much smaller than those for the solid and interface. The longest time of the gas, the convective transport time $\alpha_g/\overline{u}_g^2$, is about $O(10^2)$ times smaller than the condensed phase thermal conduction time $\alpha_c/\overline{r}_b^2$. Physically this means that the gas chemical reactions are restricted to a very thin region close to the interface. Mathematically the coupling between the two phases becomes simpler, since inertial terms in the gas equations can be dropped, and the gas can be treated as quasi-steady with respect to the condensed phase. With a different terminology, the gas is a "slave" of the solid[5]: it is determined by it, but at the same time feeds energy back to it through (2.2). With this assumption the gas phase equations reduce to:

$$m(t) = \rho_g(x,t)\,u_g(x,t) = \rho_c\,r_b(t) \qquad (2.3)$$

dim. continuity

$$p(x,t) = p(t) \qquad (2.4)$$

dim. momentum

$$p(x,t) = \rho_g(x,t) \frac{\mathcal{R}}{\overline{\overline{W}}} T(x,t) \qquad (2.5)$$

dim. state

$$\left(\frac{k_g}{k_{ref}} \frac{\partial \Theta}{\partial X}\right)_{g,s} = H_f \int_{0^+}^{X_f} \varepsilon_g \frac{\rho_g}{\rho_{ref}} \exp\left(- \frac{k_{ref}}{k_g} \frac{c_g}{c_{ref}} RX\right) dX \qquad (2.6)$$

nondim. energy, in integral form

The rate of reaction ε_g in the gas has to be modeled. For a thin flame many models are available[2,6,7,8]. While the numerical results produced by each model change somewhat, the behavior of the solution of (2.1) through (2.6) does not.

3. BIFURCATION DIAGRAMS

The set of nonlinear equations (2.1) through (2.6) does not lend itself to an analytical solution and must be integrated numerically to find Θ, as a function of X and τ, once external and internal parameters have been fixed. These solutions, however, are isolated histories, and cannot predict stability unless they are obtained for all the possible combinations of parameters. To predict the nature of the solution a different approach is needed. Firstly, since the gas is slaved to the condensed phase, take the surface temperature as the dependent variable of importance. Impose then, following Von Karman and Polhausen[9], a finite temperature disturbance $u(X,\tau)$ to the steady-state solution of (2.1). Further assume $u(X,\tau)$ monotonical with X in the interval $-\xi(\tau)$, 0 corresponding to the unknown thickness ξ of the thermal disturbance profile. In particular, exponential or polynomial dependences of the type

$$u = u_s \exp(nX/\xi) \qquad ; \qquad u = u_s(1 + X/\xi)^n$$

have been used.

Integrating (2.1) from $-\xi(t)$ to 0 eliminates X as variable, yielding an ordinary differential equation in time for Θ_s

$$\frac{d\Theta_s}{d\tau} = f(\Theta_s - \overline{\Theta}_s) + g(\tau, \Theta_s - \overline{\Theta}_s) \qquad (3.1)$$

$$\Theta_s(\tau = 0) = \overline{\Theta}_s$$

where

$$f(\Theta_s - \overline{\Theta}_s) = \frac{-R\int_{-\xi}^{0} C\, d\Theta + (K\frac{\partial\Theta}{\partial X})_{c,s} - (K\frac{\partial\Theta}{\partial X})_{-\xi}}{\int_{-\xi}^{0} C\, u_c \left\{1 - \frac{X}{G(X)}\left[\frac{(u_x)_{c,s}}{u_s} - (\frac{\partial (u_x)_{c,s}}{\partial\, u_s})_{P,F_o}\right]\right\} dX}$$

$$+ \frac{F_o\int_{-\xi}^{0}\int_{\lambda_1}^{\lambda_2} f_1(X,\lambda)\,d\lambda\, dX + H_c\int_{-\xi}^{0}\varepsilon_c(\Theta)dX - \frac{S\, d_{ref}}{A}\int_{-\xi}^{0}\dot{q}_{out}(\Theta)dX}{\int_{-\xi}^{0} C\, u_c \left\{1 - \frac{X}{G(X)}\left[\frac{(u_x)_{c,s}}{u_s} - (\frac{\partial (u_x)_{c,s}}{\partial\, u_s})_{P,F_o}\right]\right\} dX}$$

and

$$g(\tau, \Theta_s - \overline{\Theta}_s) = \frac{-\int_{-\xi}^{0} C\, u_c \frac{X}{G(X)}\left[(\frac{\partial (u_x)_{c,s}}{\partial P})_{u_s,F_o}\frac{dP}{d\tau}\right.}{\int_{-\xi}^{0} C\, u_c \left\{1 - \frac{X}{G(X)}\left[\frac{(u_x)_{c,s}}{u_s} - (\frac{\partial (u_x)_{c,s}}{\partial u_s})_{P,F_o}\right]\right\} dX}$$

$$- \frac{\left.(\frac{\partial (u_x)_{c,s}}{\partial F_o})_{u_s,P}\frac{dF_o}{d\tau}\right] dX}{\int_{-\xi}^{0} C\, u_c \left\{1 - \frac{X}{G(X)}\left[\frac{(u_x)_{c,s}}{u_s} - (\frac{\partial (u_x)_{c,s}}{\partial u_s})_{P,F_o}\right]\right\} dX}$$

being

$u_c = (1 + \frac{X}{\xi})^n$ and $G(X) = 1 + \frac{X}{\xi}$ for polynomial disturbance

$u_c = \exp(n\frac{X}{\xi})$ and $G(X) = 1$ for exponential disturbance

The autonomous part f represents the inherent response of the system to finite perturbations; the part g contains the effect of the external forcing functions driving the coupled solid/gas system. It is not necessary to find stability maps depending on time: thus the treatment will be restricted to the case of

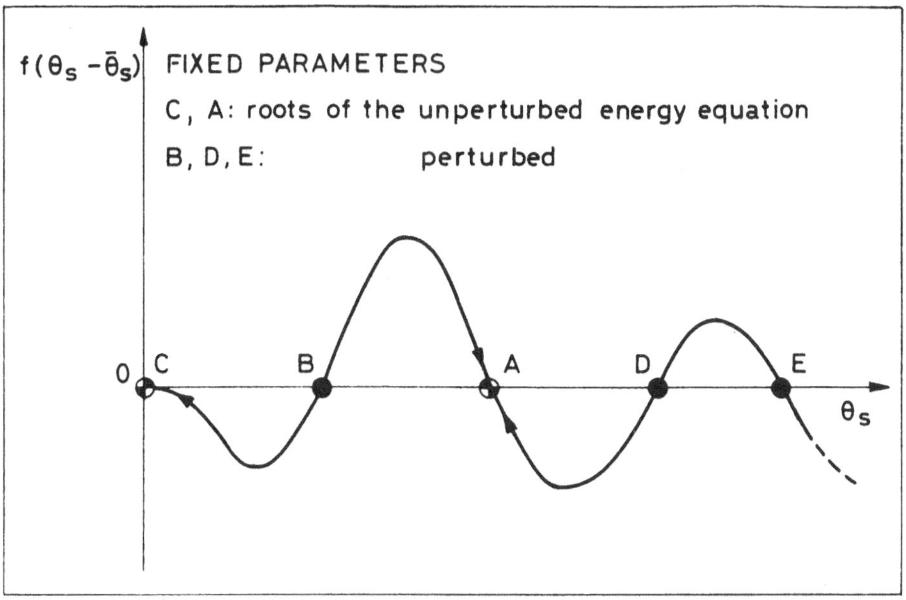

Fig. I. Nature of the static restoring function.

forcing functions steady in time. In this way $g \equiv 0$ and (3.1) reduces to

$$\frac{d\Theta_s}{d\tau} = f(\Theta_s - \overline{\Theta}_s) \qquad (3.2)$$

Notice that $\overline{T}_f(p, T_{-\infty})$ and $\overline{r}_b = a\,\overline{p}^n$ are considered known. In any event, Eqn. (3.2) does not need to be solved to extract useful information. The roots of f and the way they vary with the parameters, coupled with (2.1) through (2.6), yield the information needed to build bifurcation diagrams for Θ_s. Fig. 1 shows a typical f vs Θ_s plot; by inspection the point C is the root corresponding to the solution $\Theta_s = 0$ (no combustion). The point B is an unstable root, as any disturbance changing $\Theta_s(B)$ will cause $d\Theta_s/d\tau$ to keep going in the same direction. Therefore, for disturbances causing Θ_s to go below $\Theta_s(B)$, C is the end of the trajectory and extinction follows; for $\Theta_s > \Theta_s(B)$ the solution moves toward the root A (possibly with damped oscillations). That is the steady-state time independent solution, unless the bifurcation parameter (see below) has already reached its critical value. All the other roots (B, D and E) cannot sustain a steady combustion (with burning rate constant in time) being solutions of the solid phase energy equation only in "perturbed" conditions. Past the critical value of the bifurcation parameter, the roots A and D are reversed, and Θ_s will oscillate indefinitely between D and E (limit cycle around A).

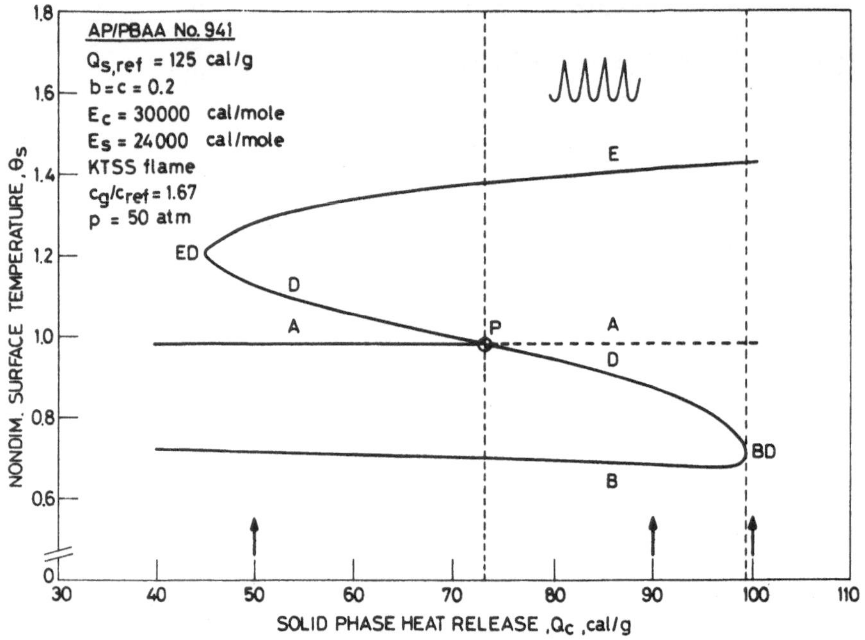

Fig. 2. Surface temperature as a function of the bifurcation parameter Q_c.

Fig. 2 is a bifurcation diagram for Θ_s with Q_c (energy release in condensed phase per unit mass of the propellant) as the bifurcation parameter and all other parameters constant. The solution bifurcates due to the fact that, for a particular range of the bifurcation parameter, the root A becomes unstable. The following information can be extracted from Fig. 2: (1) if the combustion process has resulted in Θ_s going below B, the solution goes to C (unreacting solution) and extinction follows; (2) for $Q_c < Q_c(P)$ and if Θ_s does not go below B, the stable reacting configuration A will be asymptotically reached (possibly with damped oscillations); (3) for $Q_c(P) < Q_c < Q_c(BD)$ and if Θ_s does not go below B, Θ_s will oscillate indefinitely between D and E with constant frequency and amplitude; (4) for $Q_c > Q_c(BD)$ only the unreacting solution C can be obtained under steady operating conditions.

By systematic plotting of bifurcation diagrams it was found that Q_c, $Q_{s,ref}$, p, F_0, n, n_s are bifurcation parameters whereas E_s and w are not. The energy loss term $\dot{q}_{out}$ has negligible influence on the stability properties, being several orders of magnitude less than other terms. Experimental evidence[3] of the transition from constant burning rate to self-sustained oscillatory combustion (before extinction at pressure deflagration limit) confirmed the validity of the analytical predictions at least when the pressure is implemented as bifurcation parameter. Numerical checks of the analytical predictions were performed, for all of the explo-

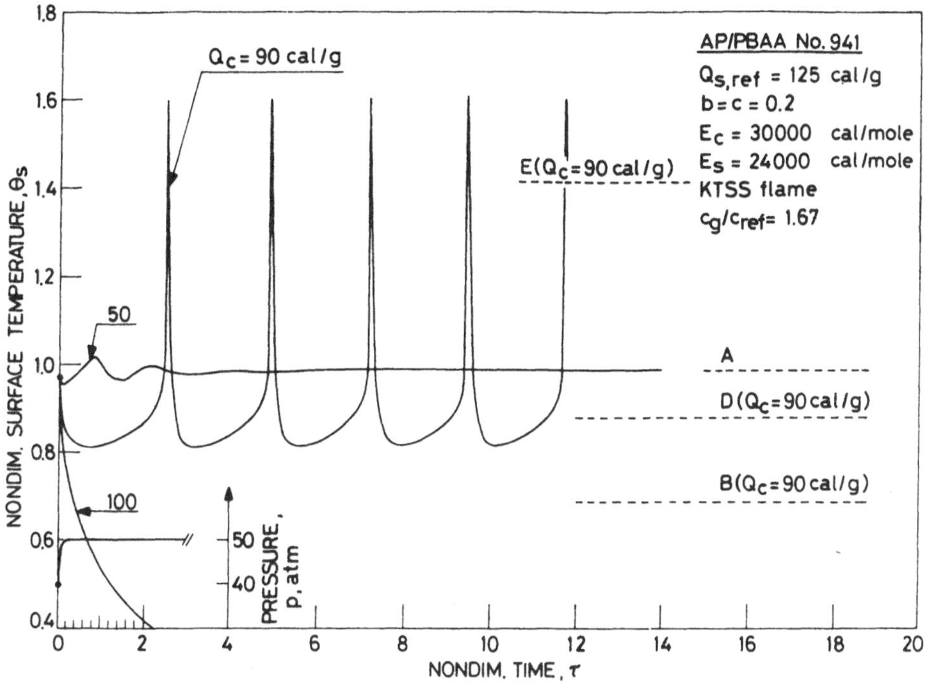

Fig. 3. Different histories of surface temperature, following a pressure increase from 40 to 50 atm, as a function of the condensed phase heat release, Q_c.

red bifurcation parameters, under a wide variety of operating conditions; results show a very good agreement. An example of numerical check is given in next section.

4. NUMERICAL SOLUTION

The analytical predictions obtained from (3.2) must be confirmed by solving the set of equations in space and time. The numerical solution of Eqns. (2.1)-(2.6) is obtained through an implicit finite difference scheme[10]. The time step is a (variable) fraction of the condensed phase characteristic time; the space step size is chosen by taking into account the requirements of both boundary conditions. The cold boundary condition at minus infinity requires a coarse grid which goes deep inside the solid phase, whereas the boundary condition at the burning surface requires a fine grid to keep the volumetric terms small compared to surface terms. Tests are performed to check that both boundary conditions are properly satisfied. The initial condition of the partial differential equation is the steady thermal profile corresponding to burning at fixed conditions. Integration with double precision requires about 3' of computational time on a Sperry-Univac 1100/80 computer. Notice that the bifurcation diagram of Fig. 2 can be obtained in about one hour using a HP-85 desk calculator.

Fig. 3 shows three solutions, for the three values of Q_c marked by an arrow in Fig. 2, corresponding to an imposed jump in pressure from 40 to 50 atmospheres. They quantitatively match the predictions of the bifurcation diagram. For Q_c = 90 cal/g the amplitude of the oscillations is somewhat larger than predicted due to dynamic effects; however, results in this range are independent on the forcing function p(t). The new steady-state A corresponding to the final p = 50 atm is instead correctly calculated for Q_c = 50 cal/g.

5. CONCLUSIONS

The results of Sections 3 and 4 indicate that the response of heterogeneous deflagrations of AP-based solid propellants can be successfully modeled by keeping the nonlinear features of the phenomenon. The analytical predictions (verified by judicious integration of the nonlinear reaction/diffusion equations for the condensed phase and the gas) yields easily bifurcation diagrams that can be used as maps to both check and investigate in detail the response of the heterogeneous burning system in time.

Success of this treatment bids well for the stability analysis of other heterogeneous combustion problems, such as catalytic and liquid drop combustion.

REFERENCES

1. Steinz, J.A., Stang, P.L., and Summerfield, M., 1969, The Burning Mechanism of Ammonium Perchlorate Based Composite Propellants, Princeton University AMS Report No. 830, Princeton, N.J.
2. Merkle, C.L., Turk, S.L., and Summerfield, M., 1969, Extinguishment of Solid Propellant by Depressurization: Effects of Propellant Parameters, AIAA Paper No. 69-176.
3. Bruno, C., Riva, G., Zanotti, C., Dondè, R., Grimaldi, C., and De Luca, L., 1983, Experimental and Theoretical Burning of Solid Rocket Propellants Near the Pressure Deflagration Limit, International Astronautical Federation Paper IAF-83-367.
4. De Luca, L., Riva, G., and Zanotti, C., 1983, Nonlinear Burning Stability of Solid Propellants, U.S. Army European Research Office, Final Technical Report, Milano, Italy.
5. Haken, H., 1978, "Synergetics", Springer-Verlag, N.Y., p. 195.

6. Krier, H., T'ien, J.S., Sirignano, W.A., and Summerfield, M., 1968, Nonsteady Burning Phenomena of Solid Propellants: Theory and Experiments, AIAA J., Vol. 6, No. 2, p. 278.
7. Kooker, D.E., and Zinn, B.T., 1974, Numerical Investigation of Nonlinear Axial Instabilities in Solid Rocket Motors, B.R.L. Contract Report No. 141.
8. Levine, J.N., and Culick, F.E.C., 1974, "Nonlinear Analysis of Solid Rocket Combustion Instability", Air Force Report AFRPL TR-74-75.
9. Goodman, T.R., 1964, Application of Integral Methods to Transient Nonlinear Heat Transfer, in "Advances in Heat Transfer", Vol. 1, Academic Press, N.Y., p. 51.
10. De Luca, L., Riva, G., and Tanturli, I., 1982, Numerical Solution of Solid Propellant Burning, L'Aerotecnica Missili e Spazio , Vol. 61, No. 2, pp. 61-68.

NOMENCLATURE

A = cross section area of propellant, cm^2
C = specific heat ratio, $c_c/c_{ref} = 1 + c\Theta$
c = specific heat, cal/gK
d = thickness, cm
E = nondimensional activation energy, $\tilde{E}/(\mathcal{R}T)$
$\tilde{E}$ = activation energy, cal/mol
$f(\Theta_s - \overline{\Theta}_s)$ = nond. static restoring function (see text)
G = nondimensional function depending on the assumed disturbance thermal profile
g = nond. function depending on nature of temperature disturbance profile (see text) and forcing functions
H = Q/Q_{ref}, nond. heat release
K = thermal conductivity ratio, $k_c/k_{ref} = 1 + b\Theta$
k = thermal conductivity, cal/cm sK
n = order of the approximating temperature disturbance profile (see text); also: ballistic exponent
p = pressure, atm
p_{ref} = reference pressure, atm
P = p/p_{ref}, nond. pressure
$\dot{q}$ = Φ/Φ_{ref}, nond. energy flux
Q = heat release, cal/g
Q_{ref} = $c_{ref}(T_{s,ref} - T_{ref})$, reference heat release, cal/g
Q_s = net surface heat release (positive if exothermic), cal/g
r_b = burning rate, cm/s
$r_{b,ref}$ = $r_b(p_{ref})$, reference burning rate, cm/s
R = $r_b/r_{b,ref}$, nond. burning rate
$\mathcal{R}$ = universal gas constant, cal/mol K
S = perimeter of propellant sample, cm
t = time, s

T	=	temperature, K
T_{ref}	=	300 K, reference temperature, K
$T_{s,ref}$	=	$T_s(p_{ref})$, reference surface temperature, K
u	=	nond. finite size disturbance of temperature defined as $u(X,\tau) = \Theta(X,\tau) - \overline{\Theta}(X)$
u_x	=	nond. finite size disturbance of thermal gradient defined as $u_x = \partial\Theta/\partial X - d\overline{\Theta}/dX$
w	=	power of pyrolysis law
x	=	space variable, cm
X	=	$x/(\alpha_{c,ref}/r_{b,ref})$, nond. space variable

Greek Symbols

ε_c	=	nond. condensed phase reaction rate
Θ	=	$(T - T_{ref})/(T_{s,ref} - T_{ref})$
ξ	=	nond. thickness of disturbance thermal layer
T_m	=	minimum temperature for occurrence of chemical reaction in the condensed phase
T_k	=	matching temperature for surface pyrolysis law
ρ	=	density, g/cm^3
τ	=	$t/(\alpha_{c,ref}/r^2_{b,ref})$ nond. time
t_{ref}	=	$\alpha_{c,ref}/r^2_{b,ref}$ = 1.059 ms, reference time
T	=	$T()/T()$, ref., nond. temperature
Φ_{ref}	=	$\rho_c c_{ref} r_{b,ref}(T_{s,ref} - T_{ref})$ reference heat flux, cal/cm^2s

Subscripts and Superscripts

b	=	burning
c	=	condensed phase
ξ	=	penetration depth
f	=	flame
g	=	gas phase
p	=	pressure
s	=	surface
c,s	=	surface from condensed phase side
g,s	=	surface from gas phase side
out	=	lost by the system
ref	=	reference
$-\infty$	=	far upstream
$+\infty$	=	far downstream
$-$	=	steady state
$=$	=	average value

THERMOELASTICITY AND MECHANICAL INSTABILITIES

C. E. Bottani and G. Caglioti

Istituto di Ingegneria Nucleare
CESNEF-Politecnico di Milano
20133 MILANO, Italy

INTRODUCTION AND SUMMARY

The objectives of these lectures are:

- To suggest that time is ripe to revisit the whole field of the mechanical behavior of materials. This is a field relevant for its engineering and technological applications as well as for its economical implications: but the conventional approach to it turns out to be inadequate to face the complexity of the process occurring during the transformations promoted by external mechanical action on a material. The fundamental physical fact is too often ignored that the "mechanical" transformation of materials are not purely machanical, but belong rather to the domain of non-equilibrium cooperative phenomena. Therefore, a comprehensive understanding of these transformations cannot be achieved on a purely mechanical basis, focussing the attention solely on mechanical quantities such as deformation and stress. It requires, for instance, that also temperature charges, associated with the production and fluxes of entropy contestually accompanying these transformations, should be carefully measured and considered.
- To indicate that
  i) the mathematical methods used to handle non-equilibrium cooperative phenomena can be used to develop theoretical models of dynamical instabilities relevant in the mechanical behavior of solids: at least one case, that of the strain driven thermoelastic-plastic instability (yield) has been worked out satisfactorily; and

ii) recent progresses in the development of sensitive temperature sensors and in the construction of sophisticated systems of acquisition, elaboration and presentation of stress, deformation and temperature data, allow to follow in real time the behavior of materials under stress. Both the Hooke-Kelvin regime of the thermoelastic deformations occurring in the thermodynamic branch, and plastic flow developing after the thermoelastic-plastic dynamical instability, can now be kept under experimental control and more comprehensively characterized.

These lectures are divided in two parts.

The first part highlights the thermoelastic properties of materials. To understand the nature of the processes occuring in a material undergoing an elastic deformation is a necessary step toward the interpretation of more complex non-equilibrium cooperative phenomena subsequently occurring in materials under stress, such as yielding, plastic flow, creep, fracture, etc. But the thermoelastic branch ideally belongs to the field of linear irreversible thermodynamics of systems just slightly displaced from equilibrium, while this meeting is focussed instead on non-equilibrium phenomena. Only a very schematic account on thermoelasticity is thus proposed in the first part below, so that these proceedings don't reflect the emphasis actually given during the oral presentation to the thermoelastic effect. In any case the interested reader can find an up to date review on the statistical mechanics and thermodynamics of the elastic deformation, and the thermoelastic effect in e.g. Ref. [I.1]

The second part of these lectures is more directly concerned with the main objectives of this NATO Advanced Study Institute. It contains a detailed account of a mechanical instability specifically studied since several years, the thermoelastic-plastic transition.An up-dating of our views on the nature of this instability it presented, together with recent experimental results obtained by "thermal emission", a new method to explore the mechanical properties and behavior of materials.

I. THE THERMOELASTIC EFFECT

In synopsis I the conceptual scheme of the first principles derivation of the thermoelastic effect in a quasi harmonic solid is outlined.

As indicated in the general introduction, the actual derivation of the balance equations governing the thermoelastic behavior of materials has been presented elsewhere [1,2]. The reader is referred to e.g. the review in [1] and the references therein for additional information including the experimental findings.

In the present context the thermoelastic effect is especially important since it allows to define the reference state, $\mathcal{R}_o$, of uniform thermoelastic cooling associated with a uniformly increasing imposed deformation, around which critical thermoacoustic fluctuations may arise as soon the thermoelastic-plastic limit (yield) is reached.

SYNOPSIS I

Conceptual scheme for a first principle derivation of thermoelasticity

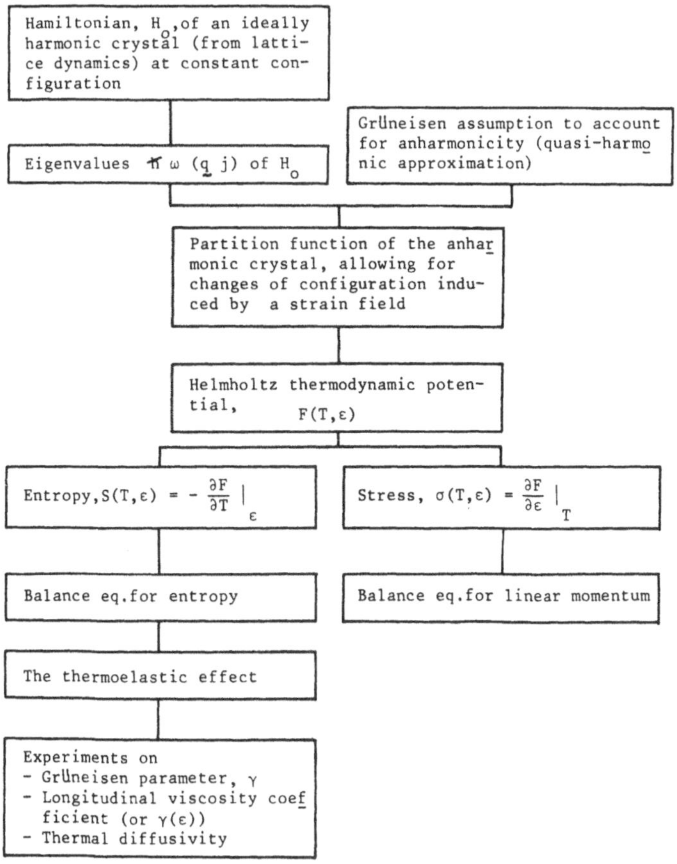

II. THE THERMOELASTIC-PLASTIC TRANSITION IN METALS(*): Thermal Emission as a Probe to Identify the Yield Point

1. Introduction

Our objectives are

i) to offer some suggestions for an interpretation of yielding (the onset of plastic flow) as a dynamical instability occurring at the end point of the Hooke-Kelvin thermodynamic branch;

ii) to suggest that a new method, thermal emission, can help ascertain the existence and location of a sharp division between the elastic and plastic regimes of deformation, even in the case of metals exhibiting an apparently vanishing ductility.

In what follows, the definition of yield, currently given in purely mechanical terms, is revised so as to include the relevant thermodynamic features of the irreversible transformation leading from the thermoelastic regime to that of plastic flow. Some recent experimental results on the thermal emission are then presented and discussed in brief.

2. The Yield Point as the Critical Point of a Dynamical Instability

A comprehensive definition of the elastic limit -the commencement of yielding- cannot be given in purely mechanical terms [1]. A metal subject to a non-isothermal tensile test undergoes complex thermodynamic transformations . Initially, it is macroscopically elastic: its behavior is essentially reversible. Its state varies along the thermodynamic branch. The entropy produced is minimal. Beyond the elastic limit, plastic flow sets in, involving irreversible deformations. Various kinds of dissipative structures can arise. Dislocations are created and the effects of their interaction become increa-

---

(*) Work supported in part by a grant from the Consiglio Nazionale delle Ricerche, under the Progetto Finalizzato Metallurgia. The experimental data reported here will be submitted for publication in Materials Letters. Part of these data have been presented also at the Workshop on Statics and Dynamics of Nonlinear Systems, Erice, 1983, at the Ettore Majorana Centre for Scientific Culture

singly significant. "Coherent" dissipation occurs, while a conspicuous amount of entropy is produced per unit time.

On the thermodynamic branch, where dislocations are by and large immobile and the system is only slightly removed from equilibrium, the thermodynamic state may be defined [2] via a local Helmholtz potential that is a functional of only the local temperature T (conjugate to the local entropy S) and the strain tensor $\underset{\approx}{\varepsilon}$ (conjugate to the stress $\underset{\approx}{\sigma}$).
The situation is different for a dissipative structure undergoing plastic flow. As dislocations are created anew or become mobile, the dislocation density should be an additional state variable. However, since the system is far removed from equilibrium, its description in terms of a thermodynamic state is questionable. In this perspective, a precise definition of the yield point would be simply the end point of the thermodynamic branch, beyond which the behavior of the system can no longer be described in conventional thermodynamic terms [3]. Of course, in order to pinpoint the departure of the system from the thermodynamic branch, the nature of the physical phenomena occurring while the state of the system changes along this branch should be understood clearly.

A theory of the thermoelastic instability has been presented in Ref.[2][*]. This is based on the balance equations for entropy and momentum in a sample deformed at a constant applied tensile strain rate. These coupled nonlinear partial differential equations allow one to describe the space and time dependence of the fluctuations of the temperature and deformation fields about a thermoelastic reference state, $\mathcal{R}_0$, that corresponds to a uniformly increasing deformation accompanied by a uniform adiabatic thermoelastic cooling. The linear stability of $\mathcal{R}_0$ is studied by means of a normal mode analysis of these fluctuations, using the Hurwitz criterion. The component of the fluctuation field that becomes unstable first as the imposed strain (the control parameter) is increased is found to be the thermoacoustic mode characterized by the fundamental wavevector. This mode can be visualized as a damped phonon, bearing two polarization components - the fields of the displacements around $\mathcal{R}_0$ of the deformation and temperature -. The amplitude of this mode (the order parameter) begins to diverge exponentially in time and "thermal emission" and acoustic emission set in as soon as the control parameter exceeds a certain threshold.

(*)The theoretical predictions concerning a time-dependent state should be taken with caution (editor's note).

During the initial elastic stage of deformation, which may be regarded as an ideal isentropic transformation, stability is realized through thermoelastic coupling. The sample shrinks, thus opposing the tensile transformation ε_{zz}^{ℓ}. The relative drop in temperature is proportional to the relative change in volume, so that

$$\Delta T / T_o = -\gamma(1 - 2\nu)\ \varepsilon_{zz}^{\ell} \tag{1}$$

Here T_o is the reference temperature of the specimen (i. e., the temperature of the thermal bath it is in contact with), ν is the Poisson ratio, and the Grüneisen parameter

$$\gamma = -\ d(\ln\omega_D) / d(\ln V) \tag{2}$$

is related to the anharmonicity of the interatomic potential in the material concerned : ω_D is the Debye frequency and V the specimen volume. At the end of the thermoelastic branch, a destabilizing tendency begins to dominate, manifesting itself through substantial viscous heating. Plastic flow commences. In the balance equations of Ref. 2 , a deformation-dependent phenomenological Grüneisen parameter, $\gamma(\varepsilon)$, was introduced in order to describe in a convenient manner the behavior of the entropy and the stress in terms of the state variables T and ε. The strain dependence of ε has been estimated from experimental data in the case of α-Ti, using Eq.(1) with γ replaced by $\gamma(\varepsilon)$ [4]. $\gamma(\varepsilon)$ was found to decrease rapidly to a vanishing value with increasing ε. This enables us to identify the critical value $\varepsilon = \varepsilon_\Theta$ of the control parameter as that value of the imposed deformation at which $\gamma(\varepsilon_\theta) = \emptyset$. When $\varepsilon = \varepsilon_\theta$, the thermoelastic cooling is exactly balanced by the thermoplastic heating. As $\varepsilon \rightarrow \varepsilon_\theta$, the coupling between the strain and the temperature in the Helmholtz potential vanishes. This situation corresponds to the onset of a dissipative structure, in which the motion of dislocations is activated and sustained by the external mechanical work. However, as already stated, such a dissipative structure is far removed from equilibrium and cannot be described in simple thermodynamic terms. A kinetic model is necessary to account for the several possible mechanisms of dissipation, and, in particular, for the rate of production of entropy and momentum by the multiplication and motion of dislocations. A fully satisfactory description of the thermoelastic-plastic instability and the dissipative structure beyond the bifurcation is not available at present.

Our main interest here is in thermal emission, and

so we focus on the equation for the balance of entropy. We derive this equation below, with the additional purpose of exhibiting the nature of the information required to replace the phenomenological parameter $\gamma(\varepsilon)$ by the relevant entropy production term associated with the multiplication and motion of dislocations, in an improved theory. The entropy $S(T,\varepsilon)$ of an elastic body slightly removed from equilibrium can be written as [5]

$$S(T,\varepsilon) = S(T_o, \emptyset) + \gamma(1 - 2\nu)C_v\varepsilon^{e\ell}_{zz} + C_v(T-T_o)/T_o \qquad (3)$$

where $S(T_o,\emptyset)$ is the equilibrium entropy of the undeformed body of temperature T_o. The continuity equation for entropy is

$$\partial S/\partial t + \mathrm{div}\,(\underset{\sim}{q}/T) = P[S] \qquad (4)$$

where

$$\underset{\sim}{q} = -\,K\,\mathrm{grad}\,T \qquad (5)$$

is the heat flux density vector, K is the thermal conductivity and P[S] is the entropy produced per unit time and unit volume.This in turn is of the form

$$P[S] = \underset{\sim}{q}\,.\,\mathrm{grad}\,(1/T) + P^d[S]. \qquad (6)$$

The first term on the right is the rate of entropy production in irreversible heat flow induced by a temperature gradient.
Since

$$\mathrm{div}(\underset{\sim}{q}/T) = \underset{\sim}{q}\,.\,\mathrm{grad}(1/T) + (1/T)\,\mathrm{div}\,\underset{\sim}{q}\,, \qquad (7)$$

on using Eqs.(6) and (7) in Eq.(4) we see that the thermal component of the entropy production is compensated for by a part of the incoming entropy flux, $\mathrm{div}(\underset{\sim}{q}/T)$. The second term on the right in Eq.(6), P[S], arises from the irreversible motion of dislocations. It depends in a complex manner on the microstructure and kinetics of the dislocation field -on the instantaneous density of mobile dislocations, their average velocity, their "mean free path", and the "viscosity" of the medium in which they move.

The foregoing equations thus yield

$$\partial T/\partial t - \chi\nabla^2 T = -\gamma T_o(1-2\nu)\partial\varepsilon^{e\ell}_{zz} / \partial t + T_o P^d[S]/C_v, \qquad (8)$$

where $\chi = K/C_v$ is the thermal diffusivity. All the terms but the last one in Eq.(8) are measurable or are readily modelled. They are also independent of the details of the defect dynamics. Thus Eq.(8) can be used [6],[7] as a starting point for testing models of defect kinetics. Further work on this program is presently underway in our laboratory.

We now turn to a brief discussion of some experimental results on the variation of T vs ε and of σ vs ε.

3. "Thermal Emission" in 100Cr6 steel

Figure 1 shows the σ vs ε characteristic of a specimen of 100Cr6 steel subjected to a tensile test. This steel exhibits an extraordinarily extended range of apparent elasticity. It is pertinent to ask whether this entire range corresponds to genuine elastic behavior. The answer is contained in Fig.2, which shows the temperature versus deformation characteristic obtained by means of instrumentation described elsewhere [1],[3]. An abrupt change in the variation of T as a function of ε occurs at the critical strain ε_θ of the thermoelastic-plastic transition, which suggests the phrase "thermal emission" for the T-ε characteristic. As discussed earlier, although ε_θ marks the onset of highly irreversible dissipative motion of defects, there is little indication of a sharp separation between the elastic and (macro)plastic regimes in a purely mechanical experiment such as the one that yields the σ-ε characteristic. We remark that the apparent deviation from the Kelvin law (Eq.(1)) of the T - ε curve even within the thermodynamic branch should be ascribed in part to heat diffusion and in part, especially near ε_θ , to the entropy production associated with dislocation motion, P[S].

Also the behavior of temperature versus stress has been investigated.

An example of the T-σ characteristic is presented in Fig.3. An abrupt change in the variation of T as a function of σ occurs at the critical stress of the termoelastic-plastic transition, σ_θ.According to the definition of yield proposed here, yield occurs at ε_θ and σ_θ namely when the thermoelastic cooling is exactly balanced by the thermoplastic heating. ε_θ and σ_θ have been measured also on other specimens of the same 100Cr6 steel, and are reported in Table I.

The values of ε_θ and σ_θ are distributed around their averages with a standard deviation of 5% and 6.9% respec-

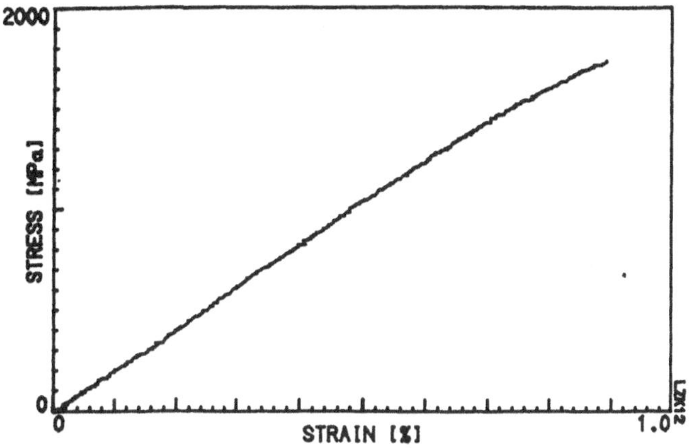

Fig.1 The stress-strain characteristic of a tensile specimen of 100Cr6 steel. While for most steels the upper limit of the range of elastic deformations seldom exceeds 0.2%, this sample seems to behave in a linear elastic way for deformations as high as 0.8%. Nevertheless, as it will become apparent from Fig.2, the thermoelastic branch extends only up to a thermoelastic-plastic limit deformation ε_θ = 0.61%

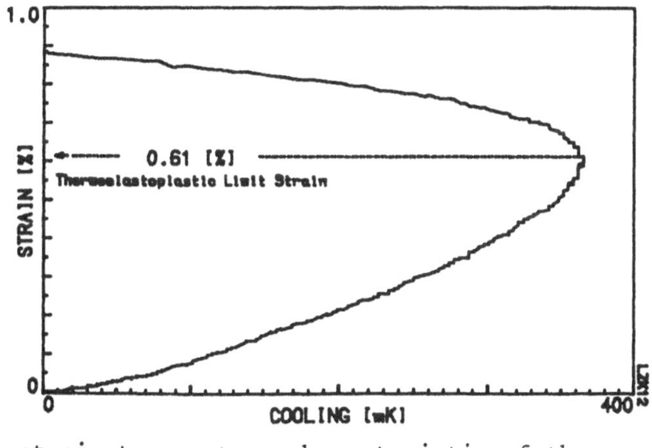

Fig.2 The strain-temperature characteristic of the same sample of 100Cr6 steel of Fig.1. While this sample is deformed within an elastic regime, it undergoes a thermoelastic cooling partly compensated by thermal diffusivity and incipient plastic heating. At the thermoelastic-plastic limit strain ε_θ = 0.61%, a conspicuous thermal emission occurs, marking the onset of irreversible plastic flow

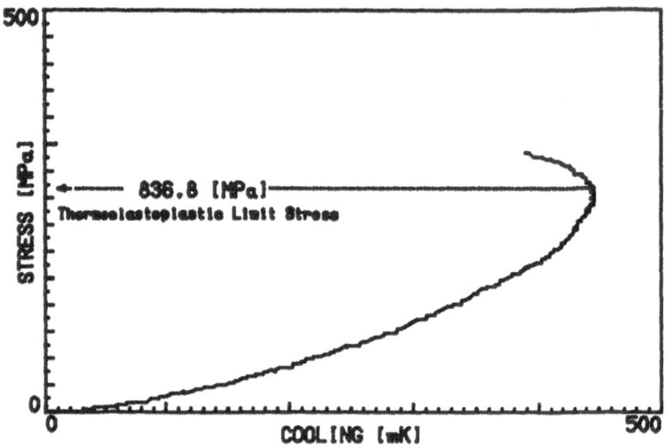

Fig.3 The stress-temperature characteristic of sample No.05 of 100Cr6 steel (see also Table I). This sample exhibits an anomalous value of σ_θ, which is lower than the average for the samples belonging to the same ingot

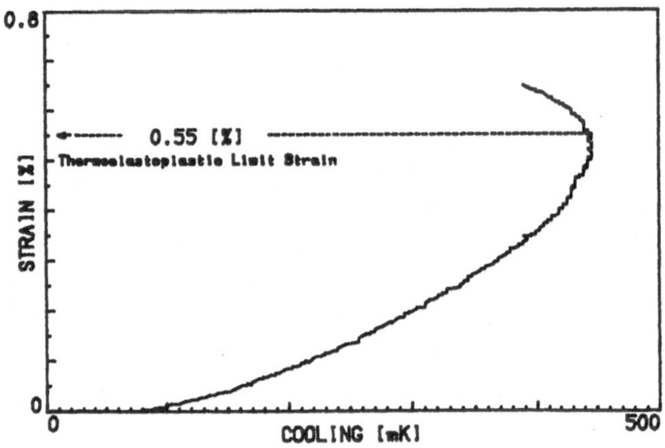

Fig 4. The strain-temperature characteristic of sample No. 05 of 100Cr6 steel (see also Fig. 3 and Table 1). This sample exhibits a comparatively low value of both σ_θ and ε_θ

TABLE I

The thermoelastic-plastic limit strain and stress, ε_θ and σ_θ respectively, for several 100 Cr6 steel samples. The average values, and their standard and relative standard deviations are presented inthe bottom of the Table

Sample Number (LZK=100 Cr6)	ε_θ (%)	σ_θ (MPa)	Note
01	0.63	1180	
02	0.61	1203	
03	0.65	970	
05	0.55	837	Sample N.05 has not been included in the statistical computations below
07	0.57	1196	
09	0.61	1214	
10	0.61	1278	
11	0.61	1283	
12	0.61	1191	
14	0.59	1231	
15	0.57	1212	
16	0.56	1174	
Average value	0.60	1194	
Standard deviation	0.03	82	
Standard deviation / Average value	5%	6.9%	

tively.
The specimen No.05 has been excluded from the above statistical analysis since it exhibits a value of σ_θ lower than the average by more than three times the standard deviation (837 MPa against (1194 ± 82) MPa). As indicated also in Fig.4, the sample showing this anomaly suddendly breaks shortly after the thermoelastic-plastic transition has occurred.
Nevertheless, while the stress at which fracture occurs (968 MPa) is lower than the average value of σ_θ by about three times the standard deviation, the strain at fracture (0.65%) is within twice the standard deviation: the apparent elastic modulus of this sample is considerably lower than the average.

The above results indicate that:

- by thermal emission, both ε_θ and σ_θ can be defined even for steels exhibiting an apparently vanishing ductility; these quantities could be conveniently adopted as a reference in the so called elastic design.
- a material sample whose σ_θ is lower than the average by more than two standard deviations could manifest anomalous and dangerous behavior during use
- σ_θ and ε_θ can be defined without the ambiguity inherent in current standard definitions of the yield strength.

4. Conclusions

The experimental results reported above and elsewhere [1,3,8] would seem to vindicate our contention that thermal emission is a valuable and reliable technique to locate the point of transition between thermoelastic and plastic behavior, even in an extreme case of apparent null ductility. The next step is to sharpen the physical description beyond the transition, using among other inputs specific microscopic models for the entropy production term P[S] that occurs in Eq.(8). P[S] is a measure of the irreversible nature of the processes occurring in materials under stress. We thus feel that, via P[S], essential features of these processes can be revealed, measuring thermal emission and correlating it with suitable theoretical models.

On the basis of our experience, this suggested procedure should prove especially useful when irreversibility is an important characteristic of the transformation under study. In particular we have in mind situations where a conspicuous fraction of the mechanical work done in deforming a material is at once converted into heat: including, of course, mechanical instabilities such as fatigue failure, fracture etc. Work in these directions is in progress

ACKNOWLEDGMENTS

A critical discussion with V.Balakrishnan on the contents of this paper is gratefully acknowledged. We thank the Centro Sperimentale Metallurgico, Roma Eur, for permission to publish the experimental results reported in Sec.3.

REFERENCES I

1. G.Caglioti,The Thermoelastic Effect: Statistical Mechanics and Thermodynamics of the Elastic Deformation in"Mechanical and Thermal Behavior of Metallic Materials" , Proc.of the Int.School of Physics, E. Fermi,LXXXII Course, Varenna 1981; G.Caglioti and A.Ferro (Editors),North Holland,Amsterdam (1982)
2. S.Boffi,C.E.Bottani,G.Caglioti,P.M.Ossi, Zeit.Phys. B, 39, 135 (1980)

REFERENCES II

1. C.E.Bottani,G.Caglioti,A.Novelli,P.M.Ossi,F.Rossitto and G.Silva, Met.It. 74,530 (1982)
2. S.Boffi,C.E.Bottani,G.Caglioti and P.M.Ossi, Zeit.Phys B, 39, 135 (1980)
3. C.E.Bottani and G.Caglioti, Phys.Scrip. T1, 119,(1982)
4. C.E.Bottani, G.Caglioti and P.M.Ossi,J.Phys,F 11 541 (1981)
5. L.D.Landau and E.M.Lifshitz, Theory of Elasticity, Pergamon Press, London (1970)
6. M.Beghi, Nuovo Cim, D 1, 778 (1982)
7. M.Beghi, Analisi Termomeccanica del Comportamento Elastoplastico dei Metalli, Thesis, Politecnico di Milano (1979) (unpublished)
8. C.E.Bottani and G.Caglioti, Mat.Letters 1, 119 (1982)

NON-EQUILIBRIUM EFFECTS SEEN IN MOLECULAR DYNAMICS CALCULATIONS OF SHOCK WAVES IN SOLIDS*

Franklin E. Walker, Arnold M. Karo, and John R. Hardy**

Lawrence Livermore National Laboratory
Livermore, Calif. 94550
and
Behlen Laboratory of Physics
University of Nebraska
Lincoln, Nebr. 68588

INTRODUCTION

In a number of experiments[1-3] carried out to study the effects of shock waves in condensed materials (particularly in chemical explosives), we found evidence for the mechanical fracture of covalent bonds in or very near the shock fronts which produced free atoms and free radicals, as well as other thermally-activated atomic and molecular species. Scrutiny of the streak and framing camera records obtained in these experiments led us, with other analysis, to the formulation of a new concept of the shock initiation of explosives.[4,5] To obtain some corroboration of this concept and to elucidate the microscopic processes occurring in the shock, we completed computer modeling and molecular dynamics analysis of the experiments.

We have now carried out a large number of molecular dynamics calculations on the behavior of shock waves in condensed systems. The first of these studies was reported at the Sixth International

*Work performed under the auspices of the U.S. Department of Energy by the Lawrence Livermore National Laboratory under contract W-7405-ENG-48 and the Office of Naval Research under contract E00014-82-F-0094.

**Consultant to Lawrence Livermore National Laboratory.

Colloquium on Gasdynamics of Explosions and Reactive Systems[6] and demonstrated that, for nonuniform materials, shocks alone could produce free atoms and atomic and molecular clusters. Possibly the most important result of this early work was a clear demonstration of shock integrity as manifested by the highly nonergodic behavior of shock-loaded perfect lattices: specifically, shock energy imparted to a perfect lattice is not significantly degraded into random thermal motion but remains localized in the shock front until it encounters a surface or other boundary feature. Our calculations also demonstrated that shocks are localized on a scale of atomic dimensions. These two properties enable shocks to deposit their energy selectively into lattice boundaries. This leads to the production of energetic microscopic spalled fragments of the lattice. For perfect crystalline materials it is by this mechanism that shock energy is most readily degraded.

In subsequent studies[7,8] we demonstrated that the presence of imperfections and crystallographic irregularities could provide additional channels for the decay of shock energy. The interplay between the various channels available in any given system can produce very complex behavior.

We have now developed general and flexible computer codes which enable us to replace these general qualitative observations by detailed quantitative studies. Specifically, we have introduced a sophisticated neighborhood look-up procedure and the ability to monitor selectively and in detail the energy flux in any "region" of the lattice. The size of this region is arbitrary. It can be as small as a specific atom or atomic pair or can encompass the entire lattice.

In the present paper we shall present the first results using these codes. We have chosen to examine the manner in which energy is absorbed from the shock by single diatomic molecules in a monatomic host lattice. Our central concern in these initial studies was to determine the "rise-time" of the shock as it transits the diatomic molecule. Specifically, we wished to obtain information about the time required for a major part of the shock energy to be converted into internal molecular energy.

DISCUSSION

The technique of molecular dynamics has been described previously.[6,8] To study energy flux into and out of a region such as we have described in the Introduction, our new codes continuously monitor the center-of-mass translational energy for the region, its rotational energy (about the center-of-mass), its vibrational energy, and its total energy. The potential energy of the region can be calculated in two ways. One can include only

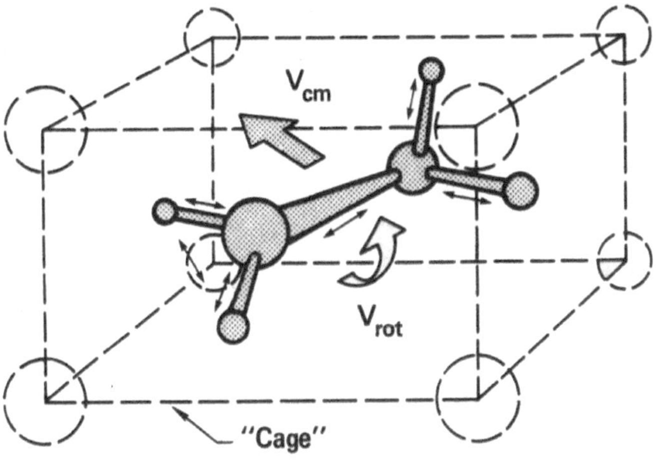

The molecular unit can be treated as isolated or as interacting with the lattice

Partitioning of the total molecular energy:

- **Center of mass energy**
- **Total internal energy: isolated or interacting**
- **Total rotational energy**
- **Total vibrational energy: isolated or interacting**

Fig. 1. Schematic illustration of the energy partitioning used in the present work.

bonds between atoms identified as being within the region, or one can include all these internal interactions together with the interactions of the atoms within the region with atoms lying outside the region - these exterior atoms forming the so-called "cage." Since, for the present we are more interested in the storage and flux of energy within the region, the first quantity is more appropriate for our purposes. For clarity these two types of partitioning are illustrated in Fig. 1, and the problem parameters and output are shown in Fig. 2.

The number of possible studies can be seen to be virtually unlimited. However, as indicated in the Introduction, we have restricted ourselves to the simplest region that can represent a molecular entity, i.e., nearest-neighbor pairs or diatoms. Computational economy dictated that our initial studies be restricted to two-dimensional simulations. Our codes are fully three dimensional; however, the size of a three-dimensional system which would be free of spurious surface effects is such that studies of the present variety would be computationally very expensive. Finally, two-dimensional calculations are also

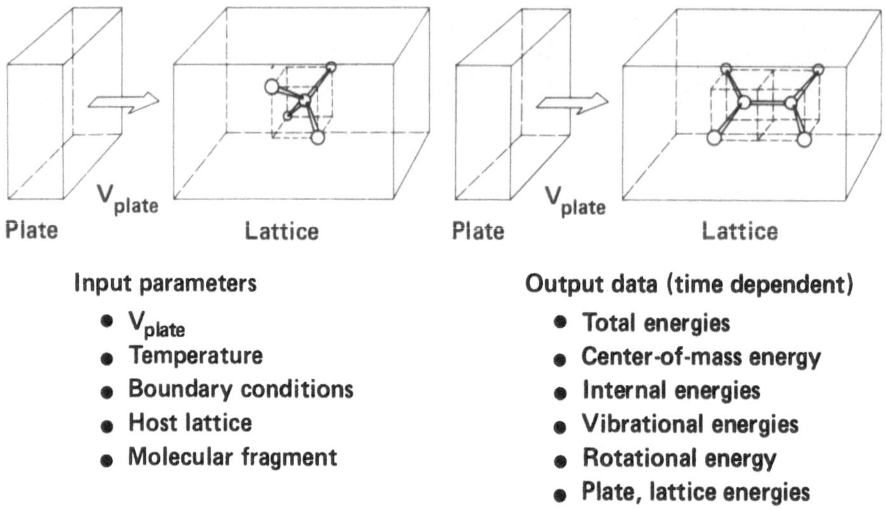

Plate
V_{plate}
Lattice
Plate
V_{plate}
Lattice
Input parameters
• V_{plate}
• Temperature
• Boundary conditions
• Host lattice
• Molecular fragment
Output data (time dependent)
• Total energies
• Center-of-mass energy
• Internal energies
• Vibrational energies
• Rotational energy
• Plate, lattice energies

Fig. 2. Problem parameters and output for the molecular dynamics simulations.

reasonable, since past experience has shown that at least the qualitative nature of our conclusions should be unaffected by this reduction in dimensionality.

We shall now present results for two different situations that we have studied:

(a) a light symmetric diatom, equivalent to an H_2 molecule, except for bond length, embedded in a heavy (iron) host lattice; and

(b) a light but highly asymmetric diatom, equivalent to a CH group, except for bond length, in a heavy (iron) host lattice.

In both cases a shock is generated by the impact of a plate of the host lattice from the left. Initial and final configurations are shown in Fig. 3 together with the location of the diatomic impurity. This location is the same in both studies and has been chosen so that after spall has occurred the impurity remains well within the host lattice structure. To highlight the effects of differences between the molecules only the impurity masses and the potential between them were changed. In order to ensure that all effects be clearly visible we deliberately used very large shock loading (shock pressures $\sim$ 0.5 - 2.0 Mbar). This procedure "overdrives" the rather soft Fe-Fe bonds to the point where their stiffness approaches that of typical organic bonds.

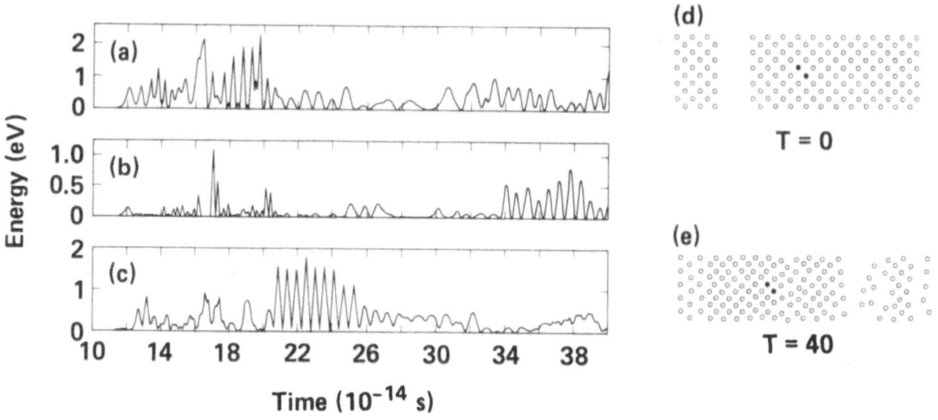

Fig. 3. Energy flux for the H_2 molecule in an iron lattice. (a) Center-of-mass energy versus time. (b) Rotational energy versus time. (c) Vibrational energy versus time (includes only H_2 potential energy). (d,e) Initial and final system configurations. Units of time are in 10^{-14} s.

We shall now describe each of the simulations in turn.

<u>H_2 in an Iron Lattice</u>

Figure 3 shows the vibrational, rotational, and center-of-mass energies for this system as functions of time. The first quantity was calculated including only the potential energy of the H-H bond. We also show the initial and final configurations of the ensemble. In this and the subsequent simulation it will be observed that in the final configuration the diatom is well removed from the region of spall induced by the shock.

The behavior observed is both interesting and highly complex. However, certain broad features are clearly apparent. The first is that there is an extremely rapid rise of both rotational and center-of-mass energy during times of the order of 1 - 2 femtoseconds (fs) as the shock reaches the diatom. This is followed by an equally sharp rise in the vibrational energy of the order of 1 fs later. Subsequently, all energies but the rotational appear to build to values in the eV range and hold at these levels for some 60-70 fs. Following this, it appears that some local transient restructuring of the system "bleeds" all forms of the internal energy out of the system with high efficiency. Probably the most significant feature is that the H_2 molecule spends a long period of time in a very hot vibrational state with an energy having an equivalent temperature of more than twenty times the melting temperature of the host lattice. From the lattice configuration at T = 40 it can be clearly seen, however, that the host has certainly not melted. This type of highly efficient and rapid energy transfer can be explained most easily by the highly nonergodic character of the system's evolution, particularly during the onset of shock loading. The result is a startlingly efficient coupling of shock energy to internal molecular motion. Should the same situation be present for larger molecules, it would appear likely that shock loading could very readily input to the whole molecule sufficient energy to rupture at least a single bond.

CH in an Iron Lattice

The time evolution and final configuration for this system can be seen in Fig. 4. The history is broadly similar to that for H_2 in an iron lattice. However, it appears that the asymmetric mass distribution of the diatom considerably enhances the efficiency

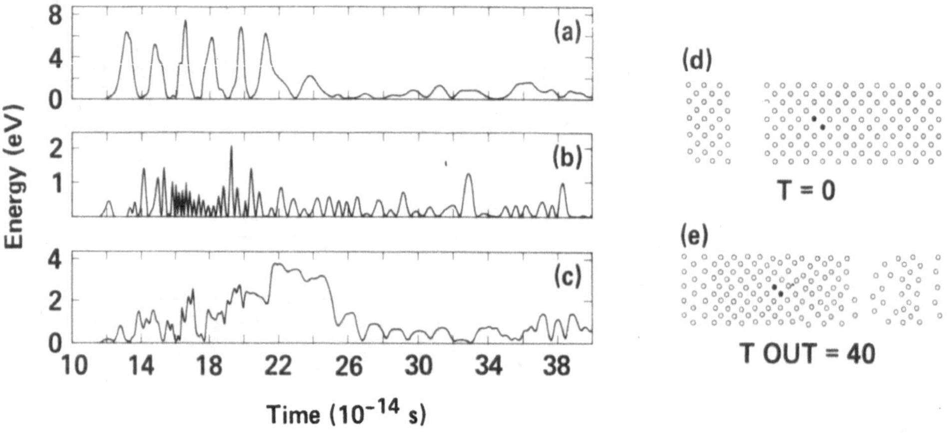

Fig. 4. Energy flux for the CH molecule in an iron lattice. (a) Center-of-mass energy versus time. (b) Rotational energy versus time. (c) Vibrational energy versus time (includes only CH potential energy). (d,e) Initial and final system configurations. Units of time are in 10^{14} s.

of the coupling of shock energy to internal motion. Consequently, for a given shock loading, or initial plate velocity, the CH group becomes vibrationally even hotter. Once again at a later stage during the evolution the internal energy appears to be dissipated by local restructuring.

The Order-Disorder-Order Transition

It was seen clearly in the photographs from the shock initiation experiments of nitromethane[2,3] that the shock front as it enters the nitromethane sample is very smooth to the naked eye. However, in our work with non-initiating shock waves[1] and in the studies of shock rise times in nitromethane completed by Presles and Harris[9] it appears there is some interesting microscopic structure due to low levels of chemical reaction in or very near the front.

When the initiation to detonation starts, there is much macroscopic disorder which is soon (a few microseconds) self-organized to a relatively smooth and ordered wave front again. It is possible that the "shock barrier"[10] related to the vibrational velocities of the atoms in the molecules of the explosive, as well as some kinetic effects, adds a constraint that assists this self-organization in what could otherwise be a very chaotic system.

CONCLUSIONS

The present work clearly demonstrates that strong shock fronts, apparently because of their extreme sharpness on an atomic scale, can impart to internal molecular motion startlingly large amounts of energy ($\sim$ eV's per bond) over dramatically short times ($\sim 10^{-12} - 10^{-13}$ s). As a consequence, for a considerable period of time during and after shock transit, the internal energy of the molecule is one to two orders of magnitude larger than the thermal energy of the host lattice which remains after the shock has passed. Although the present studies concern only diatomic systems, it would appear that the uptake of energy per bond would remain largely unchanged for polyatomic systems. Thus, large molecules should readily and rapidly be excited to the point of dissociation by shock loading. For such dissociation to occur it is obviously necessary for such molecules to be free to escape from their host environment. If this is not the case, the excess energy in the molecule will dissipate and may in fact do so rapidly. Escape mechanisms are obviously provided by defects such as surfaces, interfaces, and microscopic voids. Defects such as these could then serve as sources of free radicals which, should the shock-loaded material be an explosive, provide a mechanism for initiating detonation.

REFERENCES

1. F. E. Walker and R. J. Wasley, Initiation of nitromethane with relatively long-duration, low-amplitude shock waves, Combustion and Flame 15:233 (1970).
2. F. E. Walker and R. J. Wasley, Initiation patterns produced in explosives by low-pressure, long-duration shock waves, Combustion and Flame 22:53 (1974).
3. F. E. Walker, Initiation and detonation studies in sensitized nitromethane, Acta Astronautica 6:807 (1979).
4. F. E. Walker and R. J. Wasley, A general model for the shock initiation of explosives, Propellants and Explosives 1:73 (1976).
5. F. E. Walker, Quantum mechanics and molecular dynamics calculations provide new evidence for a free radical shock initiation model, Propellants, Explosives, Pyrotechnics 7:2 (1982).
6. A. M. Karo, J. R. Hardy, and F. E. Walker, Theoretical studies of shock-initiated detonations, Acta Astronautica 5:1041 (1978).
7. J. R. Hardy, A. M. Karo, and F. E. Walker, The molecular dynamics of shock and detonation phenomena in condensed matter, Progress in Aeronautics and Astronautics 75, J. R. Bowen, ed., American Institute of Aeronautics and Astronautics (1981).
8. A. M. Karo and J. R. Hardy, The study of fast shock-induced dissociation by computer molecular dynamics, Proceedings of the NATO Advanced Study Institute on Fast Reactions in Energetic Systems, C. Capellos and R. F. Walker, eds., D. Reidel Publishing Co., Dordrecht-Holland, Boston, USA (1981).
9. P. Harris and H. N. Presles, Comparison of the optical reflectivity of a shock front in liquid water and in liquid nitromethane, U.S. Army Armament Research and Development Command, Dover, N.J., Technical Report ARLCD-TR-82025 (November 1982).
10. F. E Walker, Description of a shock-wave velocity barrier, Propellants and Explosives 6:15 (1981).

CURRENT TOPICS IN REACTION-DIFFUSION SYSTEMS

Paul C. Fife*

Department of Mathematics, University of Arizona

1. INTRODUCTION

In these lectures, I want to outline various topics of high current interest in the theory of reaction-diffusion systems. I say "theory", but in fact they all reside in the applied side of the theory, as in all cases they are motivated by applications in the natural sciences. RD systems, in fact, provide an extremely fertile source of models in all these sciences. In accordance with the theme of the conference, the applications will by and large be related to dissipative structures of one kind or another.

Those interested in the "purer" side of the theory of these systems might consult the recent book by Smoller (1983).

The two broad areas I will dwell upon are singular perturbation methods and bifurcation methods. In the former category (which occupies the lion's share of the material, secs. 3-8), I will discuss traveling and stationary front phenomena. For a single equation, the theory of fronts is quite complete, but in the case of systems with more equations, some kind of perturbation method is usually employed. Perturbative methods require the existence of suitably placed small or large parameters, which is a limitation on the repertoire of problems amenable to this class of techniques. Despite this fact, the versatility of these methods is remarkable. In the case of the singular perturbation approach to wave fronts (secs. 4-8), for

*Supported by the National Science Foundation and by the Mathematics Department and Institute for Fundamental Physics, Kyoto University.

instance, the applications to target and rotating spiral patterns involve, in my opinion, very realistic models.

Bifurcation theory has been the most popular route by which theorists have deduced the existence and nature of dissipative structures in reaction-diffusion systems; there have been an enormous number of papers on this subject. Again, in principle the method depends on the existence of parameters with values near certain critical values. And again, this limitation has not seemed to have dampened the versatility of the method in illuminating various sorts of structural phenomena. In the simplest cases, the underlying theory is well known (see, for example, the above-mentioned book by Smoller or many other references) and will not be repeated here. There has been a lot of interest in the study of possible large-scale structures enveloping smaller scale bifurcating patterns. This field of investigation is promising but in a very primitive stage, so I shall focus attention there in sec. 9. In particular, I derive, for illustration, the equation governing the large scale structures in the case that the smaller scale ones are periodic traveling waves. Envelop equations of this general nature (including so-called λ-ω systems) have been quite popular as a vehicle for the construction and study of target patterns and rotating spirals.

The final section (10) of the paper serves in some sense as a bridge between singular perturbation results and bifurcation results. It surveys the work a group of mathematicians in Kyoto and Hiroshima have done in developing a remarkably rich and complete study of the 2-parameter branching behavior of steady solutions of a model system.

The original motivation for the study of the model problem in sec. 10 was from population biology, but (for some parameter ranges) the equations are essentially the same as those on which the material on target and spiral patterns in secs. 6 and 7 is based. These latter objects are seen in certain chemical reagents and in certain physiological tissues. The study of stationary structures, such as in sec. 8, is typically motivated by issues in developmental biology. So a rich range of applications are in the background of the material presented here. Each of the applications has been and is currently the subject of a great deal of research. I do not intend to give an adequate survey of this work; for example, many important studies of spirals will not be mentioned. Rather, the object of the lectures is to attempt an understandable explanation of certain important tools for the study of dissipative structures. At the same time, a few items apparently appear here for the first time: the formation process for spiral patterns in the context of sharp wave front analysis, the compound layer in sec. 8, and the formalization of sharp wave front dynamics, including trigger and phase types, in sec. 4.

In preparing these notes I benefited a great deal from insight gained in discussions with Y. Kuramoto.

I. DYNAMICS AND STATICS OF WAVE FRONTS

2. The Simplest Wave Fronts

As mentioned before, traveling fronts play a crucial role in the constructions I intend to elaborate. These fronts can, in turn, be best understood by examining their properties in the case of a scalar reaction-diffusion equation. Moreover, these simple fronts can serve as building blocks for more complicated phenomena.

So let me give a brief outline of the salient facts about traveling front solutions of the equation

$$u_t = u_{xx} + f(u). \qquad (2.1)$$

By their strictest definition, they are solutions of the form $u = U(x-ct)$ for some constant c (the velocity) which connect one stationary state u_1 at $x-ct = -\infty$ to another one, u_0, at $x-ct = +\infty$. By "stationary state" I mean a value of u such that $f(u) = 0$. For nontrivial fronts to exist, it is therefore necessary that f have at least two distinct zeros. Moreover, this condition proves to be nearly sufficient as well, since if f has two distinct zeros, is nonzero between them, and is continuously differentiable, then it has been shown that fronts exist connecting these two states. The smoothness assumption, in fact, can be relaxed considerably. The most careful construction of such fronts was possibly that in (Aronson and Weinberger 1978), and a very simple less rigorous argument based on phase plane analysis can also be given. We categorize the case according to the properties of f.

A. The Fisher Case:

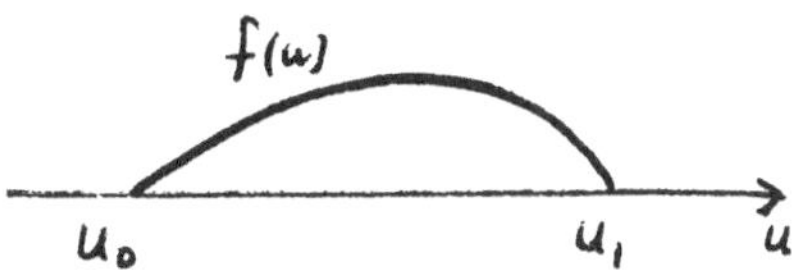

Fisher (1937) first proposed (2.1) with this nonlinearity in connection with a problem in population genetics. There exists a positive number c^* such that a front exists for every velocity $c > c^*$. These fronts, though stable to perturbations restricted to bounded intervals, do not enjoy the stability properties of fronts in the bistable case below. In the case when f is negative between the two rest states, there again exists a continuum of possible velocities for fronts connecting u_1 at $-\infty$ to u_0 at $+\infty$, but this time they are all negative.

B. <u>The Bistable Case:</u>

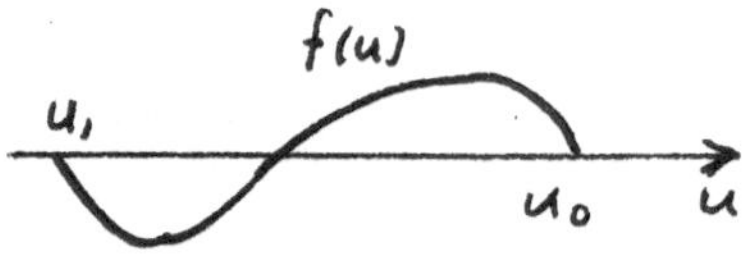

If $f'(u_0) < 0$, $f'(u_1) < 0$, and f has only one zero between them, (and in some other cases as well), then there exists a unique wave front connecting u_0 and u_1. Therefore only one value of c is possible. The sign of c is determined as follows. The state u_0 is called "dominant" if

$$\int_{u_1}^{u_0} f(u)du > 0,$$

and u_1 is dominant if the inequality is reversed. (More or less, the graph of f forms two lens-shaped regions; the one with the larger area is adjacent to the dominant state.) Then the traveling front always moves in the direction which will enlarge the "territory" of the dominant state. Thus if u_0 is dominant and $U(+\infty) = u_0$, then necessarily $c < 0$. Of course $c = 0$ if and only if $\int_{u_1}^{u_0} f(u)du = 0$. Bistable fronts were proved in Fife and McLeod (1977) to be very stable structures.

C. The Ignition Case:

In this case, fronts connecting u_0 to u_1 are also unique, with unique velocity, and enjoy a limited degree of stability. See Kanel' (1962), Fife and McLeod (1981).

3. Slowly Varying Fronts

Consider the equation like (2.1) but with f depending weakly on x and t (as well as on u). Specifically, let v be a given slowly varying function of x and t:

$$v = v(\varepsilon x, \varepsilon t) \qquad (\varepsilon \ll 1),$$

and consider the equation

$$u_t = u_{xx} + f(u,v), \qquad x \in \mathbb{R}. \tag{3.1}$$

Assume that for each v, f is in the bistable case, and denote the two stable zeros by $u = h_{\pm}(v)$, $h_+(v) > h_-(v)$.

Intuitively, one can picture a progressing front-like solution, slowly changing its speed and shape. To bolster intuition, the following formal construction is possible:

Denote the center of the front, where u equals some fixed value between $h_-(v)$ and $h_+(v)$ (assuming there is such a value for all v), by $\frac{y}{\varepsilon}$. Now look for y as a function of εt, and u as a function of $z \equiv x - \frac{y(\varepsilon t)}{\varepsilon}$ and $\tau = \varepsilon t$:

$$u = U(z,\tau). \tag{3.2}$$

Setting $\xi = \varepsilon x$, we have $v = v(\xi, \tau)$, and substituting (3.2) into (3.1), we get

$$-y'(\tau)U_z + \varepsilon U_\tau = U_{zz} + f(U, v(y(\tau) + \varepsilon z, \tau)) \tag{3.3}$$

<u>Lowest order approximation</u>: Set $\varepsilon = 0$ to obtain

$$-y'(\tau)U_z = U_{zz} + f(U, v(y(\tau), \tau)). \tag{3.4}$$

Now if v were fixed, (3.1) would have a unique traveling front solution $u = U(x-ct)$, $U(-\infty) = h_+(v)$, $U(+\infty) = h_-(v)$, $U(0) =$ the above-mentioned intermediate value. Here c would depend on v: $c = c(v)$, and U would satisfy the equation

$$-cU' = U'' + f(U, v) \tag{3.5}$$

Comparing this with (3.4), we see immediately that

$$y'(\tau) = c(v(y(\tau), \tau)) \equiv Y(y, \tau). \tag{3.6}$$

If one integrates this with any assumed initial position $y(0)$, one finds the trajectory of the front, its velocity $y'(\varepsilon t)$ being slowly varying. At each instant of time, the profile $U(z, \tau)$ will be the same function of z as the function $U(z)$ in (3.5).

If we wish to obtain higher order approximations, call the one just obtained (y_0, U_0), and seek an approximate solution of (3.3) in the form $y = y_0 + \varepsilon y_1$, $U = U_0 + \varepsilon U_1$. To this order, we find

$$LU_1 = -f_v(U_0(z,\tau), v(y_0,\tau))v_\xi(y_0,\tau)(y_1(\tau) + z) - y_1'(\tau)U_{0z} + U_{0\tau}, \tag{3.7}$$

where L is the linear differential operator

$$LU \equiv U_{zz} + y'_0(\tau)U_z + f_u(U_0(z,\tau), v(y_0(\tau),\tau))U.$$

The fact is that this operator has a one-dimensional nullspace (on $_2(-\infty, \infty)$) and the equation (3.7) has a solution $U_1(z,\tau)$, for any given τ, if and only if the right side is orthogonal (in the sense of $_2$), to the nullvector, which happens to be $U_{0z}(z,\tau)$. (See Fife (1974), for example.) Applying this orthogonality condition results in an equation of the form

$$y_1'(\tau) + a(\tau)y_1(\tau) + b(\tau) = 0,$$

which may be integrated (with $y_1(0) = 0$) to obtain the correction $y_1(\tau)$. And the correction U_1 is, for each τ, the solution of (3.7) satisfying $U_1(0) = 0$. There is exactly one such solution.

This procedure could, of course, be continued.

4. Coupling with another reactant: propagator-controller systems

This section establishes the framework for handling solutions of reaction-diffusion systems with two equations such that one component propagates an abrupt front, this propagation (its speed and amplitude, that is) being modulated by the second component. We call such a pair a "propagator-controller" system. Systems with these general properties are very common in applications (we shall see a couple of them), and a great variety of phenomena can be fit into this framework. The idea of such a modulated propagation event was possibly first put forward by Ostrovskii and Yahno (1975), followed independently by Ortoleva and Ross (1975) and Fife (1976a). Some of the material on wave front dynamics in this section represents an extension of that presented in Fife (1976c).

We now interpret the function $v(\xi,\tau)$ in sec. 3 not as a given fixed function, but rather as a solution of a differential equation:

$$v_\tau = g(u,v)$$

Let us also write (3.1) in terms of the variables $\xi = \varepsilon x$, $\tau = \varepsilon t$. The result is a system

$$\varepsilon u_\tau = \varepsilon^2 u_{\xi\xi} + f(u,v) \tag{4.1}$$

$$v_\tau = g(u,v) \tag{4.2}$$

This scale change from (x,τ) to (ξ,τ) compresses the front studied in sec. 3 to a narrow layer. Inside the layer, u is expressed in the form

$$u = U\left(\frac{\xi - y(\tau)}{\varepsilon}, \tau\right),$$

where y satisfies (3.6), and to lowest order, the function U is the profile constructed in the preceding section.

Outside this frontal layer, the function u will not suffer abrupt changes, and it is safe to neglect the terms with ε and ε^2 in (4.1), so that $f(u,v) = 0$. The solutions of this latter equation which were provided are $u = h_\pm(v)$, and of course to be compatible with the profile in the layer, which approaches values $h_\pm(v)$ to the left and right, respectively, we must choose the specific outer solutions

$$u = h_+(v), \quad \xi < y \quad \text{(left of the layer)},$$

$$u = h_-(v), \quad \xi > y \quad \text{(right of the layer)}.$$

In both outer regions we continue to have (4.2), so substituting for u in that equation its possible values as given above, we obtain an equation for v which does not depend on u directly; only on the front's position y:

$$v_\tau = G(v,\xi-y) \equiv \begin{cases} g(h_+(v),v), & \xi < y \\ g(h_-(v),v), & \xi > y, \end{cases} \tag{4.3}$$

$$y_\tau = c(v(y(\tau),\tau)). \tag{4.4}$$

One problem is that v may develop discontinuities and hence (4.4) may be difficult to interpret. We deal with this problem later.

Let me illustrate how initial value problems for (4.3,4) can be solved. Let y_o be a given number and $v_o(\xi)$ a given continuous function. We attempt to solve (4.3,4) under the initial conditions $v(\xi,0) = v_0(\xi), y(0) = y_0$. The steps are as follows:

1. For each $\xi > y_o$, integrate the following equation, to obtain $v(\xi,\tau)$ for $\tau > 0$:

$$v_\tau = g(h_-(v),v); \quad v(\xi,0) = v_0(\xi). \tag{4.5}$$

2. Do the analogous thing for $\xi < y_0$, using $h_+(v)$. This establishes a function $v(\xi,\tau)_1$ which is expected to agree with the desired solution in some but not all parts of the (ξ,τ) plane. For $\tau > 0$, it will in general be discontinuous at $\xi = y_o$.

3. Assuming $c(v_0(y_0)) \neq 0$, solve the initial value problem

$$y'(\tau) = c(v(y,\tau)) \tag{4.6}$$

with the function v determined above. The function y is initially monotone, since by assumption $c \neq 0$ then. If it ceases to be monotone at some later time τ_1, then discontinue the integration at that time. Let T denote the graph, in the (ξ,τ) plane, of the function $y(\tau)$ for times up to that cessation time (Fig. 1).

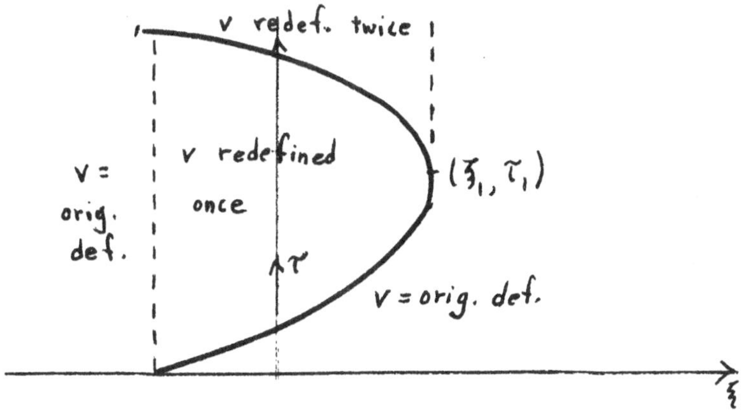

Figure 1

4. For time $\tau < \tau_1$, the desired solution v is, except for the region above T, the one constructed in 1. and 2. above.

Above T, we determine a new function $v(\xi,\tau)$ by integrating (4.5) again, but with the following change: the (sign) subscript on h is to be the opposite from what it was in the region below T. For example, if y is monotone increasing (as in Figure 1), then T lies to the right of $\xi = y_o$, and on crossing T, we change the subscript from "-" to "+". The initial values of the new v are assumed on T, and coincide there with the values of the original function v. This redefines the original function v in the region above T.

5. with the new v, continue integrating (4.6) at the point τ_1 we had previously stopped, if there is such a point (if there is not, we are done).

This is the stage at which the difficulty mentioned following (4.4) could first appear. In fact, letting (ξ_1,τ_1) be the endpoint of T, we see that for $\tau > \tau_1$, the right side of the differential equation (4.3) satisfied by v in general suffers a discontinuity as ξ crosses the value ξ_1. Hence v itself will in general be distontinuous there, and the right side of (4.4) as well. this leaves the problem ill defined, and our model is incomplete. We complete it by providing an interpretation of (4.4). This is done by cases. Note that necessarily $c(v(\xi_1,\tau_1)) = 0$. For $\tau > \tau_1$, we let

$$C_+(\xi_1,\tau) = \lim_{\xi\downarrow\xi_1} c(v(\xi,\tau)); \quad C_-(\xi_1,\tau) = \lim_{\xi\uparrow\xi_1} c(v(\xi,\tau)).$$

(a) If C_+ and C_- are both nonnegative for $\tau > \tau_1$ in some neighborhood of τ_1, then for $\xi = \xi_1$, the function v appearing in the right side of (4.4) is to be interpreted as the limit as ξ approaches ξ_1 from above. There then results a trajectory $y(\tau)$ which will not move left as τ increases by a small amount, so will not cross the line of discontinuity.

(b) If they are both nonpositive, we interpret v as its limit from below.

(c) If C_+ is nonpositive and C_- is nonnegative, we replace the right side of (4.4) by 0. This results in the trajectory remaining at the position $y = \xi_1$ for awhile. This is reasonable, since if it strayed to the right, it would enter a region where its velocity is nonpositive, which entails a contradiction. Similarly if it strayed to the left.

(d) If C_+ is nonnegative and C_- is nonpositive, then interpreting the right side of (4.4) as either C_+ or C_- would bring no contradiction; we leave the choice open and say that the solution is not unique beyond τ_1; it can be continued in two different ways.

These categories exhaust all but the pathological cases, which we shall not discuss. As we follow the trajectory beyond τ_1, it may happen, of course, that the case may change (from (c) to (b), for example); then the interpretation will also change, according to the rules above.

Step 5 is now well defined. Continue the integration until y reaches its next point of nonmonotonicity, if it ever does. Then redefine v above the new section of the graph of y, as before. Continue this process.

Let us be more specific about terminology and notation. Fronts as described above the graph T will be called trigger fronts. These serving to expand the region where $u = h_+(v)$ will be denoted by T_+; those contracting it by T_-. For completeness, we call the stationary ones T_0.

The foregoing analysis implicitly assumes that the functions $h_\pm$ and c are defined for all values of v; we now explore the complications introduced when that assumption is dropped. to be specific, let us suppose the function $f(u,v)$ has the sigmoidal nature shown in Figure 2. the interval $\underline{v} < v < \overline{v}$ contains the only values of v for which h_+ and h_- are both defined. We further assume that $f_u < 0$ for $u = h_\pm(v)$, $v \in (\underline{v},\overline{v})$ (excluding the endpoints $\underline{v}$ and $\overline{v}$). For those values of v, the function $f(u,v)$ is in the bistable category, so that $c(v)$ is uniquely defined. At the endpoints $v = \underline{v}$ or $\overline{v}$, f is in the Fisher category, and there is an unbounded interval of possible values for c. These endpoints are also, of course, the points where either h_+ or h_- ceases to exist. The fact that the endpoints can be characterized in both of these ways enables us to generalize the front-trajectory analysis described before. The generalization proceeds as follows.

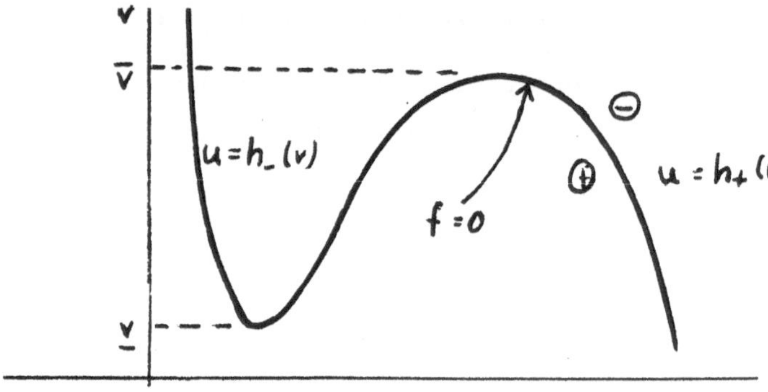

Figure 2

We assume

$$g(h_-(v),v) < 0; \quad g(h_+(v),v) > 0; \tag{4.7}$$

and

$$v_o(\xi) > \underline{v}, \quad \xi > y_o; \quad v_o(\xi) < \bar{v}, \quad \xi < y_o.$$

In step 1, we integrate the equation (4.5) only until the value of τ at which h_- ceases to be defined; i.e. until v reaches $\underline{v}$. This point in time will, of course, depend on ξ, say $\tau = \tau^*(\xi)$. Let the graph of this function be denoted by T^*_+. In the analogous way, we get a graph T^*_- for $\xi < y_o$. (See Fig. 3.)

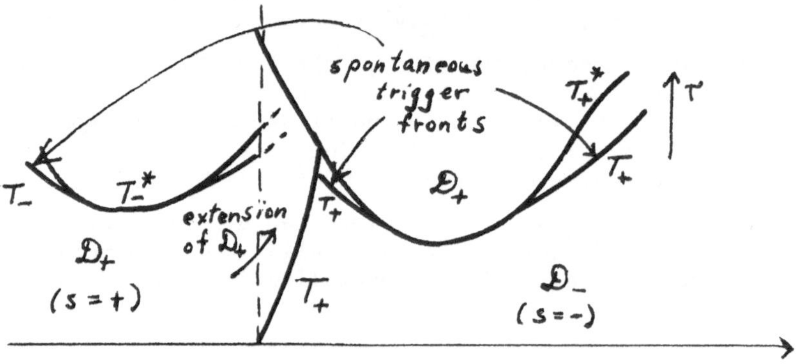

Figure 3

At this point, it is best to pass from a description of the dynamics based on the function y to a description based on the "state" function $s(\xi,\tau)$. This function takes on only the values "+" or "-", and specifies whether the pair (u,v) lies on the h_+ or the h_- branch of the S-curve in Fig. 2. For $\tau = 0$, we have $s(\xi,0) = +$ for $\xi < y_0, s(\xi,0) = -$ for $\xi > y_0$. Let $D_+(D_-)$ be the connected component of the (ξ,τ) plane in state + (state -) containing the negative (positive) half of the ξ-axis. We show how to determine these regions. If $\xi > y_0$ is fixed and time evolves until T^*_+ is reached, and if s remains "-" during this entire evolution, then s must change to + as T^*_+ is crossed. To see why this is true, we note that the pair (u,v), for that particular ξ, has, during the evolution, traveled down the h_- branch in Figure 2 to the bottom. At that point, g continues to be negative, and so the point is forced off the S-curve. Then the last term f in (4.1) is no longer 0, but rather positive. From (4.1) we see this forces εu_τ to be comparably positive, hence u_τ is very large and positive. The point (u,v) is rapidly attracted to the other branch h_+, thus changing s.

So clearly, D_- is bounded above by both T^*_+ and the trajectory T_+ constructed in step 3 above. In addition, it may be bounded above by spontaneously generated fronts. How they come about is as follows. Consider a portion of the curve T^*_+ which has positive slope. Let

$$q_+(\tau) = \frac{d\xi}{d\tau}\Big|_{T^*_+} > 0.$$

be the velocity of a point which travels along T^*. If one fixes a point on this portion, the state of the medium changes from + to - as ξ increases through that point, τ being held constant. (This change of course, represents a front.) This front may or may not generate a trigger front T_+ moving to the right from that point. The criterion for when this happens is easily seen. If such a front is generated, it will move into D_-, and into a region where $v > \underline{v}$. Therefore it will have, on first entering that region, velocity $c(\underline{v}) = \lim_{v \to \underline{v}} c(v)$. Clearly this is only possible if $c(\underline{v}) > q_+$ at that point, and $c(\underline{v}) > q_+(\tau)$ for a small time interval beyond that point. On the other hand if $c(\underline{v}) - q_+(\tau)$ changes from negative to positive values at some point, then such a trigger front, being a very stable type of structure, will indeed develop. This latter is therefore a strict criterion.

Similarly, trigger fronts T_+ can be spontaneously generated moving to the left, at places on T^*_+ with negative slope. When they first appear, they will have negative velocity $-c(\underline{v})$. (See Figure 3).

Fronts converting the medium from state + to state - may in a similar manner be generated spontaneously at points on T^*_-. In that case the criterion is:

> T_- type trigger fronts generated at points on T_- where $q_-(\tau)$ increases past the positive value $|c(\bar{v})|$, or decreases past the negative value $c(\bar{v})$.

For completeness we also restate the criterion for new T_+ fronts:

> T_+ - type trigger fronts generated at points on T^*_+ where $q_+(\tau)$ increases past the positive value $c(\underline{v})$ or decreases past the negative value $(-c(\underline{v}))$.

Now it is clear that the portion of the half-plane $\xi > y_o$ lying below all these curves T_+ and T^*_+ will be contained in D_-. And a similar statement is true about D_+. Above the minimum of these curves, we are in a different state, and a revised function v must be constructed by integrating a differential equation (4.3) with dynamics determined by the new state. In this way, it is easy to see how the rest of D_- (if any) or D_+ (if any) may be obtained; see Figure 3 again.

One may continue, in this manner, to obtain a global picture for the funtions v and s. The latter will serve to partition the plane into regions in state + or state -, separated by traveling fronts which are depicted as portions of curves $T_\pm, T_o$, or $T^*_\pm$. In accordance with a dichotomy already used by Winfree and others in the literature, we call the former "trigger fronts" and the latter "phase fronts" (Tyson and Fife 1980).

5. Accounting for diffusion of v

In (4.1), one may divide by ε to obtain ε as the diffusion coefficient of u. The first question we consider is, what is the effect of introducing a diffusion term $\epsilon v_{\xi\xi}$ of like magnitude into (4.2)?

In our construction in sec. 4, the function $v(\xi,\tau)$ obtained in the end had large gradients or discontinuities only near stationary fronts T_0. Other than this and the discontinuous derivatives on the front trajectories $T_\pm$ and $T^*_\pm$, it was smooth. It is reasonable to then suppose that the effect of the diffusion term will not alter the picture appreciably: near $T_\pm$ and $T^*_\pm$, it will only round off the corners (discontinuities in derivative) of v. And although it will smooth out v's discontinuity on T_0, replacing it by a large gradient in a small neighborhood of T_0, it will not alter the position or alignment of T_0, nor the values outside that neighborhood.

When a larger diffusion term is admitted in (4.2), however, the problem becomes more difficult. For greatest simplicity, let us return to the case when there are no phase fronts and only one trigger front T_+, given by $y = y(\tau) > y_0$. (Figure 4).

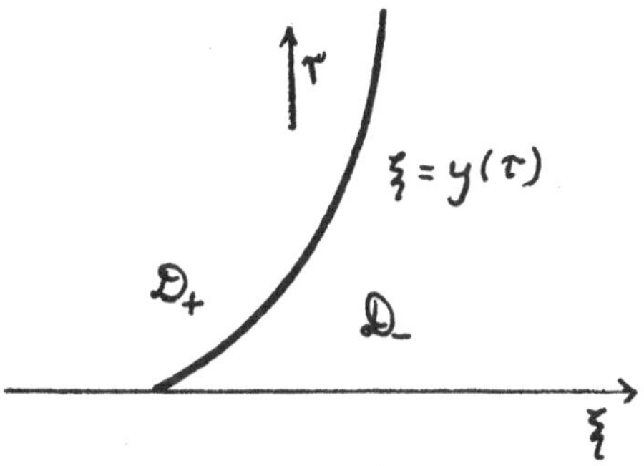

Figure 4

The problem then becomes a free boundary problem for the determination of v and y. Again, denote by $D_-(D_+)$ the region to the right (left) of the interface curve. The problem is to solve

$$v_\tau - v_{\xi\xi} = g(h_-(v),v) \quad \text{in } D_- \quad (g(h_+(v),v) \ \text{in } D_+)$$

with the interface condition that v be continuous, and satisfy

$$y'(\tau) = c(v(y(\tau),\tau)).$$

I am not aware of a mathematical theory for this type of problem, nor for its generalization when phase fronts are allowed.

6. Target patterns for the Belousov-Zhabotinsky reagent

The familiar expanding blue and red rings seen in this reagent (Zaikin and Zhabotinsky 1970) offer a fertile ground for the application of the foregoing theory. Indeed, it was shown in (Tyson 1979, Tyson and Fife 1980) that (4.1,2), with f having properties similar to those depicted in Figure 2, is a reasonable reduction of the Oregonator dynamics proposed for the BZ reaction (Field, Koros, Noyes 1972). Actually, a diffusion term $\varepsilon v_{\xi\xi}$ would be added to (4.2), but again may be neglected for small ε according to the argument in sec. 5. Tyson (1982) has further investigated the appropriateness of the Oregonator equations as a reduction of a more complex system describing the BZ dynamics more accurately. In short, the propagator-controller scenario of sec. 4 is applicable to the BZ reagent.

In modeling the reaction and diffusion problem by (4.1,2), one identifies u (the propagator) with the concentration of $HBrO_2$ in the reagent, and v (the controller) with the concentration of Ce^{iv}. Regarding the function g, it may be taken to depend on another parameter in the problem, which we called b. For $b > 1$, the nullcline $g = 0$ intersects the left branch h_- (Figure 5), and for $b < 1$, it intersects the

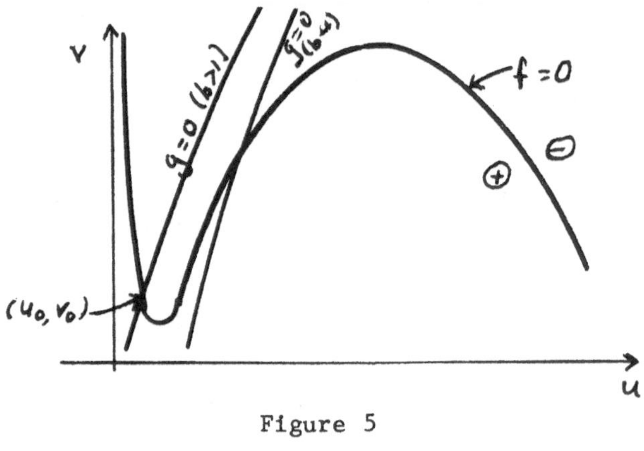

Figure 5

intermediate unstable branch. In all cases, $g < 0$ on the left of its nullcline, and $g > 0$ on its right.

I shall show (see Tyson and Fife (1980) for more details) that the one-dimensional analog of target patterns will arise naturally under homogeneous initial conditions, if b is allowed to depend on ξ, being less than 1 near the origin, and greater than one and constant further away. This inhomogeneity could represent an extraneous particle in the reagent slightly altering the chemistry in its neighborhood. Specifically, we assume (Fig. 6) that

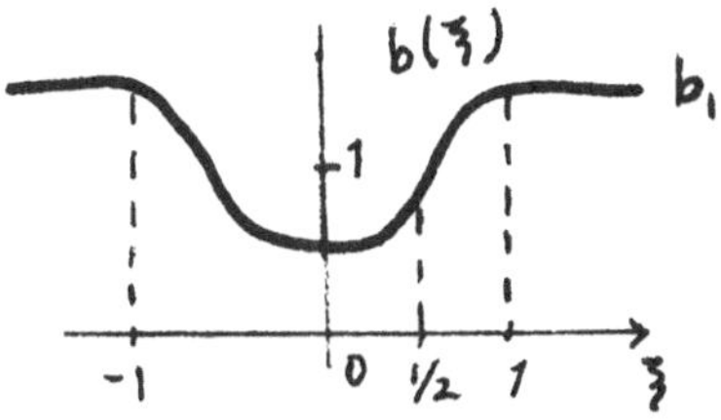

Figure 6

$$b < 1 \quad \text{for} \quad |\xi| < 1/2; \quad b \geq 1 \quad \text{otherwise};$$

$$b \equiv \text{const} = b_1 > 1 \quad \text{for} \quad |\xi| > 1.$$

For $b = b_1 > 1$, there exists a stable critical point (rest state) (u_0, v_0) on the left branch (Figure 5). Let us take the uniform initial condition $v(\xi, 0) \equiv v_0$, $s(\xi, 0) = -$.

First, we find the curve T^*_+. When $|\xi| > 1/2$ we have $b > 1$, so that the solution $v(\xi, \tau)$ of (4.5) with initial condition v_0 never reaches $\underline{v}$. This means the curve T^*_+ is not defined for those ξ; it exists only for $|\xi| < 1/2$. It is shown as the (in part) dotted line in Figure 7. This means that necessarily a forward-moving trigger front is generated at some point on T^*_+, namely where the phase velocity q_+ first drops below the trigger velocity $c(\underline{v})$. And a backward moving front is generated on the other side. These fronts soon enter the region where $v = v_0$, and at that point reach their ultimate velocity $c(v_0)$. This accounts for the birth and eventual linearity of the graph T_+ in Figure 7.

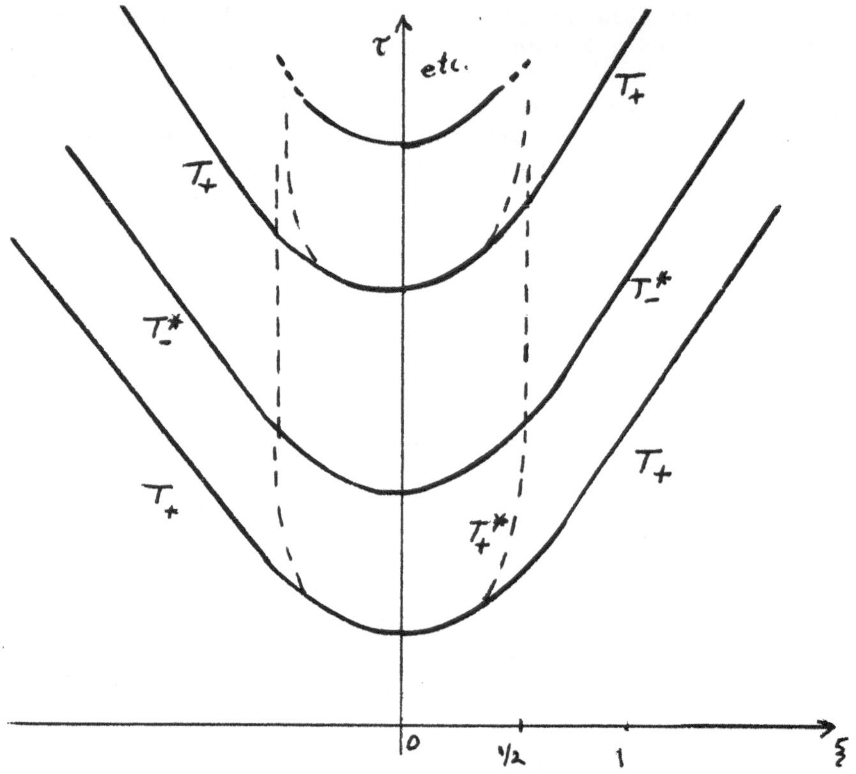

Figure 7

Immediately above this first front, the medium is in state $+$, where g remains strictly positive. This means that v (according to (4.3)) will inexorably increase, for each ξ, until its upper limit $\overline{v}$ is reached. This forms the curve T^*_- as shown above T_+. It is roughly parallel to T_+ because v takes roughly the same amount of time to reach $\overline{v}$, no matter where on the front T_+ it starts from. The question now arises as to whether or not a trigger front can be spontaneously generated from T^*_-. This will happen if q, for points on T^*_-, drops below $-c(\overline{v}) > 0$. If the S-curve is skewed as shown in Figure 5, then this will not happen, because the trigger velocity at the bottom of the S-curve is sufficiently larger than that at the top. Therefore T^*_- represents, in its entirety, an actual phase front.

Proceeding to the next stage, we see that the third expanding front is, like the first, phase-like near the origin and trigger-like elsewhere. The time interval between the 2nd and 3rd fronts will be larger than that between the first and second, etc., because the v dynamics on the h_- curve are slower than on the h_+ curve.

In this way, an infinite sequence of fronts, alternately trigger and phase, progress outward in both directions. Notice that in the region $|\xi| > 1$, v never returns to its original value v_0. Roughly speaking, since v_0 is an equilibrium point for the dynamics on h_-, v approaches v_0 very slowly on that branch. This leaves time for the trigger wave to appear and convert the state to $+$. Also notice that the trigger fronts, beginning with the third front pictured, move more slowly than the first. This is because their velocity is $c(v)$, where v here is the value of v at the front. Since, as we have seen, this is larger than v_0 and c decreases monotonically with v, their speed will be less.

Apparently another chemical regime results in a g nullcline intersecting the right hand branch h_+ near its top in Figure 5. A similar analysis shows a different kind of target pattern to appear, reminiscent of that observed by Smoes (1980) in which the colors are reversed.

In our model, the center has different properties from the rest of the medium ($b = b(\xi)$), and this heterogeneity accounts for the spontaneous generation of all the waves at that point. The same effect, however, is seen with a homogeneous model in which a third slowly reacting chemical, coupled with the first two, exists. Then the position of the pattern center is determined from some inhomogeneity in the initial conditions (Fife 1981).

Finally, conditions for the appearance of solitary waves (pulses) can easily be elucidated within the framework of our model, this time homogeneous with b = const. Suppose we are in the excitable regime ($b > 1$), so that a rest point (u_0, v_0) exists as in Figure 5. Again, take initial data $v(\xi,0) = v_0$, but this time $s(\xi,0) = +$ for $\xi < 0$, and $= -$ for $\xi > 0$. This evolution is pictured in Figure 8.

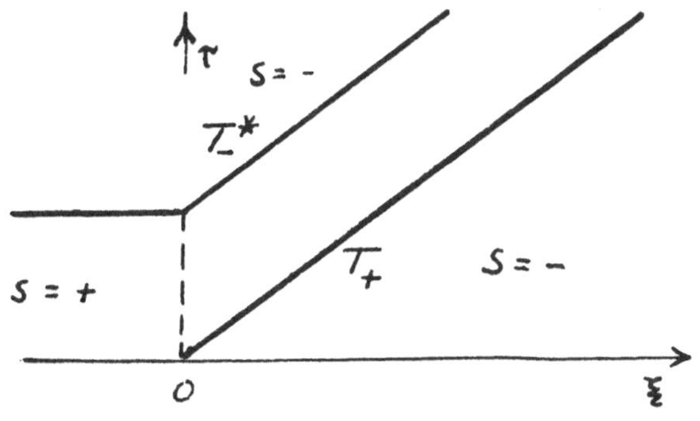

Figure 8

The trigger front T_+ has constant velocity $c(v_0)$. It is followed by a parallel phase front T^*_- for the same reason as before. But beyond that, there is no new front generated because there is no new curve T^*. In fact, the presence of the rest point on the left branch prevents v from ever reaching $\underline{v}$. There is a horizontal continuation of T^*_- to the left into the region $\xi < 0$.

7. The generation of Spirals

These striking figures (also called "rotors" by Winfree (1973, 1978)) are seen under much the same conditions as targets are seen. Unlike targets, their models must be two-dimensional, as no one-dimensional analog apparently exists (Jahno (1975) proposed one, but it proved to be very unstable structurally (Fife 1980)).

The framework in sec. 4 can readily be extended to two space dimensions. The regions $D_\pm$ are now regions in (ξ_1, ξ_2, τ)-space; fronts are now represented by surfaces in that space separating the regions. This applies to both T^* and T. Their velocity c now refers to velocity in the normal direction.

Consider now a solitary wave of the type constructed at the end of the last section. But now imbed it in a space with one higher spatial dimension. The curves T_+ and T^*_- now become oblique planes parallel to the ξ_2-axis. At one instant of time, their appearance in the (ξ_1, ξ_2) plane is shown in Figure 9. Their direction of motion is also indicated there.

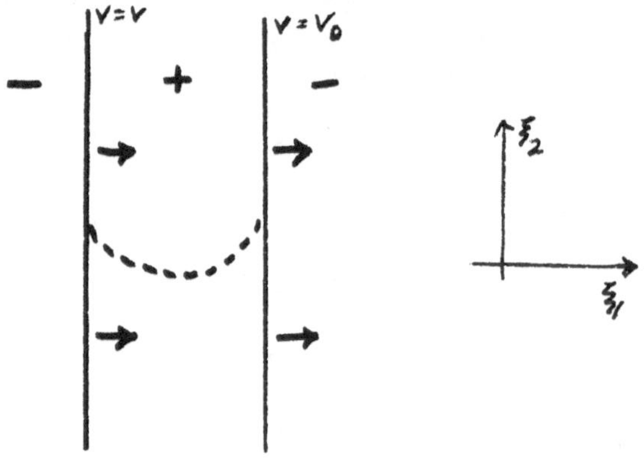

Figure 9

We now perturb this configuration by moving the variable u, in the portion of the "+" strip in Figure 9 below the dotted line, to the left toward the branch h_- (Figure 5). The system rapidly equilibrates to the state "−" in that perturbed region, resulting in the configuration shown in Figure 10. On the straight portions of the outline of the front, v remains equal to $\bar{v}$ and v_0, as in Figure 9. As we follow along the curved portion, however, the value of v on the front decreases smoothly from $\bar{v}$ to v_0. At some intermediate point (shown), v equals the value v^* for which $c(v^*) = \int_{h_-}^{h_+} f(u,v^*)du = 0$. This is the picture at time $\tau = 0$. The initial velocity of the front at each point is indicated by arrows.

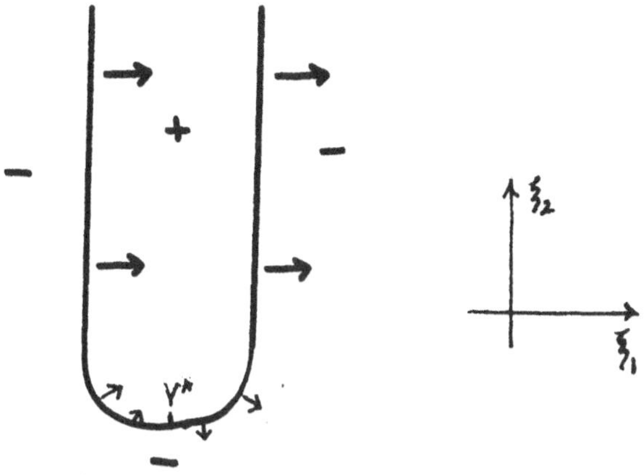

Figure 10

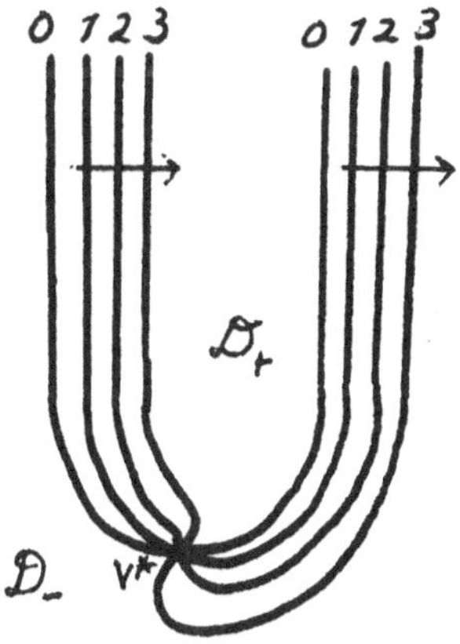

Figure 11

After a small increment in time has passed, we can imagine that the front will have moved a small distance in the direction of the arrows in Figure 10, resulting in a slightly distorted curve marked with "1" in Figure 11. During this small time increment, we have to ask what has happened in the meantime to the function v. In the region D_+ it increases, and in region D_- it either remains at v_0 (if already there) or else decreases toward that value. The latter is the case near the motionless point where $v = v^*$ (but in D_-). These two opposite tendencies in D_+ and D_- have a couple of striking effects near the point where $v = v^*$. First of all, they tend to "pull v apart", creating a discontinuity or steep gradient in the function v. Secondly, as the front on the left of that point moves into D_+ where v has increased, its normal speed, being $c(v)$, also increases. This accelerates the motion of the front in that small area, creating further distortion as indicated by the subsequent curves "2", "3",... Carried to its logical extreme, this type of analysis would predict that the front would have unbounded angular velocity and unbounded curvature as one approaches that critical point. This singular behavior is inconsistent with assumptions we used in constructing the model in the first place. For example, we neglect the diffusion of v on the basis that v would be fairly smooth. And large curvatures certainly cast grave doubt on the adequacy of a model which has normal velocity a function only of v at a given point, independent of the curvature. In short, any analysis of the subsequent development of the spiral pattern near its center must employ a different rescaling of the original problem near that point.

Nevertheless, our qualitative analysis can certainly be continued in regions away from the core. Doing so, it is seen that although the straight parts of the edges of the domain D_+ continue moving to the right with constant speed, the lower part develops a pronounced asymmetry with the bulge on the right eventually twisting and moving to the left. What we are left with is a curved tip, as in Figure 12. Of course this process continues, with a full spiral eventually developing. Keener (1980) gave a description of developed spirals outside the core in terms of frontal at dynamics.

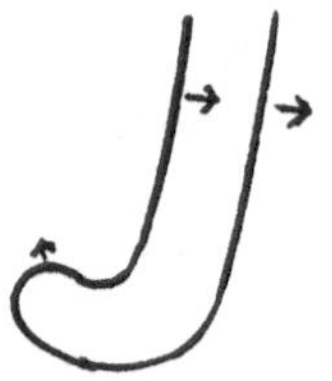

Figure 12

It is beyond my scope to survey the numerical and analytical investigations of spirals in reactive media, but there have been many. In view of its being based on a sound model of the chemistry of the BZ reaction and its marked success as applied to the target pattern problem, the framework presented here would seem worthwhile to pursue. Although the analysis of this model is far from complete, the argument presented gives a clear picture of a mechanism for the generation of spiral patterns.

One comment which should be made is that if we define the vertex of the developing rotor as the position on the front where $v = v^*$, then it does not follow that this point remains stationary. In fact, meandering has been observed, and Rossler and Kahlert (1979) attempt to link this meandering with instabilities at the vertex associated with singularities like those mentioned above. It may turn out, then, that the dynamics near the center of the spiral is essentially unstable in some sense.

8. Compound layers and stationary solutions

Let us perform another thought experiment. Take (4.1,2) with small diffusion in v:

$$u_t = \varepsilon u_{\xi\xi} + \frac{1}{\varepsilon} f(u,v) \tag{8.1}$$

$$v_\tau = \varepsilon v_{\xi\xi} + g(u,v) \tag{8.2}$$

We choose f to be as in Figure 5 or 13, and the nullcline of g as shown in Figure 13. Thus a stable critical point (u_0,v_0) is as shown.

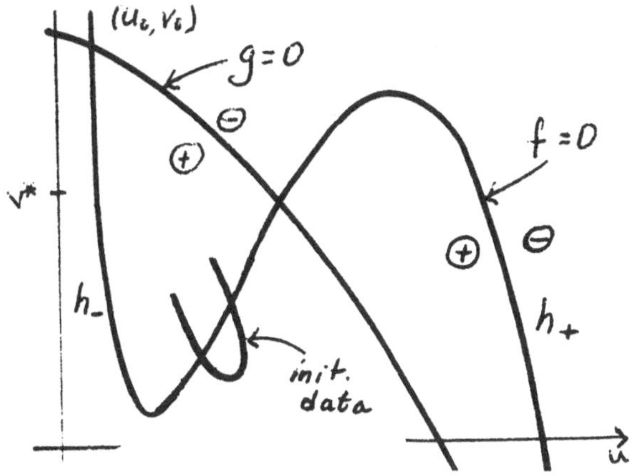

Figure 13

We impose initial conditions on our system as follows: $v(\xi,0) = v_0(\xi)$, $u(\xi,0) = u_0(\xi)$, where the (u,v)-plane trace of the curve $(u_0(\xi),v_0(\xi))$ is pictured in Figure 13. Only the middle portion, say for ξ_1, $< \xi < \xi_2$ lies below the curve $f = 0$. then the smallness of ε means the system rapidly semi-equilibrates by u moving to one or another of the branches $h_{\pm}$. After this rapid semi-equilibration, we are in the framework of the evolution problem (4.3,4). But there will be two trigger fronts to start off with: one beginning at $y(0) = \xi_1$ and the other at $y(0) = \xi_2$ (See Figure 14). Consider the one on the right Initially, + is the dominant state at that front (it is at the other one also), and $s = +$ to its left, so it moves to the right. As it so moves, the value of v at the front increases for two

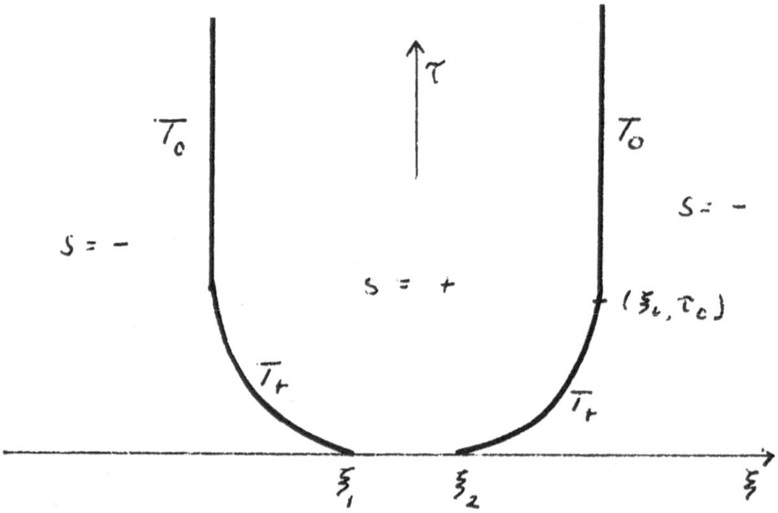

Figure 14

reasons: the initial conditions show $v_0(\xi)$ increasing as ξ increases beyond the right-hand front; and also in that region v increases with τ because $g > 0$ there. The effect of v increasing along the front's trajectory is to decelerate the front, as c is a decreasing function of v. Eventually, we reach (in finite time!) a point where $c = 0$, and the analysis changes at that point (call it ξ_0, τ_0). Here we follow the procedure outlined in sec. 4. Call the trajectory just constructed T_+. Above it v follows the dynamics of (4.3) with the + sign chosen, so decreases in time. Therefore c is always positive in that region. On the other hand, for $\xi > \xi_o$ and $\tau > \tau_o$, v follows the other dynamics (- sign) and so c is negative there. We are therefore in case (c) of sec. 4, and our T_+ is converted into a T_o. The front remains stationary from that time forward. a similar thing happens to the front issuing originally from ξ_1. So a configuration is reached in which two stationary fronts of type T_o separate space into three regions, the middle one being in state + and the outer ones in state -. In the middle region, v continues to decrease in time; and in the outer regions, it increases and approaches a limit v_o.

This gives the gross structure of the solution; this structure was obtained without reference to the internal structure near the singular line T_o, just as we were able to trace the developing spiral, in its gross outline, in the preceding section without investigating its fine structure near the center. But in the present case, it is not too difficult to give a satisfactory analysis of the inner region, near T_o. It is composed of a "layer within a layer": a narrow transition layer associated with the scale change $\xi - \xi_o = \varepsilon x$, in which v is approximately equal to the value v^* at which $c(v^*) = 0$, imbedded within a larger transition layer appropriate to the scale change $\xi - \xi_o = \sqrt{\varepsilon}\, x'$, in which $u = h_\pm(v)$ and v satisfies a nonlinear diffusion equation with discontinuous nonlinearity. I'll not give the details.

Our second thought experiment will involve only one change: in (8.2), we allow v to have greater diffusion, replacing $\varepsilon v_{\xi\xi}$ by $v_{\xi\xi}$. Then of course the front dynamics problem described in sec. 4 becomes a more difficult free boundary problem. I shall be content with examining the possible final configurations attained by this evolution problem. In contrast to the above case, stationary final states are possible. The stationary problem (neglecting the fine structure of internal layers, in the spirit of sec. 4) is the following. (For simplicity we have assumed the initial data to be even in ξ; this evenness is then preserved in the stationary problem.)

Stat. Prob.: Find $\xi_0 > 0$ and $v(\xi)$, even in ξ, such that

$$v'' + g(h_+(v),v) = 0, \quad |\xi| < \xi_0,$$

$$v'' + g(h_-(v),v) = 0, \quad |\xi| > \xi_0,$$

$$v(\xi_0) = v^*.$$

As a reasonable sample case, conforming to Figure 13, we set

$$g(h_-(v),v) = \alpha(v_0 - v^*),$$

$$g(h_+(v),v) = -\beta < 0,$$

Then the problem has a unique solution (Figure 15)

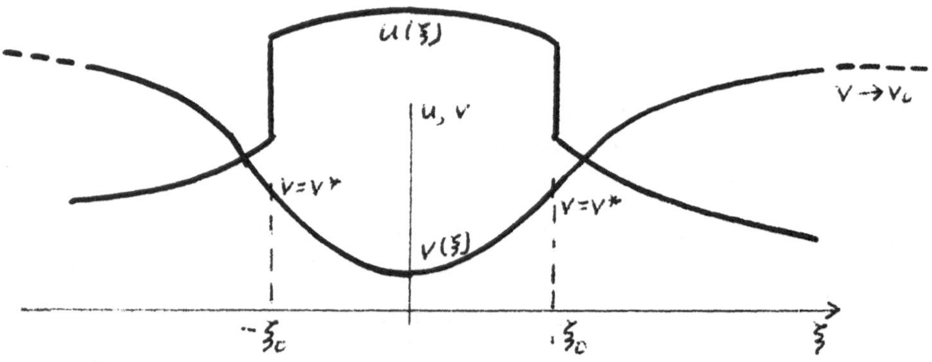

Figure 15

$$\xi_0 = (v_0 - v^*)\sqrt{\alpha}/\beta \ ,$$

$$v = v^* + \frac{1}{2}\beta(\xi^2 - \xi_0^2), \quad |\xi| < \xi_0,$$

$$v = v_0 - (v_0 - v^*)\exp[-\sqrt{\alpha}\ (\xi - \xi_0)], \quad |\xi| > \xi_0,$$

Structures like this were described by Koga and Kuramoto (1980) who also reported numerical results showing that the structure can oscillate in certain parameter regions. This intriguing phenomenon bears further investigation.

A procedure for investigating the stability or "realizability" of similar structures was given in Fife (1976a). A completely different approach to the problem of obtaining stable large amplitude stationary patterns was given in Fife (1977).

Stationary solutions of reaction-diffusion equations were proposed by Gierer and Meinhardt (1972) as a mechanism of providing for morphogen distributions in developing organisms. In these models, they proposed a faster-diffusing substance (v or (-v) in our case) to act as an inhibitor, the other one being an activator. These characterizations do not exactly fit, in the case of our model.

II. SMALL AMPLITUDE STRUCTURES

Bifurcation analysis is the most traditional approach to the discovery of structured solutions of reaction-diffusion equations, and as such needs no detailed explanation here. Suffice it to say that the theory is usually applied in the situation when the problem at hand has adjustable parameters and a constant (in space-time) solution for all reasonable values of these parameters. This constant solution may or may not be stable, and its stability is easily tested, except in borderline cases. Again excepting marginal cases, parameter space can be divided into two regions, in which this trivial solution is either stable or unstable. For parameter values near the boundary between these two regions (sometimes necessarily confined to one side or the other of it), there typically exist other solutions which are close to the constant ones. These others may be structured in space and/or time. Since they are close to being constant, we call them small-amplitude solutions. Bifurcation theory is useful in showing their existence, approximate properties, and stability or instability.

The most vexing and interesting problems arise when the spatial domain of the solutions is all space, and we shall be concerned with this case in sec. 9. The object there is to point out a research area which needs further development. Sec. 10, finally, will be a brief report on recent work drawing a connection between the two seemingly disparate approaches considered in these talks: the singular perturbation approach of Part I, and the bifurcation approach.

9. Small wave trains and associated solutions

For reaction-diffusion problems in all space, bifurcation phenomena can in general be categorized according to whether the new solutions are oscillatory or not in time, and whether their spatial structure has finite or infinite characteristic length (Fife (1978), Ermentrout (1981a,b)). Since there are no spatial boundaries present, these properties are not affected by geometry; they depend merely on the equations.

For such bifurcating solutions, it has been shown that other small structures, varying slowly in space-time and enveloping the original structures, can sometimes appear. The equation governing these large-scale envelops is

$$\frac{\partial z}{\partial t} = K \frac{\partial^2 z}{\partial \xi^2} + \alpha z + \beta z|z|^2, \tag{9.1}$$

and its generalization in higher space dimensions, where $z = z_1 + iz_2$ is a complex-valued function of (ξ, τ), and ξ and τ are large-scaled versions of space and time. The parameters K, α, β are in general complex. This equation has, in the literature, been variously called the (possibly complex) Ginsburg-Landau equation, the Newell-Whitehead equation (Newell and Whitehead (1969) and the $\lambda - \omega$ system (Kopell and Howard 1979 and earlier papers) (although the latter name refers to the case when K is real: and then a more general nonlinearity is allowed). It is universal in the sense that it can be derived by a formal two-timing analysis of bifurcating solutions for quite general, perhaps large order, systems. In the context of reaction-diffusion systems, see examples of its derivation, for various cases, in Kuramoto and Tsuzuki (1975), Kuramoto (1984), Cohen, Hoppensteadt and Miura (1977), Fife (1979). For a more general context, see Newell (1974). Of course the coefficients K, α, β depend on the original system, but the form of the equation (9.1) does not. The meaning of the function z is that it, in a way which will be made clearer later, represents possible amplitude modulations of the more finely structured bifurcating solutions. These modulated structures are then new small solutions of the original problem. Besides its role in constructing the new solutions, (9.1) sometimes sheds light on which of the many unmodulated bifurcating solutions are stable.

Although derivations similar to this are available here and there in the literature, I shall derive the envelop equation in the case of a general system for which the bifurcating solutions are oscillatory and have a finite characteristic spatial length (this can only occur if the system has at least four equations; the more usual case considered is when the characteristic length is infinite; this case is when the kinetics undergo a Hopf bifurcation and no spatially structured mode is unstable at the bifurcation point.)

This derivation is akin to that in Newell (1974). I shall take pains to interpret the coefficients K and α, as far as possible, in terms of the properties of the eigenvalue of an associated matrix which governs the stability ofthe basic constant solution. Our equation is real, has one parameter λ, and the spatial domain is $\mathbb{R}^2$:

$$u_t = D \Delta u + f(u,\lambda), \quad u \in \mathbb{R}^n. \tag{9.2}$$

Here D is a positive definite matrix. We assume that $f(0,\lambda) \equiv 0$, so $u \equiv 0$ is always a solution. The linear, quadratic, and cubic parts of f are symbolized as follows:

$$f(u,\lambda) = (A + \lambda B)u + Q(u,u) + C(u,u,u) + \text{higher order terms}.$$

Here A and B are real $n \times n$ matrices, and $Q(u,v)$ and $C(u,v,w)$ are real bilinear and trilinear forms. With no loss of generality, we assume Q and C are each symmetric (invariant when any two of the arguments are interchanged). Thus,

$$u_t = D \Delta u + (\Delta + \lambda B)u + Q(u,u) + C(u,u,u) + \ldots \tag{9.3}$$

Our basic assumption is that the solution $u \equiv 0$ loses its stability as λ increases from negative to positive values. The stability is traditionally tested by looking for solutions

$$u = \phi\, e^{i\underline{k}\cdot\underline{x} + \mu t}.$$

of the linear problem $(Q = C = 0)$ with $R_e\, \mu > 0$; if there is one, the 0 solution is unstable. Here $\underline{k} = (k_1,k_2)$ is a real vector, and $\underline{x} = (x_1,x_2)$ are the spatial coordinates. Clearly μ is an eigenvalue, and ϕ an eigenvector, of the matrix

$$H(p,\lambda) \equiv -pD + \Delta + \lambda B,$$

where $p = |k|^2$. We assume there exists a principal eigenvalue $\mu(p,\lambda)$ which is algebraically simple and has positive real part, for some p, only when $\lambda > 0$. We assume that μ becomes unstable at a characteristic value $p_0 > 0$. Specifically, if we set $\mu = \sigma(p,\lambda) + i\,\omega(p,\lambda)$, the assumption on σ is as in Figure 16. In accordance with that figure, we assume the following about the derivatives of μ:

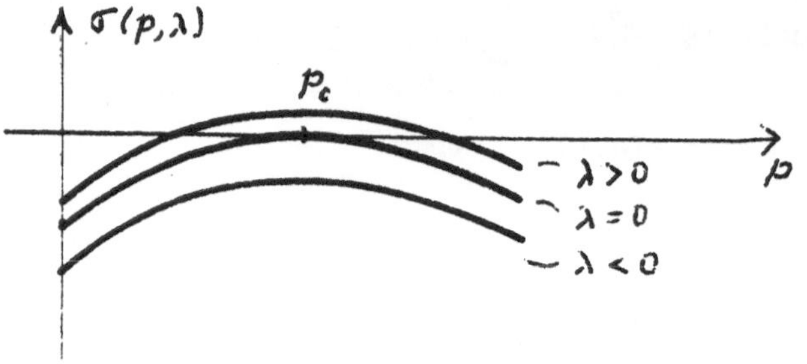

Figure 16

$$\sigma(p_o,0) = \sigma_p(p_o,0) = 0, \ \sigma_{pp}(p_o,0) < 0, \ \ \sigma_\lambda(p_o,0) > 0. \quad (9.4)$$

The statement that $p_o > 0$ implies, as we shall see, that the characteristic spatial length of the bifurcating solutions is finite. We also assume that $\omega_o \equiv \omega(p_o,0) \neq 0$; this will imply they are oscillatory.

We collect some facts about the matrix H for future reference. The simplicity of $\mu(p,\lambda)$ implies, first of all, that μ depends smoothly on p and λ and that there exists an eigenvector $\phi(p,\lambda)$ with the same smooth dependence, normalized so that

$$|\phi(p,\lambda)|^2 = 1 \quad (9.5)$$

(We use the usual complex scalar product in $\mathbb{R}^n$:

$$\langle\phi,\psi\rangle = \sum_i^n \phi_i\bar{\Psi}_i; \ |\phi|^2 = \langle\phi,\phi\rangle.)$$

Secondly, if we set $H_1 \equiv H(p_o,0) - i\omega_o I$, and understand by the symbol ϕ the vector $\phi(p_o,0)$, then the equation

$$H_1 q = h \in \mathbb{R}^n \tag{9.6}$$

has a solution if and only if $\langle h_1,\psi\rangle = 0$, where ψ is an eigenvalue of the adjoint problem:

$$H_1^* \psi = 0.$$

Finally, the algebraic simplicity of μ means there exists no solution of $H_1 q = \phi$, so necessarily $\langle\phi,\psi\rangle \neq 0$. By multiplication of ψ by a scalar, we may ensure that

$$\langle\phi,\psi\rangle = 1.$$

If the orthogonality condition is satisfied, the solution of (9.6) is determined only up to an additive multiple of ϕ. We denote by

$$q = H_1^{-1} h$$

the particular solution which satisfies $\langle q,\phi\rangle = 0$.

Differentiating

$$H(p,\lambda)\phi(p,\lambda) = \mu(p,\lambda)\phi(p,\lambda),$$

we obtain

$$H\phi_p - \mu\phi_p = D\phi + \mu_p\phi, \tag{9.7}$$

$$H\phi_{pp} - \mu\phi_{pp} = 2(D + \mu_p)\phi_p + \mu_{pp}\phi, \tag{9.8}$$

$$H\phi_\lambda - \mu\phi_\lambda = -B\phi + \mu_\lambda\phi \, . \tag{9.9}$$

Also differentiating (9.5), we get

$$\langle \phi, \phi_p \rangle = 0. \tag{9.10}$$

The right sides of (9.7-9) must all be orthogonal to ψ when $p = p_0$, $\lambda = 0$. Using this, (9.4), and (9.10) yields the following, where we assume $p = p_0$, $\lambda = 0$:

$$\langle D\phi, \psi \rangle = -i\omega_p$$

$$\phi_p = H_1^{-1}(D\phi + i\omega_p \phi), \tag{9.11}$$

$$2\langle (D + i\omega_p)H_1^{-1}(D + i\omega_p)\phi, \psi \rangle = -\mu_{pp} \tag{9.12}$$

$$\langle B\phi, \psi \rangle = \mu_\lambda . \tag{9.13}$$

We now proceed to find small solutions of (9.3) by formal expansions. We assume $\lambda > 0$ is small, and define

$$\varepsilon = \sqrt{\lambda} . \tag{9.14}$$

We seek solutions in the form

$$u = \varepsilon u^1 + \varepsilon^2 u^2 + \dots \tag{9.15}$$

We suppose them to be functions not only of x and t, but also of the scaled variables $\xi = \varepsilon x$, $s = \varepsilon t$, and $\tau = \varepsilon^2 t$. This means that in (9.3), we replace the differential operator ∂_t by $\partial_t + \varepsilon\partial_s + \varepsilon^2\partial_\tau$, and the (vector) gradient ∂_x by $\partial_x + \varepsilon\partial_\xi$, so

$$\Delta u = \sum_1^2 \left(\partial_{x_i} + \varepsilon\partial_{\xi_i}\right)^2 u = \Delta_x u + 2\varepsilon\partial_x \cdot \partial_\xi u + \varepsilon^2 \Delta_\xi u,$$

and $L \equiv \partial_t - D\Delta - (A + \varepsilon^2 B)$ can be written $L = L^0 + \varepsilon L^1 + \varepsilon^2 L^2$, where

$$L^0 = \partial_t - D\Delta_x - A, \tag{9.16a}$$

$$L^1 = \partial_s - 2D\partial_x \cdot \partial_\xi, \tag{9.16b}$$

$$L^2 = \partial_\tau - D\Delta_\xi - B. \tag{9.16c}$$

Putting (9.14-16) into (9.3) and equating coefficients of each power of ε, we obtain a sequence of equations

$$L^0 u^1 = 0 \quad (9.17)$$

$$L^0 u^2 = -L^1 u^1 + Q(u^1, u^1), \tag{9.18}$$

$$L^0 u^3 = -L^1 u^2 - L^2 u^1 + 2Q(u^1, u^2) + C(u^1, u^1, u^1). \tag{9.19}$$

These equations ((9.17), for instance) will have a great number of solutions. We should agree at this point to limit consideration only to solutions which (i) are bounded (to be compatible with the assumed validity of the expansion (9.15)) and (ii) are not transient: do not decay as $t \to \infty$. Neglecting transients simply means we are looking for permanent strucures; limiting the search this way makes things much easier.

Fourier analysis in x applied to (9.17) gives the following as a complete set of solutions satisfying these two criteria:

$$\{\phi\, e^{\pm i\,(\underline{k} \cdot \underline{x} + w_0 t)} : |\underline{k}|^2 = p_0\},$$

where $\phi = \phi(p_0, o)$ and ω_0 were defined earlier. Therefore the general solution of (9.17) takes the form

$$u^1 = \int_{|\underline{k}|^2 = p_0} [e^{i(\underline{k} \cdot \underline{x} + \omega_0 t)} da(\underline{k}) + e^{-i(\underline{k} \cdot \underline{x} + \omega_0 t)} db(\underline{k})],$$

where the measures a and b can also depend arbitrarily on ξ and τ.

It is quite tedious to push through the calculations allowing this much generality in the form for u^1. Therefore we impose another requirement in our search for structured solutions: it is that the lowest order form for the solution, namely u^1, should be real and take the form of a unidirectional traveling wave in the variables x and t. taking the direction to be that of the x_1-axis, we find that necessarily u^1 simplifies to:

$$u^1 = a(\xi,\tau)e^{i(k_0 x_1 + \omega_0 t)} \sim \qquad (k_o = \sqrt{p_o} \ . \qquad (9.20)$$

Here we use the notation "~" to denote the statement "plus the complex conjugate of the preceding term". We could equally well use the symbol for "real part of", but it will be essential in the calculations below to write u^1 as the sum of two terms.

This requirement on the form of u^1 is simply a requirement on the type of fine structure that we wish our solutions to have. Many other types of structures, such as that of standing waves, could also be handled.

For convenience, we use the notation $\theta \equiv k_0 x_1 + w_0 t$.

From (9.16b,20) we find

$$L^1 u^1 = (\partial_s - 2iD\underline{k}\cdot\partial_\xi)a\phi e^{i\theta}\sim = (\partial_s - 2iDk_o\partial_{\xi_1})a\phi e^{i\theta} \sim,$$

and

$$Q(u^1,u^1) = a^2 Q(\phi,\phi)e^{2i\theta}\sim + 2a\bar{a}\, Q(\phi,\bar{\phi}).$$

Therefore the right side of (9.18) assumes the form of a quadratic polynomial in the variables $e^{i\theta}$ and $e^{-i\theta}$.

Here we digress to determine solvability conditions for equations of the form

$$L^o u = Ve^{im\theta} \qquad (m \text{ an integer}, \ V\in \mathbb{C}^n).$$

It is easily checked that the only bounded nontransient solutions u are $H_m^{-1} V e^{im\theta}$ plus the general solution of (9.17) given above. In this expression we use $H_m \equiv H(m^2 k_o^2, 0) - im\omega_o I$. By our original assumption (Figure 16), this is invertible for $m \neq \pm 1$, and for $m = 1$, is invertible on the orthocomplement of ψ. It can be shown in the same way that for $m = -1$, H_m is invertible on the orthocomplement of $\bar{\psi}$. From all this, it follows that an equation of the form

$$L^o u = V^o + V^1 e^{i\theta} \sim + V^2 e^{2i\theta} \sim + \ldots\ldots$$

is solvable in the set of functions we are using if and only if $\langle V^1, \psi\rangle = 0$; and if this condition is satisfied, the solution is

$$u = H_o^{-1} V^o \sim + H_1^{-1} V^1 e^{i\theta} \sim + H_2^{-1} V^2 e^{2i\theta} \sim + \cdots +$$

(gen. sol'n. of (9.17)).

Applying this to (9.18), we find the criterion to be

$$\langle(\partial_s - 2iDk_o\partial_{\xi_1})a\phi, \psi\rangle = 0$$

But in view of (9.11) and $\langle\phi,\psi\rangle = 1$, the condition reduces to

$$a_s - 2k\omega_p a_{\xi_1} = 0 \tag{9.21}$$

Note that (9.21) means that when applied to a or to a function of a, the differential operator ∂_s may be replaced by $2k_o\omega_p\partial_{\xi_1}$. We shall make use of this fact. In particular, we have $(\partial_s - 2iDk_o\partial_{\xi_1}) = -2ik(D + i\omega_p)\partial_{\xi_1}$, hence

$$L^1 u^1 = -2ik\, a_{\xi_1}(D + i\omega_p)\phi e^{i\theta} \sim . \tag{9.22}$$

The solvability condition (9.21) being satisfied, we may obtain the solution u^2 of (9.18) by use of (9.22):

$$u^2 = -2ika_{\xi_1} H_1^{-1}(D+i\omega_p)\phi e^{i\theta} \sim + a^2 H_2^{-1} Q(\phi,\phi) e^{2i\theta} \sim + \tag{9.23}$$
$$+ 2a\bar{a} H_0^{-1} Q(\phi,\bar{\phi}) + u_g .$$

where u_g is a solution of (9.17).

The next step is to put (9.20) and (9.23) into (9.19) and apoply the orthogonality condition once more. Of course the condition relates only to the terms on the right of (9.19) involving $e^{\pm i\theta}$. The only place where u_g plays a role in this condition is in the first term $-L^1u^2$ on the right of (9.19). And, as before, the condition's only implication for u_g is an analog of (9.21)--which merely serves to specify how u_g depends on s. With this fact duly noted, we may now disregard u_g in discovering the other implications of the orthogonality condition. Doing so, we calculate

$$L^1u^2 = -4k^2\partial^2_{\xi_1} a(D+i\omega_p)H^{-1}_1(D+i\omega_p)\phi e^{i\theta}\sim + \text{ terms}$$

$$\text{in } e^{\pm 2i\theta} + \text{ terms independent of } \theta;$$

$$L^2u^1 = (\partial_\tau a\phi - \Delta_\xi aD\phi - aB\phi)e^{i\theta}\sim;$$

$$Q(u^1,u^2) = Q(a\phi, 2a\bar{a}H^{-1}_o Q(\phi,\bar{\phi}))e^{i\theta}\sim + Q(\bar{a}\bar{\phi}, a^2H^{-1}_2 Q(\phi,\phi))e^{i\theta}\sim$$

$$+ \text{ terms in } e^{im\theta},\ m \neq \pm 1;$$

$$C(u^1,u^1,u^1) = a^2\bar{a}[C(\phi,\phi,\bar{\phi}) + C(\phi,\bar{\phi},\phi) + C(\bar{\phi},\phi,\phi)]e^{i\theta}\sim$$

$$= 3a^2\bar{a}\, C(\phi,\phi,\bar{\phi})e^{i\theta}\sim + \text{ terms in } e^{im\theta},\ m \neq \pm 1.$$

Now applying the orthogonality condition to the right side of (9.19) yields (using (9.13))

$$4k^2_o\partial^2_{\xi_1} a\langle(D+i\omega_p)H^{-1}_1(D+i\omega_p)\phi,\psi\rangle - [\partial_\tau + i\omega_p\Delta - \mu_\lambda]a +$$

$$4a^2\bar{a}\langle Q(\phi,H^{-1}_o Q(\phi,\bar{\phi})],\psi\rangle + 2a^2\bar{a}\langle Q(\bar{\phi},H^{-1}_2 Q(\phi,\phi)),\psi\rangle +$$

$$3a^2\bar{a}\langle C(\phi,\phi,\bar{\phi}),\psi\rangle = 0,$$

which can be written in the form (recall (9.12))

$$\partial_\tau a = -k_o^2 \mu_{pp} \partial_{\xi_1}^2 a - i\omega_p \Delta a + \mu_\lambda a + a^2 \bar{a} \beta,$$

where

$$\beta \equiv 4\langle Q(\phi, H_o^{-1} Q(\phi,\bar{\phi})),\psi\rangle + 2\langle Q(\bar{\phi}, H_2^{-1} Q(\phi,\phi)),\psi\rangle$$
$$+ 3\langle C(\phi,\phi,\bar{\phi}),\psi\rangle.$$

This is the generalization of (9.1) in the present case. Notice that in this equation, the second order spatial differential operator is a sum of second derivatives without mixed derivatives, with coefficients which are purely imaginary, except for the term $\partial_{\xi_1}^2 a$, whose coefficient has a positive real part $-k^2 \sigma_{pp}$. The most obvious but least exotic solutions are the plane waves $a = R\exp(i(\kappa \cdot \xi + \Omega\tau))$ and conditions relating R, κ and Ω which provide the existence of such solutions can be seen from the equation.

The equation (9.1) in its various forms has been used as a vehicle for the study of target patterns, rotating spirals, and chaotic solutions of reaction-diffusion systems. For example, see Cohen, Neu and Rosales (1978), Greenberg (1978,80), Hagan (1980), Kopell and Howard (1981 and previous papers), and Kuramoto (1984 and previous papers); the last mentioned is a review of the subject.

Nevertheless, much work of a basic nature needs to be done on (9.1), particularly in the cases when K is complex. In this connection, I should mention that very recently a mathematician by the name of Chang in the Institute of Applied Mathematics, Beijing, has given a global existence theory for the initial value problem for equations like (9.1) (unpublished manuscript).

10. Piecing together a global picture for a model problem

In this final section I wish to indicate at least the flavor of some very attractive and thorough work which has been done over the last several years on stationary solutions of

$$\begin{aligned} u_t &= d_1 u_{xx} + f(u,v); \\ v_t &= d_2 v_{xx} + g(u,v); \end{aligned} \qquad (10.1)$$

specifically, of homogeneous Neumann problems on a finite interval, $x \in [0,1]$. This work is contained in a number of papers by Fujii, Hosono, Mimura, Nishiura, and Tabata; see the references for some of them. The function f in (10.1) is assumed to have the same properties, depicted in Figure 17, as the one we were concerned with earlier (Figure 5). The function g is assumed to be negative on the left branch (h_-) in Figure 17, and positive on the other. (Compare Figure 5 with $b < 1$).

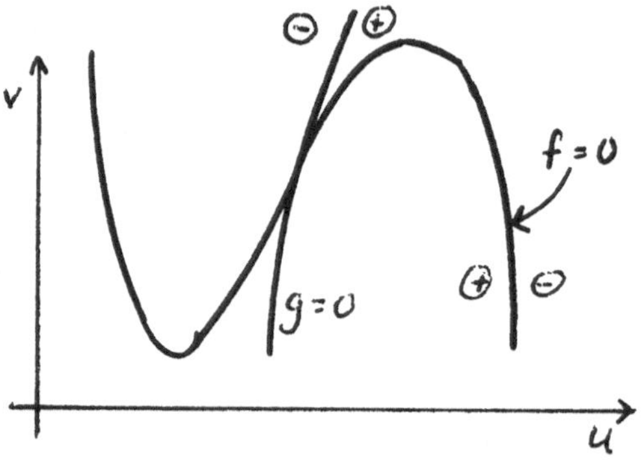

Figure 17

The goal has been a global analysis of all solution branches with respect to the two positive variable parameters d_i. By combining rigorous results about various limiting cases with computational results, an understanding of the behavior of the solution branches which, though not complete, is apparently very near to being complete, has been obtained. The most interesting features of these results are these:

(1) solutions in the parameter regime appropriate to singular perturbation analysis, namely solutions with interior and boundary layers, are shown to connect with solutions in the parameter regimes where primary bifurcations take place. A characterization of the solutions in the singular perturbation limit and near simple bifurcation points can be made according to the number of nodes of u, i.e. the minimal number of intervals on which u is monotone. This characterization permeates the global picture to some extent. The solution sheets connecting the two regimes mentioned have, in general, quite complex behavior, due to the presence of bifurcations from double eigenvalues, and secondary bifurcations.

(2) "Recovery and loss of stability" curves on the solution sheets are indicated. For example, suppose $d_2 > 0$ is fixed. As d_1 decreases, new solutions bifurcate from the constant solution. If the first bifurcation is supercritical, for example,

then the first new solution is stable, and subsequent bifurcating solutions will be unstable. If d_2 is large enough, however, then apparently these other solution branches recover their stability when d_1 becomes small enough, and keep it as $d_1 \to 0$. Results of this type appear from numerical simulation, and for d_2 very large the recovery property has been proved by Nishiura (1983). When d_2 is smaller, however, it may happen tht solution branches recover stability and then lose it again as d_1 decreases.

There are two kinds of asymptotic analyses which are relevant to this study: (a) $d_2 > 0$ fixed, $d_1 \to 0$, and (b) $d_1 > 0$ fixed, $d_2 \to \infty$.

Type (a) involves the kind of singula perturbation arguments (but made more rigorous) that I brought out in secs. 4,5, and 8. In fact in sec. 8, a stationary solution with two layers on an infinite interval was put together using the idea of a limit problem. A theory, including existence proofs, for analogous solutions of the Dirichlet problem on a finite interval with perhaps many layers was given by Fife (1976b) and extended to the Neumann problem by Mimura, Tabata, and Hosono (1980). These solutions appear in Figure 18.

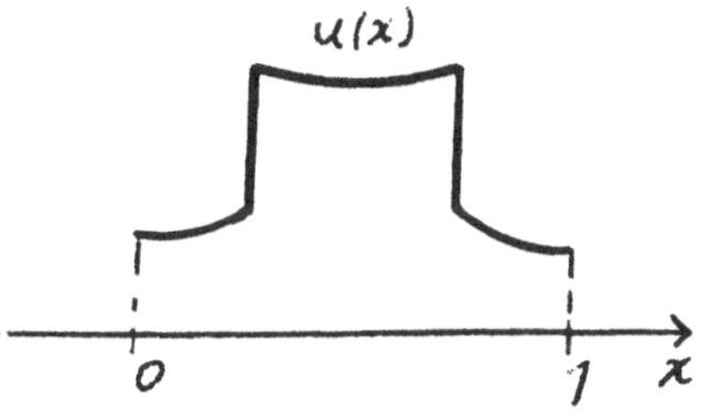

Figure 18

Type (b) involves a limit problem called the "shadow" system, formally obtained by arguing that for very large d_2, v should be approximately constant, say $v \sim \eta$. Then integrating (10.1b), one obtains

$$\int_0^1 g(u,\eta)dx = 0,$$

which, when coupled with (10.1a):

$$d_1 u_{xx} + f(u,\eta) = 0,$$

completes the definition of the shadow system. The global theory of this system is complete.

An especially intriguing singular limit is when the limits in (a) and (b) occur simultaneously; i.e. in the words of the above authors, the "singular shadow edge" is approached. Analysis remains to be done in this region.

References

D. G. Aronson and H. F. Weinberger 1978, Adv. in Math. 30, 33-76.

D. S. Cohen, F. C. Hoppensteadt, and R. M. Miura (1977), SIAM J. Appl. Math. 33, 217-229.

D. S. Cohen, J. C. Neu, and R. R. Rosales 1978, SIAM J. Appl. Math., 35, 536.

G. B. Ermentrout 1981a, Quart. Appl. Math., Apr. 1981, 61-86.
1981b, preprints on stationary homogeneous media.

R. J. Field, E. Koros, and R. M. Noyes 1972, J. Amer. Chem. Soc. 94, 8649.

P. C. Fife 1974, J. Differential Equations 15, 77-105.
1976a, J. Chem. Phys. 64, 854-864.
1976b, J. Math. Anal. and Appls. 54, 497-521.
1976c, SIAM-AMS Proceedings, Symp. on Asymptotic Methods and Singular Perturbations, New York, 23-49.
1977 J. Math. Biol. 4, 358-362.
1978, Bull. Amer. Math. Soc. 84, 693-726.
1979, pp. 143-160 of Applied Nonlinear Analysis, V. Lakshmikantham, ed., Academic Press, New York.
1980, Math. Research Center TSR No. 2110.
1981, pp. 45-56 in Analytical and Numerical Approaches to Asymptotic Problems in Analysis, Axelsson, Frank, and Van der Sluis, ed., Mathematics Studies 47, North-Holland, Amsterdam.

P. C. Fife and J. B. McLeod 1977, Arch. Rational Mech. Anal. 65, 335-361.
1981, Arch. Rational Mech. Anal. 75, 281-314.

R. A. Fisher 1937, Ann. of Eugenics 7, 355-369.

H. Fujii, M. Mimura and Y. Nishiura 1982, Physica D-Nonlinear Phenomena 5D, 1-42.

H. Fujii and Y. Nishiura 1982, Proceedings, U.S.-Japan Seminar on Nonlinear Partial Diff. Eqns. July 1982, Tokyo.

A. Gierer and H. Meinhardt 1972, Kybernetika 12, 30-39.

J. M. Greenberg 1978, SIAM J. Appl. Math. 34, 391-397.
1980 SIAM J. Appl. Math 39, 301-309.

P. Hagan 1980, manuscript on spiral waves.

V. G. Jahno 1975, Biofizika 20, 669-674.

Ya. I. Kanel' 1962, Mat. Sbornik 59, 245-288.
J. P. Keener 1980, SIAM J. Appl. Math. 39, 528-548.
S. Koga and Y. Kuramoto 1980, Prog. Theor. Phys. 63, 106-121.
N. Kopell and L. N. Howard 1973, Stud. Appl. Math. 52, 291-328.
N. Kopell and L. N. Howard 1979, Stud. Appl. Math. 64 (8), 1-56.
Y. Kuramoto 1984, Chemical Oscillations, Waves, and Turbulence, Springer Series in Synergetics #19.
Y. Kuramoto and S. Koga 1981, Prog. Theor. Phys. 66, 1081-1085.
Y. Kuramoto and T. Tsuzuki 1975, Prog. Theor. Phys. 54, 687.
M. Mimura, M. Tabata and Y. Hosono 1980, SIAM J. Math. Anal. 11, 613-631.
A. C. Newell 1974 Lectures in Appl. Mathematics, Vol. 15, Nonlinear Wave Motion, A. C. Newell, ed., Amer. Math. Soc., Providence.
A. C. Newell and J. A. Whitehead 1969, J. Fluid Mech. 38, 279.
Y. Nishiura 1982, SIAM J. Math. Anal. 13, 555-593.
1983, Proc., Workshop on modelling of patterns in space and time, Heidelberg, July, 1983.
P. Ortoleva and J. Ross 1975, J. Chem. Phys. 63, 3398-3408.
L. A. Ostrovskii and V. G. Yahno 1975, Biofizika 20, 489-493.
O. E. Rossler and C. Kahlert 1979, Z. Naturforsch. 34a, 565-570.
M.-L. Smoes 1980, pp. 80-96 in Dynamics of Synergetic Systems, H. Haken, ed., Springer-Verlag, Berlin.
J. Smoller 1983, Shock Waves and Reaction-Diffusion Equations, Springer-Verlag, Berlin.
J. Tyson 1979, Ann. N. Y. Acad. Sci. 36, 279-295.
J. Tyson 1982, preprint on reducing the F-K-N mechanism.
J. Tyson and P. C. Fife 1980, J. Chem. Phys. 73, 2224-2237.
A. T. Winfree 1972, Science 175, 634-636.
1978, Theor. Chem. Vol. 4, Academic Press, New York, pp. 1-51.
A. N. Zaikin and A. M. Zhabotinsky 1970, Nature 225, 535-537.

DISCRETE NONLINEAR DYNAMICS

Siegfried Grossmann

Fachbereich Physik, Philipps-Universität
Renthof 6, D-3550 Marburg, F.R.G.

1. INTRODUCTION

If external stress is applied to macroscopic systems they leave the equilibrium state and develop new orderings. One observes symmetry breaking with respect to space as well as to time. In this latter case the system either oscillates or shows even more complicated temporal evolution, characterized by the notion of chaos. It is this irregular behavior for which discrete nonlinear dynamics is a useful discipline. Although a very simplified description of realistic systems it retains one of their most essential features, namely, their nonlinearity. Nevertheless, it allows to calculate many properties analytically (and not only numerically). These lectures will deal with several aspects, methods, and results of discrete nonlinear dynamics which partly have been found recently in many experiments with dissipative systems.

2. WHY DISCRETE DYNAMICS?

If a parameter, say a, which measures the distance from equilibrium, caused by changes in the boundary conditions (b.c.) or by flows through an open system, is increased, the physical system only hesitatingly develops new macroscopic degrees of freedom, also called order parameters x_α, $\alpha = 1,2,\ldots,d$. If $d = 1$, the only possible long time behavior is $x_1 = \text{const}$ (or ∞), describing a *steady state.* - For d = 2, besides the steady state another dynamical quality may appear, a *limit cycle.* The Fourier spectrum then shows peaks at the inverse cycle period and, depending on the nonlinearity in the equation of motion, higher harmonics. - If the phase space has d = 3 dimensions *chaotic motion* may show up as another dynamical quality.

Chaos is described as follows. $x_\alpha(t)$ is

i) permanently t-dependent, despite of static b.c.;
ii) bounded;
iii) nonperiodic, irregular;
iv) attracted by strange subsets in phase space;
v) sensitively depending on the initial conditions, so
vi) unpredictable in the long run although determined by a set of coupled o.d.e.'s;
vii) the Fourier spectrum $S(\omega)$ is broad (with or without additional sharp peaks);
viii) correlations decay.

To understand this selfgenerated quasi-stochastic, quasi-noisy behavior the nonlinearity in the equations of motion

$$\dot{x}_\alpha = f_\alpha(x,a) \quad , \quad \alpha = 1,2,\ldots,d \tag{1}$$

is essential. Analytic solutions are not possible in general. By a stroboscop-like discretization in time one considers discrete trajectories $x_\alpha(t_0)$, $x_\alpha(t_0+\Delta t)$,... instead of continuous ones, $x_\alpha(t)$. If the discretization is performed in accordance with the system's dynamics by a Lorenz or Poincaré mapping, it is sufficient to consider anyone of the variables, say x, at successive times τ, starting with some initial value: x_0, x_1, $x_2,\ldots,x_\tau,\ldots$. The deterministic character is represented by a discrete version of (1), the *dynamical law* $f_a(x)$,

$$x_{\tau+1} = f_a(x_\tau) \quad , \quad \tau = 0,1,2,\ldots \quad . \tag{2}$$

Only a few, characteristic features of $f_a(x)$ are important. Due to the boundedness of the motion $f_a(x)$ cannot be monotonous. Thus $f_a(x)$ is expected to have one (or a few) maxima and/or one (or a few) discontinuities. Several examples will be considered later. - The existence of a dynamical law $f_a(x)$ describing chaos can be expected, if the continuous motion's phase space has dimension ≥ 3 but the system's attractor being small. In any case, (2) is only an approximation to the correct equations of motion, since the $f_a(x)$ of interest are non-invertible while (1) is. Nevertheless, the predictions of (2) have proven not only to be qualitatively correct but to some extent even quantitatively. This is why discrete dynamics is not only a fascinating mathematical discipline (iteration theory) but of relevance for physics, chemistry, ecology,... .

If $d > 3$ also other routes to chaos are possible, which only in part are mimicked by discrete dynamics. I am aware of the following:

1)Via quasi-periodicity: D.Ruelle, F.Takens, 1971 [1]. A few bifurcations lead from steady state through periodic state, quasi-period motion to a strange attractor. - A subroute, which is possible although not "generic", is Landau-Hopf's route, 1944 [2], with ever increasing number of non-commensurate periods.

2)Via frequency-locking (or sinusoidal external driving) followed by period doubling: M.Feigenbaum, 1978 [3], P.Coullet, C.Tresser, 1978 [4], and S.Grossmann, S.Thomae, 1977 [5]. Experimental realizations are reported in sect. 3.

3)Via intermittency: Y.Pomeau, P.Manneville, 1979 [6]. A steady or oscillatory state is irregulary interrupted by bursts of irregular, chaotic motions.

There are abbreviated routes or mixtures of those, depending on the particular system together with its particular b.c.. It is period-doubling and (in part) intermittency which is successfully treated by discrete nonlinear dynamics.

3.EXPERIMENTS

The period-doubling route to chaos is characterized by a sequence of a-intervals I_n of geometrically decreasing length, $I_{n+1} = I_n/\delta$ (n not too small). Within $I_n = a_{n+1} - a_n$ the trajectory is or after a transient approaches a $p = 2^n$ periodic state $x_0, x_1, x_2, \ldots, x_{p-1}, x_p = x_0$. This period is born at a_n by slope-type bifurcation; the difference $|x_{p/2} - x_0|$ starts to grow from zero. At a_{n+1} the period looses stability in favour of twice that period. Another bifurcation happens, $|x_p - x_0|$ starts to grow from zero. The magnitudes of successive bifurcations scale by α (considering equivalent values of a in successive intervals). The numbers $\delta \cong 4.67$ and $\alpha \cong 2.50$ ([5],[3]) only depend on the order z of the $f_a(x)$-maximum but not on details of the dynamical law. In this sense they enjoy the attribute *universal* [3]. The experiments referred to below all seem to show the quoted values for δ, α, belonging to a quadratic maximum, $z = 2$. It is not yet known, whether there are definite deviations (indicating $z \neq 2$) or whether particular systems might miss the quadratic maximum in favour of

$f_a \propto (x-x_{max})^4$. If z = 1, i.e. in a broken linear transformation, the period doubling phenomenon is missing at all.

In the Fourier spectrum each new bifurcation shows up as a new *subharmonic* (together with its higher harmonics due to the nonlinearity). The spectral power in successively born subharmonics again reflects the scaling of the variable being governed by α. The mean intensity of the subharmonic of order n+1 together with its odd order harmonics is about 14 dB less than the corresponding one of order n [7]. The smallest subharmonic peaks of successive orders are scaled down by ≅ 18.4 dB [8]; more precisely, it is for $a_n \to a_\infty$

$$S(\omega)/S(2\omega) = [1-2\alpha\cos 2\pi\omega + \alpha^2]/4\alpha^4. \qquad (3)$$

Figure 1 summarizes the bifurcation sequence, fig.2 displays a sequence of spectra representative for many observations in many different systems.

Beyond the limit $a_\infty = \lim_{n\to\infty} a_n$ chaotic motion is possible, i.e. broad band spectra show up. There is an *inverse cascade* [5] of parameter values $\tilde{a}_n$ at which the n-th subharmonic (sharp for $a<\tilde{a}_n$) broadens, quickly looses intensity and disappears with increasing a. This disappearance of the subharmonics goes along with a gradual increase of the noise level. The period shrinks, the chaotic amplitude increases, both phenomena being governed by the same parameters δ and α [5]. The slowing down along the direct cascade $a_n \to a_\infty$ turns into a speeding up along the inverse cascade $a_\infty \ldots \tilde{a}_n$, $\tilde{a}_{n-1}, \ldots$. The critical exponent is

$$\omega_n \propto |a_n - a_\infty|^\nu \propto (\tilde{a}_n - a_\infty)^\nu \ , \ \nu = ln2 \ / \ ln\delta \cong 0.45 \ . \qquad (4)$$

At the $\tilde{a}_n$ pairs of chaotic bands merge, which below $\tilde{a}_n$ were

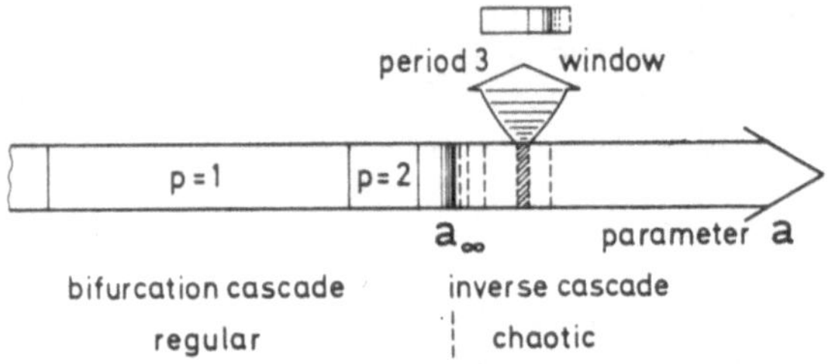

Fig.1: Bifurcation sequence (schematic). a_∞ critical attractor, onset of chaos. L.h.s direct cascade $p = 2^n$, n = 0,1,...; r.h.s. inverse cascade with example of a window showing stable periods.

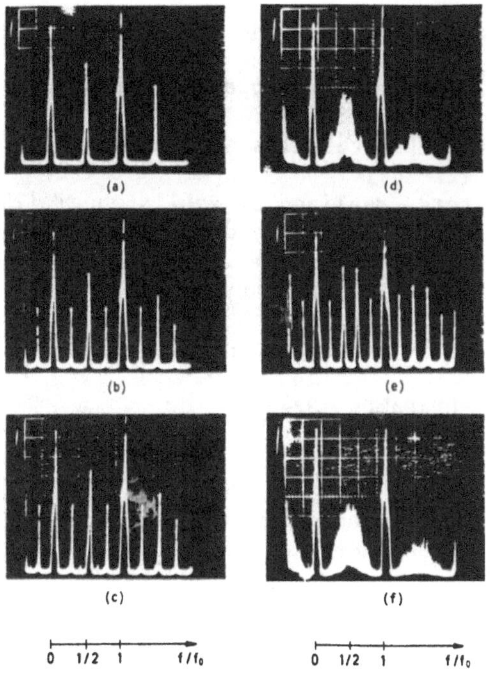

Fig.2: Spectra of a nonlinear electric oscillator which is driven with increasing voltage (Lauterborn et.al.[24]).

visited periodically, causing the corresponding subharmonic peak. Between the band merging points $\tilde{a}_n$ of the inverse cascade there are windows of regular, periodic motion in the chaos regime. Each window is characterized by some higher basic period m = 3,4,..., and shows a hierarchy of bifurcations on a reduced a-scale, for example $3, 3\cdot 2, 3\cdot 2^2, \ldots a_\infty^{(3)}, \ldots, \tilde{a}_n^{(3)}, \ldots$. The order in which these windows of stable periods show up was given by Metropolis, Stein, and Stein [9], the scaling regularity in the corresponding a-values was found by Geisel and Nierwetberg [10]. With increasing order m the windows are increasingly smaller and cannot be detected experimentally. Figure 2e shows a spectrum in the 5-window. For a method to calculate the window-width see later, section 9.

The description of the full transition region was first developed [5] for the parabolic map $f_a(x) = 4a\, x(1-x)$. Meanwhile many experiments support that it seems to be quite universally correct. Due to limitations in stabilizing a and in resolving the measured amplitudes x_τ only a limited number of bifurcations can be observed as well as only a limited number of windows along the inverse cascade. Sometimes there seem to be still unexplained irregularities in the appearance of spectral lines.

I am aware of the following large number of experiments, which show the period doubling route as well as deterministic chaos.

i) Rayleigh-Bénard heat flow in fluid layers, $a \propto \Delta T$ [11-14].

ii) Momentum flow in Taylor's concentric rotating cylinders, $a \propto \Omega$ [15-17].
iii) Concentration fluctuations in well-stirred open tank reactors, $a \propto$ flowrate, "chemical turbulence" [18-21].
iv) Nonlinear electric oscillators (RLC circuits with a varactor diode), $a \propto$ driving voltage, [22-24].
v) Nonlinear mechanical oscillators, realized by Pohl's wheel [25].
vi) Acoustic cavitation noise piezo-electrically driven, $a \propto$ driving voltage, "acoustical turbulence", [26].
vii) "Optical turbulence" in lasers and bistable systems [27-29].
viii) Faraday's shallow water surface waves [30].
ix) Quantum vortex line generation by 1st sound in superfluid He II [31].
x) Driven Josephson junctions, SUPARAMPS [32-34].
xi) "NMR-turbulence" [35].
xii) Baroclinic flow instability [36].

Further experiments are expected to have increased accuracy or new aspects rather than to add just another example.

4. TRIFURCATION ?

After the introductory survey, this and the following sections will deal with certain aspects of discrete nonlinear dynamics, elucidating some more or less connected ideas, methods, or results.

Is there only bifurcation? For maps equivalent to $f_a(x) = af_1(x)$ and f_1 having a negative Schwarzian derivative $S_f = f'''/f'-(3/2)(f''/f')^2$ the answer is yes. A p-cycle $x_0, x_1, \ldots, x_{p-1}, x_p = x_0$ looses its stability if $dx_p/dx_0 = -1$, which it had gained as a function of a if $dx_p/dx_0 = +1$. The performance starts again with $x_{2p}(x_0)$ for the *two* newly born fixed points. That is connected with the possibility of having just *two* real roots of 1.

In iteration theory Julia (1918) and Fatou (1919) and also Myrberg (1962) studied trajectories got by iterating holomorphic maps. Allowing in the holomorphic quadratic map

$$z_{\tau+1} = \lambda z_\tau(1-z_\tau) \tag{5}$$

the variable z as well as the parameter $\lambda \equiv 4a$ to be complex, a fascinating new richness appears. An existing p-cycle looses stability [37,38] if

$$dz_p/dz_0 = \exp(2\pi i m/n) \ , \tag{6}$$

m,n relative prime. Varying λ to meet (6) yields n-furcation. It is the restriction to real x,a that only *bi*furcation occurs: $m/n = 1/2$. The range of stability shrinks $\propto n^{-2}$ [38], so irrational phases do not seem to exist.

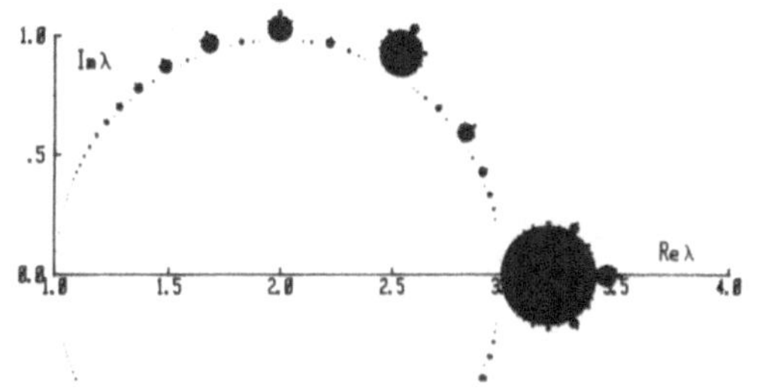

Fig.3a: Complex λ plane with regions of various stable periods obtained by n-furcations.

The parameter scaling (and the n-cycle's amplitudes splitting) along a certain m/n path in the λ-plane can be described by

$$(\lambda_j - \lambda_{j-1})/(\lambda_{j+1} - \lambda_j) = \delta_{m/n} \text{ (and } \alpha_{m/n} \text{ resp.).}$$

In the renormalization transformation (cf.sect.7) the n-th iterate is considered instead of the second one. The δ's and α's have modulus and phase, depending on m/n, describing the extension and rotation of the n^k-fold cycle. Figure 3a shows the bifurcation diagram in the λ-plane [38]. The basins of attraction are regions with fractal boundary, cf.fig.3b. Applications to lattice structures and structural phase transitions, described by area preserving maps are due to Janssen and Tjon [39].

5.DENSITY OF STATES, RENORMALIZATION, AND CHAOS MAP

Critical behavior near phase transitions is described by renormalization group equations [40-41] which are nonlinear. Thus they are suspicious to produce chaotic interactions. Up to now this does not seem to have happened; only recently [42] this possibility was discussed for spin-glass behavior in frustrated Ising models. To contribute to the physical understanding, a simple example may be useful.

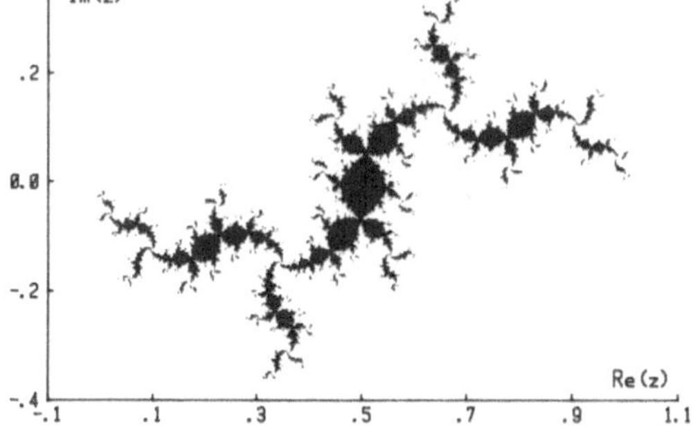

Fig.3b: Basin of attraction for the 9 cycle, after two 2/3 trifurcations. (from reference [38].)

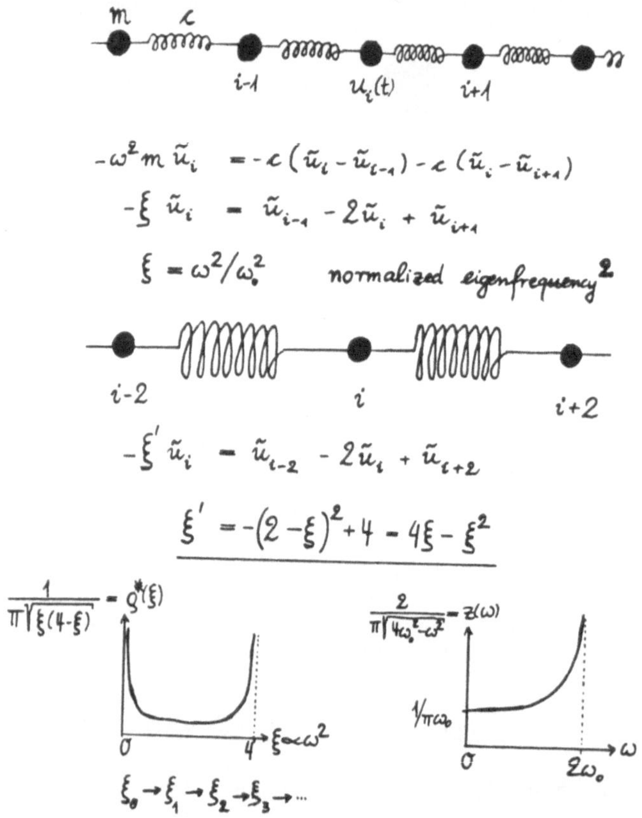

Fig.4: Viewgraph copy: linear harmonic chain and derivation of its density of states.

Consider the well-known linear harmonic chain (fig.4). The only relevant parameter is the eigenfrequency $\sqrt{\xi}$ (in terms of the bare oscillator frequency $\omega_0 = \sqrt{c/m}$). Elimination of each second mass' Fourier amplitude leads to an equation for the remaining oscillators of the same type, but with effective, renormalized parameter ξ'. This may be interpreted as renormalized mass, or spring constant, or another possible frequency ω in a linear harmonic chain. Starting with $x_0 \equiv \xi_0/4$ one finds reduced (squared) frequencies

$$x_{\tau+1} = 4x_\tau(1-x_\tau) \ . \tag{7}$$

The low frequency increase is $x_\tau = 4^\tau x_0$; but as x_τ is fenced into [0,1], too large a frequency must be followed by a smaller one again, etc. Although there are infinitely many x_0 which give rise to a periodic orbit, for almost every (a.e.) x_0 one finds a nonperiodic chaotic trajectory, which fills the whole interval [0,1] densely (although not equally dense). This can be most appropriately described by a density: $\rho^*(x)dx$ is defined as the relative number of visits to the subinterval $(x, x+dx)$.

Discrete nonlinear dynamics tells us, that for the map (7) $\rho^*(x)$ is a smooth function ($\rho^* \in L_1$) and in particular that it is *the same* for a.e. initial frequency x_0. This uniqueness is tantamount to the statement of ergodicity of the discrete dynamics which is generated by the "fully developed" (i.e. $a = 1$) parabola map (7).

We readily interpret ρ^* as the density of frequencies which a harmonic chain enjoys. How does ρ^* look like? In the case of (7) this can be answered analytically, using an old result. Equation (7) is solved by one-to-one transformation

$$x_\tau = 0.5 - 0.5 \cos\pi\Theta_\tau = \sin^2(\Theta_\tau \pi/2) \ .$$

The distribution of the real numbers $\Theta_\tau = 2^\tau \Theta_o \pmod 1$ is uniform in [0,1] for a.e. Θ_0 according to H.Weyl, 1916 [43]. From $\rho^*(x)dx = \hat{\rho}^*(\Theta)d\Theta$, the uniform distribution $\hat{\rho}^*(\theta) = 1$ of the phase Θ yields the (normalized) x-distribution

$$\rho^*(x) = 1/\pi\sqrt{x(1-x)} \ . \qquad (8)$$

By $\rho^*(x)dx = z(\omega)d\omega$ the density of states $z(\omega)$ is found, see fig.4, which is well-known, of course, from text books.

What do we learn? Even systems which are not suspicious of chaos may be tractable by iterated discrete maps. - Chaotic solutions may give rise to perfectly smooth physical properties of the system considered. - It is very useful to know the densities $\rho^*(x)$ implied by chaotic solutions of discrete maps.

This simple example is a blown-up version of a method to calculate the spectral dimension $\tilde{d}$ of the d-dimensional Sierpinski gasket by Toulouse [44] for the trivial case $d = 1$. $\tilde{d}$ is defined by the low-frequency density of states $z(\omega) \propto \omega^{\tilde{d}-1}$. It is an important quantity for random walks ($\langle R^2 \rangle \propto t^{\tilde{d}/d_H}$, d_H fractal dimension), the percolation problem, etc. [44]. Note, that it was not necessary to introduce the reciprocal lattice (which is not clear on Sierpinski gaskets).

The method to calculate the density of states can be extended [45] to a harmonic lattice with basis. Let $\kappa = M/m$ be the mass ratio of the two masses, $c/m = \omega_0^2$, ξ as before. The relevant parameter now is $\eta = (\xi-2)(\kappa\xi-2)$, which characterizes the unit cell of the lattice and which renormalizes when integrating the degrees of freedom as

$$\eta' = (\eta-2)^2 \ . \tag{9}$$

This is equivalent to (7) by $x = 1-\eta/4$. Remembering (8) one obtains the η-distribution. That, together with the two branched inverse function

$$\xi_\pm = \frac{1+\kappa}{\kappa} \pm \sqrt{\left(\frac{1+\kappa}{\kappa}\right)^2 + \frac{\eta-4}{\kappa}}$$

yields the density of states on both phonon branches (acoustical and optical) in accordance with conventional lattice dynamics.

6. RANDOM NUMBERS ?

The sequence of iterates $\{x_\tau\}_{\tau=0}^{\infty}$ generated by the parabola map (7) is chaotic in the sense defined in sect.2, with the only exception of (iv): the attractor is a whole interval. The most important properties (v) through (viii) can be easily shown, once the ergodicity property (iii) is clear. (Remark: the proof of ergodicity uses the conjugation to the hat-transformation and a theorem by Li and Yorke [46] for the latter, see e.g. [5]).

Ad (v), sensitive dependence on the initial value x_0: Consider two trajectories starting at x_0 and at $x_0 + \varepsilon_0$. The first images are x_1 and $x_1 + \varepsilon_1$ with $\varepsilon_1 \cong f'(x_0)\varepsilon_0$. The trajectories tend to separate if $|f'(x_0)| > 1$ at least in the x_0-mean. For the parabola $\Lambda \equiv \langle |f'(x_0)| \rangle = 8/\pi > 1$. Initially neighbouring trajectories separate exponentially. Correspondingly the map is *mixing*.

A consequence of mixing is that the invariant density can be calculated as the asymptotic solution of a *master equation*, $\rho^* = \lim_{\tau\to\infty} \rho_\tau$,

$$\rho_{\tau+1} = H\rho_\tau \ , \quad (H\rho)(x) \equiv \int \delta(x - f(y))\rho(y)dy. \tag{10}$$

The steady state ρ^* solves the Frobenius-Perron equation $\rho^* = H\rho^*$. The reader may verify this in particular for (8), (7). H is denoted as Frobenius-Perron operator.

Ad (vi): Because of the exponential divergence of neighbouring trajectories there is no long term prediction whenever there is a finite uncertainty ε_{min} in the initial condition. If ε_{tol} denotes the tolerable uncertainty in x_τ for a prediction of interest, the time range τ_{pre} which *can* be predicted is estimated by

$$\varepsilon_{tol} \cong \Lambda^{\tau_{pre}} \varepsilon_{min} \ , \text{ or } \ \tau_{pre} \cong \ln(\varepsilon_{tol}/\varepsilon_{min})/\ln\Lambda \ . \tag{11}$$

Usually the Lyapunov exponent

$$\lambda = \langle ln|f'|\rangle$$

is calculated instead of $ln\Lambda$. Stable periodic attractors have $\lambda < 0$, mixing systems $\lambda > 0$. For (7) it is $\lambda = ln2 > 0$.

Ad (vii): $S(\omega)$ = const, broad band spectrum for a.e. parabola trajectory, since

ad (viii): correlations decay instantaneously:

$$c_\tau = \langle \delta x \, \delta x_\tau \rangle = 0.125\delta_{0,\tau} \; . \tag{12}$$

This is concluded from δx_τ *and* ρ^* being symmetric functions of $\delta x = x - 1/2$, while δx itself is antisymmetric.

Thus the parabola generates a sequence of "*random numbers*"? It does. Were it not for the quickly growing rounding error, one could use it in practice as a random generator. But numerically one soon sees nothing but the round-off properties of the computer, as was observed already by von Neumann [47]. To overcome this, one better uses linear maps, which have to be broken or discontinuous, of course, to meet with the *non*-monotony.

Commonly used random number generators are based on integers i_τ instead of reals x_τ,

$$i_{\tau+1} = \mu i_\tau + b \pmod{i_{max}} \; . \tag{13}$$

It therefore may be useful to comment on analytical properties of such maps, transformed back to the interval [-1,1] or [0,1] (fig.5).

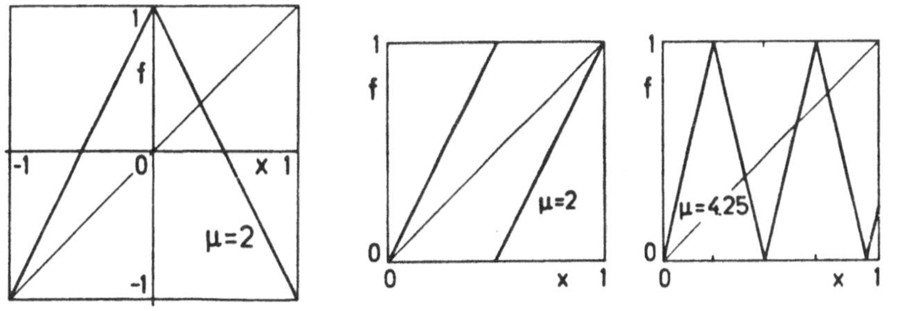

a) $f(x) = 1-\mu|x|$ b) $f(x) = \mu x \pmod 1$ c) $f(x)=(-1)^{[\mu x]}\mu x \pmod 1$

Fig.5: Broken linear or discontinuous maps. a) Hat or tent map, b) Modulo generator, c) Sawtooth map.

They are all ergodic and mixing if $\mu > 1$. The Lyapunov exponent is $\ln\mu > 0$. What distinguishes the discrete dynamics of these maps? Map (a) has $\rho^* = 1$ and instantaneous correlation decay, $c_\tau = (1/3)\delta_{0,\tau}$. Precisely all rational numbers x_0 give rise to periodic orbits, while all irrationals lead to random sequences.

Map (b) also has a constant $\rho^*(x) = 1$, but correlations are finite! I will demonstrate how correlations are calculated. The method is applicable quite generally. Write
$c_\tau = \langle xx_\tau\rangle - \langle x\rangle^2 = m_\tau - 1/4$. Then
$m_\tau = \int xLx_{\tau-1}dx = \int x_{\tau-1}Hxdx = m_{\tau-1}/2 + 1/8$. Here L is the operator adjoint to the Frobenius-Perron operator H introduced earlier. L has the meaning of the Liouvillian, the time translation operator $Lx_0 = x_1$ or $LG(x_0) = G(x_1)$ for any function G. We arrive at $c_\tau = c_0/2^\tau$, or, for arbitrary integer $\mu \geqq 2$, the correlation decays $\propto (1/\mu)^\tau$. So, the advantage of easily computing $x_{\tau+1}$ with the modulo generator is payed by finite (though exponentially decaying) correlation, disturbing the randomness. This is tolerable if μ is chosen as a large integer.

Map (c) in general does not even show a uniform distribution ρ^*. In fact, it is not even known (analytically) for arbitrary μ. But if for the μ of interest the particular trajectory $1, x_1(1), x_2(1), \ldots, x_\tau(1), \ldots$ has only a *finite* number n of different elements (containing an unstable cycle) the invariant density is a step function consisting of at most n steps. An extension to more general broken linear transformations together with a proof was given in [5]. (The Frobenius-Perron operator here is a finite dimensional matrix, being a special case of Ulam's [48] idea to approximate $\underline{H}$ by a matrix to obtain a numerically stable procedure to determine ρ^*. By Li's theorem [49] it is known, under which propositions this approximation converges with increasing matrix rank.)

7. PERIOD-DOUBLING RENORMALIZATION GROUP

If μ in map (a) is *decreased*, infinitely long lasting correlations show up. For particular μ-values the attractor consists of a set of subintervals, which are periodically visited by the phase x_τ. To this infinitely correlated periodic motion a chaotic component is superimposed. *Within* the subintervals x_τ has random positions. Therefore the spectrum consists of peaks (representing the inverse period) and a noisy broadband background (representing the random position within the subintervals).

The particular values are $\tilde{\mu}_n = 2^{2^{-n}}$. They are defined by the property that $f_\mu^{[2n]}$ is equivalent to the $\mu = 2$ hat in a small subbox, see fig.6. The reduction factors are $\alpha_n = (2^{2^{-n}}-1)^{-1} \cong 2^n/\ln 2$ and $\sqrt{2}\alpha_n$. They describe the scaling from step n-1 to n. A view on the figure will convince you that for μ slightly smaller than $\tilde{\mu}_n$ the pairwise connected boxes separate, while for μ slightly larger $\tilde{\mu}_n$

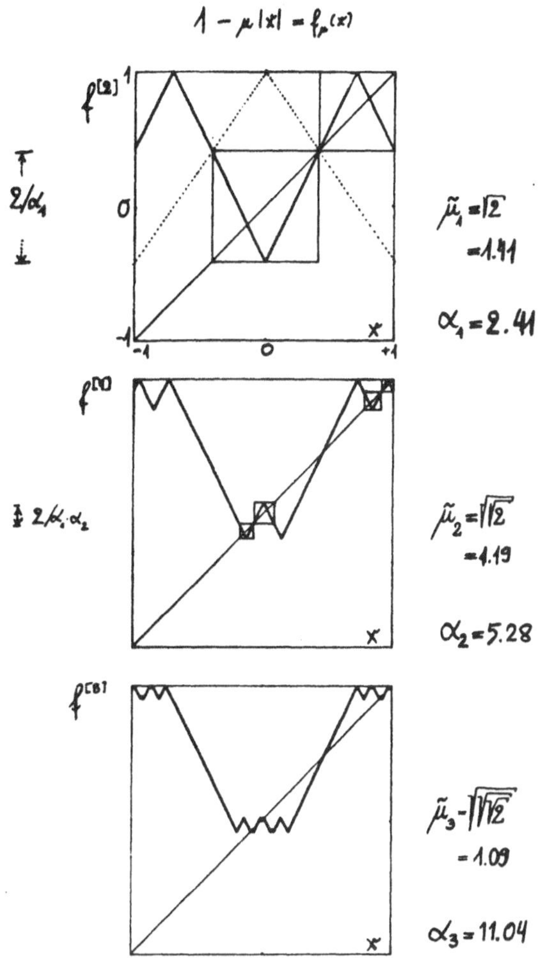

Fig.6: Higher iterates of the hat map for $\tilde{\mu}_n$, n = 1,2,3 (view graph copy).

they overlap. Thus the $\tilde{\mu}_n$ are the band merging points, forming the inverse cascade [5], cf. fig.7. One easily verifies (using the given values of $\tilde{\mu}_n$) that $(\tilde{\mu}_{n+1} - \tilde{\mu}_n)/(\tilde{\mu}_n - \tilde{\mu}_{n-1}) \to 1/2$. So the geometric reduction factor is $\delta = 2$ for the hat map; it was 4.669... for the parabolic map, the difference being caused by the different orders of the maximum (z = 1 hat, z = 2 parabola). - The reduction factors α_n have a finite limit 2.502... for the parabolic map, while for the hat map $\alpha_n \to \infty$. Due to this the bands shrink to tiny threads. The noisy period-2^n states approach the critical state at $\tilde{\mu}_\infty = 1$. The limit attractor has measure zero but cardinality continuum, as holds for the parabola map. But the fractal dimension here is $d_H \cong 1/n \to 0$ instead of $d_H \cong 0.53$ for the parabola [50]. As a function of μ the map's properties have a discontinuity at $\mu = 1$.

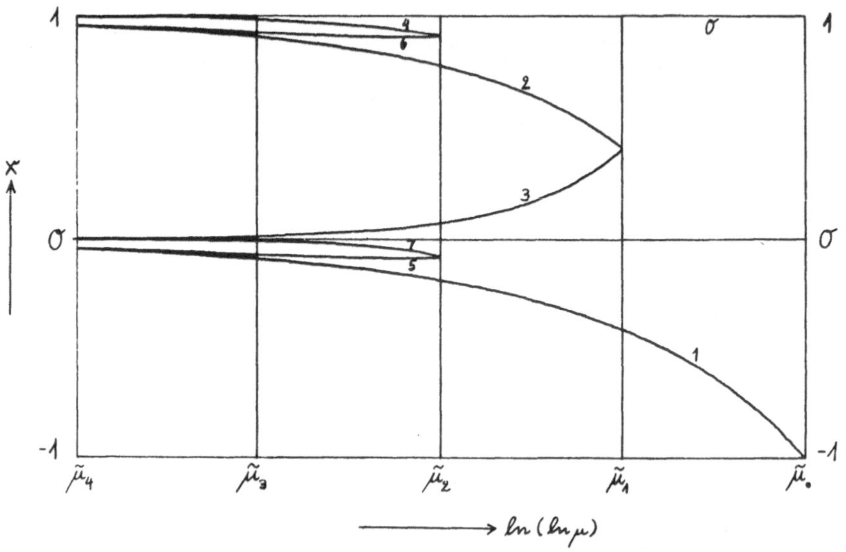

Fig.7: Band boundaries for the hat map as a function of μ. The band merging (splitting) points are indicated.

For $\mu < 1$ there remains one stable fixed point only.

The period-doubling transformation which generates the inverse cascade of band merging (splitting) states, can be given explicitly.

$$(Tf)(x) = -(1/q)f^{[2]}(-qx) \quad , \quad (Tf)(0) = 1 \tag{14}$$

was introduced by Feigenbaum [3] and by Coullet and Tresser [4] to prove the universality of the period-doubling sequence towards the onset of chaos. For the hat map considered here there is *no* such direct cascade of period-2^n states. Only the inverse, noisy-periodic cascade [5] exists and can be calculated from (14). In general (14) is a functional transformation, but here it degenerates into a parameter transformation.

$$f_\mu^{[2]}(x) = 1 - \mu|1 - \mu|x|| = 1 - \mu + \mu^2|x| \quad , \text{ x small,}$$

$$(Tf_\mu)(x) = (\mu - 1)/q - \mu^2|x| \equiv 1 - \mu'|x|.$$

Hence the renormalization group transformation generating the inverse cascade and the scaling factor α are

$$\mu' \equiv T\mu = \mu^2 \quad , \quad \alpha(\mu) = 1/q(\mu) = 1/(\mu - 1). \tag{15}$$

The fixed point is $\mu^* = T\mu^* = 1$; the unstable eigenvalue is found by linearizing, $d\mu' = 2d\mu$, so $\delta = 2$; the cascade starting from

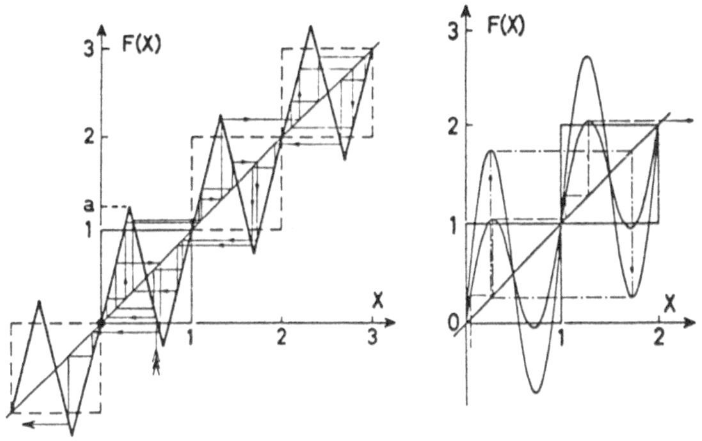

Fig.8: Examples of periodic maps, sawtooth and sinusoidal.

$\tilde{\mu}_0 = 2$ is $\sqrt{2}$, $\sqrt{\sqrt{2}}, \ldots, 2^{2^{-n}}, \ldots$; and $\alpha_n = (2^{2^{-n}} - 1)^{-1}$, all this as before.

8.DETERMINISTIC BROWNIAN MOTION?

If the dynamical map is too steep and high, the trajectory is no longer restricted to an interval. Interesting motions occur, if the map has periodic structure (fig.8). More precisely, let the dynamical law be

$$X_{\tau+1} = F_a(X_\tau) = X_\tau + f_a(X_\tau) \;, \; -\infty < X_\tau < \infty. \tag{16}$$

$f_a(X + l) = f_a(X)$ is assumed as periodic, for instance sinusoidal (r.h.s.) or broken linear (l.h.s.). All solutions are then translational invariant: $\{X_\tau + Ml\}$ is equivalent to $\{X_\tau\}$ for all integers M. l denotes the periodicity length, henceforth chosen as 1. (16) generates a Wiener-Levy process for appropriate choice of $f_a(X)$. Its basic properties are the drift velocity v and the diffusion coefficient D, which determine the mean phase and its variance by

$$\langle X_t \rangle \cong vt \;, \quad \Gamma_t = \langle (X_t - vt)^2 \rangle = 2Dt. \tag{17}$$

To clarify the conditions under which this Brownian motion happens the phase X_τ is decomposed into two random parts, $X_\tau = N_\tau + x_\tau$, uniquely defined by demanding N_τ integer and $x_\tau \in [0,1)$. Correspondingly (16) is decomposed into

$$N_{\tau+1} = N_\tau + \Delta_\tau(x_\tau) \quad , \; x_{\tau+1} = g(x_\tau). \tag{18}$$

$\Delta(x)$ is integer and describes the magnitude of the box-to-box jumps, while the statistics of the trajectories is caused by the properties of the *reduced map* $g(x) = F(X) - [F(X)] \in [0,1)$. Both Δ and g depend on the parameter a, of course. Averages are meant either as time means or (if g(x) is ergodic) as initial value means. The box numbers N are the analogs of the positions of a Brownian particle, the fractional part x within a box corresponds to the different motions of the surrounding atoms, leading to different kicks. Note that the observation of Brownian motion usually is done on a space grid and is discretized in time anyhow.

While clearly $v = \langle\Delta(x)\rangle$, one may ask whether and why the variance grows $\propto$ t and how one can calculate the diffusion coefficient D.

To answer this, write $X_t = \sum_{\tau=0}^{t-1} \Delta_\tau + x_t$. The variance then consist of three terms.

i) $\langle(\delta x_t)^2\rangle$, which is clearly bounded by 1/4.

ii) $2\sum_{\tau=0}^{t-1} \langle\delta\Delta_\tau \, \delta x_t\rangle$, which is of the order $t_{corr} \times \langle\delta\Delta\delta x\rangle$, so is also bounded.

iii) The double sum $\sum_{\tau=0}^{t-1} \sum_{\tau=0}^{t-1} \langle\delta\Delta_{\tau_1} \, \delta\Delta_{\tau_2}\rangle$; whenever the jump correlations decay sufficiently fast, the first sum converges to a t-independent value and the second sum yields the linear increase with t.

It is thus a basic mechanism which leads to normal diffusion, analogous to the well-known fact that the variance of macroscopic variables A is *not* $\propto V^2$ but increases as the variables themselves, $\langle(\delta A)^2\rangle \propto V \propto \langle A\rangle$.

Together with $\Gamma_t \propto t$ we have derived a correlation function representation of the diffusion coefficient

$$2D = \langle(\delta\Delta)^2\rangle \; (1 + 2\lim_{t\to\infty} (1/t) \sum_{\tau=0}^{t-1} \sum_{\lambda=1}^{\tau} C^o_\lambda). \tag{19}$$

Here C^o_λ is the normalized correlation function of the jump function $\Delta - \langle\Delta\rangle$. (19) is the basic formula that governs normal deterministic diffusion. It has been derived in [51].

As an application the onset of diffusion is briefly discussed. There is no diffusion, if the local height $a = \max_{0\leq x<1} F_a(x)$ of the map is less than 1 (or l). If by increasing a the height exceeds $a_c = 1$, diffusion sets on. Since there is only a small fraction of the box length, in which $F_a(x)$ exceeds 1, and since jumps can only occur if x just hits this fraction (of length $\tilde{\delta}$, say), diffusion starts continuously with $a - 1$. From (19) one concludes

$$2D = \tilde{\delta}(a) \cdot \rho^*_{a=1}(x \in \tilde{\delta}) \; . \tag{20}$$

The onset diffusion constant consists

i) of an universal factor $\tilde{\delta}(a) \propto (a - 1)^{1/z}$, determined by the maximum's order z of the tip that exceeds 1. This universality, being of geometrical origin, has been stated first by Geisel and Nierwetberg [52] together with the scaling due to diffusion enhancement by external noise. But D consists

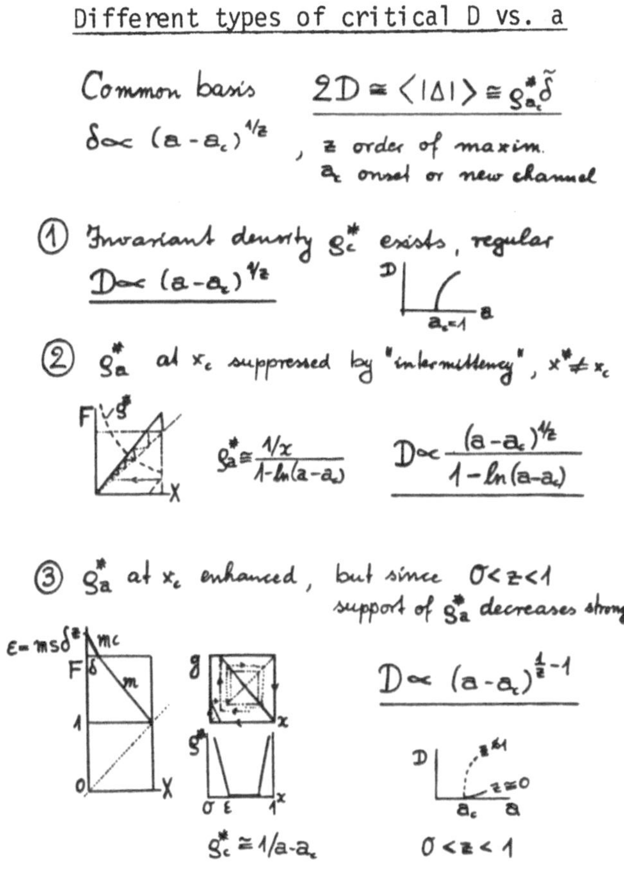

Fig.9: Examples for the diffusion constant D *vs* height a of the map near the onset (view graph copy).

ii) of a direct dynamical quantity, the invariant density $\rho^*_{a=1}$ of the dynamical map. This yields nonuniversal corrections, unless ρ^* stays regular for $a \to 1$. The corrections may even change the qualitative behavior of D near onset. For example D may jump discontinuously, like in a first order phase transition. The implications caused by $\rho^*_{a=1}$ have been extensively discussed in [51b]. Deviations from $D \propto \sqrt{a-1}$ are signals of interesting dynamics in the diffusion process. - Examples are summarized in fig.9.

9.WINDOWS OF NONDIFFUSIVE STATES

The derivation of the basic formula (19) for D tells us that the variance of a diffusing cloud varies differently if there are long range correlations in the jump function $\Delta(x_\tau)$. Those are unavoidable if the tip of $F_a(x)$ is flat, i.e. if $z > 1$. For appropriate parameter ranges the g-map or an iterate $g^{[n]}$ has a stable fixed point. This implies periodicity in the jump function $\Delta(x_\tau)$. Hence, instead of performing Brownian motion the phase travels in regular jumps into either positive or negative direction. We [51] called that "running mode". There is symmetry breaking but no diffusion. It also may happen that there are as many steps in positive direction as in negative (for instance in a 2-cycle of $x_0, x_1, x_0, x_1, \ldots$ with $\Delta(x_0) = -\Delta(x_1)$). Then the motion is localized but again nondiffusive.

By using a continuity argument one convinces oneself, that *if* g_a has a stable cycle, then there is always a whole *a-interval* where this happens. Therefore nondiffusive motion always occurs in whole parameter intervals, called *windows* or regular *bands*. These can be classified by the smallest cycle m of g(x) in the band. Together with m the whole bifurcation cascade $m \cdot 2^n$ (direct as well as inverse) shows up in the band. The width of the parameter interval decreases with the order m and soon submerges in the external noise. This, by the way, is the reason that in the sinusoidal map a continuous onset of the diffusion can be (numerically) observed at all.

Let us consider as an example, $F(X) = X + \mu \sin 2\pi X$ (fig.8 r.h.s.). Whenever μ is integer, $\mu = k = 1,2,\ldots$ a running mode band of velocity $v = +k$ or $-k$ starts [51,53], caused by a fixed point of the reduced map. The jumps are of size k. The width of the k-th band slowly decreases with increasing k. In good approximation [54]

$$\delta\mu_k = \frac{k}{2}\left(\sqrt{1+\frac{9}{2\pi^2 k^2}} - 1\right) \cong 9/8\pi^2 k \tag{21}$$

in agreement with numerical results [53]. The idea to find the band width analytically is: The onset of the band occurs if $g^{[n]}$ (here g itself) touches the angle bisector; the end of that band is reached if the maximum is mapped on the nearby unstable fixed point: $g^{[n]}(g^{[n]}(x_c)) = x^*$. Formulae (22) summarize the issue [54] for m = 1 bands' widths as a function of the order z of the maximum. Note

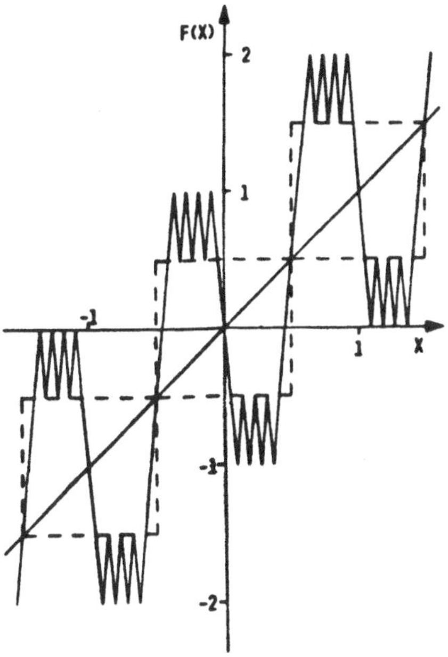

Fig.10: Broken linear map causing diffusion with long correlations between successive jumps, if the number M of peaks in a local cluster is large (here M = 4).

that the bands $k = 1,2,...$ decrease nonanalytically with $z \to 1$, i.e. approaching a broken linear transformation. Saw-tooth maps ($z = 1$) have *no* bands.

$$F_a \cong a - a4^z|x - 0.25|^z, \tag{22a}$$

$$\delta a_k(z \to 1) \cong 1/(2k + 0.5)^{1/(z-1)}, \tag{22b}$$

$$\delta a_k(z = 2) \cong 9/8(8k + 2) . \tag{22c}$$

10.ANOMALOUS DIFFUSION?

To clarify the role of the jump function's correlation decay we considered a periodic map with easily changeable correlation decay, but with certainly no regular running or localized modes [55]. The map is shown in fig.10. By constructing the phase trajectory the reader easily verifies, that for larger peak number M the phase X travels several steps in one direction until it eventually hits that small subinterval which maps it on the opposite peaks, whence it travels back. We calculated the Δ-correlation function analytically. For not too small M it is

$$C^o_\tau = [(2M - 1)/(2M + 3)]^\tau \cong \exp(-\frac{2}{M}\tau) . \tag{23}$$

Consequently if $t \lesssim M/2$ the strong correlation between successive jumps implies (cf. sect.8) $\Gamma_t \propto t^2$, enhanced diffusive broadening.

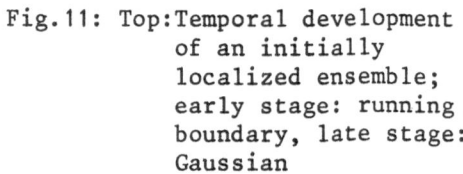

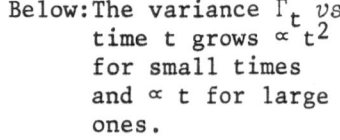

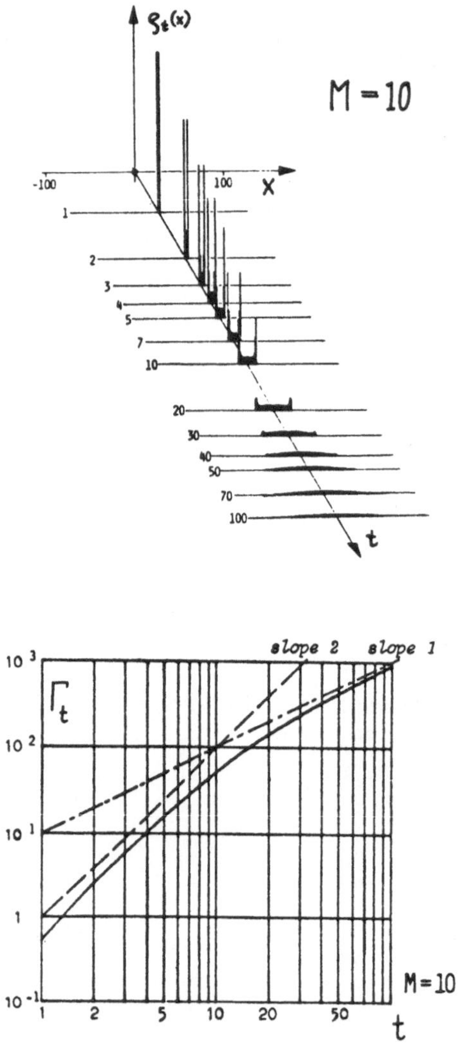

Fig.11: Top: Temporal development of an initially localized ensemble; early stage: running boundary, late stage: Gaussian

Below: The variance Γ_t *vs* time t grows $\propto t^2$ for small times and $\propto t$ for large ones.

But for $t \gg M/2$ one again has normal diffusion, $\Gamma_t \propto t$, although with enhanced diffusion coefficient. Figure 11 displays the cross-over from anomalous to normal Brownian motion and also the qualitative change of the distribution function, if one starts with a cloud of phase points localized in one box with constant density. More details on this correlation enhanced anomalous diffusion are published elsewhere [55].

11. ORNSTEIN-UHLENBECK PROCESS

I finally would like to explain that stochastic processes subject

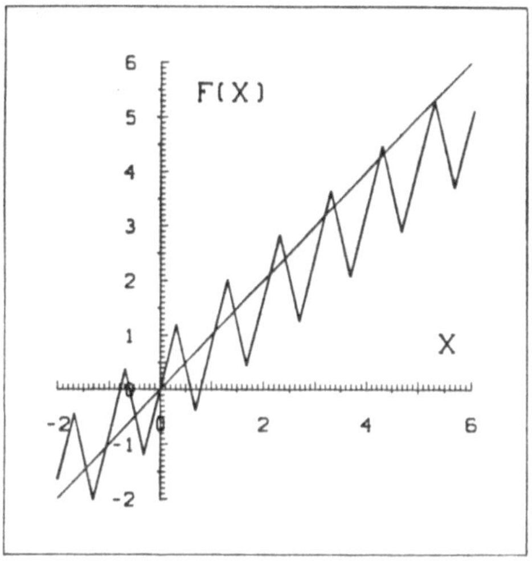

Fig.12: Sawtooth map climbing along a direction slightly less than 45°.

to a systematic restoring force in addition to the stochastic one also have their counterparts in deterministic diffusion generated by discrete nonlinear maps [56]. Consider the map in Fig.12, which slightly deviates from the climbing sawtooth map with strict translational invariance. It is defined by

$$X_{\tau+1} = e^{-\gamma} X_\tau + f_\alpha(X_\tau) \quad , \quad f_\alpha \text{ periodic,} \tag{24}$$

γ small. Due to the deviation $(e^{-\gamma} - 1)X \cong -\gamma X$ the global motion of the phase X_τ now is restricted to a finite interval, approximately given by $(-1/\gamma, +1/\gamma)$. The probability distribution to find the phase in the box with number n at time t ($\gg \gamma^{-1}$) is [56]

$$P_t(n) \cong \frac{1}{\sqrt{2\pi\sigma_t}} \exp\left\{-\frac{(n-n_0 e^{-\gamma t})^2}{2\sigma_t}\right\} , \tag{25a}$$

$$\text{with } \sigma_t = (D_0/\gamma)\,(1-e^{-2\gamma t}) \,. \tag{25b}$$

$P_t(n)$ approaches a Gaussian steady state with equilibrium width $\sigma_\infty = D_0/\gamma$, where D_0 is the diffusion constant according to $\gamma = 0$ in (19). The form of the distribution demonstrates that (24) models an Ornstein-Uhlenbeck process, i.e. white noise diffusion in an external potential $-\gamma X^2/2$.

Helpful discussions with Stefan Thomae and Ulrich Brosa are gratefully acknowledged, in particular their support in preparing some of the figures.

REFERENCES

1. D.Ruelle, F.Takens, Comm. Math. Phys. 20,167 (1971)
2. L.D.Landau, C.R.(Dokl.) Acad. Sci. USSR 44,311 (1944)
   E.Hopf, Comm. Appl. Math. 1,303 (1948)
3. M.J.Feigenbaum, J. Stat. Phys. 19,25 (1978)
4. P.Coullet, C.Tresser, J. Physique (Paris), Colloque 39,C5-25(1978)
5. S.Grossmann, S.Thomae, Z. Naturforsch. 32a,1353 (1977)
6. Y.Pomeau, P.Manneville, Phys. Lett. 75A,1 (1979); Comm. Math. Phys. 74,189 (1980)
7. M.J.Feigenbaum, Phys. Lett. 74A,375 (1979)
   M.J.Feigenbaum, in: Nonlinear Phenomena in Chemical Dynamics, Eds. C.Vidal, A.Pacault, Proc. Int. Conf. Bordeaux, Sept. 7-11, 1981, Springer: Berlin etc., 1981, p. 95
8. S.Thomae, S.Grossmann, Phys. Lett. 83A,181 (1981)
9. N.Metropolis, M.L.Stein, P.R.Stein, J.Comb. Theor. A15,25 (1973)
10. T.Geisel, J.Nierwetberg, Phys. Rev. Lett. 47,975 (1981)
11. G.Ahlers, Phys. Rev. Lett. 33,1185 (1974)
12. A.Libchaber, J.Maurer, J.Physique (Paris), Colloque, C3-51 (1980)
13. J.P.Gollub, S.V.Benson, J. Fluid Mech. 100,449 (1980)
14. M.Giglio, S.Musatti, U.Perini, Phys. Rev. Lett. 47,243 (1981)
15. J.P.Gollub, H.L.Swinney, Phys. Rev. Lett. 35,927 (1975)
16. R.P.Fenstermacher, H.L.Swinney, J.P.Gollub, J. Fluid Mech. 94, 103 (1979)
17. Yu.N.Belyaev, A.A.Monakhov, S.A.Shcherbakov, I.M.Yavorskaya, JETP - Lett. 29,295 (1979)
18. O.E.Rössler, K.Wegmann, Nature 271,89 (1978),see also O.E.Rössler, Z. Naturforsch. 31a,1168; 1664 (1976)
19. L.F.Olsen, H.Degn, Nature 267,177 (1977)
20. R.A.Schmitz, K.R.Graziani, J.L.Hudson,J.Chem.Phys. 67,3040(1977)
21. J.C.Roux, A.Rossi, S.Bachelard, C.Vidal, Phys. Lett. 77A,391 (1980); J.S.Turner, J.C.Roux, W.D.Mc Cormick, H.L.Swinney, Phys. Lett. 85A,9 (1981)
22. P.S.Linsay, Phys. Rev. Lett. 47,1349 (1981)
23. J.Testa, J.Pérez, C.Jeffries, Phys.Rev.Lett. 48,714 (1982)
24. W.Lauterborn, W.Meyer-Ilse, Diplomarbeit, Göttingen (1982)
25. G.Mayer-Kress, Bild der Wissenschaft, April 1983, p.14-15
26. W.Lauterborn, E.Cramer, Phys. Rev. Lett. 47,1445 (1981)
27. H.Haken, Phys. Lett. 53A,77 (1975); 62A,133 (1977)
28. H.M.Gibbs, F.A.Hopf, D.L.Kaplan, R.L.Shoemaker, Phys. Rev. Lett. 46,474 (1981)
29. F.T.Arecchi, R.Meucci, G.Puccioni, J.Tredicce, Phys. Rev. Lett. 49,1217 (1982)
30. R.Keolian, I.Rudnick, L.A.Turkevich, S.J.Putterman, J.A.Rudnick, Phys. Rev. Lett. 47,1133 (1981)
31. C.W.Smith, M.J.Tejwani, D.A.Farris, Phys.Rev.Lett. 48,492(1982)
32. Prediction by B.A.Huberman, J.P.Crutchfield, N.H.Packard, Appl. Phys. Lett. 37,750 (1980)
33. M.Cirillo, N.F.Pedersen, Phys. Lett. 90A,150 (1982)
34. R.Y.Chiao, P.T.Parrish, J. Appl. Phys. 47,2639 (1976)

35. D.Meier, R.Holzner, B.Derighetti, E.Brun, in: Evolution of Order and Chaos, Ed. H.Haken, Synergetics Symposium April 26-May 1,1982, Springer: Berlin, etc., 1982
36. J.D.Farmer, J.Hart, P.Weidman, Phys. Lett. 91A,22 (1982)
37. B.B.Mandelbrot, Ann. N.Y.Acad. Sci.357,249 (1980)
38. P.Svitanović, J.Myrheim, Phys. Lett. 94A,329 (1983)
39. T.Janssen, J.A.Tjon, Phys. Lett. 87A,139 (1982); J. Phys. A: Math. Gen. 16,673; 697 (1983)
40. K.G.Wilson, Phys. Rev. B4,3174,3184 (1971) F.Wegner, Phys. Rev. B5,4529 (1972)
41. L.P.Kadanoff, Phys. Rev. Lett. 34,1005 (1975) M.E.Fisher, Rev. Mod. Phys. 46,597 (1974)
42. S.R.Mc Kay, A.N.Berker, S.Kirkpatrick, Phys.Rev.Lett. 48,767(1982)
43. H.Weyl, Math. Ann. 77,313 (1916)
44. R.Rammal, G.Toulouse, J. Physique-Lett. 44,L-13 (1983)
45. S.Grossmann, S.Thomae, to be published
46. T.-Y.Li, J.A.Yorke, Trans. Am. Math. Soc. 235,183 (1978)
47. J.von Neumann, J. Res. Nat. Bus. Stand.Appl.Math.Ser. 3,36(1951)
48. S.M.Ulam, A Collection of Mathematical Problems, Interscience Tracts in Pure and Applied Mathematics 8 pp. 73 (1960)
49. T.-Y.Li, J.Approx. Theory 17,177 (1976)
50. P.Grassberger, J.Stat. Phys. 26,173 (1981)
51. S.Grossmann, H.Fujisaka, Phys. Rev. A26, 1779 (1982); Z. Phys. B48,261 (1982)
52. T.Geisel, J.Nierwetberg, Phys. Rev. Lett. 48,7 (1982)
53. M.Schell, S.Fraser, R.Kapral, Phys. Rev. A26,504 (1982)
54. Details of the results in this section will be published elsewhere
55. S.Grossmann, S.Thomae, Phys.Lett. 97A (1983)
56. H.Fujisaka, S.Grossmann, S.Thomae, to be published

DETERMINISTIC DIFFUSION - A QUALITY OF CHAOS

T. Geisel *

Institut für Theoretische Physik
Universität Regensburg, D-8400 Regensburg
Federal Republic of Germany

ABSTRACT

This paper reports on chaotic systems that can exhibit a deterministic "random" walk as an additional chaotic quality. This diffusive motion is generated within the dynamical system and not by external random forces. It is governed by a master equation, which is derived from an exact equation. The onset of diffusion is analogous to a phase transition and is described by a universal scaling function. Under certain circumstances the power spectra show excess noise at low frequencies and the mean-square displacements have an anomalous asymptotic behavior. In 2-dimensional systems there is a crossover in the critical behavior at the onset of diffusion.

1. INTRODUCTION

It is well known by now that dynamical systems in the form of nonlinear differential equations or iterative mappings may exhibit seemingly irregular and random motions although they are not exposed to external random forces as in models for Brownian motion. The "randomness" which we call chaotic behavior is generated within

* Heisenberg fellow

the deterministic system. The simplest examples are iterative 1-dimensional mappings [1], where a time series x_t is generated by repeated application of a function $x_{t+1}=f(x_t)$. It was found by Feigenbaum that the onset of chaos via period-doubling has strong analogies with a phase transition. The succession of period-doubling bifurcations is described by a universal bifurcation rate $\delta=4.66920$, which plays the role of a critical exponent [2]. Also the fine structure of the chaotic regime is governed by bifurcation rates [3] which converge to a universal constant $\gamma=2.94805$. Other maps can produce intermittent chaos [4], where universal behavior in analogy to a phase transition has also been found [5,6].

It is clear that chaotic behavior may be distinguished by different qualities which may be expressed by certain physical or mathematical quantities. Examples are period-doubling and intermittent chaos, which were already mentioned above. Other criteria are mere non-periodicity, ergodicity, mixing and unpredictability (measured by the Lyapunov exponent) [7]. One may also distinguish systems which are as random as a coin toss [7]. We have recently found some 1-dimensional systems which fulfil another criterion [8]. They are as random as a random walk. Their long time behavior can be characterized by a diffusion coefficient, which measures the degree of randomness and plays the role of an order parameter. The onset of diffusion has universal scaling properties and thus represents another case with analogies to phase transitions. Such a deterministic diffusion may occur in certain physical systems; it has been found e.g. in simulations of driven Josephson junctions [9,10] as diffusion of the phase difference.

In section 2 a simple example will first be presented. Then starting from an exact equation a master equation is derived describing the statistics of the random walk. The diffusion coefficient is calculated and expressed in terms of a universal scaling function and critical exponents are obtained. In section 3 the power spectrum is calculated for diffusion with intermittent characteristics. It shows excess noise at low frequencies. Anomalous asymptotic diffusion is described in section 4, and results for 2-dimensional systems are mentioned in section 5.

2. THE ONSET OF DIFFUSION

2.1 AN EXAMPLE

We will consider dynamical systems in the form of iterative maps

$$x_{t+1} = f_\mu(x_t) \tag{2.1}$$

where x_t will be the diffusing variable, t is a discrete time and μ is a parameter. The model which we have proposed originally as a simple example exhibiting a deterministic "random" walk is [8]

$$x_{t+1} = x_t - \mu \sin(2\pi x_t) \tag{2.2}$$

Its graph is shown in Fig. 1.

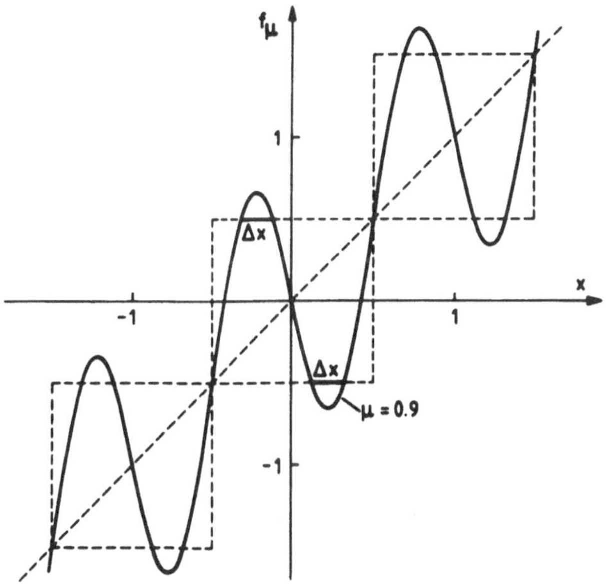

Fig. 1 Dynamical system $x_{t+1}=f_\mu(x_t)$ for the example of Eq. (2.2). Intervals with successors in neighbouring cells are denoted by Δx.

In a computer experiment we have measured the mean-square displacements $<(x_t-x_0)^2>$ averaged over 2000 orbits x_t with initial values $-{}^1/_2<x_0<{}^1/_2$. The result is shown as a function of time in Fig. 2.

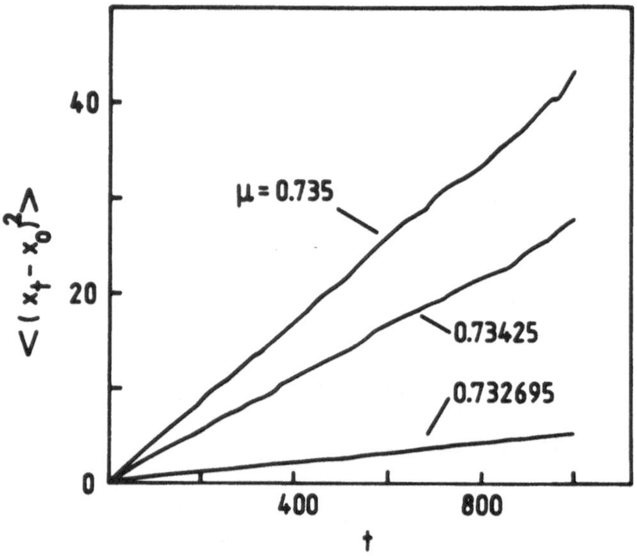

Fig. 2 Mean-square displacements as a function of time generated by Eq. (2.2).

The mean-square displacements diverge linearly in time. In any experiment this is the criterion to verify diffusive motion. The diffusion coefficient is given by half the slope of the lines in Fig. 2. It vanishes when the parameter μ is below a critical value μ_c=0.732644. Diffusive motion is usually explained by random forces, e.g. due to random collisions of a heavy Brownian particle with the molecules of a surrounding liquid. Here, however, it is generated intrinsically in the absence of random forces.

From Fig. 1 we may get a heuristic idea of how the diffusive motion arises. The real axis x and the axis of the successors $f_\mu(x)$ can be partioned into unit cells as shown by the dashed lines. Most of the values x in a unit cell have successors $f_\mu(x)$ in the same unit cell. Only points in the intervals denoted Δx have successors in neighbouring cells. When the parameter μ is lowered the intervals Δx become smaller and even vanish. In the latter case there is no diffusion because an orbit x_t starting in a given cell will always remain in it and cannot spread along the x-axis. When Δx is small it takes many iterations until an orbit x_t falls into Δx and is transferred into a neighbouring cell. I.e. the residence time within a cell is large compared with the time of a jump into a neighbouring cell (which is one time step). The jump rate is given by the probability of hitting an interval Δx, which is expected to be proportional to Δx. In this way one might phenomenologically construct a master equation for the hopping motion from cell to cell. However, it is possible to rigorously derive a master equation starting from an exact equation as discussed below.

2.2 MASTER EQUATION

We will henceforth deal with models more general than Eq. (2.2). As universal statements will be made we will consider universality classes of models defined as [8]

$$x_{t+1} = f_\mu(x_t) + \sigma\, \xi_t \tag{2.3}$$

where a random variable ξ_t has been added that is assumed to have zero mean, standard deviation 1 and a Gaussian distribution $v(\xi)$. The parameter σ measures the strength of the external noise. The following theory can be carried out analogously in the absence of noise (σ=o), i.e. for a purely deterministic system. However, as we are also interested in the effect of small external fluctuations (σ<<1) we immediately treat the external noise case for convenience. The deterministic case may be recovered for $\sigma \to 0$. The map $f_\mu(x)$ is assumed to have the following properties: $f_\mu(x)$ is an odd function

$$f_\mu(-x) = - f_\mu(x) \; , \tag{2.4}$$

$f_\mu(x)-x$ is periodic with period 1

$$f_\mu(x+n) = n + f_\mu(x) \tag{2.5}$$

and $f_\mu(x)$ has a relative maximum per period at x_c+n, in particular with $-{}^1/_2<x_c<0$. The vicinity of the maximum is assumed to be of the form

$$f_\mu(x) = a(\mu) - b(\mu)|x-x_c|^z \tag{2.6}$$

where a and b are coefficients depending on a single parameter μ and the exponent $z>0$ determines the type of maximum and distinguishes the universality classes.

The statements on diffusion and universal properties will only be made on the set of those μ for which an invariant distribution function $q(x)$ exists. Considering unit cells of length 1 we define a conditional probability $\rho_t(x)dx$ of finding a value between x and x+dx at time t if the initial value was in the 0th cell $(|x_0|<{}^1/_2)$. This probability must satisfy the following (exact) equation due to conservation of probability

$$\rho_{t+1}(y) = \iint\limits_{-\infty}^{+\infty} \rho_t(x)v(\xi)\delta[y-f_\mu(x)-\sigma\xi]dxd\xi \; . \tag{2.7}$$

It means that the probability of finding a value y at time t+1 is equal to the sum of probabilites that at time t the system had values x and ξ such that $y=f_\mu(x)+\sigma\xi$. In the noise-free case Eq. (2.7) reduces to the Frobenius-Perron equation, which is investigated e.g. in ref. [11]. An equation like (2.7) was also used by Haken and Mayer-Kress [12] to derive a Chapman-Kolmogorov equation. Here we carry out some integrations in order to derive a master equation. The probability $p_t(l)$ for a transition from cell 0 to cell l in time t is

$$p_t(l) = \int_{l-{}^1/_2}^{l+{}^1/_2} \rho_t(x)dx \; . \tag{2.8}$$

Consider now the map $f_\mu(x)$ reduced to the interval $-^1/_2 \leq x \leq ^1/_2$ where all unit cells are identified with the zero'th cell. If this map has an invariant distribution q(x) we can write

$$\rho_t(x) = q(x')p_t(l) \tag{2.9}$$

for $l-^1/_2 < x < l+^1/_2$ and $x'=x-l$.
Carrying out some integrations in Eq. (2.7) and using Eq. (2.9) we have derived a master equation describing the statistics of the diffusive motion in the critical region [8]

$$p_{t+1}(l) - p_t(l) = - wp_t(l) + \frac{w}{2}p_t(l+1) + \frac{w}{2}p_t(l-1) \tag{2.10}$$

with the jump rate

$$w = \int_{-^1/_2}^{+^1/_2} q(x)\ \mathrm{erfc}[(1/2-f(x))/\sqrt{2}\,\sigma]\ dx \tag{2.11}$$

As mentioned before all statements on diffusion of course are true only on the set of those parameters μ for which an invariant distribution q(x) exists. The existence may be investigated for individual maps. Let us consider the example of Eq. (2.2) and the critical point μ_c where diffusion sets in. We perform a topological conjugation of this map

$$u_{t+1} = h^{-1}[f(h(u_t))] = g(u_t) \tag{2.12}$$

using the homeomorphism

$$x = h(u) = -\frac{1}{2}\cos(\pi u) \quad (0 \leq u \leq 1)\ . \tag{2.13}$$

The conjugated map g(u) is

$$u_{t+1} = \frac{1}{\pi}\arccos\{\cos(\pi u_t)-2\mu_c\sin[\pi\cos(\pi u_t)]\} \tag{2.14}$$

This is a piecewise C^∞ function and with the exception of the extrema its slope is everywhere $|g'(u)|>2$ as illustrated in Fig. 3. Applying a theorem of Lasota and Yorke [13] it follows that g(u) has an invariant distribution $q_g(u)$ and therefore f(x) has an invariant distribution denoted by q(x).

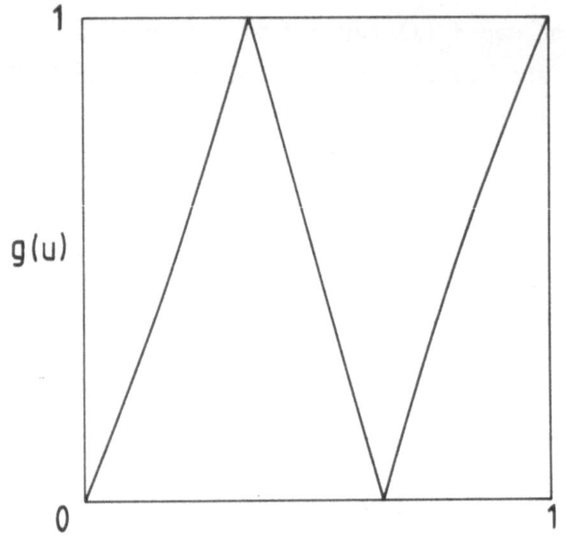

Fig. 3 Map g(u) Eq. (2.14) topologically conjugated to f(x).

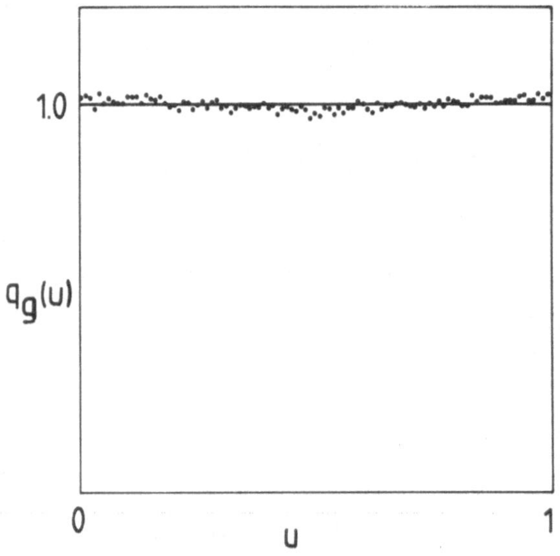

Fig. 4 Invariant distribution $q_g(u)$ belonging to the conjugated map g(u) in comparison with the approximation $q_g(u) \approx 1$.

The topological conjugation also allows us to calculate the distribution function q(x). The map g(u) (see Fig. 3) is very close to a piecewise linear map which would have a constant distribution (following from the Frobenius-Perron equation). Therefore as an approximation we may use $q_g(u)=1=\text{const}$. Comparison with numerical results in Fig. 4 shows that this approximation is actually very accurate. Conservation of probability under conjugation then gives us an approximation for the distribution q(x) (belonging to the map f)

$$q(x) = q_g(u)\left|\frac{du}{dx}\right| \tag{2.14}$$

$$q(x) \approx \left|\frac{du}{dx}\right| = \frac{1}{\pi}\left(\frac{1}{4} - x^2\right)^{-1/2} . \tag{2.15}$$

These considerations referred to $\mu=\mu_c$. In addition we know [3,14] that μ_c is an accumulation point for ergodic parameters μ. I.e. in every neighbourhood of μ_c there are parameters where the invariant distribution q(x) exists.

2.3 CRITICAL PROPERTIES OF THE DIFFUSION COEFFICIENT

The master equation (2.10) can be solved and used to calculate statistical quantities of interest like correlation functions and mean-square displacements. As the correlation functions are trivial we concentrate on the critical behavior of the diffusion coefficient D. The critical point μ_c is determined by the condition

$$f_{\mu_c}(x_c) = \frac{1}{2} . \tag{2.16}$$

In its vicinity we expand

$$a(\mu) = \frac{1}{2} + a'(\mu-\mu_c) \tag{2.17}$$

The diffusion coefficient was obtained from Eqs. (2.10-11) [8]

$$D = \sigma^{1/z}\, d\left(\frac{\mu-\mu_c}{\sigma}\right) \tag{2.18}$$

$$d\left(\frac{\Delta\mu}{\sigma}\right) = \frac{q(x_c)}{z}\left(\frac{\sqrt{2}}{b}\right)^{1/z} \int_{-\frac{a'}{\sqrt{2}}\frac{\Delta\mu}{\sigma}}^{\infty} \operatorname{erfc}(u)\left[u + \frac{a'}{\sqrt{2}}\frac{\Delta\mu}{\sigma}\right]^{\frac{1}{z}-1} du \tag{2.19}$$

where erfc(u) denotes the complementary error function. D has thus been expressed in terms of a universal scaling function in analogy to a phase transition. It is universal for a class defined by the exponent z (except for nonuniversal prefactors); it is a scaling function because simultaneous scaling of $\Delta\mu$ and σ leaves d invariant. The parameters μ and σ thus play the roles which e.g. in a ferromagnetic phase transition are played by the temperature T and the applied magnetic field H. Eqs. (2.18-19) can be used to determine critical exponents. In the absence of noise ($\sigma \to o$) these equations become

$$D = 2q(x_c)\left(\frac{a'}{b}\right)^{1/z} (\mu-\mu_c)^{1/z} + O\left(\Delta\mu^{3/z}\right) . \qquad (2.20)$$

For $\mu=\mu_c$ the noise dependence is $D=\sigma^{1/z} d(o)$.

In order to illustrate and to test the accuracy of the scaling behavior a computer experiment was carried out for the map

$$f_\mu(x) = \mu\sqrt{27}\,\left(2x^3-\frac{1}{2}x\right) \qquad \left(|x|<{}^1/_2\right) . \qquad (2.21)$$

We have measured the diffusion coefficient D for 3 different noise levels σ and 100 values of $\Delta\mu/\sigma$ between -1 and +1. In Fig. 5 D and $\Delta\mu$ are scaled in such a way that according to Eq. (2.18) the scaling function d should results. Indeed the data points for the 3 experiments form a single curve and thereby experimentally show the existence of the scaling function. The line in Fig. 5 is the analytic result which is obtained without adjustable parameters.

The critical behavior investigated above refers to the first onset of diffusive motion when the maximum of the map is increased. Under special circumstances deviations from Eq. (2.20) like logarithmic corrections may be encountered. This was investigated by Grossmann and Fujisaka [15]. Maps like Eq. (2.2) can also have other transitions to diffusion e.g. via a tangent bifurcation. This happens for larger values of μ and was noted by Schell et al. [16].

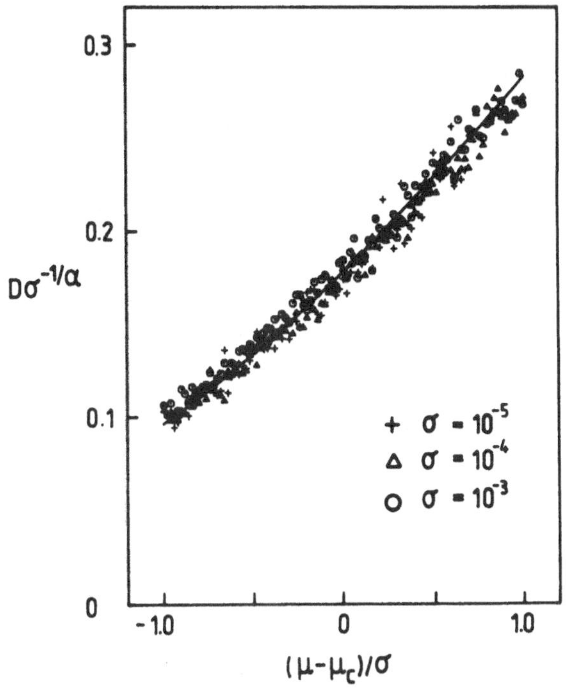

Fig. 5 Measurement of the diffusion coefficient D in a computer experiment demonstrating the existence of a scaling function. Full line: Analytic expression for the scaling function Eq. (2.19).

3. EXCESS NOISE FOR INTERMITTENT DIFFUSION

The power spectra for the situation treated so far are trivial. A more interesting case is when the map comes close to a tangent bifurcation for a drifting orbit and diffusion becomes intermittent. Schell et al. have shown that the critical law becomes $D \sim 1/\sqrt{\Delta\mu}$. I will now present results for the power spectrum which were obtained in a work with J. Nierwetberg [17].

It is easy to understand that intermittent diffusion involves correlated jumps. In the example of Eq. (2.2) there are tangent bifurcations to running modes at $\mu=1$ and more generally $\mu=m$. The initial value $x_o=x_c=-1/4$ generates $x_1=x_c+m$ and $x_t=x_c+tm$. There is also a solution drifting to the left for $x_o=-x_c$. At parameters slightly below $\mu=m$ the tangent bifurcation is preceded by intermittent chaos as usually [4]. In the present case, however, the laminar motion is coupled with jumps over m unit cells. These jumps are correlated as long as the laminar motion lasts.

The quantities of interest are the velocity correlation function

$$\langle v(o)v(t)\rangle = \lim_{T\to\infty} \frac{1}{T} \sum_{\tau=1}^{T} v_\tau v_{\tau+t} \tag{3.1}$$

(where $v_\tau = x_{\tau+1} - x_\tau$) and its power spectrum

$$S_{vv}(\omega) = \sum_{t=-\infty}^{+\infty} e^{i\omega t} \langle v(o)v(t)\rangle \ . \tag{3.2}$$

We have obtained these quantities in a continuum approximation [17] for $\mu\nearrow m$. The correlation function is

$$\langle v(o)v(t)\rangle = m^2(1-t/t_m) - \frac{m^2}{x_o t_m c} \ln \frac{\cos(\alpha - 2\alpha t/t_m)}{\cos\alpha} \tag{3.3}$$

where $t \le t_m$, $2x_o \ll 1$ is the width of the laminar region, $c=2\pi^2 m$,

$$\alpha = \mathrm{arctg}\{x_o[c/(m-\mu)]^{1/2}\} \ , \tag{3.4}$$

and

$$t_m = \frac{2\,\alpha}{[c(m-\mu)]^{1/2}} \tag{3.5}$$

is the maximum laminar time.

The power spectrum was obtained numerically and analytically. Fig. 6 shows the numerical spectral analysis of the time series v_t generated by Eq. (2.2) for $\mu=1-5\cdot10^{-7}$. Obviously there is a 1/f-regime for $\omega>10^{-1}$ and a $1/f^2$-regime for $\omega<10^{-1}$. For $\omega<2\pi/t_m$ the spectrum must saturate at a finite value. If we want to compute the spectrum by Fourier-transforming Eq. (3.3) we must expect artifacts because the approximation (3.3) is identically zero for $t \ge t_m$ and because the spectrum Eq. (3.2) is a Fourier sum and not a Fourier integral. These can be partly avoided by considering only the limit

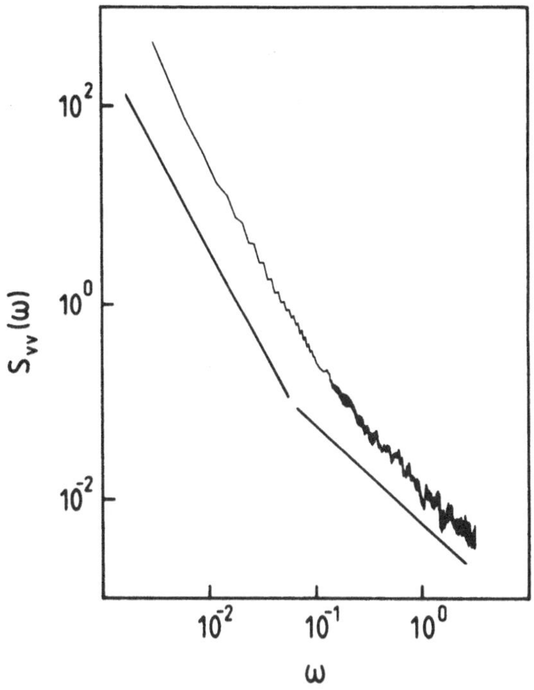

Fig. 6 Velocity power spectrum for Eq. (2.2) with $\mu=1-5\cdot 10^{-7}$. The straight lines with slopes 2 and 1 are included for comparison with ω^{-2} and ω^{-1} behavior.

$\mu\to m$, i.e. $t_m\to\infty$, where the Fourier sum (3.2) becomes an integral. For that purpose we must introduce the scaled variables $\tau=t/t_m$, $\omega'=\omega t_m$ and $S'=t_m S(\omega)$. The sum over t becomes an integral over $d\tau$. For odd frequencies $\omega'=(2n+1)\pi$ only the first term of Eq. (3.3) contributes

$$\lim_{\mu\to m} t_m \; S(\omega) = 4m^2 \frac{1}{\omega^2} \tag{3.6}$$

for even frequencies $\omega'=2n\pi$ only the second term contributes

$$\lim_{\mu\to m} t_m \; S(\omega) = \frac{m}{\pi x_o} \frac{1}{\omega} \quad . \tag{3.7}$$

This difference between odd and even frequencies arises because our approximation (3.3) for the correlation function becomes identically zero at $t=t_m$.

Eqs. (3.6-7) cannot literally reproduce the spectrum measured in Fig. 6, however, they can explain the $1/f^2$ and 1/f-regimes. Because $x_0(m)=x_0(m=o)/m$ there is a crossover frequency $\omega_{cr}=4\pi x_0(m=o)$. Above ω_{cr}, the 1/f-behavior (Eq. (3.7)) dominates, and below ω_{cr}, the $1/f^2$-behavior (Eq. (3.6)) dominates.

In this section we found that instead of being white the power spectrum may exhibit excess noise at low frequencies. This is due to correlated jumps, which arise when the diffusion process becomes intermittent.

4. ANOMALOUS DIFFUSION

Besides producing excess noise, intermittent diffusion may also give rise to anomalous diffusion. I.e. the asymptotic behavior of the mean-square displacements may be different from the standard case $\langle\Delta x^2(t)\rangle\sim t$. I will now resume the results of a collaboration with S. Thomae [18].

Consider as an example maps of the type

$$x_{t+1} = x_t + a x_t^2 \qquad (x \le x_c \le \tfrac{1}{2}) \tag{4.1a}$$

where $x_c=[(1+6a)^{1/2}-1]/2a$, and

$$x_{t+1} = \frac{3}{2} - 2\,\frac{x-x_c}{1-2x_c} \qquad (x_c \le x \le \tfrac{1}{2}) \tag{4.1b}$$

The remaining part of the map $x_{t+1}=f_a(x_t)$ along the real axis is defined by symmetry requirements

$$f_a(-x) = -\,f_a(x) \tag{4.1c}$$

$$f_a(x+n) = n + f(x)\ . \tag{4.1d}$$

Eq. (4.1b) is introduced in order to bring down the map to an unstable fixed point at x=1/2. Its explicit form is not essential for the following results. The essential feature of the map is its behavior close to the origin $x_t=o$ in Eq. (4.1a). This functional form generates intermittent chaos, such that an orbit may stay arbitrarily long close to the origin without performing a diffusive jump into a neighbouring unit cell. This leads to a distribution of residence times which is responsible for an anomalous asymptotic behavior.

We have obtained the distribution $\psi(T)$ of residence times T in a continuum approximation [18]

$$\psi(T) = 2a[2 + aT]^{-2} . \tag{4.2}$$

Note that although $\psi(T)$ is normalized, it does not have any finite moments (when T is treated continuously). Eq. (4.2) leads to the following long time limit for the mean-square displacements

$$\langle\Delta x^2(t)\rangle = \frac{a}{2}\frac{t}{\ln[\frac{a}{2}te^{-\gamma}]} \qquad (t\to\infty) \tag{4.3}$$

where $\gamma=0.577216$ is Euler's constant. Eq. (4.3) shows that the mean-square displacement never reaches the standard diffusive behavior (linear in time) but instead has a logarithmic correction.

5. DIFFUSION IN TWO DIMENSIONS

We have recently considered 2-dimensional generalizations of Eq. (2.1). At the same time they are generalizations of Chirikov's standard map [19]. Consider the following 2-d map

$$x_{t+1} = f_\mu(x_t) + (1-\gamma)v_t \tag{5.1}$$

$$v_{t+1} = x_{t+1} - x_t \tag{5.2}$$

where $f_\mu(x_t)$ is assumed to have the same properties as in section 2.2. We have included a dissipation parameter γ. In the conservative limit ($\gamma=0$) and with $f_\mu(x)=x-\mu\sin(2\pi x_t)$ we recover Chirikov's standard map. In the strong dissipation limit ($\gamma=1$) we recover the 1-d map (2.2) treated before.

In the large friction regime ($1-\gamma\ll 1$) it is possible to treat Eqs. (5.1-2) perturbatively [20]. It turns out that there is a crossover from a critical exponent 1/z to an exponent 1/2. This is illustrated in Fig. 7 for a map characterized by an exponent z=4. The critical exponent 1/z=1/4 goes over into an exponent 1/2 close to the transition. The analytic results will be published elsewhere [20].

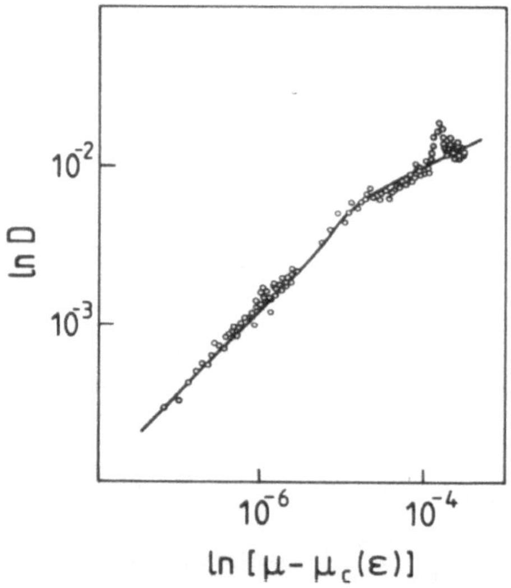

Fig. 7 Diffusion coefficient for a map whose maximum is characterized by an exponent z=4. The full line shows the analytic result [20].

ACKNOWLEDGEMENTS

I thank J. Nierwetberg and S. Thomae for many discussions, who were my co-workers for the yet unpublished work of the last three sections. I also acknowledge financial support by Deutsche Forschungsgemeinschaft.

REFERENCES

1. R.M. May, Nature 261, 459 (1976).
2. M.J. Feigenbaum, J. Stat. Phys. 19, 25 (1978) and 21, 669 (1979).
3. T. Geisel and J. Nierwetberg, Phys. Rev. Letters 47, 975 (1981).
4. P. Manneville and Y. Pomeau, Phys. Letters 75A, 1 (1979) and Physica 1D, 219 (1980).
5. J.E. Hirsch, B.A. Huberman and D.J. Scalapino, Phys. Rev. A 25, 519 (1982).
6. J.-P. Eckmann, L. Thomas and P. Wittwer, J. Phys. A 14, 3153 (1981).
7. see e.g. R. Shaw, Z. Naturforschung 36a, 80 (1981).
8. T. Geisel and J. Nierwetberg, Phys. Rev. Letters 48, 7 (1982).
9. B.A. Huberman, J.P. Crutchfield and N.H. Packard, Appl. Phys. Letters 37, 750 (1980).
10. A. Reithmayer, diploma thesis, Universität Regensburg 1982.
11. S. Grossmann and S. Thomae, Z. Naturforschung 32a, 1353 (1977).
12. H. Haken and G. Mayer-Kress, Z. Phys. B 43, 185 (1981).
13. A. Lasota and J.A. Yorke, Trans. Am. Math. Soc. 186, 481 (1973).
14. T. Geisel and J. Nierwetberg, in Dynamical Systems and Chaos, Lecture Notes in Physics 179, ed. L. Garrido (Springer, Berlin 1983) p. 93.
15. S. Grossmann and H. Fujisaka, Phys. Rev. A 26, 1779 (1982) and Z. Phys. B 48, 261 (1982).
16. M. Schell, S. Fraser and R. Kapral, Phys. Rev. A 26, 504 (1982).
17. T. Geisel and J. Nierwetberg, to be published.
18. T. Geisel and S. Thomae, to be published.
19. B.V. Chirikov, Phys. Rep. 52, 263 (1979).
20. J. Nierwetberg and T. Geisel, to be published.

STOCHASTIC SPACE-TIME PROBLEMS

Ludwig Arnold and Peter Kotelenez

Fachbereich Mathematik, Universität

D-2800 Bremen 33, West Germany

1. REACTION-DIFFUSION MODELS

1.1 Introduction and Motivation

As an example let us look at the following reaction scheme in a d-dimensional reactor sitting on $\Omega \subset R^d$ with volume V :

$$A + X \underset{k_1'}{\overset{k_1}{\rightleftarrows}} 2X \ ,$$

$$C \underset{k_2'}{\longleftarrow} B + X \ .$$

Assume that the concentration of A, B and C is kept constant. Then the concentration $x = X/V$ of the reactant X satisfies the following partial differential equation (PDE) :

$$\frac{\partial x}{\partial t}(r,t) = f(x(r,t)) + D \, \Delta \, x(r,t) \ , \qquad \text{initial and boundary conditions} \ , \tag{1}$$

where $r \in \Omega \subset R^d$, $t \geqq o$, $D > o$, $\Delta = \sum_{i=1}^{d} \frac{\partial^2}{\partial r_i^2}$ the

Laplacian, $f(x) = \lambda(x) - \mu(x)$ with $\lambda(x) = k_1 ax$ the gain term and $\mu(x) = k_1' x^2 + k_2' bx$ the loss term. This is the phenomenological or macroscopic description, and (1) is the equation of reaction and diffusion.

However, as explaned by Haken [10] and Nicolis and Prigogine [18], in order to understand pattern formation one needs to take fluctuations into account. This is done by working on the mesoscopic level, where the main object is the general noisy evolution equation, or, more mathematically, a nonlinear stochastic partial differential equation (SPDE)(cf. Haken [11]).

It is the aim of these lectures to develop rigorous mathematical tools for handling certain space-time models including fluctuations. As pointed out by Van Kampen [20], the correction of (1) and its solution caused by fluctuations cannot be obtained by just adding (white) noise to (1) but only by a detailed analysis on the mesoscopic level. Infact, our final result in Section 3 will be that the concentration with fluctuations included is $x(r,t) + Y(r,t)/\sqrt{V}$, where $Y(r,t)$ satisfies an SPDE .

1.2 Deterministic Model. Sobolev Spaces

Let us fix the following set-up for (1):

$$\lambda(x) = \sum_{i=0}^{p} a_i x^i, \ a_i \geqq 0; \ \mu(x) = \sum_{i=1}^{q} b_i x^i, \ b_i \geqq 0;$$

$$f(x) = \lambda(x) - \mu(x) = \sum_{i=0}^{n} c_i x^i, \ c_0 \geqq 0 ,$$

where we assume $c_n < 0$ if $n > 1$ to avoid explosion. For the sake of simplicity we put $d = 1$, $\Omega = [0,1]$, $D = 1$ and treat zero flux boundary conditions:

$$\frac{\partial x}{\partial r}(0,t) = \frac{\partial x}{\partial r}(1,t) \quad \text{for all} \quad t \geqq 0 .$$

The initial condition is $x(r,0) = x_0(r) \geqq 0$, and $\Delta = \partial^2/\partial r^2$.

A PDE like (1) is most conveniently treated in the set-up of Sobolev spaces for the r-coordinate which we are going to introduce now.

Set

$$H^o = L_2(0,1) = \{\varphi : \int_o^1 \varphi^2(r)dr < \infty\} .$$

This is a Hilbert space with inner product $<\varphi,\psi>_o = \int_o^1 \varphi(r)\psi(r)dr$ and norm $|\varphi|_o^2 = <\varphi,\varphi>_o$. Furthermore, for any $m \geqq 1$ define the Sobolev spaces (all derivatives are in the distributional sense!)

$$H^m = \{\varphi : \varphi, \varphi', \ldots, \varphi^{(m)} \in H^o\} .$$

These are also Hilbert spaces with $<<\varphi,\psi>>_m = \sum_{i=o}^{m} <\varphi^{(i)},\psi^{(i)}>_o$, and

$$\ldots \subset H^m \subset H^{m-1} \subset \ldots \subset H^2 \subset H^1 \subset H^o = L_2(0,1).$$

Denote by $C^m[0,1]$ the m times continuously differentiable functions on $[0,1]$. Then by the Sobolev imbedding theorem (Adams [1]) H^m can be compactly imbedded into $C^{m-1}[0,1]$, $m \geq 1$, $H^m \xrightarrow{\text{comp.}} C^{m-1}[0,1]$ (in particular, H^m can be considered a subset of $C^{m-1}[0,1]$).

We now build-in our boundary conditions into H^2 by putting

$$H_2 = \{\varphi \in H^2 : \varphi'(o) = \varphi'(1) = o\}$$

(the derivative making sense by $H^2 \subset C^1[0,1]$).

Theorem (Kuiper [15]). If $x_o \in H_2$ then (1) has a unique global solution $x(r,t)$ such that $x(\cdot,o) = x_o$ and

$$x \in C^1([o,\infty),H^o) \cap C^o([o,\infty),H_2).$$

This solution x will be used as a deterministic reference solution around which we will expand the fluctuating concentration (Van Kampen's approximation).

1.3 Markov Jump Processes

A stochastic process $X(t)$, $t \geqq o$, with a countable state space E is called a (homogeneous) Markov jump process if

$$P(X(t+s)=j|X(s)=i,X(t_n)=i_n,\dots,X(t_1)=i_1)=P(X(t+s)=j|X(s)=i) =: p_{ij}(t)$$

for all $i_1,\dots,i_n,i,j \in E$; $0 \leq t_1 \leq \dots \leq t_n \leq s \leq t + s$. The transition probabilities $p_{ij}(t)$ satisfy $p_{ij}(t) \geq 0$, $\sum_{j\in E} p_{ij}(t) = 1$, $p_{ij}(t+s) = \sum_{k\in E} p_{ik}(s)p_{kj}(t)$ (Chapman-Kolmogorov equation). All finite-dimensional distributions of $X(t)$ are uniquely determined by $p_{ij}(t)$ and an initial distribution $P(X(0)=i) = p_i$. Moreover, the transition intensities $q_{ij} \geq 0$ defined by

$$p_{ij}(\Delta t) = P(X(t+\Delta t) = j|X(t) = i) = q_{ij}\,\Delta t + o(\Delta t)$$

exist and are finite for $i \neq j$. For $i = j$, $1-p_{ii}(\Delta t) = q_i\Delta t + o(\Delta t)$, where $0 \leq q_i \leq \infty$. Assume from now on $0 < q_i < \infty$ and $\sum_{\substack{j\in E\\ j\neq i}} q_{ij} = q_i$. Then we always have Kolmogorov's backward equation

$$\dot{p}_{ij}(t) = \sum_{k\in E} q_{ik}p_{kj}(t) ,$$

while Kolmogorov's forward or Fokker-Planck equation

$$\dot{p}_{ij}(t) = \sum_{k\in E} p_{ik}(t)q_{kj}$$

holds only under certain restrictions and is a more delicate object. Summing the last equation over i yields an equation for the absolute probabilities $p_j(t) = P(X(t)=j)$,

$$\dot{p}_j(t) = \sum_{k\in E} p_k(t)q_{kj},$$

the so-called <u>Master equation</u>. Notice that it only gives the one dimensional distributions of $X(t)$. If $Q = (q_{ij})$ with $q_{ii} = - q_i$, then Q uniquely determines the $p_{ij}(t)$'s provided no explosion from the state space E in finite time takes place, which we will assume from now on.

The process $X(t)$ with intensity matrix Q can be described as follows: It has piecewise constant paths, it stays in state i with an exponential holding time

τ_i with $E\tau_i = 1/q_i$, and after time τ_i it jumps to state j with transition probability q_{ij}/q_i

For more information on Markov jump processes see Gihman and Skorohod [9].

1.4 Stochastic Model of Reaction with Diffusion

Since $d = 1$ we put $V = L$ and subdivide our total volume L into N cells of size $l = L/N$. In each cell reaction is going on, while adjacent cells are coupled by "diffusion". The whole phenomenon is modeled as a Markov jump process $X(t) = (X_1(t),\ldots,X_j(t),\ldots,X_N(t))$ with state Space $E = \mathbb{N}^N = \{k=(k_1,\ldots,k_N): k_j \in \mathbb{N} = \{1,2,\ldots\}\}$, where $X_j(t)$ = number of particles of reactand X in the j-th cell.

We prescribe the following transition intensities (Nicolis and Prigogine [18], Haken [10], Arnold and Theodosopulu [3]): Let $e_j = (0,\ldots,0,\underset{\substack{\uparrow \\ \text{j-th place}}}{1},0,\ldots,0)$, $k,m \in \mathbb{N}^N$.

Then reaction is described by

$$q_{k,k\pm e_j} = \begin{cases} \lambda_{k_j} = l\lambda_l\left(\frac{k_j}{l}\right), & \lambda_l(x) = \sum_{i=o}^{p} a_i x\left(x-\frac{1}{l}\right)\ldots\left(x-\frac{i-1}{l}\right) \\ u_{k_j} = l\mu_l\left(\frac{k_j}{l}\right), & \mu_l(x) = \sum_{i=1}^{q} b_i x\left(x-\frac{1}{l}\right)\ldots\left(x-\frac{i-1}{l}\right) \end{cases},$$

(the a_i and b_i are the coefficients of the gain and loss polynomials $\lambda(x)$ and $\mu(x)$ introduced in 1.2) while diffusion is described by

$$q_{k,k+e_{j+1}-e_j} = \frac{D^*}{2}k_j = q_{k,k+e_{j-1}-e_j}, \quad D^* > o.$$

We put $q_{k,m} = o$ otherwise. This way we have automatically built-in zero flux boundary conditions, i.e. reflection in cell 1 and N.

It can be checked that those intensities together with an initial distribution uniquely define a Markov jump process.

1.5 Consistency of the Stochastic and Deterministic Model (Thermodynamic Limit)

The problem of consistency is as follows: How do the three independent parameters L, l and D^* (we have $Nl = L$) of the stochastic model have to move in order that $X(t)$ and the deterministic solution $x(r,t)$ of (1) are "close"?

To be able to compare $X(t)$ and $x(r,t)$ we turn $X(t)$ into a step function and density on [0,1] by

$$X(r,t) = \frac{X_j(t)}{l} \quad \text{for} \quad r \in (\tfrac{j-1}{N}, \tfrac{j}{N}],\ j = 1,\dots,N .$$

Obviously, H^o is the "best" space in our H^m scale which contains both $X(\cdot,t)$ and $x(\cdot,t)$.

Theorem 1 (Law of large numbers in $H^o = L_2(0,1)$, cf. Arnold and Theodosopulu [3], Kotelenez [12]). Suppose

(i) $L \to \infty$ and $N \to \infty$,

(ii) $D^*/2 = N^2$,

(iii) $\lim E|X(o)-x_o|_o^2 = 0$,

(iv) $D^*/l \to 0$, and degree $(f) = n \leqq 6$.

Then there is a sequence of uniformly bounded stopping times $\tau_m \uparrow \infty$ such that

$$\lim_{D^*,l,L} E\Big(\sup_{o\leqq t\leqq \tau_m} |X(\cdot,t)-x(\cdot,t)|_o\Big)^2 = 0.$$

The proof is a combination of martingale theory and functional analysis and will be sketched for a later version of the law of large numbers. Condition (ii) is needed to get the Laplacian in the limit. Condition (iv) expresses dominance of reaction over diffusion and entails in particular $l \to \infty$ quite fast. This condition cannot be relaxed in H^o .

1.6 Consistency of the Stochastic and Deterministic Model (continued)

Can we get rid of condition $D^*/l \to o$ if we are content with a weaker kind of convergence? What is the strongest topology in H^o weaker than the norm topology in which $X(t)$ and $x(t)$ look close under the minimal conditions $L,N \to \infty$, $D^*/2 = N^2$? A hint is given by

$$\lim_{\substack{L,N\to\infty \\ D^*/2=N^2}} E\langle X(t)-x(t),\varphi\rangle_o^2 = o$$

for any C^∞ test function φ on $[0,1]$.

We will now introduce a continuous scale of Sobolev Hilbert spaces as follows. Let I be the identitiy operator on H^o . Then $I-\Delta$ is an unbounded, selfadjoint, positive operator on H^o with domain H_2 . It has discrete spectrum $\lambda_k = 1 + k^2\pi^2$, $k = 0,1,\ldots$ with $\varphi_o = 1$, $\varphi_k = \sqrt{2} \cos k\pi r$ $(k\geq 1)$ a complete orthonormal system of eigenfunctions in H^o . Note that $\varphi_k \in C^\infty$.

Definition. For $\alpha \in R$ let

$$H_\alpha = \{\text{distributions } \varphi \text{ on } [0,1]: \sum_o^\infty \lambda_k^\alpha \, \varphi(\varphi_k)^2 < \infty \} .$$

H_α is a Hilbert space with inner product

$$\langle\varphi,\psi\rangle_\alpha = \sum_o^\infty \lambda_k^\alpha \, \varphi(\varphi_k)\psi(\varphi_k) ,$$

$H_o = H^o$, H_2 = our old H_2 with a different, but equivalent norm. We identify H_o with its dual H_o' and obtain a scale of Sobolev spaces, such that for $\beta \geq \alpha \geq o$

$$H_\beta \subset H_\alpha \subset H_o = H_o' \subset H_{-\alpha} \subset H_{-\beta}$$

with continuous injections. While for $\alpha \geq o$

$$H_\alpha = \{\varphi \in H_o: |(I-\Delta)^{\alpha/2}\varphi|_o < \infty \} ,$$

$H_{-\alpha}$ can be considered as the completion of H_o w.r.t. the weaker inner product

$$\langle\varphi,\psi\rangle_{-\alpha} = \langle (I-\Delta)^{-\alpha/2}\varphi,\ (I-\Delta)^{-\alpha/2}\psi\rangle_o,\quad \varphi,\psi \in H_o .$$

Thus $H_{-\alpha}$ $(\alpha>o)$ is a space that contains genuine distributions. E.g. for $r_o \in [0,1]$ $\delta_{r_o} \in H_{-\alpha}$ for $\alpha > 1/2$ since

$$|\delta_{r_o}|_{\alpha}^2 = \sum_{o}^{\infty} \lambda_k^{\alpha}\, \varphi_k(r_o)^2 < \infty \quad \text{iff} \quad \alpha > 1/2$$

Similarly, $\delta'_{r_o} \in H_{-\alpha}$ for $\alpha > 3/2$.

The injection $H_\beta \to H_\alpha$ is even compact for $\beta > \alpha$ it is Hilbert-Schmit if $\beta > \alpha + 1/2$, and it is nuclear for $\beta > \alpha + 1$.

Since two elements in $H_o = L_2(0,1)$ look closer in the norms of $H_{-\alpha}$, $\alpha > o$ (e.g. $|\varphi_k|_{-\alpha}^2 = 1/(1+k^2)^\alpha$) , we expect the law of large numbers to hold under weaker assumptions.

Theorem 2 (Law of large numbers in $H_{-\alpha}$, improvement of Kotelenez [12]). Suppose the reaction is linear, $\varepsilon > o$, and

(i) $L \to \infty$, $N \to \infty$,

(ii) $D^*/2 = N^2$,

(iii) $\lim E|X(0)-x_o|^2_{-\frac{3}{2}-\varepsilon,N} = 0$, $\sup_{D^*,l,N}\ \sup_{o\le r\le 1} |EX(r,o)| < \infty$,

(iv) $l \to \infty$.

Then there is a sequence of uniformly bounded stopping times $\tau_m \to \infty$ such that

$$\lim_{D^*,l,L} E\Big(\sup_{o\le t\le \tau_m} |X(\cdot,t)-x(\cdot,t)|_{-\frac{3}{2}-\varepsilon}\Big)^2 = 0 .$$

Moreover, if (i) and (ii) hold and (iii) is valid in $H_{-5/2-\varepsilon}$ then (iv) can be dispensed with, and the result holds in $H_{-5/2-\varepsilon}$. The results are best possible in the sense that they are not valid anymore for $\varepsilon = 0$.

In condition (iii) $|\varphi|_{-\alpha,N} := |(I-\Delta_N)^{-\alpha/2}\varphi|_o$, where Δ_N is the discrete Laplacian on H_o defined by

$$\Delta_N x(r)=N^3 \Big(\int\limits_{(\frac{j}{N},\frac{j+1}{N}]} x \; - \; 2 \int\limits_{(\frac{j-1}{N},\frac{j}{N}]} x \; + \int\limits_{(\frac{j-2}{N},\frac{j-1}{N}]} x \Big), \; r\in(\tfrac{j-1}{N},\tfrac{j}{N}], j=1,\dots,N.$$

A one line "proof" is as follows: If we subtract the systematic part from $X(t)$,

$$X(t) - X(o) - \int_o^t (f(X(s)) + \Delta_N X(s))ds = Z(t),$$

we obtain a martingale $Z(t)$ ("pure fluctuations"). Since $\Delta_N \to \Delta$, $Z(t) \to 0$ would entail $X(t) \to x(t)$ satisfying (the integrated version of) equation (1). $Z(t) \to o$ is assured by the conditions.

In concluding we would like to emphasize that with Theorems 1 and 2 we have different versions of the law of large numbers at our disposal. Depending on the norm in which we want to see our Markov jump process $X(\cdot,t)$ close to the deterministic $x(\cdot,t)$ we have the choice between the very restrictive condition for the cell size $N^2/l \to 0$ in H_o, the less restrictive condition $l \to \infty$ in $H_{-3/2-\varepsilon}$ and no condition whatsoever in $H_{-5/2-\varepsilon}$.

2. STOCHASTIC PARTIAL DIFFERENTIAL EQUATIONS

2.1 Motivation

Our next aim is to study the error in the law of large numbers. In Section 3 we will see that $\sqrt{L}(X(r,t)-x(r,t))$ "converges" to $Y(r,t)$, where Y satisfies the following SPDE :

$$\frac{\partial}{\partial t} Y(r,t) = (\Delta+c_1)Y(r,t) + G(r,t)\dot{\xi}(r,t) , \qquad (2)$$

where $G(r,t) = (-\frac{\partial}{\partial r}2x(r,t)\frac{\partial}{\partial r} + (\lambda+\mu)(x(r,t)))^{1/2}$ and $\xi(r,t)$ = "white noise" in function space. Here we have assumed linear reaction, i.e. $\lambda(x) = a_0 + a_1x$, $\mu(x) = b_1x$, $f(x) = \lambda(x)-\mu(x) = c_0 + c_1X$, and $(\lambda+\mu)(x) = a_0 + (a_1+b_1)x$.

In this section we would like to develop a rigorous mathematical set-up for an equation like (2). There are two problems which account for the difficulties in infinite dimensional stochastic analysis:

(A) The interpretation of "white noise" $\xi(r,t)$,

(B) The unboundedness of Δ.

(A) is solved for finite dimensions ($\xi(r,t) \equiv \xi(t)$ independent of r) by setting $\xi(t)dt = dW(t)$, $W(t)$ a finite dimensional Wiener process, and applying the Itō calculus. In infinite dimensions there are several possibilities:

- One may use a Wiener process with two dimensional "time" (r,t) (Brownian sheet) and a corresponding multiparameter Itō calculus of integration.
- One may look at $\xi(\cdot,t)$ as a function valued or distribution valued (generalized) stochastic process in t and apply an Itō calculus for some sort of function valued or distribution valued Wiener process $W(\cdot,t)$, $\xi(\cdot,t)dt = dW(\cdot,t)$. The mathematical problems involved in the definition of $W(\cdot,t)$ again give rise to different approaches:
- Abstract Wiener spaces (Kuo [16]),
- Cylinder measure valued processes (Balakrishnan [4]),
- Measure valued processes (Dawson [8]).

(B) is coped with by the same tools as for deterministic PDE:

- One can look for classical solutions:
- One can adopt the functional analysis approach where Δ is viewed as an unbounded operator from H_o into H_o etc. Now one searches for strong solutions being in the domain of Δ.
- A weaker form of solution can often be obtained when strong solutions do not exist by either

 i) the variational approach (Pardoux [19], Krylov and Rozowskij [14])

or

 ii) the semigroup approach (Da Prato [7], Curtain and Pritchard [6], Kotelenez [12]).

Our approach to (2) is a combination of the semigroup approach and the abstract Wiener space approach in connection with Hilbert-Schmidt imbeddings in the Sobolev space set-up of section 1.6

2.2 Deterministic Evolution Equations. Semigroups

We follow Curtain and Pritchard [6]. Let H be a separable Hilbert space and $A: H \to H$ an unbounded densely defined closed operator with domain $\mathcal{D} \subset H$. Assume that A is the generator of a (strongly continuous) semigroup $T(t)$, $t \geqq 0$, of bounded linear operators on H, i.e.

$$T(t) \in L(H), T(0)=I, T(t+s)=T(t)T(s), T(t)z \to z\ (t \to 0) \quad \text{for all } z \in H,$$

$$\lim_{t \downarrow 0} \frac{T(t)z-z}{t} = Az \quad \text{for all } z \in \mathcal{D} .$$

Let us now look at the following evolution equation:

$$\dot{z} = Az + f(t), \quad t \geqq 0, \quad z(0) = z_0 \in H , \tag{3}$$

with $f\colon [0,\infty) \to H$, $f \in L_p([0,T],H)$.

Theorem. If $z_0 \in \mathcal{D}$ and $f \in C^1([0,T],H)$ then (3) has a unique strong solution in $[0,T]$. This solution is given by

$$z(t) = T(t)z + \int_0^t T(t-s)f(s)ds. \tag{4}$$

The r.h.s. of (4) makes sense under much weaker conditions on f, e.g. if f is Bochner integrable. It is called the mild solution of (3). In other words, mild solutions may exist where strong solutions do not exist. But if a strong solution exists it is equal to the mild solution.

Example: Let $H = L_2(0,1)$, $A = \Delta$.

a) for $\mathcal{D} = H_2$ (zero flux b.c.)

$$T_t z = \sum_0^\infty \exp(-k^2\pi^2 t)\langle z,\varphi_k\rangle\varphi_k ,$$

$$\varphi_o = 1, \ \varphi_k(r) = \sqrt{2} \cos k\pi r , \ k \geqq 1 .$$

b) for $\mathcal{D} = H^2 \cap \{\varphi(0) = \varphi(1) = 0\}$ (Dirichlet b.c.)

$$T_t z = \sum_1^\infty \exp(-k^2\pi^2 t)\langle z,\psi_k\rangle\psi_k ,$$

$$\psi_k(r) = \sqrt{2} \sin k\pi r , \ k \geqq 1 .$$

2.3 Wiener Process and Stochastic Integrals in Hilbert space

A one dimensional Wiener process $W(t)$ is a stochastic process with continuous trajectories, $W(o) = o$ and independent increments $W(t) - W(s)$ distributed according to $N(o, |t-s|)$. Its derivative $\dot{W}(t) = \xi(t)$ is white noise, a generalized process.

A Wiener process $W(t)$ in R^n can be defined by $W(t) = \sum_{i=1}^{n} W_i(t) e_i$, where (e_i) is an orthogonal base and $(W_i(t))$ are n independent one dimensional Wiener processes. $W(t) - W(s)$ is $N(O, |t-s|I)$-distributed, I=identity operator on R^n.

We could attempt to define a Wiener process $W(t)$ on a Hilbert space H with $\dim H = \infty$ and for a CONS (φ_k) as follows: $(\langle W(t), \varphi_k \rangle)$ is a sequence of independent one dimensional Wiener processes. Its time derivative could be called "white noise in space and time". Unfortunately, there is only a finitely-additive measure on H with marginals $\langle W(t), \varphi_k \rangle \sim N(o,t)$, and the corresponding $W(t)$ is called cylindrical H-valued Wiener process. $W(t)$ has the operator $t\ I$, I being the identity operator on H, as its covariance operator.

On the other hand, C is the covariance operator of a proper probability measure on H iff C is a nuclear nonnegative selfadjoint operator, i.e. for which $\sum_k \langle C\varphi_k, \varphi_k \rangle < \infty$.

There are now two ways of arriving at a proper Wiener process on a Hilbert space:

a) Pick a CONS (φ_k) in H, $\lambda_k \geqq 0$, with $\sum_k \lambda_k < \infty$, a sequence $(W_k(t))$ of independent one dimensional Wiener processes and put

$$W(t) = \sum_k \sqrt{\lambda_k}\, W_k(t) \varphi_k .$$

Then $W(t)$ is an H-valued process with continuous trajectories, $W(o) = o$ and independent increments $W(t) - W(s)$ distributed according to $N(O, |t-s|C)$, C being a nuclear operator defined by $C\varphi_k = \lambda_k \varphi_k$. In particular, $\langle W(t),\varphi_k\rangle \sim N(0,t\lambda_k)$, $E|W(t) - W(s)|^2 = |t-s|$ trace $C = |t-s| \Sigma \lambda_k$. There is some arbitrariness due to the free choice of the nuclear operator C.

b) Abstract Wiener space set-up: Start with the cylindrical Wiener process on H (i.e. with "white noise in space and time") and complete H with respect to a (weaker) measurable inner product (Kuo [16]) resulting in a Hilbert space $\bar{H} \supset H$. The triple $(i,H,\bar{H})$, $i: H \to \bar{H}$ the injection mapping, is an abstract Wiener space, and the cylindrical Wiener process becomes a proper Wiener process in the sense of a) with a nuclear covariance in $\bar{H}$.

As an application to our set-up developed in section 1.6 $(i_\alpha,H_0,H_{-\alpha})$ is an abstract Wiener space for any $\alpha > 1/2$. Hence, our white noise $\xi(r,t)$ in space and time defines an H_0-valued cylindrical Wiener process, but for $\alpha > 1/2$ it is a proper $H_{-\alpha}$-valued Wiener process $W_{-\alpha}(t)$ with (nuclear) covariance operator $S_{-\alpha}$ = extension of $(I-\Delta)^{-\alpha}$ to $H_{-\alpha}$.

Let now H,K be separable Hilbert spaces, let $W(t)$ be a K-valued Wiener process and $G(t) \in L(K,H)$ be an operator-valued process adapted to $W(t)$ such that for $T > 0$

$$\int_0^T |G(t)|^2_{L(K,H)} dt < \infty .$$

Then the definition of the stochastic integral $\int_0^t G(s)dW(s)$ can be reduced to the one dimensional case by

$$\int_0^t G(s)dW(s) = \sum_k \sqrt{\lambda_k} \int_0^t G(s)\varphi_k dW_k(s)$$

(for more general integrands and integrators cf. Metivier and Pellaumail [17]).

If $T(t)$ is a semigroup on H, the stochastic convolution type integral

$$V(t) = \int_0^t T(t-s)G(s)dW(s)$$

can be defined similarly for fixed $t \geq o$, it is adapted to $W(t)$ and measurable. If $T(t)$ is contraction-type, i.e. if $|T(t)| \leqq \exp(\beta t)$ for some $\beta \in R$ then $V(t)$ has continuous trajectories in H (Kotelenez [13]).

2.4 Stochastic Evolution Equations

We use the set-up of section 2.3 and assume that A generates a contraction type semigroup on H . Then

$$dz = Az\, dt + G(t)dW(t) \quad , \quad z(o) = z_o \quad ,$$

has the unique mild solution

$$z(t) = T(t)z_o + \int_o^t T(t-s)G(s)dW(s),$$

which has continuous trajectories in H ; $z(t)$ is a Markov process if $G(t)$ is deterministic.

A similar theory for the following more general type of stochastic evolution equations was developed by Kotelenez [13]:

$$dz = A(t)z\, dt + G(t,z(t))dM(t), \; z(o) = z_o \; .$$

Here $A(t)$ generates an evolution operator $U(t,s)$, $G(t,z)$ is adapted $\in L(K,H)$, and $M(t)$ is a K-valued semimartingale.

3. SPDE IN REACTION-DIFFUSION MODELS

3.1 Introduction. Van Kampen's Aproximation

A simple calculation shows that if we look at $Y_L(r,t) = \sqrt{L}(X(r,t) - x(r,t))$ for $L \to \infty$ we will inevitably end up with $H_{-\alpha}$ spaces $(\alpha > o)$ since convergence in H_o is impossible.

Let $D([o,\infty),H_{-\alpha})$, $\alpha \geqq o$, be the space of all $H_{-\alpha}$-valued functions on $[0,\infty)$ which are right continuous with left hand limits. This space can be given a metric which makes it a complete and separable metric space (Billingsley [5]). $Y_L(r,t)$ defines a probability measure P_L on $D([o,\infty),H_{-\alpha})$ for all $\alpha \geq o$. The problem is to find an α (if possible, the smallest one) such that there is a probability P on $D([o,\infty),H_{-\alpha})$ with $P_L \Rightarrow P$ (i.e. $\int f dP_L \to \int f dP$ for bounded and continuous f). In other words, we want to find an $H_{-\alpha}$-valued process Y such that $Y_L \Rightarrow Y$.

3.2 Central Limit Theorem

We again restrict to linear reaction, see section 2.1. The following theorem is an improvement of results obtained by Kotelenez in his PhD Thesis [12].

Theorem 3 (Central limit theorem in $H_{-7/2-\varepsilon}$; Kotelenez [12]). Let $o < \varepsilon \leqq 1/2$ and suppose

(i) the conditions for the validity of the law of large numbers in $H_{-7/2-\varepsilon}$ are satisfied (see Theorem 2, conditions (i), (ii), (iii)),

(ii) $1/N \to 0$,

(iii) $\sqrt{L}(X(o)-x_o) \Rightarrow Y_o$ in $H_{-7/2-\varepsilon}$, where Y_o is a square integrable random variable in $H_{-3/2-\varepsilon}$ independent of $X(t)$.

Then $\sqrt{L}(X(\cdot)-x(\cdot))$ converges weakly on $D([o,\infty), H_{-7/2-\varepsilon})$ to $Y(\cdot)$, where

$$Y(t) = T(t)Y_o + \int_o^t T(t-s)G_\varepsilon(s)dW_{-1/2-\varepsilon}(s) \qquad (5)$$

is the mild solution of the SPDE in $H_{-3/2-\varepsilon}$:

$$dY = (\Delta+c_1)Y\,dt + G_\varepsilon(t)dW_{-1/2-\varepsilon}(t),\ Y(o) = Y_o ,$$

$W_{-1/2-\varepsilon}(t)$ = Wiener process in $H_{-1/2-\varepsilon}$ constructed in section 2.3, $G_\varepsilon(t) \in L(H_{-1/2-\varepsilon}, H_{-3/2-\varepsilon})$ is the dual of the square root $F_\varepsilon^{1/2}(t)$ of the operator

$$F(t) = -\frac{\partial}{\partial r}\, 2x(r,t)\frac{\partial}{\partial r} + (\lambda+\mu)(x(r,t)),$$

where $F_\varepsilon^{1/2}(t)$ is considered as an element in $L(H_{3/2+\varepsilon}, H_{1/2+\varepsilon})$. Moreover, Y is a Markov process in $H_{-3/2-\varepsilon}$, and $Y \in C([o,\infty), H_{-3/2-\varepsilon})$. Y is Gaussian if Y_o is.

The proof is again a combination of functional analysis, martingale theory and the theory of weak convergence and is omitted.

Let us emphasize that (5) gives the desired explizit form for the fluctuations of $X(r,t)$ around $x(r,t)$, so that the following intuitive expression is justified:

$$X(\cdot,t) \approx x(\cdot,t) + \frac{1}{\sqrt{L}} Y(t) .$$

While $X(\cdot,t)$ and $x(\cdot,t)$ live in H_0 (infact: in $H_{1/2-\varepsilon}$), $Y(t)$ is a distribution and lives only in $H_{-3/2-\varepsilon}$. But $\sqrt{L}(X(\cdot,t)-x(\cdot,t))$ looks close to $Y(t)$ only in $H_{-7/2-\varepsilon}$.

It is an exercise to calculate the covariance operator of Y from formula (5).

References

[1] Adams, R. A.: Sobolev spaces. Academic Press, New York 1975

[2] Arnold, L.: Mathematical Models of chemical reactions. In: Hazewinkel, M.; Willems, J. (eds.): Stochastic Systems. Reidel, Dordrecht 1981

[3] Arnold, L.; Theodosopulu, M.: Deterministic limit of the stochastic model of chemical reactions with diffusion. Adv. Appl. Prob. 12 (1980), 367-379

[4] Balakrishnan, A. V.: Applied Functional Analysis. Springer, Berlin 1976

[5] Billingsley, P.: Convergence of probability measures. Wiley, New York 1968

[6] Curtain, R.; Pritchard, A. J.: Infinite dimensional linear system theory. Springer, Berlin 1978

[7] Da Prato, G.: Some results on linear stochastic evolution equations in Hilbert space by the Semigroups method. Stochastic Analysis and Appl. 1 (1983), 57-88

[8] Dawson, D.: Stochastic evolution equations and related measure processes. J. Multivariate Anal. 5 (1975), 1-52

[9] Gihman, I. I.; Skorohod, A. V.: The theory of stochastic processes, Vol. II. Springer, Berlin 1975

[10] Haken, H.: Synergetics. Springer, Berlin 1978

[11] Haken, H.: Advanced Synergetics. Springer, Berlin 1983

[12] Kotelenez, P.: Law of large numbers and central limit theorem for chemical reactions with diffusion. PhD Thesis Bremen 1982

[13] Kotelenez, P.: A stopped Doob inequality for stochastic convolution integrals and stochastic evolution equations. Report 102 (1983), Universität Bremen

[14] Krylov, N. B.; Rozowskij, B. L.: On stochastic evolution equations. Itogi Nauki i Tehniki, Ser. Sov. Probl. Math. no. 14 (1979), 71-146 (in Russian)

[15] Kuiper, H. J.: Existence and comparison theorems for nonlinear diffusion systems. J. Math. Anal. Appl. 60 (1977), 166-181

[16] Kuo, H.-H.: Gaussian measures in Banach spaces. Springer, Berlin 1975

[17] Metivier, M.; Pellaumail, J.: Stochastic integration. Academic Press, New York 1980

[18] Nicolis, G.; Prigogine, I.: Self-organization in nonequilibrium systems. Wiley, New York 1977

[19] Pardoux, E.: Equation aux derivées partielles stochastiques nonlinéaires monotones. Etude de solutions fortes de type Itô. Thêse doct. Sci. math. Univ. Paris Sud 1975

[20] Van Kampen, N. G.: Stochastic processes in physics and chemistry. North Holland, Amsterdam 1981

STOCHASTIC THEORY OF TRANSITION PHENOMENA IN NONEQUILIBRIUM SYSTEMS

G. Nicolis and C. Van den Broeck [+*]

Université Libre de Bruxelles
Faculté des Sciences, CP 226, Campus Plaine
Bvd du Triomphe, 1050 Bruxelles, Belgium

[+]Vrije Universiteit Brussel
Dept. Nat., Pleinlaan 2
1050 Brussel, Belgium

1. INTRODUCTION

1A. General Formulation

Cooperative behaviour in nature is intimately connected with the ability of large classes of dynamical systems to undergo transitions towards qualitatively different types of behaviour. The most familiar example is primary bifurcation (see also lectures by P. Fife and L. Lugiato). Let

$$\frac{\partial \underset{\sim}{X}}{\partial t} = \underset{\sim}{F}(\underset{\sim}{X}, \underset{\sim}{\lambda}) \tag{1.1}$$

be a dynamical system in which $\underset{\sim}{X}$ is the set of state variables viewed as a vector in an appropriate space, $\underset{\sim}{F}$ a functional acting on $\underset{\sim}{X}$, and $\underset{\sim}{\lambda}$ a set of parameters through which the system can be controlled externally. Among the multitude of solutions that may be available, we isolate a particular reference state $\underset{\sim}{X}_s$, because of its simplicity or its physical importance. For instance, in an autonomous system subject to a homogeneous environment, $\underset{\sim}{X}_s$ may represent a uniform steady-state solution. We introduce the deviation $\underset{\sim}{x}$ from $\underset{\sim}{X}_s$ through :

---

[*]Aangesteld Navorser Nationaal Fonds voor Wetenschappelijk Onderzoek, Belgium.

$$\underset{\sim}{X} = \underset{\sim}{X}_s + \underset{\sim}{x} \tag{1.2}$$

and study the dynamical system :

$$\begin{aligned} \frac{\partial \underset{\sim}{x}}{\partial t} &= \underset{\sim}{F}(\underset{\sim}{X}_s + \underset{\sim}{x}, \underset{\sim}{\lambda}) - \underset{\sim}{F}(\underset{\sim}{X}_s, \underset{\sim}{\lambda}) \\ &= \left(\frac{\partial \underset{\sim}{F}}{\partial \underset{\sim}{X}}\right)_s \cdot \underset{\sim}{x} + \frac{1}{2}\left(\frac{\delta^2 \underset{\sim}{F}}{\delta \underset{\sim}{X} \delta \underset{\sim}{X}}\right)_s : \underset{\sim}{x}\,\underset{\sim}{x} + \ldots \\ &\equiv \underset{\approx}{\mathcal{L}}(\lambda) \cdot \underset{\sim}{x} + \underset{\sim}{h}(\underset{\sim}{x}, \underset{\sim}{\lambda}) \end{aligned} \tag{1.3}$$

in which $\underset{\approx}{\mathcal{L}}$ is a linear operator and $\underset{\sim}{h}$ is the nonlinear part of $\underset{\sim}{F}$.

Let us first limit ourselves to the case in which $\underset{\sim}{X}_s$ is time-independent and the initial system, eq. (1.1) is autonomous. Classical stability analysis tells us then that the behaviour of the deviations x is determined by the eigenvalue problem :

$$\underset{\sim}{x} = \underset{\sim}{u}\, e^{\omega t}$$

$$\underset{\approx}{\mathcal{L}}(\underset{\sim}{\lambda}) \cdot \underset{\sim}{u} = \omega \underset{\sim}{u} \tag{1.4}$$

At this point, it becomes necessary to make the distinction between the two fundamental classes of dynamical systems encountered in nature, namely conservative and dissipative systems. In the first class, eqs. (1.1) and (1.3) enjoy time reversal invariance. As a result ω can be either purely imaginary or, if it is real, be part of a multiplet in which the sum of the real parts of all ω's is zero. In each case, there will be no decay of $\underset{\sim}{x}$ possible or, in more technical terms, no asymptotic stability. In the second class on the other hand, the operators in eqs. (1.1) or (1.3) are dissipative operators, compatible with the second law of thermodynamics. The conditions constraining the eigenspectrum ω in the conservative case are no longer present and, as a result, Reω can be negative or positive ensuring asymptotic stability or instability. If asymptotic stability prevails, the state $\underset{\sim}{X} = \underset{\sim}{X}_s$ or $\underset{\sim}{x} = \underset{\sim}{0}$ will be an attractor.

In what follows, we will be concerned solely with dissipative systems. Obviously, if $\underset{\sim}{x} = \underset{\sim}{0}$ is asymptotically globally stable for all $\underset{\sim}{\lambda}$'s, no transition to new states will be possible. Conversely, one possibility for observing transitions capable of leading to cooperative behaviour is that shown in fig. 1. At $\underset{\sim}{\lambda} = \underset{\sim}{\lambda}_c$, we have marginal stability and, under quite general conditions, we can expect bifurcation of new branches of solution, as shown in fig. 2. We call this phenomenon primary bifurcation. Typically beyond such a bifurcation, we observe multiple steady states (e.g. in Benard convection or in the laser), sustained oscillations (e.g. chemical clocks), or the spontaneous formation of space structures (e.g. morphogenesis).

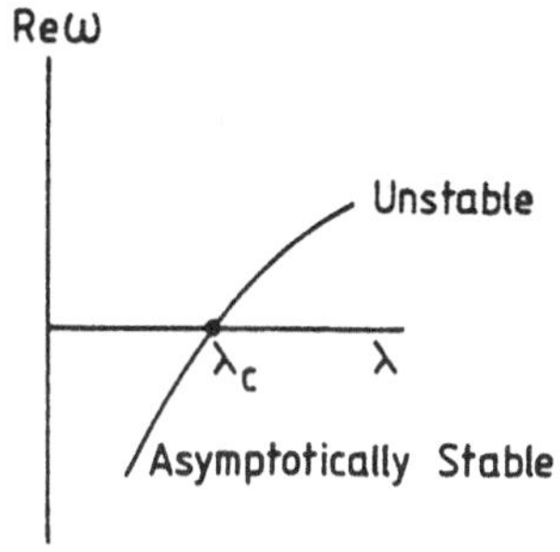

Figure 1

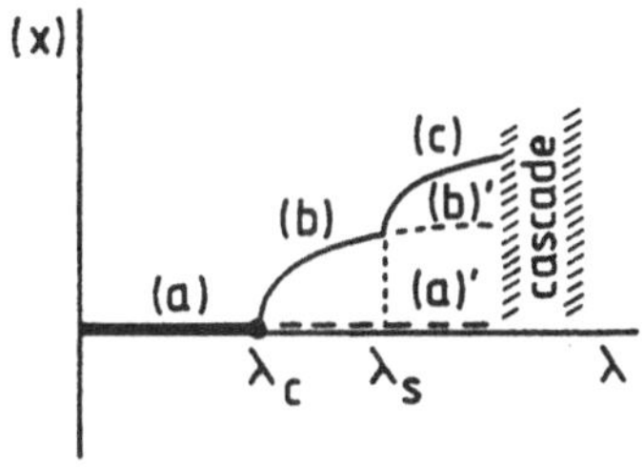

Figure 2

A second, more complex type of transition can now be introduced. Suppose that the primary branch (b) loses in turn its stability at a critical value $\underset{\sim}{\lambda} = \underset{\sim}{\lambda}_s$. New stable branches can then emerge at such a secondary bifurcation point, which can be subject to further instabilities. At present one knows several types of such cascades, some of which can lead to chaotic behaviour through intermediate steps involving, for instance, quasi-periodic solutions or period doubling. The latter has been observed in numerous experiments in fluid dynamics, chemical kinetics and optics.

The bifurcations surveyed above refer to the asymptotic behaviour of our dynamical systems, occurring in the limit of very long times, $t \to \infty$. A different class of important transition phenomena is associated to the transient evolution toward a stable attractor, which may even be the unique asymptotic solution available to the system. For instance (see fig. 3), in many cases, the state variable evolves slowly during an induction period, and then it abruptly switches in an explosive fashion to a plateau value $|x_\infty|$, at a critical ignition time t_c. Such a behaviour is observed in thermal and chemical combustion or in optical switching devices.

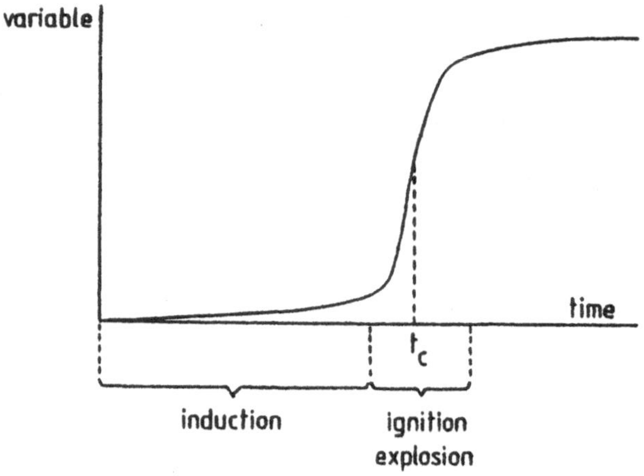

Figure 3

In what follows, we shall be concerned with the stochastic aspects of primary bifurcations and explosive behaviour. To our knowledge, cascading bifurcations have so far hardly been touched upon by stochastic methods. Their study constitutes one of the major open problems in this field. Before we give the motivation for undertaking a stochastic analysis, we briefly introduce two representative models which will recur continuously throughout these notes to illustrate the main ideas and techniques.

1B. Simple Models

The simplest example of primary bifurcation, namely bifurcation toward multiple homogeneous steady states, is nicely illustrated by the following chemical scheme, the so-called Schlögl model :

$$A + 2X \underset{k_2}{\overset{k_1}{\rightleftharpoons}} 3X$$

$$X \underset{k_4}{\overset{k_3}{\rightleftharpoons}} B \qquad (1.5)$$

Here, the chemical intermediate X is the only state variable, whereas A and B are used as control parameters.

Transient evolution in the form of explosive behaviour is best illustrated by combustion (see also the lectures by P. Clavin). A very simplified version of this problem is that of a single irreversible exothermic reaction in an adiabatically isolated vessel :

$$X \xrightarrow{k(T)} D \tag{1.6a}$$

in which k(T) is given by the Arrhenius law

$$k(T) = k_o \exp\left(-\frac{U_o}{RT}\right) \tag{1.6b}$$

T being the temperature, R the gas constant and U_o the activation energy.

The general structure of the rate laws for the ideal systems which we will consider is as follows. For systems involving chemical reactions and diffusion, one has :

$$\frac{\partial \bar{x}_i}{\partial t} = \sum_\rho \nu_{i\rho} W_\rho (\{\bar{x}_i\},\lambda) + \mathcal{D}_i \nabla^2 \bar{x}_i \tag{1.7}$$

where W_ρ is the rate of the ρ-th reaction (generally function of the concentrations $\{x_i\}$ and the constraints λ), $\nu_{i\rho}$ is the stocheometric constant for the substance i in this reaction, and $\mathcal{D}_i$ is the Fickian diffusion coefficient of species i. For example, in the case of Schlögl's model one has :

$$W_1 = k_1 a \bar{x}^2 \qquad W_{-1} = k_2 \bar{x}^3$$

$$W_2 = k_3 \bar{x} \qquad W_{-2} = k_4 b$$

$$\nu_{x,1} = 1,\ V_{x,-1} = -1,\ V_{x,2} = -1,\ \nu_{x,-2} = 1$$

In the thermochemical problem, the reaction diffusion equation (1.7) has to be supplemented by the energy balance equation :

$$c_V \frac{\partial T}{\partial t} = \sum_\rho r_{V\rho} W_\rho + \text{transport} \tag{1.9}$$

where T is the temperature of the reacting mixture, c_V its specific heat and $r_{V\rho}$ the heat of reaction.

Let us finally note that, in all these systems, transitions are associated with the existence of suitable <u>nonlinearities</u> in the kinetics. In model (1.5) the source of the nonlinearity is chemical autocatalysis, whereas in model (1.6) it is the thermal feedback arising from the increase of the rate k(T) with temperature.

It has to be emphasized however that near equilibrium none of these nonlinearities are effective : because of detailed balance, each step is cancelled by its reverse, and the ability to amplify perturbations from the reference state is compromised. It is only through the maintenance of sufficiently strong nonequilibrium constraints that the system becomes able to reveal fully the nonlinearities of its kinetics and perform transitions to new solutions.

1C. The Importance of Fluctuations.

As stressed by several lecturers at this Institute, fluctuations are universal phenomena, generated spontaneously by all physical systems. In actual fact therefore, we are always dealing in nature with stochastic dynamical systems. Nevertheless, as internal fluctuations originate in the form of localized, short scale events, their strength is expected to be small. Why then bother to incorporate them in the description ? After all, the theory of deterministic dynamical systems, which was marked by spectacular developments since the 1960's, does quite well in determining bifurcation and other transition points and in providing information on the new states arising beyond these transitions.

First, let us deal with some technical aspects of the question. The existence of multiple solutions in a dynamical system raises the problem of their relative weights and of the characteristic time scales of transitions between them. In the absence of fluctuations the problem of relative weight can simply not be addressed. As for transitions the only possibility is to impose externally a perturbation bringing the system outside the domain of attraction of one reference state and inside the domain of attraction of some other state. Such perturbations have to be provoked, otherwise the system will remain for ever at the reference state, even if the latter is not asymptotically stable. By incorporating fluctuations in the description, both problems are solved. Indeed, the relative weights are determined by the strength of maxima of the (multi-humped) probability distribution, whereas the transitions between states occur,at least in simple cases, as the result of fluctuations carrying the system over a suitably defined "potential barrier". This introduces in the dynamics a new, long time scale which is not built in the deterministic description but is rather related to phenomena similar to nucleation.

Now, a more fundamental point. Bifurcation or any other transition implies that, at some moment, a qualitative change occurs and encompasses a system of macroscopic dimensions. As a rule, in such a system, intermolecular forces are relatively short ranged and cannot thus explain the extraordinary coherence that has to be built and maintained. The study of fluctuations allows us to look for appropriate quantities, capable of characterizing coherent behaviour. In particular, as we shall discuss in section 4; the study of the spatial correlation function reveals the generation of correlations of macroscopic range in system, as soon as one deviates from thermodynamic equilibrium. We are therefore witnessing a transition of a new kind, in which the amplitude of the correlation function plays the role of an order parameter and which, therefore, has no deterministic analogue. A further transition phenomenon occurs near bifurcation, where the correlation length tends to infinity and order encompasses the entire system.

2. STOCHASTIC FORMULATION

2A. The Master Equation

A fully satisfactory approach to the study of fluctuations should be part of the microscopic theory of irreversible processes. In view of the complexity of the phenomena of interest, which involve both nonlinearities and large distance from equilibrium, a more phenomenological approach will be adopted in these lectures. We shall suppose that we can define an appropriate set of discrete state variables $\underset{\sim}{X} = \{X_\alpha\}$, which constitute a Markov process. As discussed in the lectures by L. Arnold, the probability function $P(\underset{\sim}{X},t)$ of the process obeys, under weak conditions of continuity and differentiability, the following gain-loss balance equation, also referred to as Master Equation :

$$\frac{dP(\underset{\sim}{X},t)}{dt} = \sum_{\rho} \{W_\rho(\underset{\sim}{X}-\underset{\sim}{\nu}_\rho \to \underset{\sim}{X}) \quad P(\underset{\sim}{X}-\underset{\sim}{\nu}_\rho,t) - W_\rho(\underset{\sim}{X}\to\underset{\sim}{X}+\underset{\sim}{\nu}_\rho)\ P(\underset{\sim}{X},t)\} \tag{2.1}$$

Here $W_\rho(\underset{\sim}{X}\to\underset{\sim}{X}+\underset{\sim}{\nu}_\rho)$ is the transition probability per unit time for the occurence of the elementary process ρ which changes the state of the system from $\underset{\sim}{X}$ tot $\underset{\sim}{X}+\underset{\sim}{\nu}_\rho$. The form of these transition probabilities has to be derived from the specific information characterizing the processes that take place in the system considered (e.g. elastic or reactive collisions, diffusion, heat conduction, etc. ...) and has to be chosen in accordance with the known equilibrium results. In the following section, we give their explicit form for the model systems (1.5) and (1.6).

2B. Birth and Death Processes

In the case of a well stirred chemical system in thermal and mechanical equilibrium, the state variables $\{X_\alpha\}$ are taken to be the total number of particles of the chemical species $\{\alpha\}$, and the elementary processes ρ correspond to the different chemical reactions that can take place in the system. We will introduce for illustrative purposes two further simplifications : only one chemical intermediate is considered and its number of particles X can change by amounts $\nu_\rho=\pm 1$. The Master Equation (2.1) takes the following simple form :

$$\frac{d}{dt} P(X,t) = \lambda(X-1)\ P(X-1,t) + \mu(X+1)\ P(X+1,t) - [\lambda(X) + \mu(X)]\ P(X,t) \qquad X\geq 0$$

$$\frac{d}{dt} P(0,t) = -\lambda(0)\ P(0,t) + \mu(1)\ P(1,t) \tag{2.2}$$

$\lambda(X)$ and $\mu(X)$, the so-called birth and death rates respectively, are given by

$$\lambda(X) = \sum_\rho W_\rho(X\to X+\nu_\rho)\ \delta^{Kr}_{\nu_\rho,1} \tag{2.3a}$$

$$\mu(X) = \sum_\rho W_\rho(X\to X+\nu_\rho)\ \delta^{Kr}_{\nu_\rho,-1} \tag{2.3b}$$

Since the chemical reactions are localized random events, the probability for their occurrence must scale like the system size or any other appropriate extensivity parameter N :

$$\lambda(X) = N\ \lambda(\tfrac{X}{N}) \qquad\qquad \mu(X) = N\mu(\tfrac{X}{N}) \tag{2.4}$$

This important scaling law introduces a small parameter $\varepsilon = \frac{1}{N}$ in the problem on the basis of which perturbative solutions of (2.2) can be obtained.

As an example of a birth and death process we mention the Schlögl reaction (1.5) in a homogeneous system. The birth and death rates take the following form :

$$\lambda(X) = k_1\ A\ \frac{X(X-1)}{N^2} + k_4\ B \tag{2.5a}$$

$$\mu(X) = k_2\ \frac{X(X-1)(X-2)}{N^2} + k_3 X \tag{2.5b}$$

A and B being the number of particles of the externally controlled species.

An example of a pure death process, i.e. $\lambda(X) \equiv 0$, is provided by the thermochemical model (1.6) of adiabatic explosion. One has :

$$\mu(X) = k_o X\ \exp\left[-\frac{U}{k_B T}\right] \tag{2.6}$$

The temperature variable can be eliminated by invoking energy conservation, the rise in temperature being entirely due to the exothermal chemical process :

$$c_V(T-T_o) + r_V\frac{(X-X_o)}{N} = 0 \tag{2.7}$$

T_o and X_o are the intial values of temperature and number of molecules, c_V is the specific heat of the reactive mixture and r_V the heat of reaction. The maximal temperature T_{max} is reached after complete combustion:

$$c_V(T_{max}-T_o) = \frac{r_V X_o}{N} \tag{2.8}$$

Hence, from (2.6)-(2.8) one obtains:

$$\mu(X) = k_o X \exp\left[-\frac{U}{k_B(T_{max} - \frac{r_V}{c_V}\frac{X}{N})}\right] \tag{2.9}$$

2C. Spatially Distributed Systems

In order to tackle the problem of spatial correlations or to discuss transitions towards inhomogeneous states, it is necessary to formulate a Master Equation approach for spatially distributed systems. To this end we partition real space into cells of size ΔV and assume that the numbers of particles $\{X_{i,r}\}$, of species i in the cells centered at r, constitute a Markov'process. Inside each cell, chemical reactions can take place. They are described as in the homogeneous case by the same birth and death operator $L_{ch}(r)$. On the other hand particles can diffuse from one spatial cell to an adjacent cell. This exchange is modelled as a random walk. The resulting Multivariate Master Equation takes the form :

$$\frac{dP}{dt} = \sum_{\underset{\sim}{r}} \left[L_{ch}(\underset{\sim}{r}) + \sum_{\underset{\sim}{l}} L_d(\underset{\sim}{r},\underset{\sim}{l})\right] P \tag{2.10}$$

$\underset{\sim}{r} + \underset{\sim}{l}$ are the nearest neighbours of $\underset{\sim}{r}$ and L_d is the random walk operator :

$$L_d(\underset{\sim}{r},\underset{\sim}{l})\ P(\{X_{i,\underset{\sim}{r}}\},t) =$$

$$\sum_i \frac{Di}{2d}\left[\exp\left(\frac{\partial}{\partial X_{i,\underset{\sim}{r}}} - \frac{\partial}{\partial X_{i,\underset{\sim}{r}+\underset{\sim}{l}}}\right) - 1\right] X_{i,\underset{\sim}{r}}\ P(\{X_{i,\underset{\sim}{r}}\},t) \tag{2.11}$$

d is the dimensionality of real space and the random walk rates D_i

are related to the Fickian diffusion coefficient by :

$$\mathcal{D}_i = \frac{(\ V)^{2/d}}{2d} D_i \tag{2.12}$$

Note that both L_{ch} and L_d are extensive, i.e. proportional to the cell size ΔV. In order to be consistent, the latter volume has to be chosen small enough in order to secure homogeneity and coherence within each cell.

2D. Some Important Limits

Equilibrium. We mentioned that the results obtained from the Master Equation must agree with those obtained from statistical mechanics at equilibrium. Being primarily concerned with open systems, we have to check the equilibrium results obtained from the Master Equation with the grand canonical ensemble. In the case of a unique variable constituent X in an ideal system, one has :

$$P^{eq}(X, \varepsilon_k) = Z^{-1} \exp\left[\frac{\mu X - \sum_{k=1}^{X} \varepsilon_k}{k_B T}\right] \tag{2.13}$$

where μ is the chemical potential, T the temperature, Z the partition function and ε_k the energy of the k-th particle. Summing over the energy configurations of the particles, one obtains from (2.12) a Poisson distribution for the number of particles X in a volume element ΔV :

$$P^{eq}(X) = \frac{\langle X \rangle}{X!} e^{-\langle X \rangle} \tag{2.14}$$

with average value :

$$\langle X \rangle = \sum_k \exp\left(\frac{\mu - \varepsilon_k}{k_B T}\right) \tag{2.15}$$

If one considers several non overlapping volume elements, the resulting probability distribution is a multi-Poissonian, hence fluctuations in different spatial cells are independent.

Let us consider the Schlögl model, described by the Master Equation (2.2) and (2.5). In equilibrium, the average material flux through the system must vanish. In the present case, this requires that $k_1 k_3 A = k_2 k_4 B$ and for these values of the control parameters, a Poissonian steady state distribution is consistently recovered. This result can even be extended to the multivariate problem (2.10) whose steady state solution turns out to be a multi-Poissonian, in accordance with statistical mechanics.

The above condition of zero average flux is a particular example of the fundamental property of detailed balance, characteristic for the equilibrium state. As a result of microreversibility at equilibrium, the forward rate $W_\rho(X\to X+\nu_\rho)P^{eq}(X)$ is exactly equal to the inverse rate

$W_{-\rho}(X+\nu_\rho\to X)\ P^{eq}(X+\nu_\rho)$:

$$W_\rho(X\to X+\nu_\rho)\ P^{eq}(X) = W_{-\rho}(X+\nu_\rho\to X)\ P^{eq}\ (X+\nu_\rho) \tag{2.16}$$

It is this property that guarantees the consistency of the results obtained from the Master Equation with those derived from equilibrium statistical mechanics.

Macroscopic Limit. Let us consider the equations for the first moments $\langle X_{i\underset{\sim}{r}}\rangle$ obtained from the Multivariate Master Equation (2.10) :

$$\frac{d\langle X_{i\underset{\sim}{r}}\rangle}{dt} = \langle\lambda(X_{i\underset{\sim}{r}})\rangle - \langle\mu(X_{i\underset{\sim}{r}})\rangle + \sum_{\underset{\sim}{l}} \frac{D_i}{2d}\left[\langle X_{i,\underset{\sim}{r}+\underset{\sim}{l}}\rangle - \langle X_{i,\underset{\sim}{r}}\rangle\right] \tag{2.17}$$

Using the relation (2.12) between jump rates $\mathbf{D}_i$ and Fickian diffusion coefficients $\mathcal{D}_i$, one easily verifies that the discrete random walk operator in the r.h.s. of (2.17) reduces to the Fickian diffusion term in the limit $\Delta V\to 0$ for the intensive variables

$$\bar{x}_i(\underset{\sim}{r}) = \frac{\langle X_{i,\underset{\sim}{r}}\rangle}{\Delta V} .$$

However a discrepancy with the phenomenological eqs. (1.7) subsists in the reactive terms. To understand the nature of the correction, let us consider the reaction :

$$2X \overset{k}{\to} X+E \tag{2.18}$$

for which

$$\lambda(X) \equiv 0,\ \mu(X) = k\frac{X(X-1)}{N} \tag{2.19}$$

Then we obtain (homogeneous case) :

$$\frac{d\langle X\rangle}{dt} = -\ k\,\frac{1}{N}\,(\langle X^2\rangle - \langle X\rangle) \tag{2.20}$$

Switching to the intensive variable $x = \frac{X}{N}$, one has :

$$\frac{d\langle x\rangle}{dt} = -k\langle x\rangle^2 - \frac{k}{N}\,\frac{\langle \delta X^2\rangle - \langle X\rangle}{N} \qquad (2.21)$$

The first term in the r.h.s. of (2.21) corresponds to the macroscopic reaction rate. The second term is a correction due to the fluctuations. In equilibrium, it vanishes ($\langle \delta X^2\rangle = \langle X\rangle$ for a Poisson distribution) and the macroscopic law is recovered exactly. This is not an artefact of the model considered but a general feature. In nonequilibtium the correction term is nonzero, but it is expected to be small, of order $\frac{1}{N}$. However, this in not a general feature as we shall descuss further in section 3.

Continuous Limit. Fluctuations are the consequence of the discrete nature of the microscopic processes, underlying the macroscopic evolution laws. For this reason, we have chosen to consider Markov processes in discrete state space. Nevertheless some success in describing fluctuations has been achieved, starting from the historical work of Langevin, by using continuous Markov processes. In the lectures of L. Arnold, some limit theorems were presented ensuring the equivalence of both descriptions for finite time. However, in general, this equivalence does not extend for all times. In particular, agreement on steady state results is only obtained before or at bifurcation. This raises the question of the validity of a description in terms of continuous Markov processes beyond bifurcation, e.g. in the region of bistability. We come back to this problem in section 3.

2E. Stochastic Thermodynamics

In this section, we address briefly the question of stability within the framework of the above stochastic approach. Traditionally the stability of a macroscopic system is studied at a phenomenological level by analyzing the response to external macroscopic perturbations. However, stability is an intrinsic property that is continuously being probed by the thermodynamic fluctuations present in the system. Therefore, the question arises whether stability properties are detectable from general properties of the Master Equation, without necessitating the knowledge of an explicit solution P. In particular, how are the transitions, observed at the macroscopic level, built into the stochastic description ?

In classical thermodynamics, Lyapounov functionals have been quite successful for the discussion of stability properties, at least in the linear range of irreversible thermodynamics. In stochastic theory, it is sometimes claimed that the H-functional :

$$H(t) = \sum_{\underset{\sim}{X}} P(\underset{\sim}{X},t)\,\ln \frac{P(\underset{\sim}{X},t)}{P^{st}(\underset{\sim}{X})} \geq 0 \qquad (2.22)$$

plays an analogous role. Indeed, one easily verifies that for any

P, solution of the Master Equation (2.1), the following inequality holds :

$$\frac{dH}{dt} \leq 0 \tag{2.23}$$

We conclude that H is a Lyapounov functional which guarantees the relaxation of any initial probability distribution to its final steady state value P^{st}. However, being satisfied identically, this H-theorem makes no statement about the macroscopic stability properties of the system. A finer measure is needed. Motivated by the important role played by entropy in classical thermodynamics, we introduce the information entropy of the Markov chain :

$$S(t) = - \sum_{\underset{\sim}{X}} P(\underset{\sim}{X},t) \ln P(\underset{\sim}{X},t) \tag{2.24}$$

For the entropy variation, one obtains, using the Master Equation (2.1) and invoking the conservation of probability, $\frac{d}{dt} \sum_{\underset{\sim}{X}} P(\underset{\sim}{X},t) = 0$:

$$\frac{dS}{dt} = - \sum_{\rho,\underset{\sim}{X}} \Big[W_\rho(\underset{\sim}{X}-\underset{\sim}{\nu}_\rho \to \underset{\sim}{X})\, P(\underset{\sim}{X}-\underset{\sim}{\nu}_\rho,t) - W_{-\rho}(\underset{\sim}{X} \to \underset{\sim}{X}-\underset{\sim}{\nu}_\rho)\, P(\underset{\sim}{X},t)\Big] \ln P(\underset{\sim}{X},t) \tag{2.25a}$$

By rearranging the sums, the following equivalent expression is obtained :

$$\frac{dS}{dt} = - \sum_{\rho,\underset{\sim}{X}} \Big[W_{-\rho}(\underset{\sim}{X} \to \underset{\sim}{X}-\underset{\sim}{\nu}_\rho)\, P(\underset{\sim}{X},t) - W_\rho(\underset{\sim}{X}-\underset{\sim}{\nu}_\rho \to \underset{\sim}{X})\, P(\underset{\sim}{X}-\underset{\sim}{\nu}_\rho,t)\Big] \ln P(\underset{\sim}{X}-\underset{\sim}{\nu}_\rho,t) \tag{2.25b}$$

Summation of (2.25a) and (2.25b) gives :

$$\frac{dS}{dt} = \frac{1}{2} \sum_{\rho,\underset{\sim}{X}} \Big[W_\rho(\underset{\sim}{X}-\underset{\sim}{\nu}_\rho \to \underset{\sim}{X})\, P(\underset{\sim}{X}-\underset{\sim}{\nu}_\rho,t) - W_{-\rho}(\underset{\sim}{X} \to \underset{\sim}{X}-\underset{\sim}{\nu}_\rho)\, P(\underset{\sim}{X},t)\Big] \ln \frac{P(\underset{\sim}{X}-\underset{\sim}{\nu}_\rho,t)}{P(\underset{\sim}{X},t)} \tag{2.26}$$

It is now natural to define the following stochastic fluxes and forces :

$$J_\rho(\underset{\sim}{X},t) = W_\rho(\underset{\sim}{X}-\underset{\sim}{\nu}_\rho \to \underset{\sim}{X})\, P(\underset{\sim}{X}-\underset{\sim}{\nu}_\rho,t) - W_{-\rho}(\underset{\sim}{X} \to \underset{\sim}{X}-\underset{\sim}{\nu}_\rho)\, P(\underset{\sim}{X},t) \tag{2.27a}$$

$$A_\rho(\underset{\sim}{X}) = \ln \frac{W_\rho(\underset{\sim}{X}-\underset{\sim}{\nu}_\rho \to \underset{\sim}{X}) P(\underset{\sim}{X}-\underset{\sim}{\nu}_\rho, t)}{W_{-\rho}(\underset{\sim}{X} \to \underset{\sim}{X}-\underset{\sim}{\nu}_\rho) P(\underset{\sim}{X}, t)} \tag{2.27b}$$

Note that they both vanish at equilibrium, according to the property of detailed balance, eq. (2.16). The entropy variation can now be written as the sum of two terms :

$$\frac{dS}{dt} = \frac{d_i S}{dt} + \frac{d_e S}{dt} \tag{2.28}$$

where

$$\frac{d_i S}{dt} = \frac{1}{2} \sum_{\rho, \underset{\sim}{X}} J_\rho(\underset{\sim}{X}) \, A_\rho(\underset{\sim}{X}) \geqq 0 \tag{2.29}$$

is the stochastic analogue of the internal entropy production, and :

$$\frac{d_e S}{dt} = -\frac{1}{2} \sum_{\rho, \underset{\sim}{X}} \ln \frac{W_\rho(\underset{\sim}{X}-\underset{\sim}{\nu}_\rho \to \underset{\sim}{X})}{W_{-\rho}(\underset{\sim}{X} \to \underset{\sim}{X}-\underset{\sim}{\nu}_\rho)} J_\rho(\underset{\sim}{X}) \tag{2.30}$$

plays the role of the stochastic entropy flux. In order to single out the contribution due to the fluctuations, we expand the internal entropy production (2.29) around the macroscopic state :

$$\frac{d_i S}{dt} = \left(\frac{d_i S}{dt}\right)_{\text{macr.}} + \left(\frac{d_i S}{dt}\right)_{\text{fluct.}} \tag{2.31}$$

$\left(\frac{d_i S}{dt}\right)_{\text{macr.}}$ being the classical Gibbs entropy production. As to the entropy production $\left(\frac{d_i S}{dt}\right)_{\text{fluct.}}$, associated with the fluctuations, the following properties can be established :

- $\left(\frac{d_i S}{dt}\right)_{\text{fluct}} \equiv 0$ for a Poissonian probability distribution.
- $\left(\frac{d_i S}{dt}\right)_{\text{fluct}} < 0$ at steady states close to equilibrium. This property was proven for a large class of chemical models.
- $\left(\frac{d_i S}{dt}\right)_{\text{fluct}} > 0$ at the steady state just beyond the bifurcation point of Schlögl's model.

These preliminary results indicate that fluctuations have a stabilizing influence close to equilibrium but promote the destabilization of the reference state when a bifurcation point is crossed. So far, we have not been able to relate in a general way the properties of the entropy production of the fluctuations to the bifurcation properties of the system.

3. PRIMARY BIFURCATION

3A. Critical Behaviour

Since no general stability criterium could be formulated on the basis of general properties of the Master Equation, we resort to explicit solutions, which will give us the stochastic analog of the bifurcations occurring in the macroscopic system. What do we expect? Let us consider a transition to multiple steady states, such as the transition displayed by Schlögl's model. Before bifurcation, $\lambda<\lambda_c$, we expect that the probability will be centered around the unique macroscopic steady state (see fig. 4a). In particular, at equilibrium, the probability distribution is a multi-Poissonian :

$$P^{eq}(\{X_{\underset{\sim}{r}}\}) = \prod_{\underset{\sim}{r}} e^{-\langle X_{\underset{\sim}{r}}\rangle} \frac{\langle X_{\underset{\sim}{r}}\rangle^{X_{\underset{\sim}{r}}}}{X_{\underset{\sim}{r}}!} \tag{3.1}$$

where $\frac{\langle X_{\underset{\sim}{r}}\rangle}{\Delta V} = \langle x_{\underset{\sim}{r}}\rangle$, the average number density in cell $\underset{\sim}{r}$, is equal to the unique macroscopic number density, constant throughout the entire system. The fluctuations, whose strength is measured by the variance $\langle\delta X^2\rangle$, are equal to $\langle X\rangle$. Thus, their importance relative to the average is small, as expected. Beyond bifurcation, $\lambda>\lambda_c$, (see fig. 4b), we expect that the probability distribution will be two-humped having maxima close to the two new stable macroscopic solutions. As a result, the fluctuations will be of a macroscopic order and the average value, will loose its physical meaning. What then will be the form of the probability in the immediate vicinity of the bifurcation point, which is at the borderline of the above two cases (fig. 4c) ? And what will be the relative weights of the two peaks in the probability beyond bifurcation ? Clearly, these questions are beyond the scope of the phenomenological equations (1.7), and have to be answered on the basis of the Multivariate Master Equation (2.10). Unfortunately, exact solutions can only be obtained in the uninteresting case of linear reactions, or in the case of well stirred systems for which the inverse volume size can be used as perturbation parameter. However, if we anticipate the existence of a transition, the system must display a markedly coherent behaviour at the approach of the bifurcation point. One can therefore blow up the cell size ΔV without affecting the required property of homogeneity and coherence in each cell. Of course this working hypothesis has to be verified on its consequences. Let us then briefly review the main ingredients of the

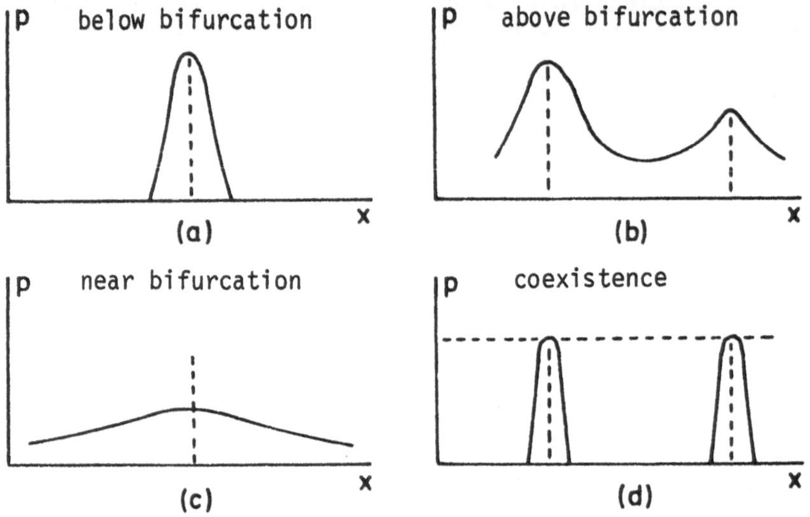

Figure 4

resulting perturbation expansion :

. For technical convenience, one writes the Multivariate Master Equation in the generating function representation :

$$F(\{S_{\underset{\sim}{r}}\},t) = \sum_{\{X_{\underset{\sim}{r}}\}} \prod_{\underset{\sim}{r}} S_{\underset{\sim}{r}}^{X_{\underset{\sim}{r}}} P(\{X_{\underset{\sim}{r}}\},t) \tag{3.2}$$

For example, the reaction step $A+2X \xrightarrow{k_1} 3X$ in the Schlögl model gives rise to the term

$$k_1 A(S_{\underset{\sim}{r}}^{\ 3} - S_{\underset{\sim}{r}}^{\ 2}) \frac{\partial^2 F}{\partial S_{\underset{\sim}{r}}^{\ 2}} \tag{3.3}$$

Note also that F, $\frac{\partial F}{\partial S_{\underset{\sim}{r}}}$, ... at $S_{\underset{\sim}{r}} \equiv 1, \forall \underset{\sim}{r}$, generate the moments of P, and that the deterministic behaviour :

$$P = \prod_{\underset{\sim}{r}} \delta^{Kr}_{X_{\underset{\sim}{r}},N\bar{x}_{\underset{\sim}{r}}} \tag{3.4}$$

corresponds to

$$F = \prod_{\underset{\sim}{r}} S_{\underset{\sim}{r}}^{N\bar{x}_{\underset{\sim}{r}}} \tag{3.5}$$

. One extracts the deterministic behaviour from P or F, by considering the function Ψ :

$$\Psi(\{S_{\underset{\sim}{r}}\},t) = \prod_{\underset{\sim}{r}} S_{\underset{\sim}{r}}^{-N\bar{x}_{\underset{\sim}{r}}} F(\{S_{\underset{\sim}{r}}\},t) \tag{3.6}$$

The resulting equation for Ψ defines a singular perturbation problem since the smallness parameter $\varepsilon = \frac{1}{\Delta V}$ multiplies higher order derivatives. This is reflected in the behaviour of the generating function which, in the limit $\varepsilon \to 0$, takes values different from zero only in the immediate vicinity of $S_{\underset{\sim}{r}} \equiv 1$, $\forall \underset{\sim}{r}$, hence its derivatives can become very large.

. To smooth out this boundary layer, we blow up the vicinity of $S_{\underset{\sim}{r}} = 1$ by setting :

$$S_{\underset{\sim}{r}} = 1 + \varepsilon^a \xi_{\underset{\sim}{r}} , \quad \forall \underset{\sim}{r} \tag{3.7}$$

Since :

$$S_{\underset{\sim}{r}}^{X_{\underset{\sim}{r}} - N\bar{x}_{\underset{\sim}{r}}} = (1+\varepsilon^a \xi_{\underset{\sim}{r}})^{\delta X_{\underset{\sim}{r}}} \cong e^{\xi_{\underset{\sim}{r}} u_{\underset{\sim}{r}}} \tag{3.8}$$

where

$$u_{\underset{\sim}{r}} = \varepsilon^a \, \delta X_{\underset{\sim}{r}} \tag{3.9}$$

the above scaling amounts to the study of the scaled fluctuations $u_{\underset{\sim}{r}}$. The value of the scaling parameter a has to be chosen such that $u_{\underset{\sim}{r}}$ is of order $\varepsilon^0 = 1$, in the limit $\varepsilon \to 0$. In the case of Poissonian or Gaussian extensive fluctuations, one has $a = \frac{1}{2}$. On the other hand, in the bistable region, fluctuations can carry the system from one macroscopic state to the other and are therefore of macroscopic order, $\langle \delta X_{\underset{\sim}{r}}^2 \rangle \sim \Delta V^2$, i.e. a=1. For the bifurcation point, at the threshold of these two situations, we expect a scaling exponent $\frac{1}{2} < a < 1$.

. In the regularised equation for the function Ψ_a :

$$\Psi_a(\xi_{\underset{\sim}{r}},t) = \Psi(1+\varepsilon^a \xi_{\underset{\sim}{r}},t) \tag{3.10}$$

one can, for a<1, identify the dominant terms. The remaining problem can be exactly solved and transformed back into the original variables.

Let us summarize the most important results. To dominant order in the inverse cell size ε, the Multivariate Master Equation is equivalent to the following Langevin equation with nonlinear drift, but process independent noise ($x_{\underset{\sim}{r}} = X_{\underset{\sim}{r}}/\Delta V$) :

$$\frac{\partial x_{\underset{\sim}{r}}}{\partial t} = f(x_{\underset{\sim}{r}}) + \frac{D}{2d} \sum_{\underset{\sim}{l}} (x_{\underset{\sim}{r}+\underset{\sim}{l}} - x_{\underset{\sim}{r}}) + \varepsilon^{1/2} F_{\underset{\sim}{r}}(t) \qquad (3.11)$$

where $F_{\underset{\sim}{r}}$ is a multi-Gaussian white noise with correlation function :

$$\langle F_{\underset{\sim}{r}}(t) F_{\underset{\sim}{r}'}(t')\rangle = \delta(t-t') \quad \{Q(\bar{x}_{\underset{\sim}{r}})\, \delta^{Kr}_{\underset{\sim}{r},\underset{\sim}{r}'} \qquad (3.12)$$

$$+ \frac{D}{2d} \sum_{\underset{\sim}{l}} \left[(\bar{x}_{\underset{\sim}{r}+\underset{\sim}{l}} - \bar{x}_{\underset{\sim}{r}})\, \delta^{Kr}_{\underset{\sim}{r},\underset{\sim}{r}'} - (\delta^{Kr}_{\underset{\sim}{r}+\underset{\sim}{l},\underset{\sim}{r}'} - \delta^{Kr}_{\underset{\sim}{r},\underset{\sim}{r}'})(\bar{x}_{\underset{\sim}{r}} + \bar{x}_{\underset{\sim}{r}'}) \right] \}$$

$f(x_{\underset{\sim}{r}})$ is the macroscopic chemical rate and $Q(\bar{x}_{\underset{\sim}{r}})$ is the chemical birth plus death rate, evaluated at the deterministic trajectory $\bar{x}_{\underset{\sim}{r}}$. We have thus established a central limit theorem, ensuring the convergence of a discrete Markov process to a continuous one. The particular status of this theorem is revealed when we calculate from it the stationary probability distribution.
In the case of Schlögl's model :

$$P_{st}(\{x_{\underset{\sim}{r}}\}) = \exp\{ -\frac{2\varepsilon^{-1}}{Q_{st}} \qquad (3.13)$$

$$\sum_{\underset{\sim}{r}} \left[\frac{\lambda-\lambda_c}{2} (x_{\underset{\sim}{r}} - \bar{x}_{st})^2 + \frac{1}{4} (x_{\underset{\sim}{r}} - \bar{x}_{st})^4 + \frac{D}{8d} \sum_{\underset{\sim}{l}} (x_{\underset{\sim}{r}+\underset{\sim}{l}} - x_{\underset{\sim}{r}})^2 \right] \}$$

This probability does not belong to the class of infinitely divisible laws, even in the homogeneous case (D=0), due to the presence of a quartic term in the exponential. This reflects the coherence associated with bifurcation : the system can no longer be partitioned into a collection of weakly correlated subsystems.
On the other hand the exponential in (3.13) displays the familiar Landau Ginzburg type of potential in discrete (position) space. This establishes the connection between the Master Equation approach and the theory of equilibrium critical phenomena. Relevant properites such as the spatial correlation, critical dimensionality and critical exponents can be obtained from renormalization group theory.
In particular, the correlation length is divergent at the approach of the bifurcation point for dimensions d>1, validating our perturbative expansion in these cases. For d=1, the system exhibits no

critical behaviour at the crossing of the bifurcation point.

Let us finally note that the critical dimensionality can also be obtained directly from the Multivariate Master Equation. Indeed, as seen from eq. (2.12), the jump rate D_i slows down as the cell size ΔV gets larger in a manner which depends on the dimensionality d. This decrease becomes less pronounced as d increases. On the other hand, near bifurcation, the chemical reaction rate is also slowed down. Moreover, as chemical reactions are local events, this "critical slowing down" is independent of the dimensionality and is entirely determined by the parameters and the dominant nonlinearities. The critical dimensionality will then be such that the rates of diffusion and reaction are comparable. This provides a relation between the number of physical dimensions d and the degree of the dominant nonlinearities. For the Schlögl model, where the dominant nonlinearity is cubic, $d_c=4$. For $d_c>4$, diffusion dominates and mean field behaviour results, whereas for $1>d>d_c$, nonclassical critical behaviour is observed.

3B. Nucleation

The perturbative approach discussed above, breaks down beyond bifurcation where fluctuations, being of macroscopic order, can no longer be distinguished from the macroscopic evolution. For this reason no analytic results are known concerning the behaviour of the inhomogeneous system beyond bifurcation, even in the simplest case of a two box model. This of course reflects the richness of behaviour displayed by bistable inhomogeneous systems (see e.g. the lectures by P. Fife). Among recent numerical results, we report here some Monte Carlo simulations concerning the relaxation of the probability distribution in a system of two coupled boxes obeying the Schlögl kinetics. In the case of a homogeneous system, the probability distribution is known and the relative weights of the two peaks, corresponding to the coexisting macroscopic steady states, can be evaluated. Equally stable states (see fig. 4d) correspond to peaks of equal weight. We now prepare the inhomogeneous system in a region of parameter space with unequal weight of these peaks and start in both boxes from the less stable state. For a small coupling (small diffusion coefficient D), the system will relax to the most stable state in each box separately (see fig. 5a). On the other hand, for a very strong coupling (large diffusion coefficient), the system essentially behaves as a homogeneous whole with a volume twice as large as the volume of an individual box (fig. 5b). The interesting case is the region of intermediate coupling. One then observes a transient inhomogeneous state, in which only one of the boxes has transited to the most stable state (fig. 5c). If one plots the mean first passage time from the initial homogeneous less stable to the final homogeneous more stable state versus diffusion (fig. 5d) one notes that this relaxation via an inhomogeneous transient is associated with a relative decrease in the first passage time. We conclude that new faster routes of relaxation reminiscent of nucleation, can occur in the case of inhomogeneous systems.

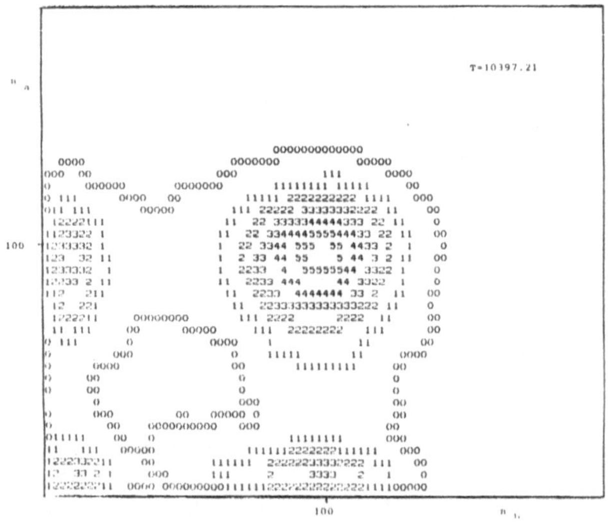

a

b

Figure 5a : Probability contour for D = 0.005.

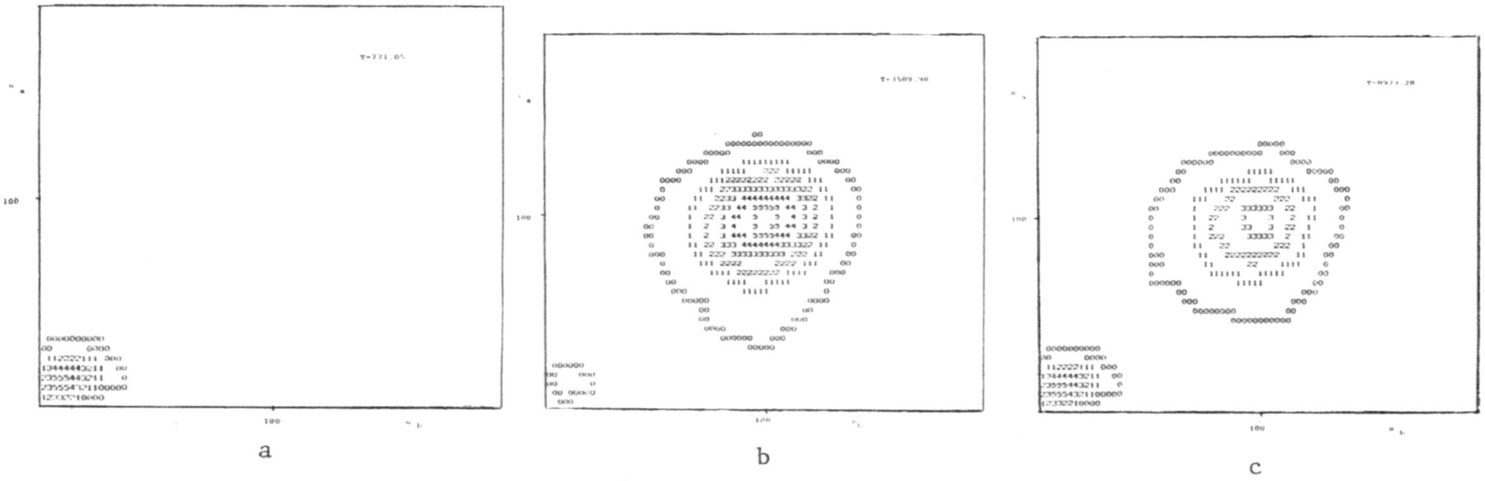

Figure 5b : Probability contour for D = 0.06. The system behaves like the homogeneous one.

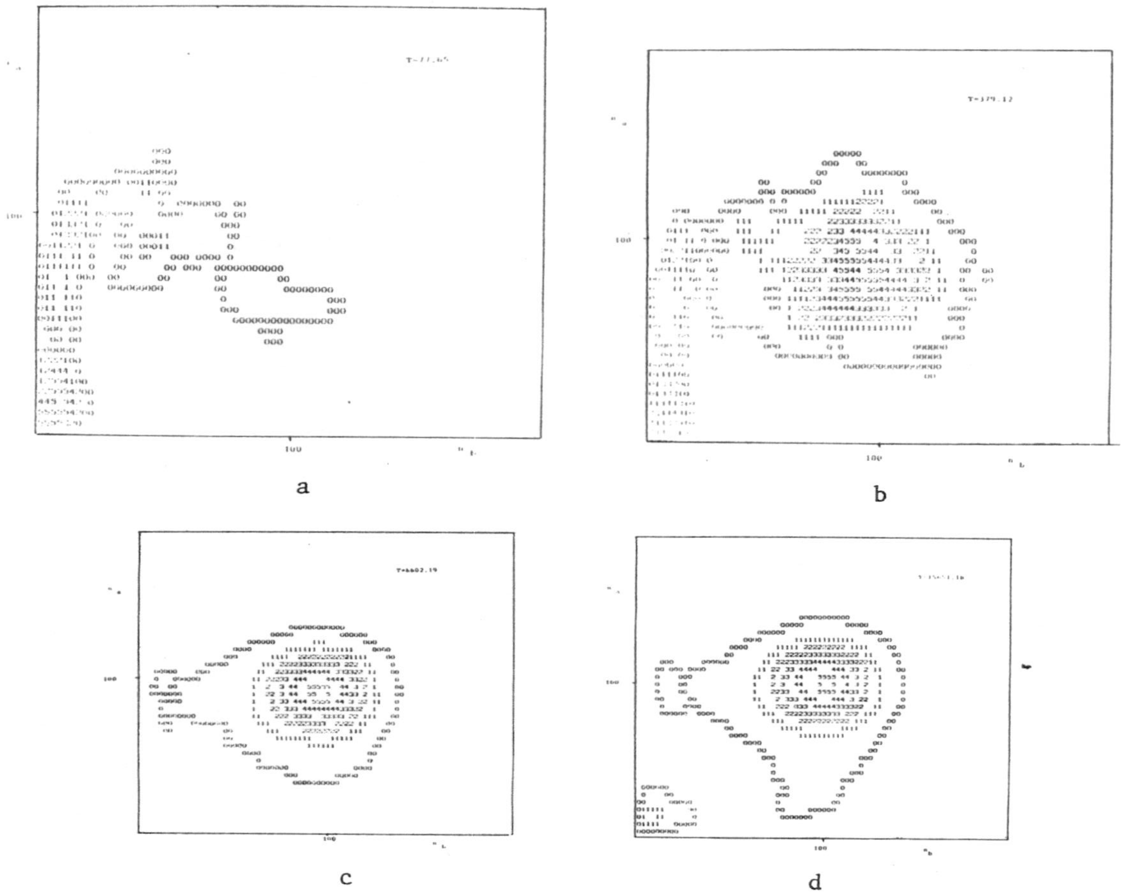

a b c d

Figure 5c : Probability contours for D-0.025. The transition between the inhomogeneous states occurs via the transient inhomogeneous states.

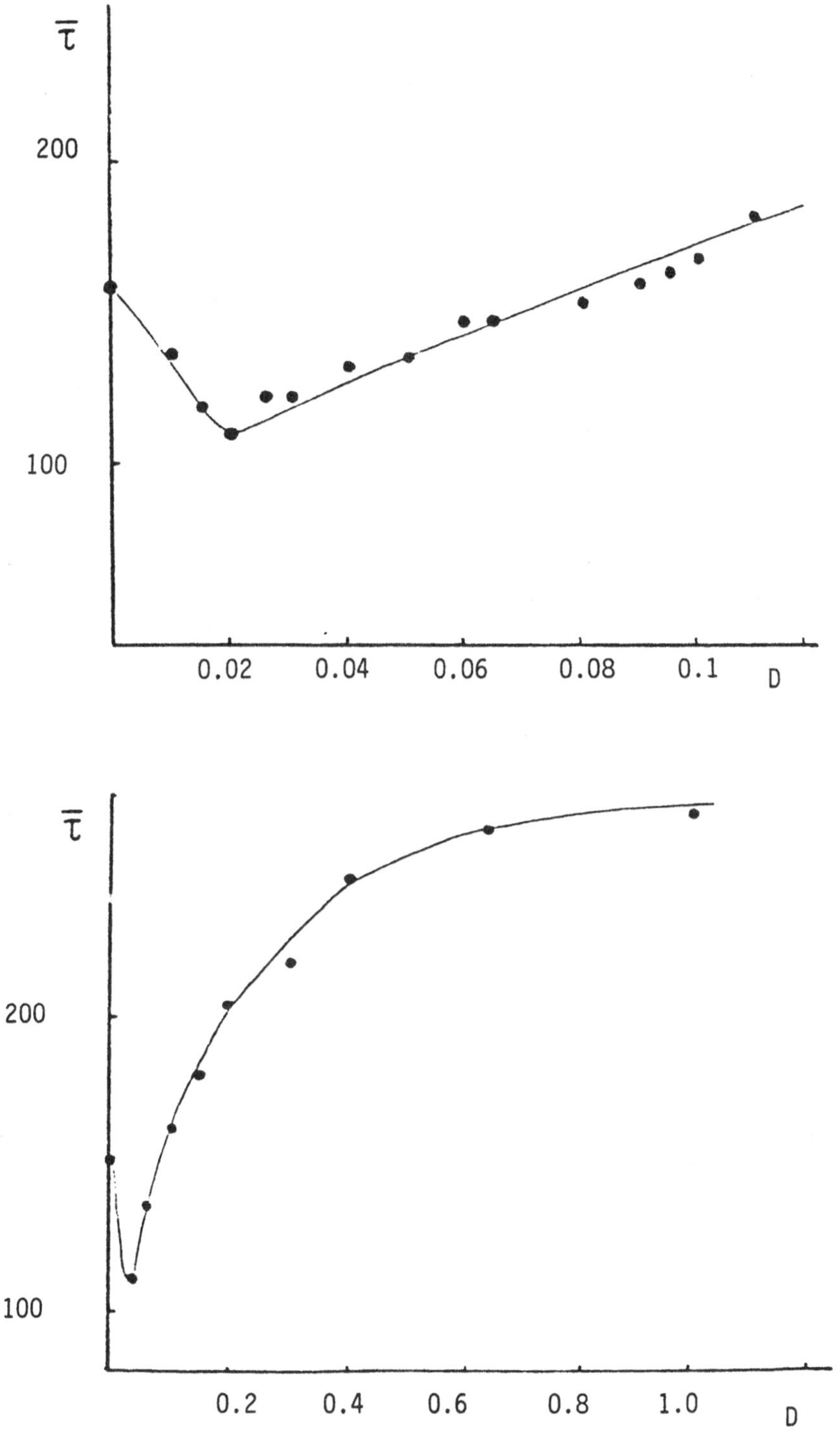

Figure 5d : Mean first passage time in function of the diffusion coefficient.

4 THE ONSET OF SPATIAL CORRELATIONS

4A. Reaction Diffusion Systems

In the previous section, we established the critical properties of reaction diffusion systems in the vicinity of a bifurcation point. How can one relate such a cooperative behaviour to the strikingly different properties of the equilibrium state ? In equilibrium systems, the competition between internal energy, which is determined by the interactions between the particles, and thermal energy can induce equilibrium phase transitions characterised by divergent fluctuations and correlation length. But how can one explain critical phenomena in ideal chemical system for which no equilibrium phase transition can occur ? The explanation must be found in the nonlinear nonequilibrium dynamics of the system. To make this point clear, let us consider a system consisting of thermally agitated particles with repulsive, attractive and no interactions (see fig. 6). In the latter case, particles are distributed at random and the number of particles X in a given volume obeys the Poisson law $\langle \delta X^2 \rangle = \langle X \rangle$.

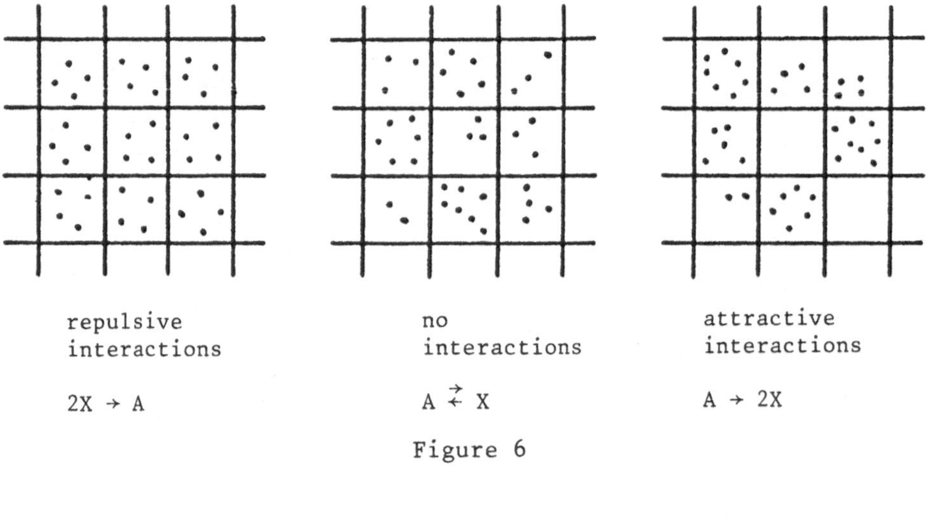

repulsive interactions — $2X \rightarrow A$

no interactions — $A \rightleftarrows X$

attractive interactions — $A \rightarrow 2X$

Figure 6

In the case of attractive interactions, particles tend to form clusters, and larger deviations from the average value are expected, whereas for repulsive interactions, the opposite is true. On the other hand, let us consider an ideal nonequilibrium chemical system.

If the system is described by linear chemical kinetics, particles will be created ($A \to X$) or destroyed ($X \to A$) at random places in space. However, in the case of nonlinear chemical kinetics, particles will either be created in clusters (for instance $A \to 2X$ gives pair of particles) or particles close to each will be destroyed reactively (the reaction $2X \to A$ depletes the system from pairs of particles). In other words, the nonlinear chemical kinetics may give rise to a similar effect as attractive or repulsive interactions. In equilibrium however, the property of detailed balance ensures that any reaction, such as $A \to 2X$, will be, in average, as frequent as its inverse, $2X \to A$, and we recover the familiar behaviour of an ideal system. In non-equilibrium, this balance is disturbed and deviations from the Poisson law are observed. How can one relate these deviations to the existence of spatial correlations ? Let us consider again the reaction step $A \to 2X$. When such a reaction occurs, the two X-particles are obviously correlated, being created together. Due to diffusion, this correlation extends in space until one of the particles again undergoes a reaction. This will happen in average at a given rate k. The resulting correlation length is thus given by :

$$l_c = \sqrt{\frac{\mathcal{D}}{k}} \tag{4.1}$$

where $\mathcal{D}$ is the Fickian diffusion coefficient of the considered chemical species.

To go beyond these qualitative arguments, we turn again to the Multivariate Master Equation (2.10). We consider the equation for the correlation function :

$$g_{\underset{\sim}{r},\underset{\sim}{r}'} = \langle \delta x_{\underset{\sim}{r}} \, \delta x_{\underset{\sim}{r}'} \rangle - \langle x_{\underset{\sim}{r}} \rangle \, \delta^{Kr}_{\underset{\sim}{r},\underset{\sim}{r}'} \tag{4.2}$$

obtained by neglecting third and higher order moments. This Gaussian type of approximation is expected to be valid well before bifurcation where the system essentially behaves as a linear system. One obtains :

$$\frac{D}{2d} \sum_{\underset{\sim}{l}} (g_{\underset{\sim}{r}+\underset{\sim}{l},\underset{\sim}{r}'} - g_{\underset{\sim}{r},\underset{\sim}{r}'}) + f'(\overline{x}_{st}) \, g_{\underset{\sim}{r},\underset{\sim}{r}'}$$

$$+ \left[\frac{1}{2} Q(\overline{x}_{st}) + x_s \, f'(\overline{x}_{st})\right] \frac{1}{\Delta V} \, \delta^{Kr}_{\underset{\sim}{r},\underset{\sim}{r}'} = 0 \tag{4.3}$$

$\overline{x}_{st}$ is the homogeneous macroscopic steady state of the system, f and Q being the macroscopic chemical rate and chemical noise strength respectively (see also (3.11) and (3.12)). The last term in the l.h.s. of (4.3) is a source term for spatial correlations. It is instructive to perform the explicit calculations for the Schlögl model :

$$f(x) = -k_2x^3 + k_1ax^2 - k_3x + k_4b \tag{4.4a}$$

$$Q(\bar{x}_{st}) = k_2\,\bar{x}_{st}^{\,3} + k_1a\,\bar{x}_{st}^{\,2} + k_3\,\bar{x}_{st} + k_4\,b \tag{4.4b}$$

Using the results (4.4), the source term in (4.3) can be rewritten as follows :

$$\frac{5}{2}\,(k_1a\,\bar{x}_{st}^{\,2} - k_2\,\bar{x}_{st}^{\,3}) + \frac{1}{2}\,(-k_3\,\bar{x}_{st} + k_4\,b) = 2J_{st} \tag{4.5}$$

where J_{st} is the net material flux through the system at the steady state :

$$J_s = k_1a\,\bar{x}_{st}^{\,2} - k_2\,\bar{x}_{st}^{\,3} = -\,k_3\,\bar{x}_{st} + k_4b \tag{4.6}$$

The equation for g in the continuum limit takes the form :

$$\nabla^2_{\underset{\sim}{r}-\underset{\sim}{r}'}\,g(\underset{\sim}{r}-\underset{\sim}{r}') + f'(\bar{x}_{st})\,g(\underset{\sim}{r}-\underset{\sim}{r}') = -2J_s\,\delta(\underset{\sim}{r}-\underset{\sim}{r}') \tag{4.7}$$

Its solution reads :

$$g(\underset{\sim}{r}-\underset{\sim}{r}') = \frac{J_s}{2\pi\mathcal{D}}\,\frac{1}{|\underset{\sim}{r}-\underset{\sim}{r}'|}\,\exp\left\{-\left[\frac{f'(\bar{x}_{st})}{\mathcal{D}}\right]^{1/2}|\underset{\sim}{r}-\underset{\sim}{r}'|\right\} \tag{4.8}$$

The exponential in the r.h.s. of (4.8) features a correlation length of the form (4.1) with $k = f'(\bar{x}_{st})$. For the number of particles in a volume element ΔV, $X = \int_{\Delta V} X_r\,dr$, one has from (4.8) :

$$\langle\delta X^2\rangle - \langle X\rangle = J_s.\Delta V \tag{4.9}$$

These results complete our picture of nonequilibrium fluctuations in chemical systems. We conclude that such <u>nonlinear</u> systems are characterized by an intrinsic correlation length l_c of macroscopic range, which diverges at the approach of a primary bifurcation (see previous section), but is otherwise a smooth function of the distance from equilibrium (see fig. 7a). However, the amplitude of the correlation function and, concomittantly, the deviation from the Poisson law, can be considered as an "order parameter" characterising the "transition" from equilibrium to nonequilibrium (see fig. 7b). Although the correlation length is an intrinsic property of the chemical system, it will not be perceived in the equilibrium state where direct and inverse elementary processes cancel out exactly due to detailed balance. Therefore the transition to nonequilibrium witnesses the sudden arousal of long range spatial correlations.

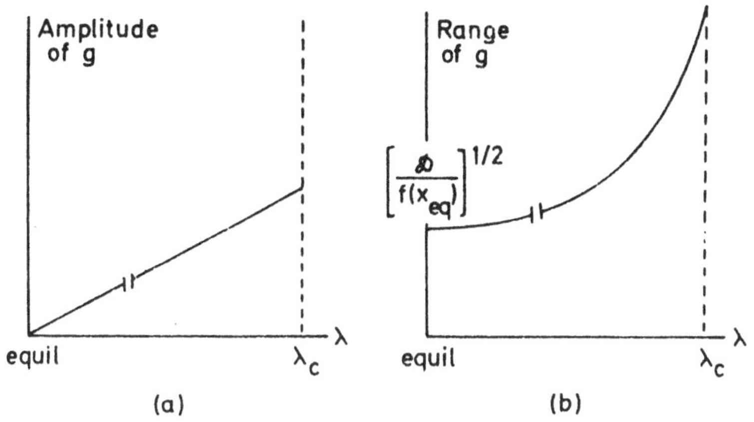

Figure 7

4B. Heat Conduction in Nonequilibrium

The chemical system discussed above is subject to a scalar nonequilibrium constant, namely the nonequilibrium concentrations of the reservoir variables A, B, ... This allows to consider an infinite system, for which no parameter related to the system size enters into the problem. Let us now briefly discuss the more complicated case of a nonequilibrium constraint which breaks the translational symmetry of the system. An example is provided by the diffusion of heat through a one-dimensional system, parallel to the z-axis, whose end points a and b are maintained at different temperatures T_a and T_b. If one supposes that heat conduction following Fick's law is the only transport mechanism taking place in the system, one verifies that the correlation function :

$$g(z,z') = \langle\delta T(z)\delta T(z')\rangle - \frac{k_B \overline{T}_z^{\,2}}{C_v}\,\delta(z-z') \tag{4.10}$$

obeys the following equation :

$$K\left(\frac{\partial^2}{\partial z^2} + \frac{\partial^2}{\partial z'^2}\right) g(z,z') = -\,2\,k_B\,\frac{\gamma^2}{C_v}\,\delta(z-z') \tag{4.11}$$

Here K is the thermal diffusivity coefficient (supposed constant), c_V the specific heat of the system, $\gamma = T_b - T_a / (z_b - z_a)$ the temperature gradient and $\overline{T}_z = T_a + \gamma(z - z_a)$ the average temperature at the position z. As in the case of the reaction-diffusion system, the source term of the spatial correlations is related to the nonequilibrium flux (in the present case the heat flux) throughout the system. However, no intrinsic length scale l_c appears in the present problem. Therefore, it is not surprising that the correlation length is determined by the total length $L = z_b - z_a$ of the system, i.e. it encompasses the entire system. Indeed, the solution of (4.11) reads :

$$g(z,z') = g(z',z) = \frac{k_B \gamma^2}{L\, c_V} z' \,(L-z) \qquad z' \leq z \tag{4.12}$$

to be compared with the result (4.8) for chemical systems. For the average

$$T = \int_{z_a}^{z_b} T(z)\, dz, \text{ one has :}$$

$$\langle \delta T^2 \rangle - \frac{k_B \overline{T}^2}{c_V L} = \frac{k_B}{12 c_V L} (T_b - T_a)^2 \tag{4.13}$$

In short, the occurrence of spatial correlations of macroscopic range is a general feature for nonequilibrium systems, independently of bifurcation phenomena. The correlation length may be either intrinsic or determined by the system size, but the amplitude of the correlation function is always related to the strength of the nonequilibrium constraint, e.g. the material or energy flux through the system. At equilibrium, this amplitude is zero masking the presence of the inherent long range coherence in the system.

5. TRANSIENT PHENOMENA

As an example of transient behaviour let us consider the model of adiabatic explosion, eq. (1.6). The deterministic equations for this problem read :

$$\frac{d\overline{x}}{dt} = - k(T)\, \overline{x} \tag{5.1a}$$

$$c_V \frac{dT}{dt} = r_V k(T) \overline{x} \tag{5.1b}$$

in which k(T) is given by Arrhenius' law (1.6b). The temperature variable T can be eliminated using energy conservation (cfr. eqs. (2.7) and (2.8)) :

$$\frac{d\bar{x}}{dt} = k(T_{max} - \frac{r_v}{c_V}\bar{x})\,\bar{x} \qquad (5.2)$$

Numerical integration of this highly nonlinear evolution equation predicts for concentration and temperature a typical explosive behaviour as depicted in fig. 8.

The effect of the fluctuations can be tackled on the basis of the theory of birth and death processes, introduced in section 2 (cf. eqs. (2.2) and (2.8)). A numerical solution of the Master Equation is represented in fig. 9. The initial condition is chosen to be a deterministic state lying in a part of fig. 8 in which the evolution is predicted to be very slow. One observes then transient bimodality in time, one of the maxima of the probability distribution being centered close to the initial state and the other maximum close to the state of complete combustion. This "bifurcation behaviour in time" becomes even more apparent if one plots the maxima of the probability distribution in the course of time (see fig. 10).

Let us outline a qualitative explanation of this unexpected phenomenon. Remember that we deal with a process involving two widely separated time scales, and that our system is initially prepared in a state in which the deterministic rate is very small. The maximum of the underlying probability distribution, whose motion roughly follows the deterministic one, will therefore move very slowly toward the region of lower values of the composition variable X. Meanwhile, because of the fluctuations, the probability will develop a width proportional to the length of the induction period and inversely proportional to the root of the number of particles. If the length of the induction period is large, this width will be appreciable. Thus, a part of the probability mass will reach the ignition point well before the maximum does so. At this moment it will be quickly entrained by the fast motion toward the region of low values of X. This leak of probability will go on continuously, but since the system cannot have negative values of X, a "traffic jamm" will arise as a result of which a new probability peak will emerge in the region of small X. Eventually the primary peak, which by then will be considerably diminished, will reach the ignition point and this will mark the end of transient bimodality.

For a typical thermo-chemical explosion in a volume of the order of $(0.1\ mm)^3$ containing about $N \sim 10^{12}$ particles, the duration of the above described transient behaviour is estimated to be about 10^{-3} sec.. It should therefore show up in the form of fluctuations of the ignition time of the order of 10^{-3} sec., which should be in the observable range for realistic systems. Moreover, since the fluctuations are predominantly inhomogeneous, one expects the formation of randomly distributed "nuclei" within the system, which can be thought of as the precursors of the hot spots or flames familiar from combustion.

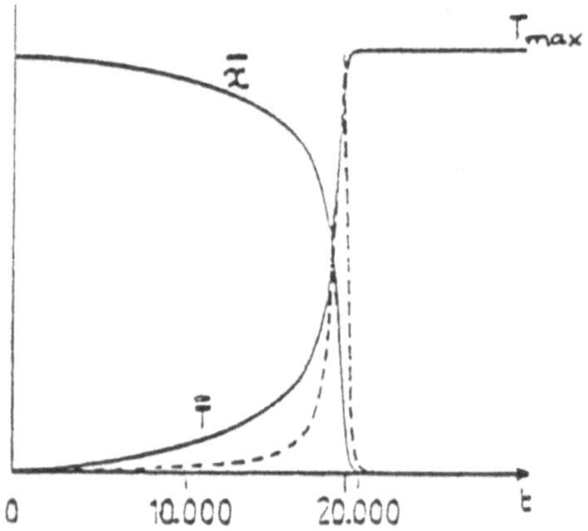

Figure 8

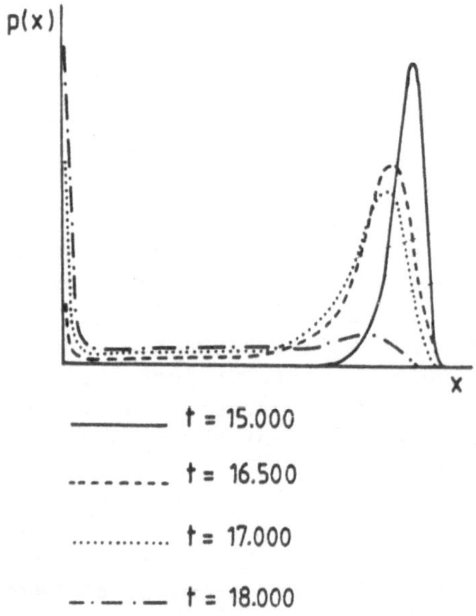

Figure 9 : Numerical solution of the Master Equation.

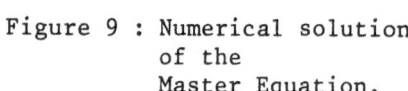

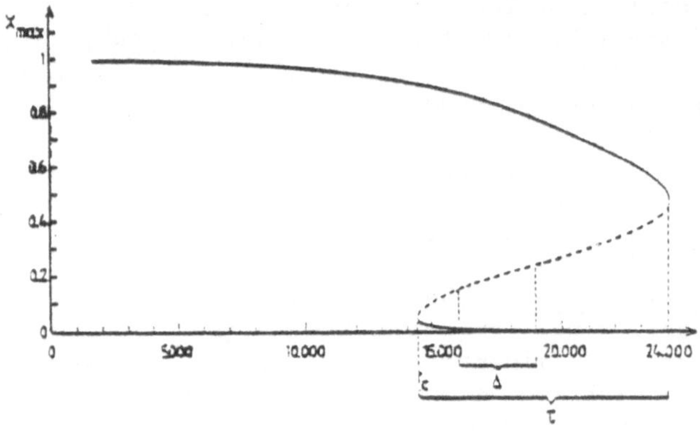

Figure 10 : Maxima of the probability distribution versus time exhibiting transient bimodality.

REFERENCES

Introduction

-G. Nicolis and I. Prigogine, *Self-Organization in Nonequilibrium Systems* (Wiley, New York, 1977).
-D. Sattinger, *Topics in Stability and Bifurcation Theory*,(Springer, Berlin,1973).
-F. Schlögl, Z. Phys. *253*, 147 (1972).
-L. Garrido, Ed., *Systems Far from Equilibrium*,(Springer,Berlin,1980).
-G. Nicolis and F. Baras Eds., *Chemical Instabilities*, (Reidel Pub., Dordrect, 1983).

Stochastic Formulation

-W. Feller, *An Introduction to Probability Theory and Its Applications*, Vol 1 and 2 (Wiley, New York, 1959).
-D. Mc. Quarrie, Suppl. Rev. Ser. Appl. Prob. (Methuen, London,1967).
-G. Nicolis and I. Prigogine, *Self-Organization in Nonequilibrium Systems*, Chpt. 9-12 (Wiley, New York, 1977).
-N.G. Van Kampen, *Stochastic Processes in Physics and Chemistry* (North-Holland, Amsterdam, 1983).
-J. Schnakenberg, Rev. Mod. Phys. *48*, No. 4, 571 (1976).
-Luo Jui-li, C. Van den Broeck and G. Nicolis, Stability Criteria and Fluctuations around Nonequilibrium States, to be published.

Primary Bifurcation

-G. Nicolis and J.W. Turner, Physica *89A* , 326 (1977).
-M. Malek Mansour, C. Van den Broeck, G. Nicolis and J.W. Turner, Ann. Phys. *131* , 283 (1981).
-M. Frankowicz and E. Gudowska-Nowak, Physica *116A* , 331 (1982).

Onset of Spatial Correlations

-C.W. Gardiner, K.J. Mc Neil, D.F. Walls and I.S. Matheson, J. Stat. Phys. *14*, 307 (1976).
-H. Lemarchand and G. Nicolis, Physica *82A*, 251 (1976).
-M. Malek Mansour and C. Van den Broeck,in *Instabilites,Bifurcation and Fluctuations in Chemical Systems* (L. Reich, Ed., Texas Press, Austin, U.S.A., 1982).
-N.G. Van Kampen, *Stochastic Processes in Physics and Chemistry*, Chapt XII (North-Holland, Amsterdam, 1983).
-G. Nicolis and M. Malek Mansour, Onset of Spatial Correlations in Nonequilibrium Systems:A Master Equation Description,to be published.

Transient phenomena

-F. Baras, G. Nicolis, M. Malek Mansour and J.W. Turner, J. Stat. Phys. *32* № 1, 1 (1983).
-M. Suzuki, in *Order and Fluctuations in Equilibrium and Non-Equilibrium Statistical Mechanics* (Wiley, New York, 1983).

NONEQUILIBRIUM SYSTEMS WITH RANDOM CONTROL PARAMETERS

M. San Miguel and J.M. Sancho

Departamento de Física Teórica, Universidad de Barcelona

Diagonal 647, Barcelona 28, Spain

INTRODUCTION

Most of the lectures in this school have been concerned with general properties and specific applications of nonlinear equations for a set of relevant variables. These equations describe the cooperative behavior of nonequilibrium systems in the framework reviewed by Haken[1]. There are two levels of discussion of these equations, the deterministic and the stochastic level. At the first level one usually studies the existence of the different homogeneous stationary states and also the spatial and temporal stationary structures which correspond to the different values of the control parameters. At this level of description fluctuations are neglected. The inclusion of fluctuations in the stochastic description allows the study of additional dynamical properties such as the relative stability of stationary states, decay of unstable and metastable states, transitions between stationary states for a fixed value of the control parameters etc. At both levels of description it is generally assumed that the control parameters have well defined constant values. In this seminar a stochastic description of nonequilibrium systems is considered in the case in which a control parameter becomes a random function of time. This describes the external noise situation defined more precisely below. A related situation which we do not discuss here is the one in which a control parameter is a periodic function of time[2]. We first discuss briefly the concept and relevance of external noise. Secondly we summarize several aspects of the mathematical description of external noise.

EXTERNAL NOISE

When discussing noise and fluctuations in a system it is convenient to distinguish between internal[3] and external noise[4,5]. Internal noise is originated in the system and reflects the effect of microscopic degrees of freedom averaged out in a mesoscopic description. As a consequence, its properties are linked to the system: In general it scales with the system size, satisfies a fluctuation dissipation relation, etc. On the other hand, external noise is not self-originated. It enters in the description of the system through external control parameters which take random values either due to a random environment or because the system is stochastically driven on purpose. From this definition it is clear that the properties of external noise are independent of those of the system and that to a certain extent they can be controlled in an experimental situation. When external noise is present in a macroscopic system,internal fluctuations are often negligeable. The different origins of fluctuations is for example clear in the optical bistable system reviewed by Lugiato[6]. Internal noise includes in this case quantum and thermal noise. The most important source of fluctuations is the external noise identified in this system with the amplitude and phase fluctuations of the incident laser.

The effect of external noise has been studied theoretically and experimentally in a variety of systems. These include electrical circuits[7], photochemical reactions[8], fluctuations of the pump parameter in a single mode dye-laser[9], magnetic ramdom fields in plasmas[10], Rayleigh-Bernard systems[11], surface catalized reactions[12], fluctuations of the magnetic field in Freederickz's transition in a nematic layer[13], fluctuations of the applied voltage in the electrohydrodynamic transition in liquid crystals[14,15], etc. In several of these cases important changes in the instabilities of the system due to the presence of external noise have been predicted or observed. For example, an instability point can be shifted by changing the noise intensity[14]. The general question one would like to answer is "What is the response of the nonlinear system to external noise?", and in particular to what extent stabilization by noise is possible. A general unique answer does not seem to exist. It rather depends in the peculiarities of the system. Motivated by the large number of systems in which external noise is of relevance we summarize in the following several methodological aspects of the common mathematical description of external noise situations.

MATHEMATICAL DESCRIPTION

The standard description of external noise[5] starts with a deterministic equation for a relevant variable q in which internal noise is neglected: $\dot{q} = f(\alpha,q)$, $\dot{q}$ indicates the time derivative of q and α is a control parameter. External noise is modelled by substituting α by $\alpha + \xi(t)$ where $\xi(t)$ is a random process of zero mean. This leads to a stochastic differential equation (SDE) for q(t) which now becomes a random process.

Linear white noise

The simplest situation is the one in which f is linear in α, $f(\alpha,q)=v(q)+\alpha g(q)$, and $\xi(t)$ is assumed to be a gaussian white noise characterized by an intensity D. In this case the resulting SDE defines a markovian process whose probability density $P(q,t)$ obeys a well known Fokker Planck equation (FPE)

$$\dot{P}(q,t) = L_q P(q,t) \equiv \left(-\frac{\partial}{\partial q} f(\alpha,q) + D\frac{\partial}{\partial q} g(q) \frac{\partial}{\partial q} g(q)\right) P(q,t) \quad (1)$$

This equation is the basis of the statistical description of the system. A prototype model often considered is defined by the SDE

$$\dot{q} = \alpha q - q^3 + q\,\xi(t) + \eta(t) \quad (2)$$

$\xi(t)$ represents the fluctuations of the parameter α and $\eta(t)$ is an independent gaussian white noise of intensity D'. This noise is added so that the variable q is not restricted to take only positive or negative values. It can be interpreted as a small thermal noise or the external noise associated with a control parameter analogous to an external field of vanishing mean value. From the FPE associated with (2) it is possible to calculate the stationary distribution $P_{st}(q)$. For fixed D' and $\alpha > o$, $P_{st}(q)$ has for $D>1$ a single maximum at $q_o = o$, while it has two well defined symmetric maxima for $D<1$. Therefore a smooth transition between a monostable and a bistable situation is possible by only changing the intensity of the external noise

Linear Colored Noise

The white noise assumption is obviously an idealization. A more realistic assumption corresponds to take $\xi(t)$ as an Ornstein-Uhlenbeck process characterized by its intensity D and correlation time τ : $\langle \xi(t)\ \xi(t')\rangle = (D/\tau)\cdot e^{-|t-t'|/\tau}$. For D fixed and $\tau \to o$ the white noise limit is recovered. The problem is now that $q(t)$ is no longer a markov process. Nevertheless an approximated FPE still exists for $P(q,t)$ when D is small enough[16]

$$\dot{P}(q,t) = L_q(\tau) \equiv \left(-\frac{\partial}{\partial q} f(q) + D\frac{\partial}{\partial q} g(q) \frac{\partial}{\partial q} H(q)\right) P(q,t) \quad (3)$$

where $H(q) = f(q)(1+\tau f(q)\frac{\partial}{\partial q})^{-1}(g(q)/f(q))$. This equation contain terms to all orders in τ. An analysis of (3) for the model (2) (with $\eta = o$) leads to the conclusion that for D fixed, $D>1$,

$P_{st}(q)$ changes its shape when increasing τ from $\tau = 0$, developing a relative maximum at $q_0 \neq 0$ which becomes dominant as $\tau \to \infty$[16]. This prediction has been checked by analogical[7] and numerical simulation[16].

In the white noise case the joint probability distribution $P(q,t;q',t')$ also obeys eq. (1) and the correlation function $\langle q(t)q(t')\rangle$ can be calculated with the FPE operator L_q. For a colored noise $\xi(t)$ this is no longer true. Moreover, due to the lack of a fluctuation dissipation relation the process $q(t)$ is nonstationary besides of being nonmarkovian[17]. As a consequence, the correlation function in the steady state cannot be calculated prescribing stationary initial conditions and using translational invariance in time. An approximate equation satisfied by the stationary correlation function can be calculated directly from the SDE[17] or from the equation for $P(q,t;q',t')$[18]

$$\frac{d}{dt}\langle q(t)q(t')\rangle_{st} = \langle (L_q^+(\tau)q)(t)q(t')\rangle_{st} + D\, e^{-(t-t')/\tau} \langle g(q(t))H(q(t'))\rangle \qquad (4)$$

The last term in the r.h.s. of (4) is a pure nonmarkovian effect which produces important changes (with respect to the case of $\tau = 0$) in the value of the relaxation time T of the system. For the model defined by eq. (2) with $\eta = 0$ it follows from eq. (4) that T increases as a function of τ for a fixed value of D [19].

Nonlinear Noise

If $f(\alpha,q)$ is nonlinear in α, the white noise assumption is completely ill-defined. In the particular case in which $f(\alpha,q) = v(q) + \alpha^2 g(q)$ and $\xi(t)$ is taken as the Ornstein-Uhlenbeck process an approximate FPE exists for the probability density $P(q,t)$[15]:

$$P(q,t) = \left(-\frac{\partial}{\partial q}\left(v(q) + (\alpha^2 + \frac{D}{\tau}\right)g(q) + \left(4\ D + \frac{D}{\tau}\right)\frac{\partial}{\partial q} g(q) \frac{\partial}{\partial q} g(q)\right)P(q,t) \qquad (5)$$

This equation is valid for $D \ll 1$, $\tau \ll 1$ and $D/\tau \sim 1$. The main effect of this form of nonlinear noise is the existence of the D/τ term in the drift of the FPE (5). This equation is of relevance for the study of external noise effects in some instabilities in liquid crystal systems[13-15]. A simplifying assumption to deal with nonlinear noise is to take $\xi(t)$ as a dichotomous noise[13,20].

Joint Description of Internal and External Noise

Internal fluctuations are usually neglected when studying the effect of external noise. Nevertheless they may become important for small systems. To calculate such finite size effects in a stochastically driven systems a joint description of internal and external noise is needed. From a fundamental point of view such a unified description is also desirable. A unified description has been recently discussed[21,22] for the case in which internal noise is modelled by a birth and death master equation, for example for a chemical system[3].

$$P(N,t) = Q(N-1,t)P(N-1,t) + R(N+1,t)P(N+1,t) - \{Q(N,t)+R(N,t)\}P(N,t) \quad (6)$$

The standard description of external noise starts at the deterministic level obtained in the thermodynamic limit of (6). Here we introduce the external noise in (6) itself. We assume that there exists an external parameter in $R(N)$ such that when it becomes random $R(N) = R_0(N)+R_1(N)\,\xi(t)$. For simplicity $\xi(t)$ is taken as a gaussian white noise. It is convenient to go to the generating function $F(s,t)$ representation of eq. (6) . Since $R(N)$ depends on $\xi(t)$, $F(s,t)$ satisfies a stochastic partial differential equation which defines $F(s,t)$ as a functional of $\xi(t)$. The average of $F(s,t)$ over the realizations of $\xi(t)$ is an effective generating function $\bar{F}(s,t)$ which satisfies the following equation

$$\frac{\partial \bar{F}(s,t)}{\partial t} = \left[(s-1)Q\left(s\frac{\partial}{\partial s}\right)+\left(\frac{1}{s}-1\right)R_0\left(s\frac{\partial}{\partial s}\right)\right]\bar{F}(s,t) + \left[D\left(\frac{1}{s}-1\right)R_1\left(s\frac{\partial}{\partial s}\right)\left(\frac{1}{s}-1\right)R_1\left(s\frac{\partial}{\partial s}\right)\right]\bar{F}(s,t) \quad (7)$$

Solving (7) for $\bar{F}(s,t)$ it is possible to calculate the statistical properties of the system with simultaneous consideration of internal and external fluctuations. In particular, an expansion around the thermodynamic limit allows the calculation of finite size effects. The analysis of (7) shows[22] that there exist crossed fluctuation terms which couple internal and external noise. As an example we consider the following nonequilibrium chemical model.

$$A + X \underset{k_1'}{\overset{k_1}{\rightleftharpoons}} 2X \quad ; \quad B + X \underset{k_2'}{\overset{k_2}{\rightleftharpoons}} C$$

in which $\alpha = K_1 a$ becomes a random parameter: $\alpha \to \alpha \pm \xi(t)$. Here a,b,c denote respectively the fixed concentration of the reactants A,B,C and, $\beta = k_2 b$, $\gamma = k_2' c$, $\delta = k_1'$. Eq. (7) becomes for this model and in the steady state

$$\left[\frac{D}{\beta} - s\left(\frac{D}{\beta} - \frac{\delta}{V\beta}\right)\right]\frac{d^2}{ds^2}\bar{F}_{st}(s) + \left[\left(1-\frac{D}{\beta}\right)-\frac{\alpha}{\beta}s\right]\frac{d}{ds}\bar{F}_{st}(s) - \frac{\gamma V}{\beta}\bar{F}_{st}(s)=0$$

(8)

The solution of this equation is known in terms of confluent hypergeometric functions[22]. From eq. (8) we obtain an equation for the second moment $\langle x^2 \rangle$ $(x=N/V)$ in which the crossed-fluctuation effect appears explicitly as a term proportional to D/V. This term vanishes in the thermodynamic limit or in the absence of external noise.

$$\frac{d}{dt}\langle x^2\rangle = 2\gamma\langle x\rangle + 2(\alpha-\beta)\langle x^2\rangle - 2\delta\langle x^3\rangle + \frac{1}{V}(\gamma + (\alpha+\beta)\langle x\rangle + 3\delta\langle x^2\rangle) - \frac{\delta\langle x\rangle}{V^2} + 4D\langle x^2\rangle - \frac{3D}{V}\langle x\rangle \tag{9}$$

Another intersting application of eq. (7) is in the study of density fluctuations in nuclear reactor models[23].

REFERENCES

1. See the lectures by H. Haken.
2. F. de Pasquale, Z. Racz, M. San Miguel, P. Tartaglia (1983).
3. See the lectures by L. Arnold and G. Nicolis.
4. See the papers by W. Horsthemke, R. Lefever, M. San Miguel and J.M. Sancho in "Stochastic Nonlinear Systems", Synergetic Series, vol. 2. Springer Verlag (1981).
5. W. Horsthemke and R. Lefever, "Noise Induced Transitions", Springer Verlag (1983).
6. See the lectures by L. Lugiato.
7. S. Kabashima, S. Kogure, T. Kawakubo and T. Okada, J. Appl. Phys. 50:6296(1979);
   J.M. Sancho, M. San Miguel, H. Yamazaki and T. Kawakubo, Physica A116:560(1982).
8. P. de Kepper and W. Horsthemke, C.R. Acad. Sci. Paris, 287C:251(1978).
9. S. N. Dixit and P. S. Sahni, Phys. Rev. Lett. 50:1237(1978).
10. H. A. Rose, Phys. Rev. Lett. 48:260(1982).
11. J.P. Gollub and F. Steinman, Phys. Rev. Lett. 45:551(1980).
12. F.J. de la Rubia, J. Garcia Sanz and M.G. Velarde, in "Nonlinear Stochastic Problems", Eds. R.S. Bucy and J.F. Moura, Reidel (1983).
13. R. Lefever and W. Horsthemke, in Proceedings of 7th International Conference on Noise in Physical Systems. Montpellier(1983).
14. S. Kai, T. Kai, M. Takata and H. Hirakawa, J. Phys. Soc. Jpn. 47:1379(1979);
    T. Kawakubo, A. Yanagita and S. Kabashima, J. Phys. Soc. Jpn. 50:145(1981).
15. M. San Miguel and J.M. Sancho, Z. Phys. B43:861(1981).
16. J.M. Sancho, M. San Miguel, S. Katz and J.D. Gunton, Phys. Rev. A26:1589(1982);
    J.M. Sancho and M. San Miguel, Z. Physik B36:357(1980).

17. A. Hernández-Machado and M.San Miguel, J. Math. Phys. (1983)
18. A. Hernández-Machado, J.M. Sancho, M. San Miguel and L. Pesquera, Z. Phys. B. (1983).
19. A. Hernández-Machado, M. San Miguel and J.M. Sancho, Preprint (1983).
20. J.M. Sancho and M. San Miguel, Prog. Theor. Phys. 69:1085(1983).
21. M. San Miguel and J.M. Sancho, Phys. Lett. 90A:455 (1982)
22. J.M. Sancho and M. San Miguel, Preprint (1983).
23. M.A. Rodriguez, M. San Miguel and J.M. Sancho, Ann. Nucl. Energy, 10:263 (1983).

NON-WHITE NOISE TRIGGERED OSCILLATIONS IN A NONLINEAR CHEMICAL PROCCESS

F.J. de la Rubia , J.J. García Sanz and M.G. Velarde

Departamento de Física Fundamental, U.N.E.D.

Apartado 50.487, Madrid (Spain)

1.- INTRODUCTION

In the last few years there has been an increasing interest in the study of the effect of external noise on the behavior of non linear dynamical systems, showing the possibility for the system to undergo *noise induced transitions* not predicted by a deterministic analysis /1-4/ . Although for one variable systems analytical work is possible both for the white and non-white noise cases /5,6/, at present such study is not yet feasible for more complicated , i.e., two-dimensional , dynamical systems exhibiting sustained oscillations (limit cycle behavior) /7,8/. Recently, we have illustrated /9,10/, the effect of noise with a non-zero correlation time on the temporal behavior of a two-dimensional system with plausible experimental relevance.

2.- DETERMINISTIC MODEL

The model, firstly introduced and studied from a purely deterministic viewpoint by Takoudis et al /11/, refers to the following pair of equations here written in dimensionless from

$$d\,\theta_i/dt = \alpha_i\,\theta_s - \gamma_i\,\theta_i - r(t) \qquad (2.1)$$

with $i = 1,2,$ $\theta_s \equiv 1 - \theta_1 - \theta_2$ and where

$r(t) \equiv \theta_1 \; \theta_2 \; \theta_s$ is the *production rate* .

In this model, which is a caricature of surface catalyzed chemical reactions /12/, θ_i $(i = 1,2)$ accounts for the ratio of chemisorbed molecules on a surface of Pt, say, to the number of available adsorption sites, therefore satisfying the conditions $\theta_i > 0$ and $0 < \theta_1 + \theta_2 < 1$. The kinetic constants α_i, γ_i are non negative parameters that take into account both static and dynamic features of the surface reaction problem.

Depending on the values of α_i, γ_i the model system (2.1) allows the possibility of one or three steady states as well as the possibility of sustained oscillations . For illustration we just consider the values $\alpha_1 = 0.016$, $\gamma_1 = 0.001$ and $\gamma_2 = 0.002$ and two values $\alpha_2 = 0.0270$ for which we have an unstable steady state leading to a limit cycle and $\alpha_2 = 0.0295$ with a stable steady state (stable focus).

It should be pointed out that the most relevant quantity in a surface reaction problem is the production rate, $r(t)$. Obviously oscillations in either of the θ_i yield oscillations in $r(t)$ /12/.

3.- STOCHASTIC MODEL

For simplicity we assume all stochastic aspects of the problem concentrated in the stochastic variation in α_2 in the form

$$\tilde{\alpha}_2 = \alpha_2 + \eta(t) \tag{3.1}$$

where α_2 is the systematic part, i.e., the deterministic value, and $\eta(t)$ is the noisy part. Keeping in mind that both α_2 and $\tilde{\alpha}_2$ must be non negative quantities we set

$$\eta(t) = (2a/\pi) \text{ arc tan } \beta_t \tag{3.2}$$

where β_t is an Ornstein-Uhlenbeck process /13/ that satisfies the equation

$$d\beta_t = -\rho\beta dt + \sigma dw_t \tag{3.3}$$

with

$$\beta_t \sim N(0, \sigma^2/2\rho) \tag{3.4.a}$$

$$\langle\beta_t \beta_{t'}\rangle = (\sigma^2/2\rho) \exp(-\rho|t-t'|) \tag{3.4.b}$$

From (3.2) follows that $-a \leq \eta \leq +a$ and therefore

$$\alpha_2 - a \leq \tilde{\alpha}_2 \leq \alpha_2 + a \tag{3.5}$$

The constant $\underline{a}$ delineates the domain of variation of the noisy parameter $\tilde{\alpha}_2$ and we take $a = \alpha_2$ to ensure the nonnegative character of $\tilde{\alpha}_2$. From (3.2) and (3.3) it follows that $\eta(t)$ is a non-gaussian diffusion process satisfying the following stochastic differential equation (here written in Stratonovich's sense /13/)

$$d\eta = f(\eta)\, dt + g(\eta)\, dw_t \tag{3.6}$$

with $f(\eta) = -\rho(a/\pi) \sin(\pi\eta/a)$ (3.7)

and $g(\eta) = (2a\sigma/\pi) \cos^2(\pi\eta/2a)$ (3.8)

The stationary probability distribution, normalized in the interval (-a, a) is

$$P_{st}(\eta) = (\pi^{1/2}/2a\varepsilon)\{\cos^2(\pi\eta/2a)\}^{-1} \exp\{-\varepsilon^{-2}\tan^2(\pi\eta/2a)\} \tag{3.9}$$

where $\varepsilon = \sigma / \rho^{1/2} \geq 0$. From (3.9) we get

$$\langle \eta \rangle = 0 \tag{3.10}$$

and

$$\langle \eta^2 \rangle = (4a^2/\pi^{5/2})\, I(\varepsilon) \tag{3.11}$$

with

$$I(\varepsilon) = \int_{-\infty}^{\infty} \text{arc} \tan^2(\varepsilon x)\, e^{-x^2}\, dx \tag{3.12}$$

$I(\varepsilon)$ is a monotonic increasing function of ε , which controls the variance of the noise η . A plot of the stationary distribution (3.9) can be seen in Fig 1 for different values of ε . To be consistent with the problem at hand we restrict consideration to the case where the most probable value of η is also the mean value , thus leading to the condition $\varepsilon \leq 1$. Note that from (3.1) this condition means that α_2 is not only the mean value of $\tilde{\alpha}_2$ but also its most probable value.

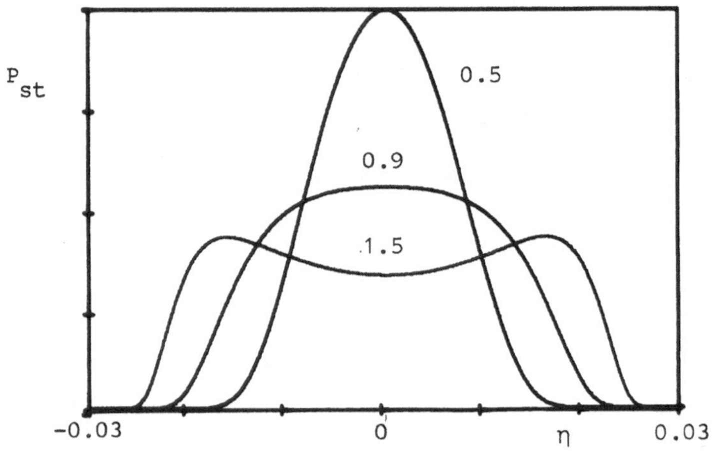

Fig. 1. Stationary probability density (3.9) for a= 0.03 and three values of ε . The value ε =1 separates the unimodal and bimodal situations

Moreover, the correlation function can be written (to a good approximation) as

$$\langle \eta(0)\eta(t) \rangle \simeq \langle \eta^2 \rangle \exp(-t/\tau) \qquad (3.13)$$

with $\tau = 1/\rho$ the time scale in the noise correlation.

As a consequence of the above given properties it follows that the triplet $\{\theta_1, \theta_2, \eta \text{ or } \beta\}$ defines a Markov process with θ_i $(i = 1,2)$ never leaving the physically relevant region $\{0 \leq \theta_i \leq 1, \ \theta_1 + \theta_2 = 1\}$. It should be stressed that this condition , essential to account for the underlying physics of the problem, can not be established for a white and thus unbounded noise.

4.- NUMERICAL RESULTS

For illustration we consider the values $\varepsilon = 0.5$ and $\tau = 50$. It is worth noting that for this value of ε, the parameter $\delta \equiv (\mathrm{Var}\ \tilde{\alpha}_2)^{1/2} / \langle \tilde{\alpha}_2 \rangle$, which measures the relative importance of the fluctuations, becomes 0.2 so that fluctuations are kept always small. For the correlation time of the noise the choice of the value taken has been dictated by our interest in illustrating the role of the noise with a correlation time of the same order (in a region to be specified below) of the evolution time for the deterministic model.

Depending of the above mentioned values for α_2, two different regions in the purely deterministic problem can be considered:

<u>Case i</u> : For $\alpha_2 = 0.0270$: the deterministic system (2.1) possesses an unstable focus surrounded by a stable limit cycle. Figure 2 shows the results found for the production rate in both the deterministic situation (broken line) and the noise-induced deformation of the limit cycle (solid line). It should also be noted that the greater ε or τ the larger is the amplitude of the oscillations.

<u>Case ii</u> : For $\alpha_2 = 0.0295$: we have a stable focus and therefore in a deterministic analysis any disturbance must decay

asymptotically in a time given by the real part of the linear eigenvalue, which defines a deterministic time scale τ_d.

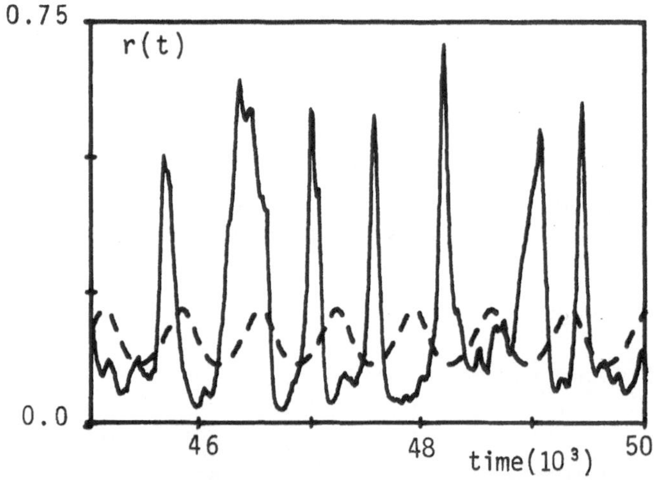

Fig. 2. Time-dependent production rate (in units of 10^{-2}) for a_2=0.0270 and ε=0.5, τ=50. *Broken line:* deterministic limit cycle. *Solid line:* effect of non-white noise.

Figure 3 depicts the noise induced oscillations for the same stochastic parameters as before. Note the striking similarity with Figure 2 thus showing that in the presence of noise it may be impossible to distinguish between the two, deterministically different modes of operation.

For the sake of completeness Figure 4 depicts the power spectrum showing a major frequency component around $T \sim 10^3$

which corresponds to the characteristic period near the steady state (inverse of the imaginary part of the linear eigenvalue).

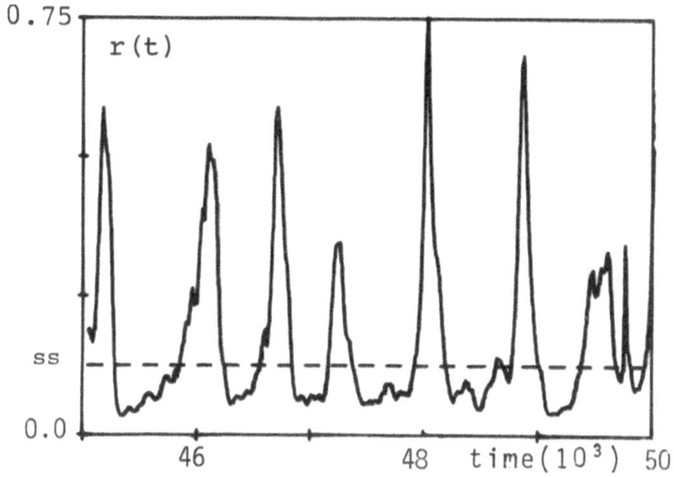

Fig.3. Same as Fig 2 for α_2 = 0.0295, ε = 0.5, τ =50. *Broken line:* stable deterministic steady state (ss). *Solid line* : effect of non-white noise. Note the striking similarity of the noise-driven production rate in Figures 2 and 3.

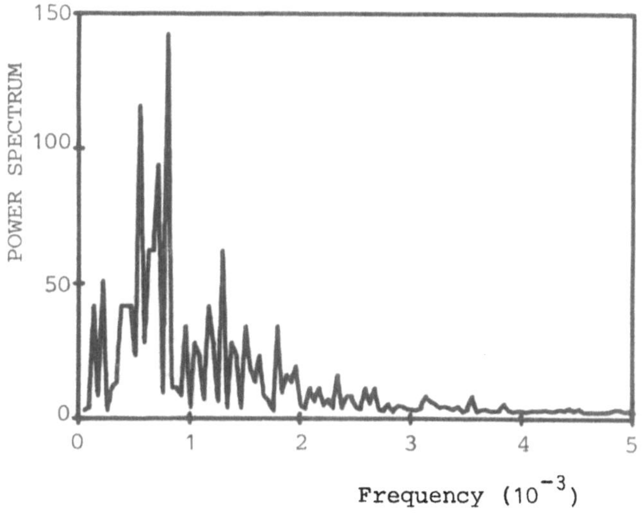

Fig.4. Power spectrum of the coverage signal θ_1 for α_2 = 0.0295, ε = 0.5, τ = 50. Note the main component close to the intrinsic deterministic frequency ($\sim 10^{-3}$)

5.- DISCUSSION AND CONCLUSIONS

The results found demand some qualification. Fig 5 shows the variation of the deterministic time scale τ_d as a function of parameter α_2. Note that for $\alpha_2 \simeq 0.016$ the correlation time of the noise takes the order of magnitude of the deterministic time. Thus although we start with a noise correlation time much shorter than the deterministic time (for $\alpha_2 \simeq 0.0295$) and the white noise aproximation could be taken as adequate, there are accesible regions where the white noise approximation breaks down and one cannot consider the noise as a mere disturbance to the dynamics of the system. Rather θ_i and η (or β) are quantities whose evolution must be studied together. Finally Fig 6, provides the steady values of the deterministic system (2.1) as functions of α_2. One sees how the temporal behavior shown earlier depends crucially on the interaction between the deterministic dynamics and the noise.

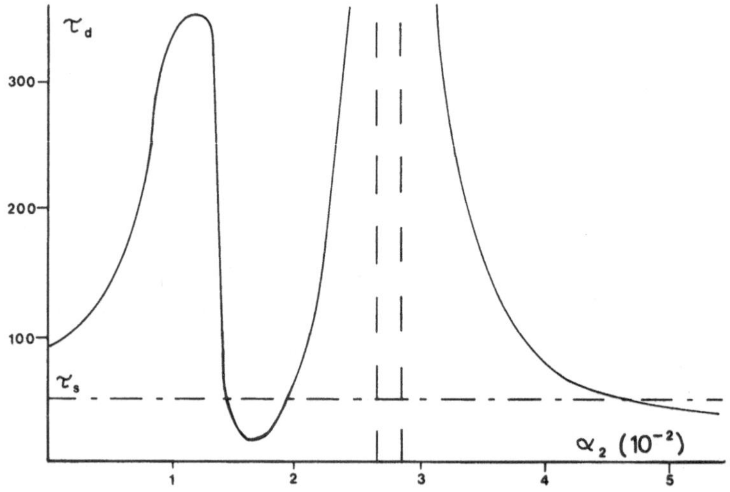

Fig.5. Variation of the deterministic time scale, τ_d, as a function of α_2. Also shown is a tipical correlation time of the noise, $\tau = 50$. The two vertical broken lines delineate the only region where the steady state is unstable and a stable limit cycle exists.

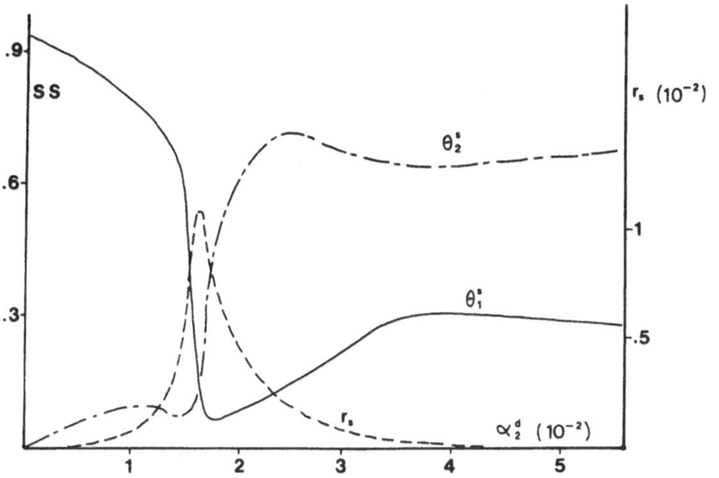

Fig.6. Steady values of coverage , $\theta_i (i=i,2)$ and production rate, r_s, illustrating the sensitivity of chemical dynamics in the region where according to the preceding Figure 5 the non-white noise cannot be considered as a mere disturbance to the deterministic model.

To summarize we have shown the effect of a non white noise on the temporal evolution of a dynamical system . It should be stressed that to explain the behavior observed no use has been made of the bifurcation diagram indicating a much deeper interaction between the system and the noise than previously considered in the literature . This result,due to the fact that both correlation time and deterministic evolution time may be of the same order of magnitude shows indeed that care must be taken when using the white noise approximation to account for the stochastic variability of control parameters in chemical reactions and related problems .

ACKNOWLEDGMENTS

Fruitful discussions with Prof. L. Arnold are acknowledged. This work has been sponsored by the Stiftung Volkswagenwerk.

REFERENCES

1. H. Haken, *Advanced Synergetics*, Springer Verlag, Berlin, 1983.
2. W. Horsthemke and R. Lefever, *Noise Induced Transitions* , Springer Verlag, Berlin, 1983.
3. A. Schenzle and H. Brand, Phys. Rev, *A20*, 1628 (1979).
4. F. J. de la Rubia and M. G. Velarde, Phys. Lett., *A69*, 304 (1978).
5. See for instance the contribution to this *ASI* by San Miguel and Sancho.
6. L. Arnold and R. Lefever (eds.), *Stochastic non Linear Systems* , Springer Verlag, Berlin, 1981 and references therein .
7. W. Ebeling and H. Engel - Herbet, Physica, *A104* , 378 (1980).
8. L. Arnold, W. Horsthemke and J. Stucki, Biometrical J., *21* , 451 (1979).
9. F. J. de la Rubia, J. García- Sanz and M.G. Velarde, in *Non Linear Stochastic Problems*, R.S. Bucy and J.M.F. Moura (eds.), 151, 1983.
10. F. J. de la Rubia, J. García- Sanz and M.G. Velarde, Surface Sci, *143*, 1 (1984).
11. C. G. Takoudis, L. D. Schmidt and R. Aris, Surface Sci. *105*, 325 (1981).
12. J. E. Turner, B. C. Sales and M. B. Maple, Surface Sci. *103*, 54 (1981).
13. L. Arnold, *Stochastic Differential Equations*, John Wiley, New York 1974.

ABOUT SOME SIMPLE FOKKER-PLANCK MODELS

J.J. Brey

Departamento de Física Teórica
Facultad de Física. Universidad de Sevilla
Apdo. Correos 1065. Sevilla. Spain

INTRODUCTION

In spite of its great interest, little is known about the derivation of stochastic equations, such as Langevin and Fokker-Planck equations, starting from a microscopic description of a system. Only the linear case seems to be quite well understood, thanks to the fundamental works of Green, Mori, Zwanzig and many others. Nevertheless, most of the real systems exhibit a nonlinear behavior. Usually, in practical applications, stochastic equations are constructed from phenomenological equations by giving a random interpretation to the variables, and by introducing some pure stochastic quantities or random forces. In order to give a meaning to the equations, one has to specify some of the properties of the random forces. This procedure requires a lot of physical intuition and, furthermore, what is meant by noise depends on the level of description[1].

Here, we will review a few results obtained with simple nonlinear models, perhaps the simplest ones, and we will point out some of the open questions and possible generalizations that we consider may be relevant.

A VERY SIMPLE MODEL

In 1973 Zwanzig[2] showed that certain kinds of evolution equations for a set of variables may lead to a nonlinear Langevin equation for one of those variables. The main idea is to formally integrate the evolution equations in such a way that the time derivative of the relevant variable is expressed in terms of the initial conditions and the values of the variable itself. The price to be paid is that the equation is in general non-Markovian, though a Markovian approximation is obtained in some cases. Once the Langevin equation is known, one can write the equivalent Fokker-Planck equation.

It is also possible to proceed in a different way, and consider from the beginning the evolution equation for the probability distribution function. This procedure has the advantage that in many cases it is easier to make approximations on the evolution of the d.f. than on the evolution of the own variables. As an example, we have considered[3] a Brownian particle interacting with a bath of harmonic oscillators. The Hamiltonian of the system is assumed to be

$$H(Q,P,q,p) = \frac{P^2}{2M} + U(Q) + \sum_{j=1}^{N} \frac{1}{2} p_j^2 + \sum_{j=1}^{N} \frac{1}{2} \omega_j^2 (q_j - \frac{\gamma}{\omega_j^2})^2 , \quad (1)$$

where we have used small letters for the variables of the bath particles and capital letters for the Brownian one. Here, $U(Q)$ is an external field acting on the B particles and γ is a coupling parameter. This Hamiltonian belongs to the class considered by Zwanzig, and the Langevin approach to the problem has been nicely discussed in Ref. 1.

The d.f. of the system ρ obeys the Liouville equation

$$\frac{\partial}{\partial t} \rho(Q,P,q,p;t) = L \rho(Q,P,q,p;t). \quad (2)$$

Here $q \equiv \{q_i\}$, $p \equiv \{p_i\}$ and L is the Liouville operator. We are interested in the reduce d.f. for the B particle, $f(Q,P,t)$, defined as

$$f(Q,P;t) = \int d\,q^N\, dp^N \rho\,(Q\,,P,\,q,\,p;\,t)\,. \quad (3)$$

Integration of Eq. (2) yields after some manipulations

$$\frac{\partial}{\partial t} f(Q,P;t) = -\frac{P}{M} \frac{\partial}{\partial Q} f(Q,P;t) + U'(Q) \frac{\partial}{\partial P} f(Q,P;t)$$
$$+ \frac{\partial}{\partial P} \int_0^t d\tau\, K(\tau) \left[\frac{P}{M k_B T} + \frac{\partial}{\partial P} \right] f(Q,P;t-\tau)\,, \quad (4)$$

where the kernel $k(\tau)$ has a very complicated expression that will not be given here. The following points must be stressed:
a) In Eq. (4) an initial value term has been omitted. The reason is that it vanishes under certain (plausible) initial conditions and it is expected to go to zero on a very short time scale for more general situations.
b) The kernel $k(\tau)$ depends on the number of oscillators N, i.e., no thermodynamic limit has been taken yet.
c) The equation is non-Markovian. This is a consequence of the reduction in the number of variables when going from Eq. (2) to Eq. (4).

As the explicit expression of $k(\tau)$ is known, one can analyze the form of Eq. (4) for specific situations. In particular, in the thermodynamic limit and when the characteristic parameters of the system verify certain conditions, one gets the usual F-P equation

$$\frac{\partial}{\partial t} f(Q,P;t) = -\frac{P}{M}\frac{\partial}{\partial Q} f(Q,P;t) + U'(Q)\frac{\partial}{\partial P} f(Q,P;t) + \xi \frac{\partial}{\partial P}\left[\frac{P}{M k_B T} + \frac{\partial}{\partial P}\right] f(Q,P;t). \tag{5}$$

The details of this derivation can be seen in Ref. 3, where the expression for ξ in terms of microscopic quantities is given, and a generalized fluctuation-dissipation theorem is formulated.

Similar studies have been carried out for more complex systems. What happens in these cases is that the expression of the memory kernel is too complicated to allow explicit calculations.

As Eq. (5) is only valid in the thermodynamic limit, we have eliminated the fluctuations associated with the size of the system. Fluctuations in our equation are due to thermal motion. This situation is the opposite of the one considered in other models, where thermal fluctuations are neglected while the effect of size fluctuations is taken into account.

The model can be generalized to include more general kinds of noise. For instance, a modification of the interaction term in (1) leads to the appearance of multiplicative noise[2,4]. It is also possible to incorporate some term of external noise in the Hamiltonian, and to study the effect of this noise on the evolution of the system. In this case, a coupling between internal and external noise shows up, giving rise to a more involved evolution of the system.

EXPANSION IN THE NONLINEARITY PARAMETER[8]

As it is well known, getting information from a nonlinear F-P equation is a difficult task. If $U'(Q)$ in Eq. (5) is nonlinear, there exists a dynamical coupling between averages and fluctuations. This coupling is important for many physical systems and, therefore, it must be taken into account when deriving nonlinear transport equations[5]. If we neglect fluctuations in Eq. (5) we get

$$\frac{d\,\tilde{Q}(t)}{dt} = - U'\left[\tilde{Q}(t)\right] . \tag{6}$$

This equation can not be understood as a macroscopic equation. The identification is only possible when the fluctuations we are dealing with are just size fluctuations. In actual experiments one measures the evolution of the averages. So, one needs theories to derive renormalized (by fluctuations) equations.

For size fluctuations, van Kampen[6] has developed a method based in a expansion in powers of the inverse of the system size. His method may be easily adapted to the case of a F-P equation, performing then an expansion in the diffusion constant[7]. Of course, van Kampen's expansion is only useful in the limit of small fluctuations. Other approximations, i.e., cumulant expansions, have been discussed in detail by Nordholm and Zwanzig[5]. Here we propose an alternative version of van Kampen's expansion. We will expand in powers of the parameter characterizing the nonlinearity in the system. To be more concrete, let us rewrite Eq. (5) with an explicit expression for $U'(Q)$:

$$\frac{\partial f}{\partial t} = - v\,\frac{\partial f}{\partial Q} + (Q + b\,Q^3)\,\frac{\partial f}{\partial v} + \gamma\frac{\partial}{\partial v}\,(v\,f) + \alpha\,\frac{\partial^2 f}{\partial v^2} . \tag{7}$$

For convenience we have also redefined the constants in the equation. Now, we introduce the following transformation of variables

$$Q = \phi(t) + b^{1/2}\,\xi \qquad\qquad v = \psi(t) + b^{1/2}\,\eta . \tag{8}$$

From the distribution function f, we construct a distribution in the new variables $\Pi(\xi\,,\eta\,;\,t)$ obeying an evolution equation that follows from (7). Then, identification of the coefficients of the different powers of b allows us to obtain a set of equations for $\phi(t)$, $\psi(t)$ and the moments of ξ and η. We have numerically solved the equations to first order in b and the results for the evolution of the average of Q fit quite well with the values obtained by means of Monte Carlo simulation, at least for times not too large[8].

Another quantity of interest, that has been extensively studied by analog computer simulation, is the spectral density $S(\omega)$, defined as

$$S(\omega) = \int_0^\infty dt\, C(t) \cos \omega t, \tag{9}$$

where $C(t)$ is the equilibrium time correlation function

$$C(t) = \langle Q(t)\, Q(o) \rangle_{eq} . \tag{10}$$

A few years ago, Keizer[9] proposed a phenomenological theory to describe fluctuations in nonlinear systems. One of the main implications of his theory is that the evolution equation for $C(t)$ can be directly written from the equation for the evolution of the average of Q. If this quantity, $\langle Q \rangle_t$, satisfies

$$\frac{\partial}{\partial t} \langle Q \rangle_t = \Phi\left[\langle Q \rangle_t\right], \tag{11}$$

the equilibrium time correlation function verifies

$$\frac{\partial}{\partial t} C(t) = \frac{\delta \Phi\left[\langle Q \rangle_{eq}\right]}{\delta \langle Q \rangle_{eq}} C(t) . \tag{12}$$

The accuracy of the spectral density obtained in this way checked against the simulation data is also quite good[8].

SOME COMMENTS

The calculations presented in the previous section correspond to a stable potential ($b > 0$). Nevertheless, an expansion in b may still be useful in certain cases of bistable potentials. Of course, the smaller b is, the deeper the potential wells are.

When studying the critical behavior, thermal fluctuations can not be neglected in general. In some cases, their effect is only to produce a shift in the position of the critical points, but for other systems, they change the nature of the bifurcation. Even more, some models have been studied where the parameter measuring the amplitude of the fluctuations plays the role of a critical parameter. In this context, it is of particular interest the model introduced by Kometani and Shimizu[10], and later studied by Desai and Zwanzig[11] and by Dawson[12]. The critical point in this model is characterized by the fact that the distribution function of the system splits into different and separate distribution functions. In other words, at a given value of the critical parameter, not only the order parameter bifurcates, but it is the whole distribution function which does it. Calculations about this model will be published elsewhere.

REFERENCES

1. R. Zwanzig in Systems Far From Equilibrium (Ed. L. Garrido) Springer-Verlag, Berlin (1980) 199.
2. R. Zwanzig, J. Stat. Phys. 9 (1973) 215.
3. J. J. Brey and M. Morillo, Nuovo Cimento 70 (1982) 187.
4. K. Lindenberg and V. Seshadri, Physica 109A (1981) 483.
5. K. S. J. Nordholm and R. Zwanzig, J. Stat. Phys. 11 (1974) 143.
6. N. G. van Kampen, Stochastic Processes In Physics and Chemistry, North-Holland, Amsterdam (1981).
7. R. F. Rodríguez and N. G. van Kampen, Physica 85A (1976) 347.
8. J. J. Brey, J. M. Casado and M. Morillo, Physica, to appear.
9. J. Keizer, J. Chem. Phys. 63 (1975) 398. For a more fundamental approach see: J. J. Brey in Systems Far From Equilibrium (Ed. L. Garrido) Springer-Varlag, Berlin (1980) 244.
10. K. Kometani and H. Shimizu, J. Stat. Phys. 13 (1975) 473.
11. R. C. Desai and R. Zwanzig, J. Stat. Phys. 19 (1978) 1.
12. D. A. Dawson, J. Stat. Phys. 31 (1983) 29.

DYNAMICS OF SYMMETRY BREAKING: PHASE COHERENCE IN FINITE AND RANDOM SYSTEMS

F. de Pasquale

Dipartimento di Fisica
Università di Roma, "La Sapienza"
I-00185-ROMA, Italy

INTRODUCTION

Purely relaxational dynamics for an n-component vectorial order parameter has been extensively used to study critical dynamics of Heisenberg magnetic systems (1) and statistical properties of transient laser radiation (2). The dynamical model we are referring to is the TDGL (time-dependent Ginzburg-Landau) model, i.e., non linear Langevin equation describing the evolution of the order parameter associated with the system. Non linearity is usually associated with a coupling between the degrees of freedom typical of the system. Only in the laser case it is possible to select a single degree of freedom (the unimode operation). In such a limit phase coherence phenomenon appears as a consequence of an instability of the system. In the opposite limit of an infinite number of degrees of freedom we have the thermodynamic limit in which the same instability can give rise to a phase transition.

The model is drastically simplified if we disregard the coupling of modes and the coupling with the external thermal bath. In such a limit we obtain a deterministic system which exhibits an instability phenomenon. Above the critical temperature the equilibrium state for the order parameter is $\psi = 0$. Below the critical temperature, in the equilibrium configuration, a degeneracy appears as far as the orientation of the vector ψ is concerned. The vector ψ will maintain its initial direction. The rotational symmetry of the theory in the n-dimensional space of the vector ψ is spontaneously broken.

If symmetry is not restored when the coupling with the thermal bath and other modes is switched on, a phase transition appears. The coupling with the thermal bath will basically produce a diffusion of the end point of the vector ψ on the surface of an hypersphere in the $n-1$ dimensional space of radius equal to the modulus of ψ . This diffusion is enough to restore the symmetry but the relaxation toward equilibrium is typically very slow. The very slow change of the phase is known as "phase coherence" in the laser's physics terminology (3). In order to have a "phase transition" the coupling between an infinite number of degrees of freedom is essential in order to stop diffusion among equivalent equilibrium states. This is the case for the three-dimensional system in which the transition to a ferromagnetic phase is possible. This transition does not occur in Heisenberg magnetic system under suitable conditions. We shall study the following cases: the size of the system is finite, the system is infinite but unidimensional , in the infinite size case and in any dimensionality in the presence of quenched local random magnetic fields.

We want to emphasize that, even when phase transition does not appear, a symmetry breaking phenomenon can be observed if the observation times are shorter than the typical time the system needs to leave a given region of phase space (5). For instance in the case of bistable systems the "escape time" from one equilibrium state to the other can be astronomical due the presence of an energy barrier. Even in the case of a continuous symmetry, where no barrier separates equilibrium states, the orientational diffusion is very slow because it is equivalent to an almost free thermal diffusion on a surface of macroscopic radius.

This lecture will be devoted to describe a simple theory of the relaxation from a given far from equilibrium initial configuration in all the quoted cases. We shall see that a very slow rate of approach to equilibrium is a common feature of all these systems.

The time behavior of the averaged order parameter (the electric field in the laser case, the magnetization in the magnetic case) and of the correlation length will be studied to support the previous conclusion.

Results are quite simple in the $n \to \infty$ limit and with suitable restrictions on the initial configuration and allowed randomness.

Within such limits the relaxation process turns out to be a

linear process with a time dependent restoring force.

The magnetization and correlation length can be calculated once a single integro-differential equation for the so called "non linear relaxation function" is solved.

Let us finally comment on the type of randomness we shall consider in the following. The coupling with the random local magnetic field will be considered to be of zero order in n. It is interesting to note that such a small randomness is enough to destroy ferromagnetic ordering. Finally it must be stressed the interest in achieving a better understanding of the non linear dynamics of disordered systems. Very slow relaxation phenomena are indeed a general and characteristic feature of such systems (6). A valuable further step in this direction should be the extension of the present approach to systems exhibiting an unaxial random magnetic field (7) or systems with random interactions which show a spin glass ordering phenomenon.

1. SPHERICAL LIMIT OF TDGL

In the TDGL model the local magnetization is associated with the n-component vectorial process (x,t)
It s time evolution is determined by the TDGL equation i.e.

$$\frac{\partial}{\partial t} \psi (x,t) = - \frac{\partial H}{\partial \psi} + \sqrt{\varepsilon}\, \xi (x,t) \tag{1.1}$$

where the effective Hamiltonian is

$$H = \int d^d x \, \{ \frac{1}{2} r \, |\psi|^2 + \frac{1}{4} \frac{u}{n} |\psi|^4 - b.\psi \} + \frac{1}{2} \int d^d x \, |\nabla \psi|^2 \tag{1.2}$$

Eq. (1.1) is the result of a coarse graining procedure which naturally introduces a lower bound for distances at which the continuous description is valid (9). Such a distance is of the order of magnitude of the interaction length in the underlying microscopic model. Stochastic terms are the gaussian local white noise and the quenched (time independent) gaussian local magnetic fields b(x) characterized by the following properties

$$<\xi> = < b > = 0 \qquad < \xi^{\alpha}(x,t)\, \xi^{\beta}(x',t') > = \delta^{\alpha\beta}\, \delta(x-x')\, \delta(t-t')$$

$$< b^{\alpha}(x).b^{\beta}(x') > = \delta^{\alpha\beta}\, \delta(x-x')(\Delta b)^2 \tag{1.3}$$

The main features of the model are non linearity and the coupling between different degrees of freedom generated by the second part of. (1.2). For $n=1$ eq. (1.2) is associated with Ising magnetic systems, for $n > 1$ to the XY ($n=2$) and to Heisenberg magnetic systems (1). For $n=2$ the same equation describes the slowly varying part of the electric field associated with the laser radiation in a cavity. In this context and independent derivation of eq. (1.2) can be found in Ref. (2). Quite accurate experiments have been performed on the statistical properties of the laser radiation during the decay from an unstable state (8). The results are in excellent agreement with (1.2) (9). A derivation from a microscopic model of the TDGL which does not assume close to equilibrium conditions can be found in Ref. 10.

If we neglect the second term of eq.(2) and the interaction with local random fields we set the local effective Hamiltonian which exhibits a single equilibrium state for $\psi = 0$ if $r > 0$ and a degeneracy of equilibrium states for $r < 0$ corresponding to different orientations of the vector ψ whose modulus is fixed by the following condition:

$$\frac{1}{n} |\psi|^2 = \frac{r}{u}$$

For $n = 1$ we have two equilibrium states

$$\psi^0 = \pm \frac{|r|}{u}$$

separated by a potential well which becomes infinitely large in the so-called hard spin (HS) limit (11), i.e., for $r,u \to \infty$ with $r/u = 1$.

For $n = 2$ the degeneracy can be described by the arbitrary phase of the complex number $z = \psi_0 + i\,\psi_1$. For $n > 2$ degeneracy is associated with the surface of an hypersphere in the $(n-1)$ dimensional space.

The qualitative change of the local potential associated with the transition from $r > 0$ to $r < 0$ is the instability phenomenon which drives a possible phase transition. A positive or negative value of r corresponds to a temperature above or below the critical temperature. The temperature determines also the coupling constant $\varepsilon = 2KT$. In order to take into account the non-local part of the effective local Hamiltonian we found very convenient to introduce the $n \to \infty$ limit i.e. to study the spherical limit of the TDGL model. In fact it has long been recognized (12) that the resulting time dependent gaussian model has relatively simple relaxational properties (13). We define the Fourier transform in a finite volume as usual as

$$f(x,t) = \frac{1}{V} \Sigma_q \quad e^{i\,q.x} \quad f(t)$$

where V is the volume of the system. Periodic boundary conditions in a box of linear dimension L are assumed: $q = \pm n^{\alpha} \frac{2\pi}{L}$

with n^{α} ($\alpha \forall$ 0, n-1) integer positive numbers. The new form of eq. (1.2) is:

$$\frac{\partial \psi_q(t)}{\partial t} = - (r + q^2)\ \psi_q - \frac{u}{V}\ \Sigma_{q'}\ A_{q-q'}\ \psi_{q'} + \sqrt{\varepsilon}\xi_q(t) \qquad (1.4)$$

$$A_q = \frac{1}{nV} \quad \Sigma_{q'} \quad \psi_{q-q'} \quad \psi_{q'} \qquad (1.5)$$

It is convenient to split the process ψ in a "macroscopic" homogeneous part oriented in $\alpha = 0$ direction of order $\sqrt{n}$ and a microscopic fluctuating part of order n^0.

In the Fourier space

$$\delta(q) = V\,\delta_{q,0}$$

$$\psi_q^{\alpha}(t) = \delta^{\alpha,0} \sqrt{n}\ \ m(t)\ \ \delta_q V + \phi_q^{\alpha}\ (t) \qquad (1.6)$$

with the initial condition

$$\phi_q^{\alpha}\ (0) = 0 \qquad (1.7)$$

A homogeneous initial configuration is an essential requirement for the simplicity of the following calculation. Expanding the evolution eq. (1.5) we obtain respectively to $O(\sqrt{n})$ and $O(1)$

$$\dot{m}(t) = - (r + u\,m^2(t) + \frac{u}{V} A_0)\ m(t) \qquad (1.8)$$

$$\dot{\phi}_q(t) = - (r+q^2 + u\,m^2(t))\,\phi_q(t) - \frac{u}{V}\ \Sigma_{q'}\ A_{q-q'}\ \phi_{q'} + \sqrt{\varepsilon}\ \xi_q(t)$$

(1.9)

The main point is to show that the nonlinear part of the restoring force is spatially homogeneous and does not fluctuate in the spherical limit

$$A_q = \frac{1}{n\,V}\ \Sigma_{q'}\ \phi_{q-q'}\ \phi_{q'}$$

$$\lim_{n\to\infty}\ A_q \ =\ a(t)\ \ V\delta_q \qquad (1.10)$$

If eq. (1.10) is true the process ϕ becomes a linear process with a time dependent restoring force to be determined selfconsistently. We shall limit ourselves to justify eq. (1.10) on the basis of a self-consistency argument. Let us suppose that eq (1.10) is valid. We obtain for

$$\phi_q(t) = \int_0^t \exp\ [\ -\ \int_{t'}^t (r+q^2\ +\ u\,m^2(t'')+u\ a(t''))dt''](\sqrt{\varepsilon}\,\xi_q(t')+b_q)dt' \qquad (1.11)$$

if we substitute eq. (1.11) in eq. (1.9) and use the following characteristic properties of the spherical limit

$$\lim_{n\to\infty}\ \frac{1}{n}\ \xi_q(t).\ \xi_{q'}(t') =<\xi_q^{\alpha}(t).\ \xi_{q'}^{\alpha}(t')> = \delta(q+q')\,\delta\,(t-t') \qquad (1.12a)$$

$$\lim_{n\to\infty}\ \frac{1}{n}\ \xi_q(t).b_{q'}\ \ =<\ \xi_q^{\alpha}(t).b_{q'}^{\alpha}\ >\ =\ 0 \qquad (1.12b)$$

$$\lim_{n\to\infty}\ \frac{1}{n}\ b_q\ .\ b_{q'}\ \ =\ \ (\Delta\,b)^2\ \ \delta\,(q\ +\ q') \qquad (1.12c)$$

we obtain that selfconsistency is verified. The following equation determines the self-consistent part of the restoring force a(t).

$$a(t)= \frac{1}{V}\ \Sigma_q\{\varepsilon \int_0^t dt'\ \exp\{\ -2\ \int_{t'}^t dt''[\ r+q^2+\ u\ m^2(t'')+u\ a(t'')]\ \}$$

$$+(\Delta b)^2\ \ (\int_0^t dt'\ \exp\{-\int_{t'}^t dt''\ [\ r+q^2+u\,m^2(t'')+\ u\ a(t'')]\})^2\} \qquad (1.13)$$

Once eqs(1.8) and (1.13) are solved, the stochastic process for the fluctuating part of the local magnetization defined by eq. (1.11) is perfectly known. We obtain a single equation instead of the coupled eqs. (1.8) and (1.13) introducing the following

$$y(t) = \frac{m^2(0)}{m^2(t)} \quad ; \quad y(0) = 1 \tag{1.14}$$

The function y describes how an initial homogeneous magnetization relaxes toward its equilibrium value. We shall refer in the following to the function y as the "inverse non linear relaxation function". As a consequence of eqs. (1.8) and (1.13) we set the following integro-differential equation for y

$$\frac{1}{2}\dot{y} = ry + \varepsilon u \int_0^t y(t')\left[\frac{1}{V}\Sigma_q\, e^{-2q^2(t-t')}\right] dt' + u\, m^2(0) + u\,\frac{(\Delta b)^2}{V}\Sigma_q \left[\int_0^t \sqrt{y(t')}\; e^{-q^2(t-t')} dt'\right]^2 \tag{1.15}$$

with the initial value given by eq. (1.14)

Finally it is easily shown that the knowledge of the relaxation function allows also to calculate the correlation length. The time dependent correlation length $l(t)$ is defined by means of the following equation

$$l^2(t) = \frac{\int d^d x \int d^d y\, |x-y|^2 < \phi(x,t).\ \phi(y,t) >}{\int d^d x \int d^d y < \phi(x,t).\ \phi(y,t) >} \tag{1.16}$$

Using eq. (1.11) we obtain

$$\lim_{n\to\infty} < \phi(x,t).\ \phi(y,t) > = V^{-1}\, \Sigma_q \{\varepsilon \int_0^t dt'\, e^{-2q^2(t-t')}[y(t')/y(t)] + \{\int_0^t dt'\, e^{-q^2(t-t')}[y(t')/y(t)]^{1/2}\,\Delta b\}^2\}.\ \exp iq.(x-y) \tag{1.17}$$

By straightforward manipulations we finally get the following relation for the time dependent correlation length .

$$l^2(t)=4d\,\frac{\varepsilon\int_0^t (t-t')y(t')dt'+\int_0^t dt'\sqrt{y(t')}\int_0^t dt'(t-t')\sqrt{y(t')}\,(\Delta b)^2}{\varepsilon\int_0^t y(t')dt' \;+\; \left(\int_0^t dt'\;\sqrt{y(t')}\;\Delta b\right)^2} \tag{1.18}$$

2. SIZE AND DIMENSIONALITY EFFECTS

Let us first consider magnetic systems with no quenched randomness present: $\Delta b = 0$. In this case eq. (1.15) simplifies a lot becoming a linear integro-differential equation which can be solved by means of Laplace transform (6). Defining the Laplace transform as:

$$y(t) = \int_0^\infty e^{-pt}\,\tilde{y}(p)\,dp$$

we set

$$\tilde{y}(p) = \frac{1 + 2u(m^2(0)/p)}{p - 2r - \frac{2\varepsilon u}{V}\Sigma_q \frac{1}{p+2q^2}} \tag{2.1}$$

For a finite volume V we consider first the "homogeneous fluctuations" case (the laser case) in which only $q = 0$ mode is taken into account. In this case we have

$$-\frac{p + 2\,u\,m^2(0)}{p^2 - 2rp - \frac{2\varepsilon u}{V}} = \frac{a_1}{p - p_1} + \frac{a_2}{p - p_2} \tag{2.2a}$$

$$p_{1,2} = r \pm \left(r^2 + \frac{2\,\varepsilon\,u}{V}\right)^{1/2} \tag{2.2b}$$

$$a_{1,2} = \frac{1}{2} \pm \frac{2\,u\,m^2(0)}{2\sqrt{r^2 + 2\frac{\varepsilon u}{V}}} \tag{2.2c}$$

The long time behavior of the inverse non linear relaxation function $y(t)$ is determined by the singularities in the positive half plane of its Laplace transform. For $r > 0$ the leading singularity is p_1 which is of the order $2r$ for small ε. The magnetization relaxes to zero very quickly for large r. For $r < 0$ the leading singularity approaches the origin $p_1 \simeq \frac{\varepsilon u}{V}$. As a consequence the magnetization decays to zero in a very long time.

There is no stable symmetry breaking solution but a drastic change appears in the relaxation time. The explicit formula for the magnetization is given by:

$$m(t) = \frac{m(0)\ \exp(-rt/2)}{\left\{ \mathrm{ch}\,(r^2 + \frac{2\varepsilon u}{V})^{1/2}\, t + \frac{2\,u\,m^2(0) + r}{(r^2 + 2\varepsilon u/V)^{1/2}}\ \mathrm{sh}\,(r^2 + \frac{2\varepsilon u}{V})^{1/2}\, t \right\}^{1/2}} \tag{2.3}$$

In the HS limit eq. 2.3 gives $m(t) = \exp(-\frac{\varepsilon t}{V})$. We note that in such a limit $m^2(0) = 1$ in order to be consistent with the initial condition $y(0) = 1$. We see that the first fast transient in which the modulus of ψ approaches its equilibrium value disappeared. What is left is the slow orientational diffusion of the vector ψ which on times $t \gg \frac{V}{\varepsilon}$ restores the rotational symmetry. As far as the contribution of modes with $q \neq 0$ to the sum appearing in the denominator of (2.1) is concerned we note that for $d > 2$ for $\frac{1}{L^d} \lesssim p \ll \frac{1}{L^2}$, i.e., for times $L^2 \ll t \lesssim L^d$ the dependence on p can be neglected. As a consequence we expect that the time evolution of the unimode case (homogeneous fluctuations case) is approximately true also for the multimode system for $d > 2$ and in the time range $L^2 \ll t$ apart form a renormalization of $r \to \bar{r}$

$$\bar{r} = r + \frac{\varepsilon u}{V} \Sigma_q \frac{1}{p + 2q^2} \tag{2.4}$$

which amounts to a shift of the critical temperature. In the same approximation it is meaningful to calculate the correlation length for a finite system using the result of eq. 2.1. apart from the substitution $r \to \bar{r}$. We obtain from eq. (1.16) and (2.1) which no initial magnetization:

$$l^2(t) = \frac{4d}{p_1} \cdot \frac{1 - p_1/p_2 \cdot \exp[-(p_1-p_2)t] + (p_1/p_2-1)\exp[-(p_1-p_2)t]}{1 - \exp[-(p_1 - p_2)t]} \tag{2.5}$$

Above the critical temperature we have

$$l^2(t) = \frac{2d}{\bar{r}} \tag{2.6}$$

for $\bar{r}^2 \gg \frac{2\varepsilon u}{V}$. Eq. (2.6) is expected to be valid according to our previous consideration for $\bar{r} \ll \frac{1}{L^2}$ i.e. in a narrow range above the critical point.

Below the critical temperature, in the HS limit we obtain

$$l^2(t) = \frac{4dV}{\varepsilon}\left(1-e^{-\frac{\varepsilon}{V}t}\right) \tag{2.7}$$

Note the very slow growth with a time constant of the order of $\frac{L^d}{\varepsilon}$ toward the asymptotic value in the quite large time range $L^2 \ll t$ The system organizes itself in very large domains whose size goes to infinity in the thermodynamic limit.

For $d < 2$ homogeneous fluctuations do not have any dominant role, we are then forced to take into account all the modes of the system.

This is more conveniently done in the infinite volume limit of eq. 2.2a using the standard rule

$$\frac{1}{V} \Sigma_q \to \int \frac{d^d q}{(2\pi)^d}.$$

For dimensionality of the system less than two the cut off Λ in the integral can be disregarded because the integral is finite for $\Lambda \to \infty$. The relaxation function is determined by:

$$y(p) = \frac{p + 2 u m^2 (0)}{p^2 - 2rp - 2 \varepsilon p^{d/2} w_d} \qquad (2.8)$$

where $w_d = \frac{1}{(2\pi)^d} \int d^d x \, (\frac{1}{1+2x^2})$. Again the leading singularities are respectively:

$$r > 0 \qquad p_1 \sim 2r$$

$$r < 0 \qquad p_1 \sim [\frac{\varepsilon u w_d}{|r|}]^{\frac{1}{1-d/2}}$$

We obtain the same qualitative behavior as in the case of homogeneous fluctuations although the nature of the leading singularity is different.

Above the critical temperature $(r > 0)$ the magnetization vanishes quickly with a typical decay time of the order of $1/2r$. Below the critical temperature $(r < 0)$ the magnetization first decays quickly to a "metaestable" value of the order of $(\frac{|r|}{u})^{\frac{1}{2}}$ and then vanishes very slowly with a decay time of the order $[\frac{|r|}{\varepsilon u}]^{\frac{1}{1-d/2}}$. As before the first quick decay of the magnetization disappears in the HS limit. In such a limit we have

$$y(p) = \frac{1}{p - \varepsilon p^{d/2} w_d} \qquad (2.9)$$

For instance the HS limit of the magnetization is for $d = 1$:

$$m(t) = \frac{e^{-\tau}}{(1 + \mathrm{erf}\sqrt{\tau})^{\frac{1}{2}}} \qquad (2.10)$$

where the scaled time τ is given by $\tau = (\varepsilon w_1)^2 t$.

The explicit expression for the correlation length in the $d = 1$ case and for $r < 0$ is greatly simplified if we first start with zero magnetization and then take the HS limit only for $rt > 0$. The result is

$$l^2(t) = \frac{4}{\varepsilon^2 w_1^2} \left\{ 1 - \frac{1+\sqrt{\tau/\pi}}{e^{\tau}\,(1+\mathrm{erf}\sqrt{\tau})} \right\} \tag{2.11}$$

Eq. 2.11 shows a square root-like growth of the correlation length toward a large but finite asymptotic value $l^2(\infty) = \frac{4}{\varepsilon^2 w_1}$.

It must be noted that this square root growth law is a universal law which seems to characterize in many cases the early stage growth of a system that relaxes from an unstable initial configuration (14). For $d > 2$ we must take into account the cut-off Λ. A standard subtraction is introduced which defines a new critical temperature

$$\int \frac{d^dq}{(2\pi)^d} \frac{1}{p + 2q^2} = \int \frac{d^dq}{(2\pi)^d} \left(\frac{1}{p+2q^2} - \frac{1}{2q^2}\right) + \int \frac{d^dq}{(2\pi)^d} \left(\frac{1}{2q^2}\right) =$$

$$= - p^{d/2 -1}\, \bar{w}_d + r_c$$

The new critical temperature is at $\bar{r} = r + r_c = 0$. The relaxation function is given by

$$y(p) = \frac{p + 2u\, m^2(0)}{p^2 - 2\bar{r}\, p + \varepsilon u\, p^{d/2}\, \bar{w}_d} \tag{2.12}$$

From eq. 2.12 we see that the leading singularity is at $p = 0$ for $\bar{r} < 0$. As a consequence magnetization reaches a non vanishing equilibrium value. In systems with dimensionalities greater than two because of the interaction between different degrees of freedom the orientational diffusion of the order parameter freezes in. We shall limit ourselves to calculate the HS limit of the correlation length for $m(0) = 0$ in the $d = 3$ case

$$l^2(t) = \frac{12}{c^2}\left\{1 + \frac{2c\sqrt{t/\pi} + c^2 t}{1 - e^{c^2 t}\,\mathrm{erfc}(-c\sqrt{t})}\right\} \tag{2.13}$$

$$c = \frac{1}{\varepsilon \bar{w}_3}\left(-1 + \varepsilon \int_{q<1} \frac{d^d q}{(2\pi)^d} \frac{1}{2q^2}\right) \tag{2.14}$$

When $c > 0$ we are above the critical temperature and the correlation length reaches a finite asymptotic value $l^2(\infty) = \frac{12}{c^2}$.

Below the critical temperature $(c < 0)$ the correlation length grows without limit with a square root law for large t. It's worth noting that an infinite correlation length characterizes the new phase even in the case in which rotational symmetry is not actually broken.

3. DYNAMICS OF THE RANDOM FIELD INSTABILITY

Since the work of ref. 4 it is well known that a static local random field of arbitrarily weak strength produces an instability of the magnetically ordered phase for $d < 4$. The ordered state is unstable against the formation of domains. The size of such domains increases as the strength of the local random field decreases. Experimental evidence of such an instability has been recently established (14) in the still controversial Ising case. We shall limit ourselves to discuss the dinamical mechanism underlying the random field instability within the theory estabilished in sec.1. First of all let us consider the homogeneous fluctuations case. From eq. 1.14, selecting only the $q = 0$ mode, we set

$$\frac{1}{2}\dot{y}(t) = ry + \frac{\varepsilon u}{V}\int_0^t y(t')dt' + u\, m^2(0) + u\,\frac{(\Delta b)^2}{V}\left(\int_0^t \sqrt{y(t')}dt'\right)^2 \tag{3.1}$$

In the HS limit with $m^2(0) = 1$, eq(3.1) gives

$$y(t) = e^{\frac{\varepsilon}{2V} t}\left\{\frac{\varepsilon}{4V\omega}\,\mathrm{sh}\,\omega t + \mathrm{ch}\,\omega t\right\}^2 \tag{3.2}$$

$$\omega^2 = \left(\frac{\varepsilon}{4V}\right)^2 + \frac{(\Delta b)^2}{V} \tag{3.3}$$

The resulting magnetization is:

$$m(t) = \frac{e^{-\frac{\varepsilon}{4V}t}}{\frac{\varepsilon}{4V}\,\mathrm{sh}\,\omega t + \mathrm{ch}\,\omega t} \tag{3.4}$$

The presence of the random fields introduces a new decay mechanism which is already effective even in the time range where the pure system behaves as in a metastable state. According to eq. (3.4) for volumes large enough even a very small randomnes will always prevail over the thermal effects. As a consequence for $t \ll \frac{2V}{\varepsilon}$ and $\frac{\varepsilon}{4V} \ll \frac{\Delta b}{\sqrt{V}}$ the magnetization behaves as

$$m(t) = \frac{1}{\mathrm{ch}(t/\tau_b)} \qquad \tau_b = \frac{\sqrt{V}}{\Delta b}$$

As far as the extension to the multimode case is concerned we see that according to eq.(3.5) the typical relaxation time scales with $L^{-d/2}$. As a consequence the homogeneous fluctuations will dominate only for $d > 4$.

Also for the far from equilibrium relaxation properties we observe a dimensionality shift effect from $d = 2$ to $d = 4$. This shift has been already found in static properties (16).

In the infinite volume limit we must solve the integral version of the complete non-linear integrodifferential eq. (1.15)

At least an asymptotic solution of it is easily found. We substitute $y = \exp \alpha t$ in 1.15 and collect the leading exponential for large t. For $d > 2$ we obtain.

$$\frac{1}{2}\alpha = \bar{r} - \varepsilon u \alpha^{d/2-1}\bar{w}_d + u(\Delta b)^2 \alpha^{d/2-2}\mu_d \tag{3.5}$$

$$\mu_d = \int \frac{d^d x}{(2\pi)^d}\left(\frac{1}{1+x^2}\right)^2$$

Below the critical temperature in the HS limit and for small ε , Δb we have:

$$\alpha = (\Delta b)^{\frac{1}{1-d/4}} \quad \mu_d^{\frac{1}{2-d/2}} \tag{3.6}$$

This is the expected asymptotic decay rate of the magnetization.

From the knowledge of the asymptotic behavior of $y(t)$ we can derive the asymptotic behavior of the correlation length.

$$\lim_{t\to\infty} l^2(t) = \frac{4d}{\alpha} \left\{ \frac{\varepsilon + 8(\Delta b)^2/\alpha}{\varepsilon + 4(\Delta b)^2/\alpha} \right\} \tag{3.7}$$

The existence of a finite correlation length in the temperature range in which the pure system is ordered can be understood as the separation of the ordered phase in domains whose everage size is given by $l^2(\infty)$. It must be noted that according to our results the rate of approach to the new equilibrium state is again very slow for small Δb.

4. CONCLUSION

Finally we want to summarize the main points of this lecture. First we have shown how the difficult problem of the nonlinear relaxation associated with the TDGL models with a continuous symmetry simplifies in the spherical limit . We have considered both usual TDGL models and models with quenched random fields. In the first case we studied size and dimensionality effects on the dynamics of symmetry breaking. When the phase transition occours in the infinite volume system $(d > 2)$ homogeneous fluctuations are the slowest decay mechanism of finite system which restores the symmetry broken by the initial conditions. Due to the occourrence of a macroscopic decay time (proportional to the volume) the system shows a metastable-like behavior. For dimensionality lower than the critical one $(d< 2)$ there is still an instability but no special role is attributed to the homogeneous fluctuations. Below the instability point the decay time of the initial magnetization and the typical size of correlated regions (domains in the magnetic language) are inversely proportional to the thermal noise strength.

We finally studied the non-linear dynamics in the presence of random local magnetic fields. We found a qualitative analogy with the previous case once a shift of the critical dimensionality from 2 to 4 is introduced. The homogeneous fluctuations will dominate for $d > 4$, their characteristic decay time is still macroscopic but of the order of the square root of a volume. For $d < 4$ the ordered state is unstable against the formation of domains. Both the asymptotic decay time and the correlation length are inversely proportional to the random field strength. In all the cases in which the symmetry breaking state is unstable we found two distinct domain growth regimes. The first one, which is typical of the early stage of domains formation, is characterized by a correlation length which grows with the square root of the time. The second one correponds to domains reaching their equilibrium size. This asymptotic increase in size is typically very slow. Finally the analysis of the random field instability suggests to perform spinodal decomposition-like experiments on these systems.

REFERENCES

1. S. K. Ma , Modern theory of critical phenomena , W. Benjamin Inc. (1976)
2. H. Risken and H. D. Vollmer, Z. Phys. 204, 240 (1976)
3. F.T. Arecchi . in Order Fluctuations in Equilibrium and Non equilibrium Statistical Mechanics ed. by Nicolis, G. Dewel and J.W. Turner , Willey New York , 1981 p. 107.
4. Y. Imry and S. K. Ma, Phys. Rev. Lett. 35, 1399 (1975)
5. F. de Pasquale, Z. Racz and P. Tartaglia, Phys. Rev. B sept. 1983
6. K. Binder, W. Kinzel: in "Disordered Systems and Localization" Lecture Notes in Physics 149 (1981) ed. by C. Castellani, C. Di Castro and L. Peliti.
7. S. Fishman, A. Ahrony , J. Phys. C 12, L729 (1979)
8. F. T. Arecchi, V. Degiorgio and B. Querzola, Phys. Rev. Lett. 19, 1168 (1967)
9. F. de Pasquale, P. Tartaglia and P. Tombesi, Phys. Rev. A25, 466 (1982)
10. J.S. Langer , Ann. Phys. 65, 53 (1971)
11. D. Sherrington, the same issue of ref. 6
12. I. Halperin, P. C. Hohenberg and S.K. Ma, Phys. Rev. Lett. 29, 1548 (1972)
13. Z. Racz, T. Tel, Phys. Lett. 60A, 3 (1977)
14. J.D. Gunton, M. San Miguel and P. Sahni, "The Dynamics of firs order phase transitions" to appear in "Phase transitions and critical phenomena ed. by C. Domb and J. Levowitz (Academic Press)
15. R. J. Birgenau et al., Phys. Rev. 28, 1438 (1983)
16. G. Parisi, N. Surlas, Phys. Rev. Lett. 43, 744 (9179)

NATO-ASI-EL ESCORIAL- AUGUST 1-11,1983-PARTICIPANTS(Main picture)

NATO-ASI-EL ESCORIAL- AUGUST 1-11, 1983-PARTICIPANTS (Additional picture)

PARTICIPANTS

Arecchi, Professor F.T. Lecturer & Co-director of ASI	Istituto di Ottica Largo E. Fermi 6 50125 Arcetri-Firenze Italy
Arimondo, Professor E. Lecturer	Istituto di Física Sperimentale Pad.20 Mostra D'Oltremare I-80125 Napoli Italy
Armbruster, Mr. D.	Institut fur Informationsverarb Universitat Tubingen 74 Tubingen, Koestlinstr 6 Germany (GFR)
Arnold, Professor L. Lecturer	Forschungsschwerpunkt Dynamische Systeme Universitat Bremen Postfach 330440 2800 Bremen 33 Germany (GFR)
Arroyo de Grandes, Ms. P.	Termología-Facultad Ciencias Ciudad Universitaria Zaragoza - 6 Spain
Artiz-Cohen, Mr. J.A.	U.N.E.D.- Física Fundamental Apdo. Correos 50.487 Madrid - 3, Spain
Bestehorn, Mr. U.	Inst. Theor. Phys. Universitat Stuttgart Pfaffenwaldring 57/IV D-7000 Stuttgart 80 Germany (GFR)

Borghi, Dr. R.P.	Faculte Sciences, Univ. Rouen 76130 Mont Saint Aignan France
Borgis, Mr. D.J.	Univ. Pierre et Marie Curie Tour 16-5 4, Place Jussieu 75230 Paris Cedex-05 France
Boyer, Professor L. Lecturer	Lab. Dynamique des Fluides, Centre Saint-Jerome F-13397 Marseille Cedex 4 France
Brey, Professor, J.J.	Física Teórica -Univ. de Sevilla Apdo. Correos 1065, Sector Sur Sevilla , Spain
Bruno, Dr. C.	Politec. di Milano Depto Energetica Piazza L. Da Vinci 32 20133 Milano Italy
Bunz Mr. H.	Inst. Theor. Phys. Universitat Stuttgart Pfaffenwaldring 57/IV D-7000 Stuttgart 80 Germany (GFR)
Caglioti, Professor G. Lecturer	Istituto Di Ingegneria Nucleare Via Ponzio 34/3, I-20 20133 Milano Italy
Califano, Mr. A.	Istituto di Ottica Largo E. Fermi 50125 Arcetri-Firenze Italy
Calvo, Mr. S.	CSIC - Instituto de Plasticos MADRID - 6 , Spain
Carreras, Dr. C.	U.N.E.D.- Ciencias Apdo. Correos 50.487 Madrid-3, Spain
Casado, Dr. J.M.	Física Teórica- Univ. de Sevilla Apdo. Correos 1065, Sector Sur Sevilla , Spain

Castellanos, Professor A.	Dpto. Electricidad y Magnetismo Universidad de Sevilla Apdo. Correos 1065, Sector Sur Sevilla, Spain
Castillo, Dr. J.L.	U.N.E.D.-Física Fundamental Apdo. Correos 50.487 Madrid-3, Spain
Chou, Mr. W.	IHES 91440, Bures-Sur-Yvette France
Chyba, Mr. D.E.	Dept. of Phys., Bryn Mawr College Bryn Mawr, PA 19010 U.S.A.
Clavin, Professor P. Lecturer	Lab. Dynamique des Fluides Centre Saint-Jerome F-13397- Marseille Cedex 4 France
Crespo, Ms. E.	U.N.E.D.- Física Fundamental Apdo. Correos 50.487 Madrid- 3, Spain
Deltour, Dr. A.R.	ENSEEIHT-IMFT 2 Rue Ch. Camichel 31071 Toulouse Cedex France
Faetti, Dr. S.	Universita di Pisa Piazza Torricelli 2 56100 Pisa, Italy
Fairen, Dr. V.	U.N.E.D.- Física Fundamental Apdo. Correos 50.487 Madrid-3, Spain
Feinberg, Mr. D.	CNRS-Groupe Transt. Phases B.P. 166 Centre de Tri 38042 Grenoble Cedex France
Fife, Professor P.C. Lecturer	Mathematics Dept. University of Arizona Tucson, AZ 85721 U.S.A.

Fonseca, Ms. T.	Dept. Química,Facul. Ciencias 4000 Porto Portugal
Fox, Mr. R.O.	Department Chemical Eng. Ksu-Durland Hall Manhattan, KS 66504 U.S.A.
Fronzoni, Dr. L.	Istituto Fisica Piazza Torricelli 2 56100- Pisa Italy
Garcia, Mr. A.L.	Center Statistical Mech. University of Texas Austin, TX 78712 U.S.A.
Garcia, Mr. M.	Hospital Ramón y Cajal Carretera de Colmenar Viejo Madrid, Spain
García-Sanz, Dr. J.	U.N.E.D.- Física Fundamental Apdo. Correos 50.487 Madrid-3, Spain
García-Ybarra, Dr. P.	Colegio Universitario Arcos de Jalón Universidad Complutense Madrid, Spain
Geisel, Dr. T.	Theoretische Physik Universitaet Regensburg D-8400 Regensburg Germany (GFR)
Giusfredi, Dr. G.S.	Istituto di Ottica Largo E. Fermi 6 50125- Firenze Italy
Gonzalez, Dr. J.J.	Agder Ingeniør-Og Distriktshogskole N- 4890-Grimstad Norway
Gribben , Dr. R.J.	Dept. Math.-University Strathclyde 26 Richmond Street Glasgow G1 1XH U.K.

Grossmann, Professor S. Lecturer	Fachbereich Physik-Ag. Statist. Physik Renthof 6 D-3550 Marburg Gernany (GFR)
Grupp, Dr. J.	Physik. Institut Domagkstrasse 75 D- 4400 Munster Germany (GFR)
Guardia-Manuel, Ms. E.	Dpto. Física Teórica Universidad de Barcelona Avda. Diagonal 647 Barcelona- 28, Spain
Gutierrez-Barquín, Ms. I.	Colegio Universitario Integrado Arcos de Jalón Madrid, Spain
Haken, Professor H. Lecturer	Inst. Theor. Phys. Univ. Stuttgart Pfaffenwaldring 57/IV D-7000 Stuttgart 80 Germany (GFR)
Halas, Ms. N.J.	Dept. Phys.,Bryn Mawr College Bryn Mawr, PA 19010 U.S.A.
Hemmer, Professor, P.C.	Institutt Teoretisk Fysikk Universitet Trondheim 7034 Trondheim Norway
Hernandez-Machado, Ms. A.	Dpto. Física Teórica Universidad Barcelona Avda. Diagonal 647 Barcelona-28, Spain
Hickey, Mr. K.A.	Nuclear Engineering University Missouri-Columbia Columbia, Missouri 65211 U.S.A.
Hofmann, Mr. P.	Th-Darmstadt , I. F. Phys. Chem. I D- 6100 Darmstadt Germany (GFR)

Horowicz, Mr. R.J.	Istituto di Fisica Via Celoria 16 20133 - Milano Italy
Irvin, Dr. B.R.	St. Andrews Presbyterian College Laurinburg, NC 28352 U.S.A.
Jimenez-Fernandez, Dr. J.	U.N.E.D.-Física Fundamental Apdo. Correos 50.487 Madrid-3, Spain
Jimenez-Gonzalez, Mr. J.	U.N.E.D.- Física Fundamental Apdo. Correos 50.487 Madrid-3, Spain
Jimenez del Paso, Mr. J.D.	U.N.E.D.-Física Fundamental Apdo. Correos 50.487 Madrid-3, Spain
Jimenez-Sendin, Dr. J.	IBM -Madrid Centro Cientifico Paseo Castellana, 4 Madrid-1, Spain
Kaashoek, Dr. J.F.	Eramus Univ. Econometric (H2-25) Burg. Owdlaan 50 Rotterdam, Holland
Kantor, Dr. J.C.	Dpt. of Chem. Engineering University of Notre Dame Notre Dame, Indiana 46556 U.S.A.
Kessler, Professor J.O.	Physics Dept., Building 81 University of Arizona Tucson, AZ 85721 U.S.A.
Kotelenez, Dr. P.M.	Forschungsschwerpunkt Dynamische Systeme Universitat Bremen-Postfach 330440 2800 Bremen 33, Germany (GFR)
Liñan, Professor A. Lecturer	ETS Ingenieros Aeronaúticos Ciudad Universitaria Madrid-3, Spain

Lugiato, Professor L. Lecturer	Istit. Science Fis. Univ. Milano Via Celoria, 1 I-20133 - Milano Italy
Mancini, Mr. H.L.	Ceilap-Citefa/Zufriategui Varela 1603-Villa Martelli Prov. de Buenos Aires Argentine
Marco, Dr. R.	Depto Bioquímica, Facultad de Medicina UAM Arzobispo Morcillo S/N Madrid, Spain
Marchesoni, Dr. F.	Depto Física Piazza Torricelli 2 I-56100 Pisa Italy
Menzinger, Professor M.	Dpt. Chemistry, Univ. Toronto Toronto, Ont. M5S 1A1 Canada
Misguich, Dr. J.H.	Stgi/Fusion Cea. B.P. 6 92260-Fontenay Aux Roses France
Monteillier, Mr. J.	Lab. Dynamique des Fluides Centre Saint-Jerome F-13397, Marseille Cedex 4 France
Moran, Dr. F.	Bioquímica-Facul. Químicas Universidad Complutense Madrid-3, Spain
Morillo, Dr. M.	Física Teórica,Univ. Sevilla Apdo. Correos 1065, Sector Sur Sevilla, Spain
Morita, Dr. T.	Instituut Lorentz Nieuwsteeg 18 2311SB Leiden Holland
Muller, Dr. S.	Max Planck Inst. Ernahrungsphys. D-4600 Dortmund Germany (GFR)

Muñoz-Sudupe, Dr. A.	Física Teórica, Fac. Ciencias Universidad Complutense Madrid-3, Spain
Nicolis, Professor G. Lecturer	Chimie Physique II (ULB) C.P. 231 Campus Plaine B-1050 Bruxelles Belgium
O'Neill, Mr. G.G.	Dpt. Math.,University Strathclyde 26 Richmond Street Glasgow G1 1XH U.K.
Pasquale, Professor F. De	Facolta di Ingegneria,Univ. Roma P.Le A. Moro 2 I-00185 Roma Italy
Pelce, Mr. P.	Lab. Dynamique des Fluides Centre Saint-Jerome F-13397, Marseille Cedex-4 France
Penland, Ms. C.	Applied Research Lab. P.O. Box 8029 Austin, Texas 78712 U.S.A.
Perez-García, Dr. C.	Dpto. Termología,Fac. Ciencias Univ. Autonoma de Barcelona Bellaterra (Barcelona) Spain
Perez-Villar, Professor V.	Dpto. Termología, Fac. Físicas Universidad de Santiago Santiago de Compostela Spain
Pettini, Dr. M.	Osservatorio Astrofísico Arcetri Largo E. Fermi 5 50125 Firenze Italy
Politi, Dr. A.	Istituto di Ottica Largo E. Fermi 6 50125 Firenze Italy

Pratsinis, Mr. S.	Univ. California, Dept. Chem. Eng. Room 5531, Boelter Hall Los Angeles, California 90024 U.S.A.
Protassov, Dr. Y.	Moscow Bauman Higher Tech. School 107005 Moscow URSS
Pumir, Mr. A.	Serv. Physique Theoretique Cen-Saclay 91191-Gif-Sur-Yvette France
Quel, Dr. E.J.	Ceilap-Citefa/Zufriategui Varela 1603 - Villa Martelli Prov. Buenos Aires Argentine
Riste, Professor T. Lecturer	Institute for Energy Technology Box 40 N-2007 Kjeller Norway
Roekaerts, Dr. D.	Inst. voor Theor. Fysica Celestijnenlaan, 200 D B-3030 Heverlee Belgium
Rolon, Dr. J.C.	C.N.R.S., Ecole Centrale Grande Voie Des Vignes 92290 Chatenay Malabry France
Rubi, Dr. J.M.	Dto. Termología, Fac. Ciencias Universidad Autonoma Barcelona Bellaterra (Barcelona) Spain
Rubia , Dr. F.J. de la	U.N.E.D.-Física Fundamental Apdo. Correos 50.487 Madrid-3, Spain
Rubio, Mr. M.A.	U.N.E.D.-Física Fundamental Apdo. Correos 50.487 Madrid-3 , Spain
Salan, Dr. J.	Facultad Ciencias, C-III Universidad Autónoma, Cantoblanco Madrid, Spain

Salieri, Dr. P.	Istituto di Ottica Largo E. Fermi 6 50125 Arcetri-Firenze Italy
San Miguel, Dr. M.	Física Teórica, Univ. Barcelona Diagonal 647 Barcelona-28 Spain
Sancho, Dr. J.M.	Física Teórica, Univ. Barcelona Diagonal 647 Barcelona -28 Spain
Schlicher, Dr. R.R.	Max Planck Institut fur Quantenoptik D-8046 Garching Bei Munchen Germany (GFR)
Schnaufer, Mr. B.	Inst. Theor. Phys. Univ. Stuttgart Pfaffenwaldring 57/IV D-7000 Stuttgart 80 Germany (GFR)
Schober, Mr. W.	Physikalisches Inst. Domagkstrasse 75 4400 Munster Germany (GFR)
Schoener, Mr. G.	Inst. Theor. Phys. Universitat Stuttgart Pfaffenwladring 57/IV D - 7000 Stuttgart 80 Germany (GFR)
Scully, Professor M.O. Lecturer & Co-director of ASI	Inst. Mod. Optics-Physics Dept. University New Mexico Albuquerque, New Mexico 87131 U.S.A
Scuricini, Professor, G.B.	Enea/ Cre. Casaccia Casella Postale 2400 00100 Roma Italy
Simonelli, Mr. F.	Istituto di Ottica Largo E. Fermi 50125 Arectri-Firenze Italy

Srulijes, Dr. J.	Univ. Essen/FB12 Mechanik Schutzenbahn 70 4300 Essen 1 Germany (GFR)
Tredicce, Dr. J.	Istituto di Ottica Largo E. Fermi 6 50125 Arectri-Firenze Italy
Ulivi, Dr. L.	Istituto di Ottica Largo E. Fermi 6 50125 Arcetri-Firenze Italy
Van den Broeck, Dr. Ch.	Dept. Nat./Vrije Univ. Brussels Pleinlaan 2 1050 Brussel Belgium
Van Dyke, Professor M.	School of Eng., Mechanical Eng. Stanford University Stanford, California 94305 U.S.A.
Velarde, Professor M.G. Lecturer & Director of ASI	U.N.E.D.-Física Fundamental Apdo. Correos 50.487 Madrid-3, Spain
Villers, Mr. D.	F. Sciences, Chimie-Phys. Therm-Univ. Mons Avenue Maistriau, 21 B-700 Mons Belgium
Viñals, Mr. J.	Facul. Ciencias-Dpto.Termología Diagonal 645 Barcelona- 28, Spain
Vulpiani, Dr. A.	Facolta di Ingegneria-Univ. Roma P.Le A. Moro 2 I-00185 Roma Italy
Walker, Dr. F.E.	Lawrence Livermore Nat. Lab. P.O. Box 808 Livermore, CA 94550 U.S.A.

Weimer, Mr. W.	Inst. Theor. Phys. Universitat Stuttgart Pfaffenwaldring 57/IV D-7000 Stuttgart 80 Germany (GFR)
Yuste, Professor M.	U.N.E.D.-Física Fundamental Apdo. Correos 50.487 Madrid- 3, Spain
Zeghlache, Ms. H.	Univ. Libre Bruxelles C.P. 231- Boulevard Triomphe 1050 Bruxelles Belgium
Zuñiga, Dr. I.	U.N.E.D.-Física Fundamental Apdo. Correos 50.487 Madrid-3, Spain

AUTHOR INDEX

SUBJECT INDEX